Analytical Applications of FT-IR to Molecular and Biological Systems

NATO ADVANCED STUDY INSTITUTES SERIES

Proceedings of the Advanced Study Institute Programme, which aims at the dissemination of advanced knowledge and the formation of contacts among scientists from different countries

The series is published by an international board of publishers in conjunction with NATO Scientific Affairs Division

A	Life Sciences	Plenum Publishing Corporation
B	Physics	London and New York
C	Mathematical and Physical Sciences	D. Reidel Publishing Company Dordrecht, Boston and London
D	Behavioural and Social Sciences	Sijthoff & Noordhoff International Publishers
E	Applied Sciences	Alphen aan den Rijn and Germantown U.S.A.

Series C – Mathematical and Physical Sciences

Volume 57 – Analytical Applications of FT-IR to Molecular and Biological Systems

Analytical Applications of FT-IR to Molecular and Biological Systems

Proceedings of the NATO Advanced Study Institute held at Florence, Italy, August 31 to September 12, 1979

edited by

JAMES R. DURIG
College of Science and Mathematics, University of South Carolina, Columbia, South Carolina, U.S.A.

D. Reidel Publishing Company
Dordrecht : Holland / Boston : U.S.A. / London : England

Published in cooperation with NATO Scientific Affairs Division

Library of Congress Cataloging in Publication Data

NATO Advanced Study Institute, Florence, 1980.
Analytical applications of FT IR to molecular and biological systems.

(NATO advanced study institute series: Series C, Mathematical and physical sciences; v. 57)
"Published in cooperation with NATO Scientific Affairs Division".
Includes bibliographical references and indexes.
1. Infra-red spectrometry—Congresses. 2. Fourier transform spectroscopy—Congresses. 3. Biological chemistry—Congresses. I. Durig, James R. II. North Atlantic Treaty Organization. Division of Scientific Affairs. III. Title. IV. Series.
QD96.I5N37 1980 543'.0858 80-18138

ISBN-13: 978-94-009-9072-2 e-ISBN-13: 978-94-009-9070-8
DOI: 978-94-009-9070-8

Published by D. Reidel Publishing Company
P.O. Box 17, 3300 AA Dordrecht, Holland

Sold and distributed in the U.S.A. and Canada
by Kluwer Boston Inc., Lincoln Building,
160 Old Derby Street, Hingham, MA 02043, U.S.A.

In all other countries, sold and distributed
by Kluwer Academic Publishers Group,
P.O. Box 322, 3300 AH Dordrecht, Holland

D. Reidel Publishing Company is a member of the Kluwer Group

CONTENTS

PREFACE

In the past few years it has become apparent that Fourier Transform infrared spectroscopy is developing into an excellent technique for solving some of the very difficult problems encountered in analytical chemistry. The applications of FT-IR include the detection and identification of chemical components separated by gas chromatography techniques, determination of low concentration components in a mixture, and problems which have energy limitations such as water samples, opaque samples and biological systems.

The lectures presented in this volume will be utilized at the NATO Advanced Study Institute in Florence, Italy from August 31 to September 12, 1980. These lectures are divided into three main sections: Instrumentation and Theory, Techniques, and Applications. The first section includes a basic introduction to interferometry and the operating parameters. The Techniques section consists of several lectures on accessories used in FT-IR, software and data systems, and special handling techniques. The third section contains an abundance of information on the applications of the FT-IR technique to inorganic and organic molecules, polymers, biological systems, solids and to the determination of molecular structures and conformational analyses. The contents of this volume should provide the reader with the present applications in this field as well as an indication of possible future trends. In general the lectures are of a pedagogical nature and are not to be considered as review articles. These lectures were meant to provide a forum for the exchange of ideas between the student and lecturer and it was hoped that they would engender discussion of topics of current scientific interest in Fourier Transform spectroscopy.

First of all the editor would like to thank Dr. J. F. Sullivan for her administrative assistance and overall supervision of both the Institute and preparation of this book. Also the editor would like to thank the co-directors, Professors W. G. Fateley and A. Müller for their help in various ways with the NATO Advanced Study Institute. Thanks are due to the authors for promptly submitting their manuscripts so that this volume would be available for the students at the Institute. Additionally, the editor would like to thank his wife, Marlene, for copy-editing the manuscripts and for the preparation of the author and subject indexes. Also, thanks

J. R. Durig (ed.), Analytical Applications of FT-IR to Molecular and Biological Systems, ix–x.

are due to Miss Jan Lattimore, a student at the University of South Carolina, who typed most of the articles in the camera ready form. Finally, the editor would like to acknowledge support of the NATO Advanced Study Institute which was sponsored by the NATO Scientific Affairs Division in Brussels.

Professor of Chemistry JAMES R. DURIG
Dean, College of Science and Mathematics
University of South Carolina
Columbia, South Carolina 29208, U.S.A.

INSTRUMENTATION AND THEORY

ONE VIEW OF THE ADVANTAGES OF INFRARED INTERFEROMETRY

Joel C. Case and W. G. Fateley

Chemistry Department
Kansas State University
Manhattan, Kansas, U. S. A.

1. INTRODUCTION

Since the mid 1950's many papers have appeared emphasizing substantial advantages of interferometry over dispersive spectroscopy. While it is generally recognized that these advantages are qualitatively correct, the magnitudes of these advantages are still subject to debate.

In the beginning it was apparent that with the excitement of this "new" form of spectroscopy, the rational behind many of these arguments was clouded and that the total advantage of interferometry might not be as great as first anticipated. In this chapter we wish to review the advantages and disadvantages of interferometry and attempt to quantify them to the greatest extent possible.

2. DISCUSSION

Let us use a simple optical system for an interferometer (see Fig. 1) with an infrared light source inclined at an angle θ with respect to the optical axis striking a collimator at an angle γ, the differential throughout may be written as:

$$dE = (\cos\theta dA_S)(\cos\gamma dA_C)/r^2 \qquad (1)$$

where:

dA_S = area of the source
dA_C = area of the collimator
r = distance between source and collimator.

J. R. Durig (ed.), Analytical Applications of FT-IR to Molecular and Biological Systems, 3–10.

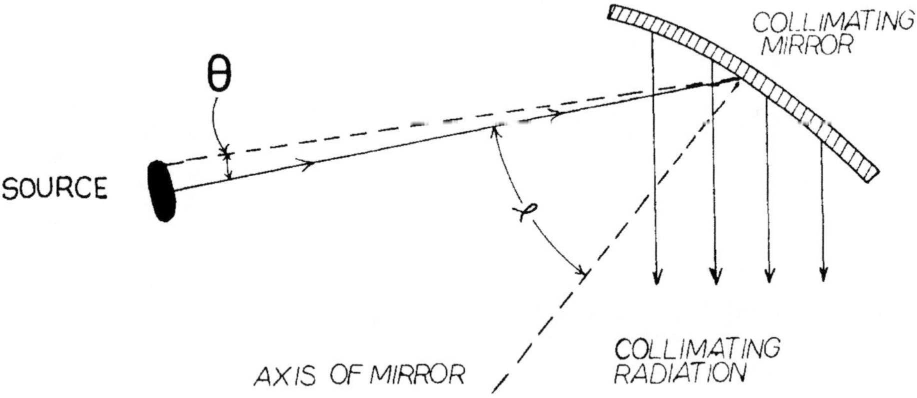

Fig. 1 Simple optical system for an interferometer.

The solid angle subtended by the source can be written as follows:

$$d\Omega = \cos\theta dA_s/r^2 \tag{2}$$

and the projected area can be expressed as:

$$dA = \cos\gamma dA_c. \tag{3}$$

Therefore, Eq. (1) can be rewritten as:

$$dE = d\Omega dA \tag{4a}$$
$$E = \Omega A \tag{4b}$$

Assuming a finite source size, the optical path difference between the two arms of a Michelson interferometer is

$$\Delta = [(\sqrt{3 + \cos^2\theta} - 1)(1 + \cos\theta) - 2]f + (\sec\theta - 1)L + 2d\cos\theta \tag{5}$$

where:

f = focal length of collimating mirror
L = distance from source to fixed mirror
d = distance from fixed to movable mirror (which is equal to the spatial displacement).

The first two terms in Eq. (5) are of different sign and are very small as compared to the third term. Thus, they tend to cancel in most cases leaving the approximation

$$\Delta \simeq 2d \cos \theta. \tag{6}$$

If the source is considered to be a point, i.e. $\theta = 0$, then Eq. (5) describing the optical path difference between the two arms of an interferometer reduces to

$$\Delta p = 2d. \tag{7}$$

It is this deviation of Δ from Δp which leads to a frequency spread, [1,2] which is

$$\Delta\bar{\nu} = \bar{\nu}_{max}\Omega/2\pi \tag{8}$$

where $\bar{\nu}_{max}$ = the highest wavenumber in the spectrum.

Substituting Eq. (8) into Eq. (4b) we obtain the relation for the throughput of a Michelson Interferometer:

$$E^I = 2\pi A^I \frac{\Delta\bar{\nu}}{\bar{\nu}_{max}} \tag{9}$$

where A^I = area of collimating mirror in the Interferometer. Superscript I will designate interferometry and D will indicate dispersive spectroscopy in our comparison.

For a <u>dispersive spectrometer</u> the solid angle may be written as

$$\Omega^D = \frac{hw}{f^2} \tag{10}$$

where: h = slit height
w = slit width
f = focal length of collimator

since:

$$w = f\frac{(d\theta)}{(d\bar{\nu})}\Delta\bar{\nu} \tag{11}$$

and

$$\bar{\nu} = \frac{m}{d(\sin i + \sin \theta)}. \tag{12}$$

The angular dispersion of the grating becomes

$$\frac{d\theta}{d\bar{\nu}} = \frac{m}{\bar{\nu}^2 \, d \cos\theta} \tag{13}$$

and the solid angle becomes

$$\Omega^D = \frac{hm}{fd \cos\theta} \frac{\Delta\bar{\nu}}{\bar{\nu}^2} \tag{14}$$

where

m = order of the grating
d = the distance between the lines of the grating.

Substituting Eq. (14) and Eq. (3) into Eq. (4b) we obtain for the throughput of a dispersive spectrometer

$$E^D = \frac{hmA^D}{fd} \frac{\Delta\bar{\nu}}{\bar{\nu}^2}. \tag{15}$$

A. Jacquinot Advantage

In 1954 Pierre Jacquinot recognized the throughput advantage of the interferometer over dispersive instruments and the ratio E^I/E^D is now usually termed "Jacquinot's advantage", [3,4].

Using Eq. (9) and Eq. (5) we obtain an equation for Jacquinot's advantage.

$$E^I/E^D = \frac{2\pi fd}{hm} \frac{A^I}{A^G} \frac{\bar{\nu}^2}{\bar{\nu}_{max}} \tag{16}$$

Substituting approximate values for the constants in Eq. (16) one obtains

$$\frac{E^I}{E^D} = \frac{0.075\,\bar{\nu}^2}{\bar{\nu}_{max}}. \tag{17}$$

It should be noted that gratings are mechanically changed several times during a complete scan of the mid infrared region and therefore d can be a function of the frequencies, as $d = d(\bar{\nu})$. Since the value of d increases stepwise with decreasing frequency the value E^I/E^D may be thought of as being "slightly less than quadratic" in frequency.

It should also be noted that Jacquinot's advantage is a measure of the amount of radiation striking the detector. Thus, this advantage is most important when the amount of radiation is minimal, e.g. such as the measurement of weak emissions, or at high resolution interferometric studies.

B. Fellgett's Advantage

If one measures a spectrum of width $\delta\bar{\nu}$ at a resolution $\Delta\bar{\nu}$ the number of spectral elements is:

$$M = \frac{\delta\bar{\nu}}{\Delta\nu} \tag{18}$$

In a dispersive instrument the time devoted to each spectral element is therefore T/M where T equals the total active time of the scan. The total active time of data acquisition may be considered to be the total time during which the sample was being irradiated.

If the noise is random, the signal-to-noise ratio becomes

$$(S/N)^D \propto \frac{T}{M}. \tag{19}$$

In an interferometer, all frequencies are impinging upon the detector during the entire active time. Therefore, for an interferometer

$$(S/N)_I \propto T, \text{ for } M = 1. \tag{20}$$

Fellgett's advantage [5,6] states that for equal sources, active times, and resolutions, the ratio of signal to noise ratios becomes

$$\frac{(S/N)^I}{(S/N)^D} = M \tag{21}$$

or that for equal signal-to-noise ratios the time required is

$$\frac{T^I}{T^D} = \frac{1}{M}. \tag{22}$$

For a scan of the entire mid infrared region at a resolution of 2 cm^{-1}, the magnitude of Fellgett's advantage may be calculated as:

$$\frac{(S/N)^I}{(S/N)^D} = \frac{(4000 - 400)^{\frac{1}{2}}}{(2)} = 42$$

It should be noted that Fellgett's advantage increases with increasing resolution, and that it becomes very important in high resolution ($\Delta\bar{\nu} < 2$ cm^{-1}) research.

It must be remembered however, that the active time of Eq. (19) is much less than the total time involved in obtaining a spectrum. In a dispersive instrument where the spectrum is usually plotted as the data are being recorded, the active time is approximately one-fourth to one-half of the total time between the start of the scan and the finished spectrum.

In an interferometer, however, the spectrum cannot be plotted until all of the interferograms have been recorded, averaged, and apodized, then the Fourier transform is performed. If only one interferogram is to be recorded, the inefficiency of the interferometer negates Fellgett's advantage. However, if several interferograms are to be averaged, the Fourier transform of previous interferograms can be averaged while the next interferogram is being collected. This improves the efficiency of the interferometer and brings the actual total time closer to the total active time of Fellgett's advantage. Simply, one can view the Fellgett advantage as one does when he multiplexes the signal.

C. Connes' Advantage

Connes' advantage may be simply stated as the ability of the interferometer to very accurately characterize the frequency. This advantage stems from the accurate measurement of the displacement of the moving mirror, by a He-Ne laser, and allows the subtraction of background spectra.

D. Cooley-Tukey Algorithm

Another way in which the efficiency of the interferometer has been improved lies in the method by which the Fourier transform is calculated. In 1966 Cooley and Tukey described an algorithm by which the total number of operations required to calculate the Fourier transform is greatly reduced, [7]. If sufficient zeros are added such that the total number of data points equals an integral power of two, theory predicts that the time required for the calculation by the "fast method", t_f, should be proportional to $\log_2 N/N$ conventional. In 1966 Forman [8] demonstrated experimentally that the proportionality constant was approximately equal to 1.8.

E. Resolution Advantage

Whereas Connes' advantage deals with the frequency accuracy of spectral data, the resolution advantage deals with the precision or resolution of the band. The resolution of an interferometer may be given as:

$$\Delta\bar{\nu} = \Delta^{-1}. \tag{23}$$

Therefore, for Δ equal to 10 cm, as is common in many commercial interferometers, $\Delta\bar{\nu}$ equals 0.1 cm^{-1}. This high resolution capability has many applications but most obvious is the elucidation of complex rotational structure in gaseous molecules.

F. Detector Performance

One factor which can be a definite disadvantage of the interferometer in our comparison to dispersive spectroscopy is the detector. Unlike a dispersive instrument where all frequencies are modulated, or chopped, at a constant frequency (ca. 10 Hz) the modulation frequency of an interferometer is dependent on the frequency of source radiation. The modulation frequency may be calculated as

$$\nu_{mod} = 2v\bar{\nu} \tag{24}$$

where v = velocity of moving mirror.

Using a typical value of 1.5 mm per sec. for v one finds that modulation frequencies may range from 3 kHz ($\bar{\nu}$ = 10,000 cm^{-1}) to 3 Hz ($\bar{\nu}$ = 10 cm^{-1}). Unfortunately, the thermocouples commonly used in dispersive spectroscopy have response times much greater than 1 ms and therefore cannot detect radiation modulated in excess of 1 kHz. To alleviate this problem most interferometers use triglycine sulfate (TGS) or deuterated triglycine sulfate (DTGS) pyroelectric bolometers. These detectors are capable of detecting radiation modulated in excess of 1 MHz but suffer from a much lower sensitivity (i.e. the D* can be less by an order of magnitude) than thermocouples. This loss of sensitivity can be very serious, reaching more than an order of magnitude difference for high frequency modulation.

If only a small limited portion of the infrared is to be observed, the TGS detector may be replaced with a photoconductive or photovoltaic detector. These detectors are much more sensitive than TGS detectors and are comparable to the thermocouples of dispersive spectroscopy.

3. SUMMARY

With the present state-of-the-art from the above optical considerations, it is evident that the number of advantages clearly favors interferometry over dispersive spectroscopy. However, the magnitude of these advantages is strongly dependent on factors such as required resolution, length of scan, available time, etc.

All of these factors must be considered before any advantage can be quantified.

It must be remembered that statements such as "poor performance largely offsets Jacquinot's advantage" are only approximations and must not be used for truly quantitative calculations, [9,10]. It must also be remembered that what might be an advantage in one instance (i.e. very high resolution) might be a needless complication in general survey work.

It is apparent that, if cost is not a factor, for most applications the interferometer is superior to the grating spectrometer. However, the requirements of each individual experiment must always be the deciding factor.

ACKNOWLEDGMENT

This work was supported by AFOSR, Grant No. 78-3617.

REFERENCES

1. W. H. Steel, J. Op. Soc. Am., 54, 151 (1964).
2. R. J. Bell, Introductory Fourier Transform Spectroscopy, Academic Press, New York, 1972, p. 149.
3. P. Jacquinot, 17^{e} Congres. du GAMS, Paris, 1954.
4. P. Jacquinot, J. Op. Soc. Am., 44, 761 (1959).
5. P. Fellgett, J. Phys. Radium, 19, 187 (1958).
6. Ibid., p. 237.
7. J. W. Cooley and J. W. Tukey, Math. Comp., 19, 297 (1965).
8. M. L. Forman, J. Op. Soc. Am., 56, 978 (1966).
9. P. R. Griffiths, Chemical Infrared Fourier Transform Spectroscopy, Wiley, New York, 1975.
10. P. R. Griffiths, H. J. Sloane and R. W. Hannah, Appl. Spectrosc., 31, 485 (1977).

BASIC THEORY AND INSTRUMENTATION FOR FT-IR SPECTROMETRY

Peter R. Griffiths

Department of Chemistry
Ohio University
Athens, Ohio, U. S. A.

1. THE BASIC MICHELSON INTERFEROMETER

A. Monochromatic Sources

The basic instrument for Fourier transform infrared (FT-IR) spectrometry is the Michelson interferometer. In its simplest form, the Michelson interferometer consists of two mutually perpendicular plane mirrors, one of which can travel in a direction perpendicular to its plane. Bisecting the planes of these two mirrors is a semi-reflecting film, or beamsplitter. An ideal beamsplitter has zero absorption, and a reflectance and transmittance both equal to 0.50.

If a collimated beam of monochromatic radiation of wavelength λ cm is passed onto an ideal beamsplitter, 50% of the incident radiation will be reflected to one of the mirrors (say, the fixed mirror) and 50% will be transmitted to the other (movable) mirror. After reflection from these mirrors, the two beams return to the beamsplitter where they recombine and interfere. Fifty percent of the beam which was reflected from the fixed mirror is transmitted through the beamsplitter while 50% of this beam is reflected back in the direction of the source. Similarly, 50% of the beam which was reflected from the moving mirror is reflected by the beamsplitter and 50% of this beam is transmitted back to the direction of the source. The beam which returns in the direction of the source is often called the reflected beam. If (as is usually the case) the input beam hits the fixed and moving mirrors normally, this beam returns along the same path as the input beam and cannot be observed. The beam which emerges from the interferometer at 90° to the input beam is called the transmitted beam,

J. R. Durig (ed.), Analytical Applications of FT-IR to Molecular and Biological Systems, 11–24.

and it is this beam that is usually detected in FT-IR spectrometry.

If the fixed and movable mirrors are equidistant from the beamsplitter, there is no path difference between the paths of the two beams when they recombine at the beamsplitter. In this case the two beams interfere constructively for the transmitted beam and destructively for the reflected beam. This is also the case for path differences equal to any integral number of wavelengths, $n\lambda$. For path differences of $(n + \frac{1}{2})\lambda$, the two beams interfere destructively for the transmitted beam and constructively for the reflected beam, and it is a simple matter to show that for a monochromatic source of intensity, $I(\lambda)$, the intensity of the transmitted beam as a function of optical path difference, or retardation, x cm, is given by:

$$I'(x) = 0.5\ I(\lambda)\ \{1 + \cos 2\pi x/\lambda\}. \qquad (1)$$

$I'(x)$ is composed of a constant component, $0.5I(\lambda)$, and a modulated component, $0.5\ I(\lambda) \cos 2\pi x/\lambda$. This modulated component is called the interferogram, and since the wavenumber, ν cm^{-1}, is simply $1/x$, the interferogram is given in terms of wavenumber as:

$$I(x) = 0.5\ I(\nu) \cos 2\pi\nu x. \qquad (2)$$

Obtaining spectral information from this sinusoidal interferogram is a trivial matter since the amplitude is directly proportional to $I(\nu)$ and its wavelength is directly proportional to the wavelength of the source. However we could obtain the same information by a series of sinusoids of unit amplitude with different wavelengths. Only when the analyzing sinusoid has the same wavelength as the measured interferogram does the product has a nonzero value. For all other wavelengths the product is zero provided that x is large enough to be considered infinite. This operation is equivalent to calculating the cosine Fourier transform of $I(x)$:

$$I(\nu) = \int_{-\infty}^{+\infty} I(x) \cos 2\pi\nu x\ dx. \qquad (3)$$

B. Polychromatic Sources

If more than one discrete frequency is emitted by the source, the interferogram is the sum of the individual interferograms due to each wavenumber, i.e.:

$$I(x) = 0.5 \sum_{i=1}^{i=N} I(\nu_i) \cos 2\pi\nu_i x. \qquad (4)$$

For a continuous source, the interferogram is the integral of the contributions from all wavenumbers in the spectrum, i.e.

$$I(x) = 0.5 \int_{-\infty}^{+\infty} I(\nu) \cos 2\pi\nu x \, d\nu. \tag{5}$$

For either of the above cases, the spectrum can be calculated by computing the cosine Fourier transform as in Eq. (3).

C. Non-ideal Systems

The discussions above were for the case of an ideal beamsplitter and an interferometer for which all wavelengths are in phase at zero retardation. A beamsplitter has never been constructed for which the reflectance, $R(\nu)$, and transmittance, $T(\nu)$, are exactly 0.50 for all wavenumbers. In this case, the spectral intensities, $I(\nu)$, in Eq. (5) must be multiplied by a wavenumber-dependent beamsplitter efficiency, $H(\nu)$, where:

$$H(\nu) = 4\, R(\nu)\, T(\nu). \tag{6}$$

For the actual measurement of $I(x)$, an infrared detector is used in conjunction with an amplifier to bring the level of the signal to the correct value for recording. The response characteristics of detectors usually vary with wavenumber to a greater or lesser extent, and also depend on the modulation frequencies present in the interferogram. Thus a measured interferogram is given by:

$$I(x) = 0.5 \int_{-\infty}^{+\infty} I(\nu)\, H(\nu)\, R(\nu) \cos 2\pi\nu x \, d\nu \tag{7}$$

where $R(\nu)$ is the response of the detector and amplifier. We can combine the product $0.5\, I(\nu)\, H(\nu)\, R(\nu)$ into one term, $B(\nu)$, so that:

$$I(x) = \int_{-\infty}^{+\infty} B(\nu) \cos 2\pi\nu x \, d\nu. \tag{8}$$

As we will see later in the discussion of rapid-scan and slow-scan interferometers, the moving mirror of certain types of interferometer moves so slowly that an external chopper must be used to modulate the beam. However, in most interferometers in use today, the moving mirror is translated with a high velocity, V cm sec^{-1}, so that the modulation frequencies due to each wavenumber in the spectrum, $f(\nu)$, are given by:

$$f(\nu) = 2V\nu \text{ Hz} \tag{9}$$

and fall in the audio-frequency range. The detector signal is then amplified by a broadband amplifier obviating the need for an auxiliary chopper. The only problem with this type of system is that a frequency-dependent phase-lag, $\theta(\nu)$, can be introduced by the amplifier and electronic filters, so that all the component waves in the interferogram are not in-phase at the zero retardation point. In this case the interferogram is given by:

$$I(x) = \int_{-\infty}^{+\infty} B(\nu) \cos [2\pi\nu x + \theta(\nu)] \, d\nu. \tag{10}$$

Whereas the ideal interferogram given in Eq. (5) is symmetrical about the x = 0 point, interferograms generated by rapid-scanning interferograms are asymmetrical about the zero retardation point. By analogy to radar terminology, this effect is called chirping. The phase-angle, $\theta(\nu)$, must be calculated and corrected before spectra can be computed with high photometric accuracy.

The cosine term in Eq. (10) may be expanded:

$$\cos[2\pi\nu x + \theta(\nu)] = \cos 2\pi\nu x \cos \theta(\nu) - \sin 2\pi\nu x \sin \theta(\nu). \tag{11}$$

The sine and cosine terms are orthogonal, and by analogy to the expression for transcendental exponential functions:

$$\exp(i\phi) = \cos \phi - i \sin \phi \tag{12}$$

we can express Eq. (10) as:

$$I(x) = \int_{-\infty}^{+\infty} B(\nu) \exp (2\pi i\nu x) \, d\nu. \tag{13}$$

The spectrum, $B(\nu)$, can be calculated from chirped interferograms by means of the complex Fourier transform:

$$B(\nu) = \int_{-\infty}^{+\infty} I(x) \exp(-2\pi i\nu x) \, dx. \tag{14}$$

Equations (13) and (14) represent a complex Fourier transform pair.

In order to compute $B(\nu)$, it is necessary to know the form of $\theta(\nu)$. In this respect we are aided by two observations: firstly, $\theta(\nu)$ usually varies slowly with ν, so that there usually exists a region close to x = 0 where most frequencies are approximately in phase, and secondly it is possible to scan the moving mirror so that the interferogram may be sampled on either side of the zero retardation point. Because of this it becomes possible to compute a low resolution phase spectrum prior to computing $B(\nu)$.

Before 1964, the Fourier transform of an N-point interferogram was performed by a series of multiplications, with N multiplications required for each output point. In 1964, Cooley and Tukey developed a matrix method for computing the Fourier transform which drastically reduced the number of computer operations needed to compute N output points. The enormous time-saving of the Cooley-Tukey (or fast Fourier transform, FFT) algorithm, together with the development of minicomputers, has directly led to the increased popularity of FT-IR spectrometry, since it became possible to compute the spectrum rapidly immediately after the measurement of the interferogram.

2. SAMPLING THE INTERFEROGRAM

The FFT is computed using the digitized values of the interferogram sampled at equal intervals of retardation. The sampling interval determines the spectral range to be covered - the larger the spectral range desired, the shorter the sampling interval. For a spectral range of $(\nu_{max} - \nu_{min})$ cm^{-1}, a sampling interval, Δx, is required, where:

$$\Delta x = 0.5(\nu_{max} - \nu_{min})^{-1} \text{ cm.} \tag{15}$$

The sampling points should be separated by precisely equal intervals, otherwise noise will be introduced into the computed spectrum.

The dynamic range of the analog-to-digital converter (ADC) is also important. The dynamic range must be great enough that the noise is greater than the least significant bit of the ADC. Since the signal-to-noise ratio (SNR) at the zero retardation point can be very large in mid-infrared absorption spectrometry, it becomes necessary to use an ADC with as large an ADC as possible, typically 15-bits and even 16-bits. Even with an ADC with such a large dynamic range, the SNR of mid-infrared interferograms using incandescent sources would still exceed the dynamic range unless the moving mirror of the interferometer is translated very rapidly. The velocity of rapid-scanning mid-infrared interferometers is usually between 1 and 10 mm sec^{-1}, whereas the mirror velocity of slow-scanning far infrared (FIR) interferometers may be two or three orders of magnitude less than this, and yet FIR sources are so weak that it is rare that even a 14-bit ADC is required.

3. FACTORS DETERMINING RESOLUTION AND LINE SHAPE FUNCTION

The ultimate factor determining the resolution of an FT-IR spectrometer is the maximum optical retardation. As an approximate rule of thumb, the resolution, $\Delta\nu$ cm^{-1}, is given by the reciprocal of the maximum retardation, X cm. Although it is fairly easy to derive this relationship intuitively, a more rigorous treatment is beneficial.

Equations (13) and (14) both have integration limits of $-\infty$ to $+\infty$. In practice no interferometer can have an infinitely long mirror travel, but we can retain the same integration limits if we multiply the theoretical (infinitely long) interferogram, I(x), by a boxcar truncation function, D(x), which is unity for values of x covered by the interferometer and zero outside of these limits, i.e.:

$$B(\nu) = \int_{-\infty}^{+\infty} I(x)\, D(x) \exp(-2\pi i\nu x)\, dx. \qquad (16)$$

The Fourier transform of the product I(x) D(x) is the convolution of the Fourier transform of each function taken individually. The Fourier transform of I(x) is the true spectrum, while the Fourier transform of D(x) is a function having the form:

$$\sigma(\nu-\nu_i) = 2\frac{\sin[2\pi X(\nu-\nu_i)]}{2\pi X(\nu-\nu_i)} = 2\ \mathrm{sinc}\ [2\pi X(\nu-\nu_i)] \qquad (17)$$

$\sigma(\nu-\nu_i)$ describes the response of the instrument to radiation of frequency, ν, when the instrument measures at a frequency, ν_i, and is known as the instrument line shape (ILS) function. The analogy to the case of a dispersive spectrometer, where the true spectrum is convolved with the slit function of the monochromator, is obvious. The sinc function in Eq. (17) has its first zero crossing at $(\nu-\nu_i) = \pm(2X)^{-1}$ and a half-width of 0.605/X cm^{-1}, so that the resolution is a little better than X^{-1} cm^{-1}.

The sinc function has very significant secondary maxima which can largely be suppressed by multiplying the interferogram by an <u>apodization function</u> which is unity at the zero retardation point and decreases monotonically until it reaches approximately zero near the maximum retardation. The most commonly used apodization function is a simple triangle, the Fourier transform of which is a sinc^2 function:

$$\sigma(\nu-\nu_i) = \frac{\sin^2[\pi X(\nu-\nu_i)]}{[\pi X(\nu-\nu_i)]^2} = \mathrm{sinc}^2\ [\pi X(\nu-\nu_i)]. \qquad (18)$$

This ILS function is identical to the slit function of a diffraction-limited grating monochromator. Other functions, including Gaussian and exponential functions, have also been used in FT-IR spectrometry, but the triangular function is by far the most common. The half-width of the primary lobe of the sinc^2 function in Eq. (18) is $0.88/X\ \text{cm}^{-1}$, and the distance between the first two zero values is twice that of the sinc function in Eq. (17). Hence a trade-off is made between sidelobe amplitude and resolution.

Certain optical criteria also have to be obeyed for the resolution allowed in theory by the ILS function to be achieved in practice. The beam must be sufficiently collimated that at the end of the scan the difference in the path of the center ray through the interferometer and the extreme ray for the same beam must be less than one half of the shortest wavelength in the spectrum. The shorter the wavelengths being investigated and the greater the retardation, the more collimated must the beam be. In practice the maximum allowed half-angle of the beam passing through the interferometer, α_{max}, is, to a good approximation, given by:

$$\alpha_{max} = \left(\frac{\Delta\nu}{\nu_{max}}\right)^{\frac{1}{2}} \text{ radians.} \tag{19}$$

This angle is usually controlled by an aperture at a focus of the beam. Sometimes this aperture is imposed by some component of the instrument such as the source, sample, or detector, but for high resolution measurements the half angle is controlled by a variable aperture situated between the source and the interferometer.

Another potential source of resolution degradation in FT-IR spectrometry is a poor mirror drive. If the moving mirror tilts to produce a path difference of more than one half of the strongest wavelength in the spectrum between the two extreme rays of the beam, the amplitude of the interferogram for that wavelength will drop to zero. Thus the higher the resolution required and the higher the wavenumber range to be studied, the higher is the precision required of the mirror drive. It is no coincidence that the first commercial interferometers were designed either for far infrared measurements or very low resolution mid-infrared measurements, since no great difficulty is encountered in designing an adequate drive mechanism. It also shows why the high resolution ($0.002\ \text{cm}^{-1}$), near infrared ($10{,}000\ \text{cm}^{-1}$) FT-IR measurements of the French school pioneered by Connes were so difficult to achieve.

4. ADVANTAGES OF FT-IR SPECTROMETRY

There are two principal advantages of Fourier spectrometry over dispersive monochromators. The first of these is the <u>multiplex advantage</u> (or Fellgett's advantage). Measurements of equal

SNR and resolution, taken on instruments with equal optical throughput and efficiency with identical sources, will take M times less time on a Fourier spectrometer, where M, the number of resolution elements, is given by:

$$M = (\nu_{max} - \nu_{min})/\Delta\nu. \tag{20}$$

For measurements taken with equal data acquisition times, the SNR of measurements taken on a Fourier spectrometer will be $(M)^{1/2}$ times better than the SNR of the same measurement taken on a dispersive spectrometer.

The other principal advantage of FT-IR spectrometry is the fact that the optical throughput of an interferometer is greater than that of a monochromator operating at the same resolution. For mid-infrared FT-IR spectrometers, this advantage (known as Jacquinot's advantage) is typically between one and two orders of magnitude. In many introductory articles on FT-IR spectrometry, one can read that Jacquinot's advantage results from the absence of slits in a Fourier spectrometer, but this is not correct since a throughput-limiting aperture (which is equivalent to the entrance slit of a monochromator) is always located somewhere in the optical train of a Fourier spectrometer. The decreased throughput of the slits of a monochromator is the cause of Jacquinot's advantage, while the lack of an "exit slit" (or spatial wavelength selector) in an FT-IR spectrometer is the cause of Fellgett's advantage.

If the total gain in SNR given by the combination of Fellgett's and Jacquinot's advantage is calculated, it might be thought that the performance of Fourier spectrometers for the mid-infrared would be greater than is actually found in practice. The reason for this is that the detectors used in commercial FT-IR and grating spectrometers are different. As noted above the scan speed of mid-infrared interferometers must be high enough that the SNR of the interferogram is kept less than the dynamic range of the ADC. Since the modulation frequencies in the interferogram ($2V\nu$ Hz) are typically from 100 Hz to 10 kHz, the response time of the detector must be very short, much shorter than the response time of thermocouple detectors usually used in grating spectrometers. The only type of room temperature, wide-range mid-infrared detectors with an adequate response in the kiloHertz range are the pyroelectric bolometers, and even the best pyroelectric (triglycine sulfate, or TGS) is between one and two orders of magnitude less sensitive than a thermocouple. Practically, this disadvantage of Fourier spectrometers effectively offsets Jacquinot's advantage, leaving only Fellgett's advantage for the FT-IR spectrometer.

If the energy passing through the interferometer is reduced, either by reducing the integrated energy from the source, or by absorption, scattering or vignetting by a sample, a more sensitive detector than TGS may be used. The most useful detector in this respect for mid-infrared measurements is the liquid nitrogen cooled mercury cadmium telluride (MCT) photoconductive detector. Depending on the cut-off wavenumber (typically 800 to 400 cm^{-1} depending on the composition), MCT detectors may be anywhere from four to forty times more sensitive than the TGS detector. For measurements at higher wavenumber, the sensitivity of photoconductive and photovoltaic detectors may be orders of magnitude greater than that of the MCT detector, so that measurements of extremely weak sources may be measured.

5. INTERFEROMETER DESIGN

The first interferometers developed for FT-IR were for far infrared measurements and were very simple mechanically. The mirror was typically driven slowly through a lubricated cylinder by a screw drive. The fringes generated by Moiré gratings were used to trigger the ADC at equal intervals of retardation. For mid-infrared measurements the accuracy required for the drive and sampling is too great to allow simple modifications to far infrared interferometers to be made, and mid-infrared interferometers are far more complex.

These interferometers must have a rapidly translated mirror which must be kept in the same plane to better than 10 μ-radians for scans up to 10 cm in length, and sampling has to be accurately performed over this entire distance. In addition, interferograms must be able to be coherently signal-averaged, so it is necessary to have some type of a marker to initiate data acquisition at precisely the same retardation for every scan.

The moving mirror of most mid-infrared interferometers is driven on an air-bearing, which allows the planarity specification to be met for retardations as long as 20 cm. The moving mirror of a smaller reference interferometer is driven by the same drive used to drive the mirror of the main interferometer, and the two mirrors are attached to each other. Through this reference interferometer is passed light from a visible white light source. The interferogram of this source has a very sharp spike at zero retardation, and data acquisition of the infrared interferogram is initiated when this signal exceeds a certain threshold. The uncertainty in this position relative to the signal interferogram is typically less than 0.1 μm from one scan to the next. The sampling points of the interferogram are usually triggered by the zero-crossings of the sinusoidal interferogram generated by passing a helium-neon laser either into the main interferometer or into the reference interferometer.

The two most popular commercial mid-infrared interferometers (those made by Digilab, Inc. and the Nicolet Instrument Corp.) have their fixed and moving mirrors mutually perpendicular, and use a beamsplitter which consists of a thin layer of germanium deposited on an optically flat KBr substrate with an identical KBr flat covering this layer to prevent excessive chirping of the interferogram. Beamsplitters made of a pellicular film of polyethylene terephthalate (Mylar, Melinex), whose thickness controls the spectral range which can be studied, are used in the far infrared. For near infrared measurements, thin layers of germanium, silicon or Fe_2O_3 on CaF_2 or quartz substrates (with identical compensators) are used.

Some interesting variations on these basic designs are also starting to appear in commercial interferometers. For example the interferometer used in the Bruker FT-IR spectrometers has a novel optical system in which both its mirrors and its beamsplitter are coplanar. The beam from the source is focused onto the beamsplitter, and the two beams from the beamsplitter are collected, collimated and passed onto each side of a moving mirror element. If the element moves a distance x, the optical retardation changes by 4x, compared with the value of 2x for a conventional interferometer design. This advantage is relatively minor compared with the advantage of focusing the beam on the beamsplitter, allowing it to be made much smaller than the beamsplitter of a conventional Michelson interferometer. Not only does this reduce the cost of the beamsplitter but it makes it possible to hold several beamsplitters optimized for different spectral regions on a wheel. Thus it becomes possible to change from one spectral region to another without breaking the purge or vacuum of the spectrometer.

A radically different type of interferometer is the design of the Analect Instrument Division of Laser Precision Corp. In this interferometer, the path difference is generated by moving a refractive wedge along the plane in which the beamsplitter is held in a conventional Michelson interferometer. Instead of a fixed and moving plane mirror, two fixed corner retroreflectors are used. It is claimed that this is a cheaper and more reliable way of generating an accurate drive than the use of a moving mirror, and this interferometer forms the basis of the least expensive mid-infrared FT-IR spectrometer currently (January, 1980) on the market.

Two other types of interferometers are each used by several workers, but are not as popular as the systems described above. These are the lamellar grating and cats-eye interferometers.

Lamellar grating interferometers are usually used for far infrared spectrometry below 100 cm^{-1}. If a conventional Michelson interferometer with a Mylar beamsplitter is used, the beamsplitter efficiency never approaches 100% and the spectral range which can be covered is quite limited; use of a lamellar grating interferometer overcomes this deficiency. In this instrument, a collimated beam of radiation is passed into a device consisting of two interleaved sets of mirrors, one set of which is fixed while the other moves creating a path difference between those parts of the incident beam reflected from the fixed and movable facets. A lamellar grating interferometer was once sold by the RIIC Division of Beckman Instruments but, to the knowledge of this author, is no longer commercially available.

For very high resolution measurements, the quality of the air-bearing drive used for medium resolution interferometers is not quite good enough, and special precautions must be taken to compensate for tilt. The most successful method of achieving this goal is through the use of cats-eye retroreflectors in the two beams from a beamsplitter. Since the throughput of very high resolution measurements is low, ADC dynamic range is rarely a limitation so that the mirror does not have to be translated very rapidly. In fact usually a stepped-scan is employed in these interferometers, where the moving mirror element is realigned each step using a system of several lasers. Interferometers of this type have been developed largely by the French, and similar systems are now being built commercially by Idealab in America.

6. DATA SYSTEMS FOR FT-IR SPECTROMETRY

In the 1960's, interferograms were usually collected on paper or magnetic tape, and spectra were subsequently computed using a remote mainframe computer. Nowadays, for all except the highest resolution measurements, data may be acquired and computed on-line using a minicomputer, often in conjunction with some peripheral memory such as a disc. A typical modern data system will have the following components:

a. minicomputer with 16K-64K work memory;
b. a moving head disc, often with 5M words of storage;
c. a 15- or 16-bit ADC capable of operating up to 50 kHz;
d. a complex input-output (I/O) interface;
e. a data terminal for inputting parameters and commands;
f. an oscilloscope with graphic and alphanumeric display;
g. a high speed digital plotter.

With commercial data systems of this type, it is possible to collect and transform interferograms up to 256K words in length, although this would be a much larger interferogram than most chemical spectroscopists would need, since a 2 cm^{-1} resolution mid-infrared spectrum would only require 8K data points.

There is a continuing controversy in FT-IR spectrometry about the optimum word-length for the minicomputer. Currently, each of the three major manufacturers of FT-IR spectrometers uses a computer with different word-lengths. Digilab uses Data General minicomputers with a 16-bit word, while Nicolet and Bruker use their own computers which have a 20-bit and 24-bit word-length, respectively. The relative advantages of each type of computer should be discussed in terms of digitization noise introduced by extensive signal-averaging of data with a high SNR digitized using a 15- or 16-bit ADC and also in terms of possible noise introduced during the Fourier transform. For interferograms with a SNR of, say, 2^{14} (a fairly typical value for mid-infrared FT-IR absorption spectrometry), a maximum of sixteen scans can be averaged using a computer with a 16-bit word-length before the SNR exceeds the word-length. With a 20-bit and 24-bit word computer, the number of scans are (approximately) two thousand and half a million, respectively. The obvious limitations of the 16-bit word computer can be overcome by the use of double-precision software, giving the computer effectively a 32-bit word-length. The 20-bit word computer only has dynamic range limitations if long data acquisition times are needed, and these limitations can be circumvented by block-averaging, i.e. averaging, say, one thousand scans, computing and storing the spectrum, and repeating the process as often as necessary, averaging the resulting spectrum each time. The 24-bit computer has essentially no dynamic range limitations for the vast majority of measurements. There is really nothing wrong with the use of double precision with a 16-bit word machine except for the increased computation time required for the FFT, but a 24-bit word computer is almost ideal for FT-IR spectrometry.

It is much less simple to discuss noise introduced during the Fourier transform quantitatively. In the only extensive study performed in this area, Foskett has shown that for any conceivable measurement, a word-length greater than 27-bits will never be required. Thus any computer used in double-precision will almost never contribute to the increase of noise during the transform.

In terms of software, all commercial data systems permit signal-averaging, apodization, phase correction, the FFT, and a variety of array arithmetic routines including ratioing, conversion of a transmittance spectrum to absorbance, and scaled subtraction. In addition, many data systems allow BASIC and FORTRAN programming, allowing the spectroscopist to manipulate and display spectral data in ways which are not possible using standard array arithmetic routines.

The end of the 1970's saw several developments which perhaps show us where developments in the data systems for FT-IR spectrometers are going to take place. Large versatile data systems of the type described above are going to become even more versatile. With the development of sophisticated hardware and software for memory mapping, it will become possible to run several interferometers from the same computer, and several tasks (data acquisition, the FFT, and plotting) for each will be able to be performed simultaneously. Array processors will be used to substantially decrease the computation time. Already, Digilab has developed a microprogrammable processor which enables a 4K FFT to be performed in double-precision in less than half a second. However it is in the field of inexpensive FT-IR spectrometers where I expect the largest advances to be made. The word-size and memory of microcomputers are now getting large enough that microcomputers will shortly be used to construct a really inexpensive FT-IR spectrometer. With such a data system, it is quite possible that most current users of small grating spectrometers will switch to FT-IR spectrometers in this decade. The 1960's was the decade of the mainframe computer for FT-IR spectrometer, the 1970's saw the predominance of the minicomputer, and the 1980's will probably see the development of microcomputer based data systems for FT-IR spectrometers.

7. BIBLIOGRAPHY

There are three comprehensive texts on the subject of FT-IR spectroscopy. These are, in order of decreasing simplicity:

P. R. Griffiths, Chemical Infrared Fourier Transform Spectroscopy, Wiley-Interscience, New York (1975).
R. J. Bell, Introductory Fourier Transform Spectroscopy, Academic Press, New York (1972).
J. Chamberlain, Principles of Interferometric Spectroscopy, Wiley-Interscience, New York (1979).

In addition to these books, one other book is devoted to various types of transform spectroscopy, including infrared:

P. R. Griffiths (ed), Transform Techniques in Chemistry, Plenum Publishing Corp., New York (1978).

The introduction to the Fourier transform by Foskett in this book is excellent, and there is also a very useful introduction to multiplex methods in spectroscopy by Comarisow and Marshall.

I have made no attempt to give references in this chapter, since they would be incomplete at best. If workers in the field of FT-IR spectrometry wish to keep up with instrumental developments in this area, perhaps the most important journal to read on a regular basis is Applied Spectroscopy. Other journals having occasional important articles include Analytical Chemistry and Applied Optics.

APODIZATION AND PHASE CORRECTION

John E. Bertie

Department of Chemistry
University of Alberta
Edmonton, Alberta, Canada

1. INTRODUCTION

Consider the Michelson Interferometer shown in Fig. 1 to be perfectly aligned. The intensity, I(x), of radiation that reaches the detector when the path-difference between the two arms of the interferometer is x is given by

$$I(x) = \int_0^\infty I_o(k) \cdot T(k) \cdot S(k) \; (1 + \cos 2\pi kx)dk; \qquad (1)$$

k is the wavenumber of the radiation in the gas, or vacuum, which fills the dashed box in Fig. 1; $I_o(k)$ is the intensity of radiation emitted by the source at wavenumber k; S(k) is the transmittance of the sample at wavenumber k; and T(k) is the part of the transmittance of the instrument that is independent of path difference. For many purposes, Eq. 1 can be abbreviated to

$$I(x) = \int_0^\infty A(k)(1 + \cos 2\pi kx)dk. \qquad (2)$$

The graph of I(x) versus x is the interferogram and A(k) = $I_o(k)$ S(k) T(k) is the desired quantity, the intensity as a function of wavenumber [1-4].

At zero path-difference,

$$I(o) = 2 \int_0^\infty A(k)dk$$

and at path differences that are very large relative to the longest radiation wavelength (1/k) in the spectrum of the source, x = ∞, the average value of I(x) is I(∞), given by

J. R. Durig (ed.), Analytical Applications of FT-IR to Molecular and Biological Systems, 25–50.

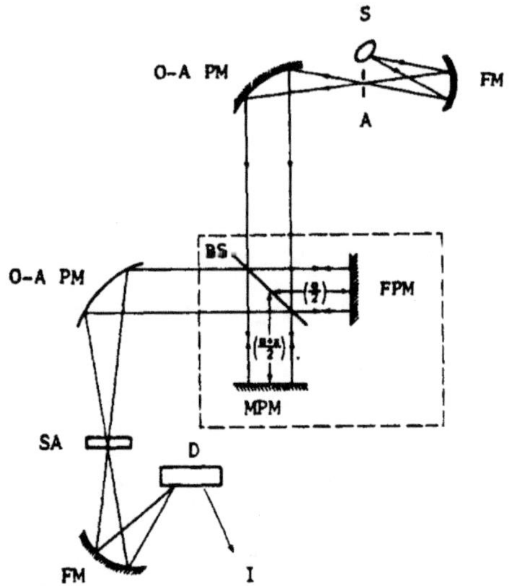

Fig. 1 A Michelson Interferometer for infrared radiation: A, aperture; S, source; FM, focussing mirror; O-A PM, off-axis parabolic mirror; FPM, fixed plane mirror; MPM, moving plane mirror; BS, beam splitter; D, detector; SA, sample; I, interferogram, i.e. intensity as a function of the path difference "x".

$$I(\infty) = \int_0^\infty A(k)\ dk = I(o)/2.$$

Thus we can calculate the interferogram function, F(x), to be

$$F(x) = I(x) - I(\infty) = \int_0^\infty A(k)\ \cos(2\pi kx)dk \tag{3}$$

and from Fourier transform theory [5,6] we can calculate the desired spectrum, i.e., the intensity as a function of wavenumber, A(k), to be

$$A(k) = 2 \int_{-\infty}^{+\infty} F(x)\ \cos(2\pi kx)dx. \tag{4}$$

I shall be concerned in this chapter with two problems that arise in the practical implementation of Eq. 4.

The first problem is that it is impossible to measure F(x) from plus to minus infinity, so that Eq. 4 can only be integrated over finite limits, which is an approximation. This approximation can be serious but can be made acceptable by the procedure called apodization. Apodization is discussed in Sec. 2.

Equation 2 describes a symmetrical interferogram such as that shown in Fig. 2, and Eq. 3 describes the corresponding sym-

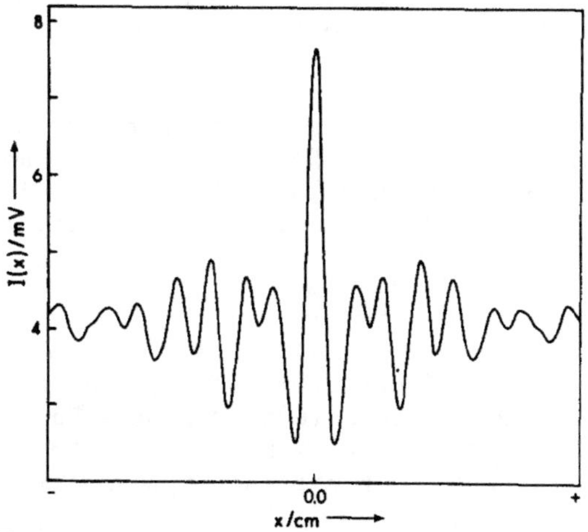

Fig. 2 A symmetrical interferogram

metrical interferogram function. The second problem arises because the interferogram function that must be Fourier transformed is rarely symmetrical. Even a symmetrical interferogram function usually yields an unsymmetrical one when it is digitized, because it is practically impossible to ensure that one data point corresponds exactly to $x = 0$. Usually digitization is done at the positions

$$x' = \delta + n\Delta x \quad \text{instead of} \quad x = n\Delta x,$$

where n is zero or integral, Δx is constant, and δ is an unknown constant. The computer regards the data point for $n = 0$ as the point for $x = 0$, δ being unknown. Thus the digitized interferogram function is unsymmetrical, being

$$F(x) = \int_0^\infty A(k)\cos[2\pi k(x + \delta)]dk \tag{5}$$

digitized at $x = n\Delta x$, where n is zero or integral, even if the analogue interferogram is symmetrical. The angle $2\pi k\delta$ is called the phase error.

Analogue interferogram functions are rarely symmetrical, particularly for the rapid-scan interferometers, and the general form of an unsymmetrical interferogram function is

$$F(x) = \int_0^\infty A(k)\cos[2\pi k(x + \delta(k))]dk \tag{5a}$$

where the $\delta(k)$ is wavenumber dependent causing the interferogram to be said to be chirped.

The procedure by which $\delta(k)$ is found and A(k) is calculated free from error is called Phase Correction. Phase Correction is discussed in Sec. 3.

The distinction between the interferogram (Eq. 2) and the interferogram function (Eqs. 3 to 5a) is not always made [2]. In the following sections it is only the interferogram function that is important and, for brevity, it will be called the interferogram.

2. APODIZATION

Consider the simplest Fourier transform, that of a symmetrical interferogram. It is never possible to evaluate Eq. 4, since it is only possible to integrate over x between finite limits, say between -X and +X. Thus we must use the approximation

$$I(k) = 2 \int_{-X}^{+X} F(x) \cos 2\pi kx \, dx, \tag{6}$$

where A(k) has been replaced by I(k), the calculated intensity, to emphasize the use of an approximation. This truncation of the interferogram is often referred to as boxcar apodization.

If the source of radiation is monochromatic with wavenumber k_a, the interferogram (Eq. 3) is simply a cosine function

$$F(x) = A(k_a) \cos 2\pi k_a x. \tag{7}$$

Equation 6 then becomes

$$I(k) = 2 \int_{-X}^{X} A(k_a) \cos(2\pi k_a x) \cdot \cos(2\pi kx) dx.$$

The integrand may be expanded as the sum of two cosine terms and integrated to yield

$$I(k) = A(k_a) \cdot 2X \cdot \left[\frac{\sin 2\pi(k_a + k)X}{2\pi(k_a + k)X} + \frac{\sin 2\pi(k_a - k)X}{2\pi(k_a - k)X}\right]. \tag{8}$$

The first term in brackets is very small as long as k_a is greater than ~10 cm^{-1}, so the spectrum of the monochromatic source obtained by this procedure is essentially

$$I(k) = A(k_a) \cdot 2X \cdot \frac{\sin 2\pi(k_a - k)X}{2\pi(k_a - k)X}$$

$$= A(k_a) \cdot 2X \operatorname{sinc} 2\pi(k_a - k)X \tag{9}$$

where the sinc function, $\frac{\sin y}{y}$, is defined by the last equality.

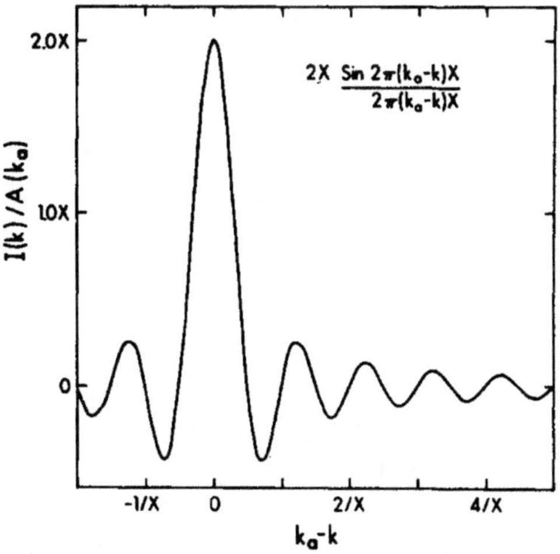

Fig. 3 Instrument line shape without apodization

The function $I(k)/A(k_a)$ given by Eq. 9 is symmetrical with respect to $k_a - k$ and is plotted in part in Fig. 3. The area under the curve is unity, the height of the central peak is 2X, the full-width of the central peak is $0.605X^{-1}$ at half-height and X^{-1} at the base where the ordinate is zero. The curve crosses zero at intervals of $(2X)^{-1}$ to either side of $k = k_a$, thus forming positive and negative side-lobes, or pods, whose heights diminish slowly, as $k_a - k$ increases, from ~20% of the height of the central peak.

Thus, if a very narrow infrared laser line at 1000 cm^{-1} illuminated a Michelson interferometer and the interferogram were recorded over the path-difference range -1 cm $\lesssim x \lesssim 1$ cm, Fourier transformation by Eq. 6 would yield a spectral line centered at 1000 cm^{-1}, with full width 0.605 cm^{-1} at half-height and 1 cm^{-1} at zero intensity, but with a series of side-lobes or pods that are strong enough to be a serious nuisance.

The shape of the spectral line given by a spectrometer when a very sharp line is studied is called the instrument line shape, and Fig. 3 thus presents the instrument line shape of a Michelson interferometer. It is clearly not satisfactory because of the strong side-lobes. Apodization is the procedure that is used to improve the instrument line shape by removing the side-lobes or pods. It can be noted in passing that this clearly makes logical nonsense of the term 'Boxcar Apodization' that is often used to describe the simple truncation of an interferogram, since the purpose of apodization is to remove the undesirable effects of the truncation. If this is understood, however, the use of the term is harmless, and sometimes convenient.

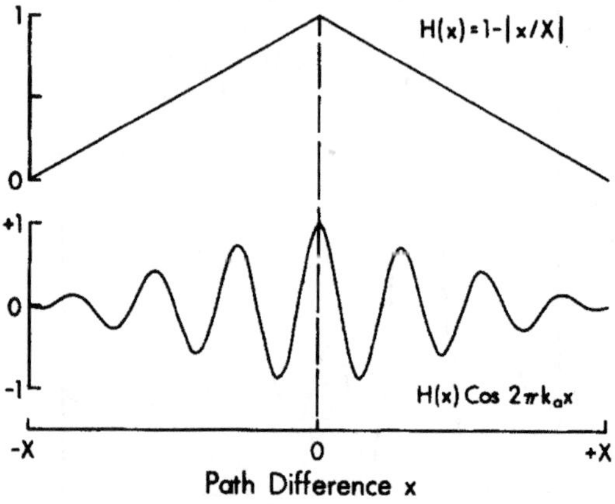

Fig. 4 Triangular apodization function and its effect on a cosine function

Apodization is achieved by multiplying the interferogram by a function, H(x), before the Fourier transform is carried out. Thus Eq. 6 becomes

$$I(k) = 2 \int_{-X}^{+X} F(x) \cdot H(x) \cos 2\pi kx \cdot dx. \tag{10}$$

If it is to be a good apodization function, the function H(x) must decay smoothly to near zero at $x = \pm X$ to reduce the discontinuity that occurs at these points when the function F(x) is truncated. A frequently used function, H(x), which has an important property is

$$H(x) = 1 - \left|\frac{x}{X}\right| \text{ if } |x| \lessgtr X, \text{ and zero otherwise.} \tag{11}$$

This function is shown, with its effect on a simple cosine interferogram, in Fig. 4. The procedure which uses this function is called triangular apodization.

For a monochromatic source of wavenumber k_a, F(x) is given by Eq. 7 and the Fourier transformation of this interferogram with triangular apodization is given by

$$I(k) = 2 \int_{-X}^{+X} A(k_a)\left[1 - \left|\frac{x}{X}\right|\right] \cos(2\pi k_a x) \cdot \cos(2\pi kx) \cdot dx. \tag{12}$$

Note that the integration limits could now be changed to plus and minus infinity without changing the result, because the integrand is zero outside of the range $-X \lessgtr x \lessgtr X$.

The integration can be done by routine methods to yield

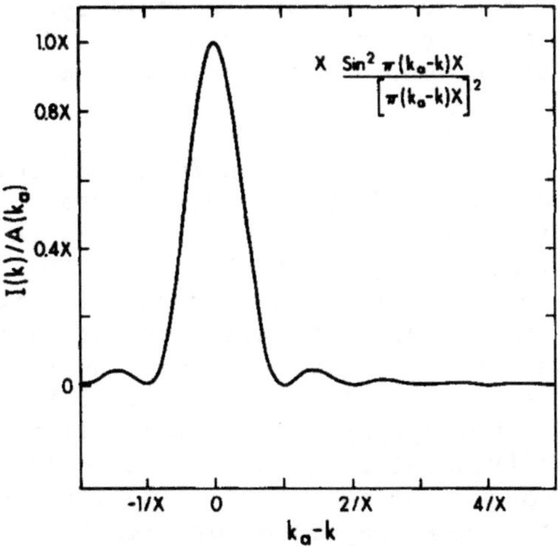

Fig. 5 Instrument lineshape with triangular apodization.

$$I(k) = A(k_a) \cdot X[\frac{\sin^2\pi(k_a + k)X}{[\pi(k_a + k)X]^2} + \frac{\sin^2\pi(k_a - k)X}{[\pi(k_a - k)X]^2}]. \tag{13}$$

As was the case for Eq. 8, the first term is extremely small for values of k_a greater than 10 cm^{-1}, so the spectrum obtained is essentially given by

$$I(k) = A(k_a) \cdot X \cdot \frac{\sin^2\pi(k_a - k)X}{[\pi(k_a - k)X]^2}$$

$$\text{i.e. } I(k) = A(k_a) \cdot X \cdot \mathrm{sinc}^2\,[\pi(k_a - k)X]. \tag{14}$$

The function $I(k)/A(k_a)$ given in Eq. 14 is plotted in Fig. 5. The area under the curve is unity, the height of the central peak is X, the full-width of the central peak is $0.88X^{-1}$ at half-height and $2X^{-1}$ at the base $(A(k)/A(k_a) = 0)$. The side-lobes are all positive and greatly reduced from those in Fig. 3, the strongest (first) pod being about 5% of the height of the main peak. Many workers use triangular apodization and, hence, accept this instrument lineshape which is, in fact, that of a diffraction-limited grating spectrometer [2,3]. This identity is not particularly significant for most infrared spectroscopists, however, because infrared grating spectrometers are usually scanning instruments operated under energy-limited conditions and have an instrument line shape that is triangular under ideal conditions.

It is clear from Figs. 3 and 5 that triangular apodization

has an important effect on Fourier transform spectra. Other apodization functions have been described [1,2,4,7], some of which are discussed in Sec. 5 with the aid of more sophisticated mathematics.

The resolution of a Fourier transform spectrometer clearly depends on the apodization function that is used and on the length of the interferogram that is transformed, 2X in the above cases. To relate these properties quantitatively the term 'resolution' must be defined, and there is [1,2] no general agreement about the most useful definition. The full-width at half-height (half-width) of the main peak of the calculated line (Figs. 3,5) is a useful measure of the resolution, and is 0.605/X without apodization or 0.88/X with triangular apodization. If the calculated spectral lines had triangular shapes, a dip would appear between the peaks of the contour of two overlapping lines of equal height when the line centers were separated by slightly more than their halfwidth. Since the lines are not triangular, this is not exactly the case, but it is nearly so. Thus, two overlapping sinc functions of equal height and width produce a single peak when their centers are separated by $\lesssim$ 1.07 halfwidths, but show a dip between the peaks of size 6% of the height of the peaks at a separation of 1.16 halfwidths, and show a 20% dip at a separation of 1.21 halfwidths. In contrast, two overlapping sinc^2 functions (Eq. 14) of equal height are just resolved when separated by their halfwidth, a 2% dip existing between the peaks; the function

$$I(k) = \frac{\sin^2[\pi(k_a - k)X]}{[\pi(k_a - k)X]^2} + \frac{\sin^2[\pi(k_a - \frac{0.88}{X} - k)X]}{[\pi(k_a - \frac{0.88}{X} - k)X]^2} \tag{15}$$

is plotted in Fig. 6 to illustrate this. Thus, a useful statement of the resolution of the interferometer is 1.16 times the halfwidth, i.e. 0.70/X, without apodization or 0.88/X with triangular apodization.

To end this section it can be noted that if all of the features in the spectrum are broad compared with X^{-1}, no significant pods are introduced into the spectrum by the truncated interferogram. This is because the modulations in the interferogram, which contain the spectral information, have faded to zero in the sampled part of the interferogram. In these cases apodization is not necessary, although it does reduce the noise level.

3. PHASE CORRECTION

The objective of phase correction is to obtain the correct Fourier transform of Eq. 5a, in which the angle $\delta(k)$ is unknown. For the moment the complication of finite integration limits will

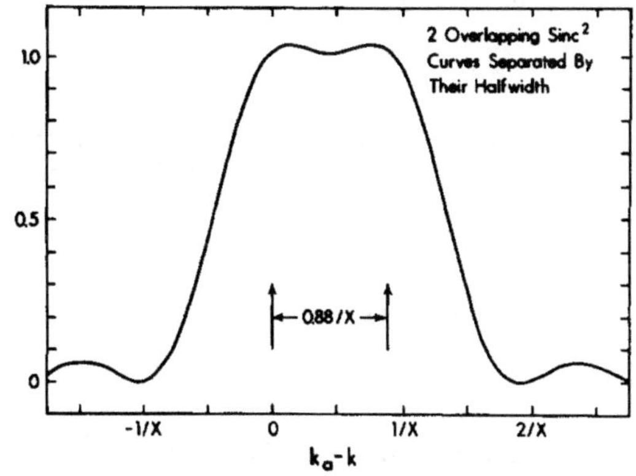

Fig. 6 The function of Eq. 15, to demonstrate the resolution of sinc^2 curves.

be ignored.

The angle $\delta(k)$ is independent of wavenumber if it arises solely because the interferogram is not sampled exactly at zero-path-difference, but for rapid-scan interferometers for the mid-infrared it is generally dependent on wavenumber. This can arise from the wavenumber dependence of the refractive index of the beamsplitter material before they are recombined. It can also arise if the electronics of the instrument, particularly electronic filters, do not have the same response for all frequencies. The moving mirror causes the path-difference between the beams to change by the wavelength, k^{-1}, of radiation of wavenumber k in the time $(2kV)^{-1}$, where V is the velocity of the moving mirror. Thus the interferometer modulates radiation of wavenumber k with a frequency $f = 2kV$, and the signal from the detector due to radiation of wavenumber k has this frequency f. If the response of the electronics is not constant for all frequencies, f, generated by the radiation, a wavenumber dependent phase shift is presented in the electronically recorded interferogram.

Equation 5a may be expanded as follows:

$$F(x) = \int_0^\infty A(k)\ \cos[2\pi k(x + \delta(k))]\ dk$$

$$= \int_0^\infty [A(k)\ \cos 2\pi k\delta(k) \cdot \cos 2\pi kx - A(k)\ \sin 2\pi k\delta(k) \cdot \sin 2\pi kx]dk.$$

The cosine Fourier transform of F(x) is [5,6]

$$C(k') = 2\int_{-\infty}^{\infty} F(x)\ \cos 2\pi k'x \cdot dx \qquad (16a)$$

$$= A(k') \cos 2\pi k'\delta(k'), \tag{16b}$$

since

$$\int_{-\infty}^{\infty} \cos 2\pi k'x \cdot \cos 2\pi kx \cdot dx = \tfrac{1}{2}\delta_{kk'} \begin{array}{l} = 1 \text{ if } k = k' \\ = 0 \text{ otherwise} \end{array} \tag{17a}$$

and

$$\int_{-\infty}^{\infty} \cos 2\pi k'x \cdot \sin 2\pi kx \cdot dx = 0. \tag{17b}$$

Further,

$$\int_{-\infty}^{\infty} \sin 2\pi k'x \cdot \sin 2\pi kx \cdot dx = \tfrac{1}{2}\delta_{kk'}, \tag{17c}$$

so the sine Fourier transform of F(x) is

$$S(k') = 2\int_{-\infty}^{\infty} F(x) \sin 2\pi k'x \cdot dx \tag{18a}$$

$$= -A(k') \sin 2\pi k'\delta(k'). \tag{18b}$$

For specified wavenumber k', therefore, the ratio of the sine and cosine Fourier transform is

$$\frac{S(k')}{C(k')} = -\tan 2\pi k'\delta(k'), \tag{19}$$

from which $\delta(k')$ can be calculated.

The phase corrected spectrum, A(k'), can then be calculated from the cosine transform via

$$A(k') = C(k')/\cos 2\pi k'\delta(k') \tag{20a}$$

or from the sine transform as

$$A(k') = -S(k')/\sin 2\pi k'\delta(k') \tag{20b}$$

or, as is usually done to make use of both transforms,

$$A(k') = C(k') \cos 2\pi k'\delta(k') - S(k') \sin 2\pi k'\delta(k') \tag{20c}$$

since, as Eq. 16b and 18b show, C(k') and -S(k') are orthogonal projections of A(k').

Alternatively, the so-called power spectrum can be calculated as

$$A(k') = [C^2(k') + S^2(k')]^{\frac{1}{2}}. \tag{21}$$

For noise free interferograms that are infinitely long A(k') should be independent of the final step of its calculation. For real, noisy interferograms the calculation of the power spectrum (Eq. 21) has the disadvantage that all of the noise appears positive in the spectrum, which can distort the spectrum with noise that is difficult to recognize as such. Nevertheless, the power spectrum calculation has been used successfully by my research group for over a decade in our far infrared spectroscopy. Further, it is the most reliable method in cases where the central fringe of the interferogram, at zero path difference, is not easy to identify because, for example, a measured emission spectrum consists of a fairly small number of sharp lines and no continuum [8].

For most infrared spectroscopy, however, Eqs. 19 and 20 are used to obtain the spectrum. Further, to reduce the amount of data to be processed and the time required to obtain and process it, the calculation is done on a one-sided interferogram, i.e. one that extends the distance required for the desired resolution on one side of the zero-path position but only a short distance on the other side. Two methods for doing this were proposed in the late sixties, by Forman, Steele and Vanasse [9] and by Mertz [10], and their relative merits have been discussed by Sanderson and Bell [11] and by Sakai, Vanasse and Forman [12]. Before these methods are described, some helpful, more sophisticated mathematics will be presented briefly.

4. MATHEMATICAL METHODS

The sine and cosine Fourier transforms can be combined into one expression by using the complex exponential notation defined by

$$e^{i\alpha} = \cos\alpha + i\sin\alpha, \tag{22a}$$

where $i = \sqrt{-1}$, and the complex conjugate

$$(e^{i\alpha})^* = e^{-i\alpha} = \cos\alpha - i\sin\alpha. \tag{22b}$$

In this notation the interferogram and the spectrum are related by [13,14]

$$F(x) = \int_{-\infty}^{\infty} B(k)\, e^{+2\pi ikx}\, dk \tag{23a}$$

and

$$B(k) = \int_{-\infty}^{\infty} F(x)\, e^{-2\pi ikx}\, dx. \tag{23b}$$

Thus

$$F(x) = \int_{-\infty}^{\infty} B(k) \cos 2\pi kx \cdot dk + i \int_{-\infty}^{\infty} B(k) \sin 2\pi kx \cdot dk$$

and the cosine transform is called the real transform while the sine transform is called the imaginary transform, following the language of complex numbers. This language can be confusing, and it should be noted that the sine transforms are imaginary only in the sense that they disappear (i.e. are zero) when all phase and sampling errors have been corrected so that F(x) is symmetric and B(k) is the correct spectrum.

Note that the integration in Eq. 23a is over negative values of the wavenumber, k, as well as positive values, although k is physically only positive. To satisfy this Fourier transform requirement, B(-k) is assumed to equal B(+k) when F(x) is calculated from B(k). If the interferogram is symmetric and the calculated spectrum is correct, both F(x) and B(k) are real quantities, i.e. the sine transforms are zero. This means that

$$F(x) = \int_{-\infty}^{\infty} B(k)\ e^{+2\pi ikx}\ dk = 2 \int_{0}^{\infty} B(k) \cos 2\pi kx\ dk$$

so that B(k) is one-half of the quantity A(k) used in the previous sections. This difference is a trivial one of definition, not an important one.

In this notation the phase correction procedure involves the transform (Eqs. 5, 16, 18)

$$F(x) = \int_{-\infty}^{+\infty} B(k)\ e^{+i2\pi kx} \cdot e^{+i2\pi k\delta(k)} \cdot dk \tag{24a}$$

$$\therefore M(k) = \int_{-\infty}^{+\infty} F(x)\ e^{-i2\pi kx}\ dx = B(k)\ e^{+i2\pi k\delta(k)}, \tag{24b}$$

$$\text{i.e. } B(k) \cdot \cos 2\pi k\delta(k) + i \cdot B(k) \cdot \sin 2\pi k\delta(k)$$

$$= \tfrac{1}{2}[C(k') - i\ S(k')], \tag{24c}$$

where C(k') and S(k') are defined by Eqs. 16a and 18a. Equations 19 to 21 follow from this using B(k) = A(k)/2. It should be noted that it is common practice to replace the negative sign in Eq. 24c by a positive sign. When this is done, the negative signs in Eqs. 19, 20b and 20c must also be replaced by positive signs.

Of particular use in the theory of apodization is the convolution theorem. It states that the Fourier transform of the product of two functions equals the convolution of the Fourier transforms of the function. Algebraically this is written (c.f. Eqs. 10 and 12):

$$I(k) = \int_{-\infty}^{\infty} F(x)\ H(x)\ e^{-2\pi ikx}\ dx = B(k) * h(k) \tag{25}$$

where B(k) is the Fourier transform of F(x) via Eq. 23b, h(k) is that of H(x), given by

$$h(k) = \int_{-\infty}^{\infty} H(x)\, e^{-2\pi ikx}\, dx, \tag{26}$$

and the $*$ means 'convoluted with', and is defined by

$$I(k') = \int_{-\infty}^{\infty} B(k) \cdot h(k' - k)dk. \tag{27}$$

5. PRACTICAL APODIZATION

The convolution theorem greatly simplifies considerations of apodization functions. Consider the case discussed in Sec. 2 of the effect of truncating an interferogram of a monochromatic radiation source. F(x) is given by Eq. 7; it is symmetric, so that B(k), which is the Fourier transform of F(x) over infinite limits (Eq. 23a), is real, and B(k) is zero except at $k = k_a$ where its value is $A(k_a)$. Thus

$$B(k) = A(k_a)\, \delta_{kk_a} \tag{28}$$

where $\delta_{kk_a} = 0$ if $k \neq k_a$ and $\delta_{kk_a} = 1$ if $k = k_a$.

For this case that H(x) is the so-called boxcar function, being 1 if $-X \lessgtr x \lessgtr +X$ and zero otherwise, Eq. 26 becomes

$$h(k) = \int_{-X}^{+X} e^{-2\pi ikx}\, dx = 2\int_{0}^{X} \cos 2\pi kx\, dx$$
$$= 2X \cdot \mathrm{sinc}\, 2\pi kX. \tag{29}$$

Equations 27 and 29 yield

$$I(k') = \int_{-\infty}^{\infty} B(k) \cdot 2X\, \mathrm{sinc}\, 2\pi(k' - k)X \cdot dk,$$

which, with Eq. 28, becomes

$$I(k') = 2X\, A(k_a)\, \mathrm{sinc}\, 2\pi(k_a - k')X, \tag{30}$$

as in Eq. 9. Note that the sinc $2\pi(k_a + k')X$ term in Eq. 8 has not appeared in this treatment. It arises from the existence, in the Fourier transform of Eq. 7, of negative frequencies which yield $B(-k_a) = B(k_a)$; because the negative frequencies have no physical meaning, they have been omitted from Eq. 28 and thus do not appear in Eq. 30.

This treatment of apodization shows that the instrument line shape is just the Fourier transform over infinite limits of the function H(x), which greatly facilitates the exploration of apodization functions. To close this discussion of apodization, a few

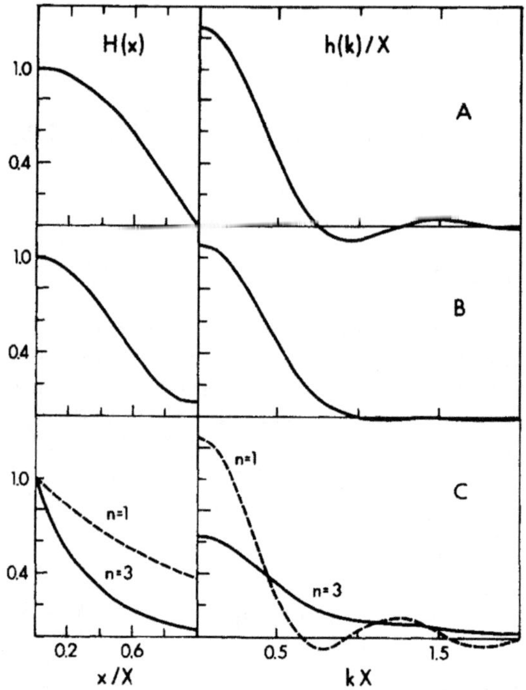

Fig. 7 Apodization functions and instrument lineshapes. All are symmetrical about x/X or kX = 0. A) $H(x) = \cos(\pi x/2X)$; B) Happ-Genzel; C) $H(x) = \exp(-n|x|/X)$.

such functions of spectroscopic interest will be described.

The quarter-wave, cosine function [2] is

$$H(x) = \cos\left(\frac{\pi}{2}\frac{x}{X}\right) \text{ for } |x| \lessgtr X \text{ and zero otherwise,} \qquad (31a)$$

where x is the path difference and X is the maximum path difference. This Fourier transforms to give the instrument lineshape

$$h(k) = \frac{\pi X}{(\pi/2)^2 - (2\pi kX)^2} \cdot \cos 2\pi kX, \qquad (31b)$$

which has the value 1.0X at $2\pi kX = \pi/2$. This apodization function and the associated instrument lineshape are symmetrical about $x = 0$ or $k = 0$, and are shown for positive x or k in Fig. 7. The halfwidth and resolution (here the separation of two lines of this shape which yields a 2% dip between the peaks of their sum) are listed in Table 1. The lineshape is intermediate between the sinc and sinc^2 shapes, being slightly sharper and easier to resolve than

Table 1. Properties of Instrument Lineshapes from some Apodization Functions

Apodization Function $H(x)$[a]		Instrument Line Shape, $h(k)/X$ FWHH	Resolution[b]	Peak height	Pod height[c] 1st	2nd	3rd
1		$0.605\ X^{-1}$	$0.68\ X^{-1}$	2.0	-21.5%	+12.5%	-9%
$1 - \lvert\frac{x}{X}\rvert$		$0.88\ X^{-1}$	$0.88\ X^{-1}$	1.0	+4.5%	+1.4%	
$\cos \frac{\pi x}{2X}$		$0.82\ X^{-1}$	$0.85\ X^{-1}$	1.27	-6.7%	+2.8%	-1.6%
Happ Genzel		$0.91\ X^{-1}$	$0.89\ X^{-1}$	1.08	-0.6%	+0.2%	-0.6%
$e^{-\frac{n\lvert x\rvert}{X}}$	$n = 1$	$0.71\ X^{-1}$	$0.73\ X^{-1}$	1.26	-5.5%	+9%	-4%
	$n = 2$	$0.85\ X^{-1}$		0.86	--	±3%	--
	$n = 3$	$1.04\ X^{-1}$	$0.88\ X^{-1}$	0.63	--	±1.6%	--
$e^{-(\frac{nx}{X})^2}$	$n = 2.24$	$\gtrsim 1.2\ X^{-1}$	$\gtrsim 1.16\ X^{-1}$		--	<0.5%	--
F_3 of Ref. 18		$0.965\ X^{-1}$			--	0.3%	--

[a] $H(x)$ is the function shown if $\lvert x\rvert \leq X$, and zero otherwise.

[b] The separation between two such functions for a dip of 2% of the height of the peaks between the peaks in their sum.

[c] As a percentage of the peak height.

the sinc^2 shape but having significantly larger pods. Lineshapes intermediate between sinc and sinc^2 functions can also be obtained by trapezoid apodization functions,

$$H(x) = 1 \text{ for } |x| \lessapprox b$$

$$H(x) = 1 - \frac{|x| - b}{X - b} \text{ for } b < |x| \lessapprox X$$

$$H(x) = 0 \text{ for } |x| > X, \text{ where b is positive.}$$

It will be seen that all apodization functions that yield a line sharper than the sinc^2 line yield much more prominent pods than those of the sinc^2 shape. One must therefore compromise between line sharpness and the size of the pods. Other functions, however, yield lines as sharp as the sinc^2 line with much smaller pods. Data on the following functions are included in Fig. 7 and Table 1.

A modification of the cosine function is the Happ-Genzel function, which is the apodization function recommended for general FTIR use by the Nicolet Instrument Corporation. The function is

$$H(x) = 0.54 + 0.46 \cos \frac{x\pi}{X} \text{ for } |x| \lessapprox X \text{ and zero otherwise} \quad (32a)$$

and its Fourier transform is

$$h(k) = \left[\frac{0.54}{\pi k} + \frac{(0.46)4\pi k X^2}{\pi^2 - 4\pi^2 k^2 X^2}\right] \cdot \sin 2\pi k X \quad (32b)$$

except at $\pi = 2\pi kX$ where $h(k) = 0.46\ X$.

Note that although $H(x)$ does not reach zero at $x = X$, it yields an instrument lineshape similar to the sinc^2 shape but with pods that do not exceed 0.6% of the peak height. These pods, however, do not decay rapidly and are still of magnitude ±0.5% at $kX = 3.25$ and 3.75.

The function

$$H(x) = Ae^{-a|x|} \quad (33a)$$

yields the Lorentzian function

$$h(k) = \frac{2A}{a} \frac{a^2}{a^2 + (2\pi k)^2} \quad (33b)$$

when Fourier transformed between infinite limits [15]. To provide an apodization function, a can be set equal to n/X, where n is a constant, and the function must be set to zero outside of the desired range, $|x| \lesssim X$, because it does not go to zero at the limit of the range. Thus the apodization function is

$$H(x) = Ae^{-\frac{n|x|}{X}} \qquad \text{for } |x| \lesssim X \text{ and} \tag{34a}$$

$$H(x) = 0 \qquad \text{for } |x| > X,$$

with A equal to one. Its Fourier transform is

$$h(k) = \frac{2AX}{n^2 + 4\pi^2k^2X^2} \cdot [n + e^{-n} \cdot [2\pi kX \sin 2\pi kX - n \cos 2\pi kX]], \tag{34b}$$

where the first term is the Fourier transform of the function of equation 33a, with $a = n/X$, over infinite limits, and the second term reflects the effect of the termination of the function at $|x| = X$. This function is, of course, interesting spectroscopically because it yields a near-Lorentzian lineshape when the effects of the termination are small enough. Fig. 7 shows H(x) and h(k) for n = 1 and 3 and Table 1 shows data for n = 1,2 and 3. When n = 1 the line is ~16% broader, but only ~7% harder to resolve, than the sinc function, but has rather smaller pods than the sinc function. When n = 3, the line is much broader than, but is as easy to resolve as, the sinc^2 and the Happ-Genzel lines. There are no negative pods when n = 3 but ripples of magnitude 1.6% of the peak height are superimposed on the extensive wings of the Lorentzian line and, like the wings, decay slowly.

A Gaussian apodization function

$$H(x) = e^{-\left(\frac{nx}{X}\right)^2} \qquad \text{if } |x| \lesssim X, \text{ zero otherwise} \tag{35a}$$

yields [15]

$$h(k) = \frac{X\sqrt{\pi}}{n}\, e^{-\left(\frac{\pi kX}{n}\right)^2} \tag{35b}$$

if H(x) is essentially zero at $|x| = X$. The FWHH of the peak of h(k)/X is then $(0.530\ n)X^{-1}$, but is larger if H(x) is not zero at $|x| = X$. H(x) = 0.02 at $|x| = X$ if n = 2, yielding a line of halfwidth $\gtrsim 1.06\ X^{-1}$. Horlick and Yuen [16] have found that when n = 2.24 the side lobes are less than 0.5% of the peak height, and although they have not reported the full width at half height it can be estimated to be $\gtrsim 1.19\ X^{-1}$. Two Gaussian peaks separated by their full width at half height show a 7.3% dip between the

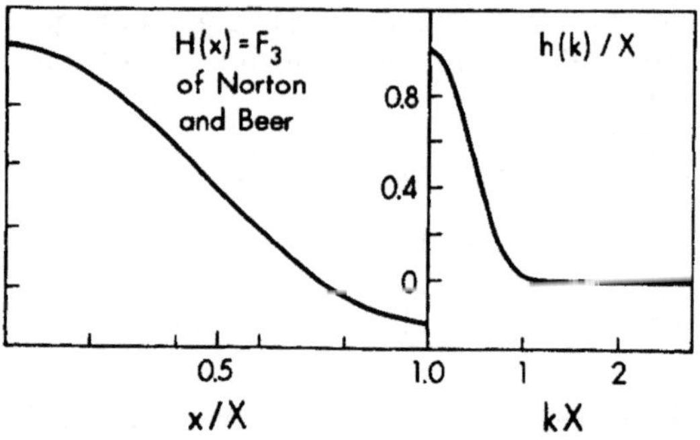

Fig. 8 Apodization function F_3 of reference 18 and its instrument lineshape.

peaks of their sum [3], so the resolution for n = 2.24 is about $1.16\ X^{-1}$. Investigation of the effect of truncating Gaussian functions at $|x| = X$ must be done by numerical methods, because there is no analytic form of the Fourier transform of a Gaussian over finite limits; I am unaware of any previous detailed investigation and will present the results of one in my talk.

A. S. Filler [17] and R. H. Norton and R. Beer [18] have investigated large numbers of apodization functions. One of these is noteworthy, the function that Norton and Beer call F_3:

$$H(x) = F_3(x) = 0.09 + 0.5875\ (1 - \frac{x}{X})^2 + 0.3225\ (1 - \frac{x}{X})^4 .$$

The analytic form of the solution is a lengthy sum of sinc, cosine and algebraic terms. It is given explicitly in reference [18], and yields a peak with FWHH = $0.965\ X^{-1}$ and the smallest side lobes that I am aware of, 0.3% of the peak height. However, as is the case for the small side lobes of the Happ-Genzel function, the side lobes decay extremely slowly. This function and instrument lineshape are shown in Fig. 8.

Finally, it should be noted that apodization is not a procedure that is peculiar to Fourier transform infrared spectroscopy. It has a long history in optics, the use of a diamond mask on a diffraction grating being one implementation of apodization, and a detailed account of this use was given by Jacquinot [19] in 1964. Further, it is one aspect of the much broader topic of correlation methods for digital filtering and smoothing, spectral identification, and line shape manipulation, which is finding extensive current application in analytical chemistry and has recently been discussed by Horlick and Hieftje [20]. One use of such digital filtering that can be of great help for qualitative chemical spectroscopy is the removal from the interferogram of the signature caused by interference fringes, with the consequent great reduction in the magnitude of such fringes in the resultant spectrum. But care should be taken with this technique, because any other information that has the same wavenumber separation as the fringes is also lost.

6. PRACTICAL PHASE CORRECTION

The methods described in Secs. 3 and 4 for the phase correction of infinitely long, double-sided interferograms are independent of whether the analogue interferogram is symmetrical or not, and require no modification, other than the inclusion of apodization, for double-sided interferograms of finite length. These methods, used with double-sided interferograms ($-X \leq x \leq +X$), yield the most accurate intensities in spectra from a Michelson interferometer. The effect of errors due to uncertainty in the zero-path position or slight asymmetry have been discussed [12].

Methods have been developed, however, to enable spectra of high intensity-accuracy to be obtained from one-sided interferograms. These interferograms extend to only very short negative path differences but to a sufficient positive path difference to provide the desired resolution. As mentioned earlier, Forman, Steele and Vanasse [9] proposed a method that involves convolution - hence 'the convolution method' - while Mertz [10] proposed the corresponding method that simply involves multiplication. After initial objections to the reliability of the multiplication method [12,21] it was shown [11] to be capable of accuracy of a few-tenths of a percent. It is the method used in most commercial mid-infrared interferometers. Both methods will be described briefly.

In the convolution method, an abbreviated interferogram, which consists of an equal, small number of data points to either side of the (estimated) zero-path position, is Fourier transformed to yield a low-resolution phase spectrum, $\delta(k)$. The phase error is a slowly varying function of frequency, so $\delta(k)$ can be calculated by interpolation at values of k between the values given by the transform. The Fourier transform of the function exp $(-2\pi ik\delta(k))$ is then calculated and convolved with the full, original one-sided interferogram. If the position of zero-path difference was well estimated, this yields an interferogram that is symmetric about the zero-path-position, apart from being much longer on one side than the other. If necessary the process can be repeated and, as Forman et al. have shown [9], very asymmetric interferograms can be made symmetrical by two applications of this procedure. Once the interferogram is sufficiently symmetrical, the spectrum can be calculated by a one-sided cosine transform (strictly one-sided now, including no negative path differences), based on Eqs. 4 and 23b:

$$B(k) = \int_{-\infty}^{\infty} F(x)\ e^{-2\pi ikx}\ dx$$

which equals

$$2 \int_{0}^{\infty} F(x) \cdot \cos(2\pi kx) \cdot dx \quad \text{if } F(x) \text{ is symmetrical.}$$

The basis for the convolution procedure is Eq. 24b. The Fourier transform of an interferogram yields a complex quantity $M(k) = B(k)\ e^{2\pi ik\delta(k)}$, where $B(k)$ is the spectrum that is sought. Thus

$$B(k) = M(k)\ e^{-2\pi ik\delta(k)} . \tag{36}$$

By the convolution theorem, the Fourier transform of the true spectrum, B(k), is the convolution of the Fourier transform of M(k) with that of the complex phase exponential. But the Fourier transform of the true spectrum is a symmetrical interferogram, while that of M(k) is the original interferogram.

In practice the limits on the integrals are finite and apodization must be used at each Fourier transformation, including that of the phase exponential [9]. This transform is ideally, using the definition that $\delta(-k) = -\delta(k)$,

$$\int_{-\infty}^{+\infty} e^{-2\pi ik\delta(k)}\ e^{2\pi ikx}\ dk$$

$$= 2 \int_{0}^{\infty} \cos 2\pi k\delta(k) \cos 2\pi kx\ dk + 2 \int_{0}^{\infty} \sin 2\pi k\delta(k) \sin 2\pi kx \cdot dk.$$

In practice the integration is between k_1 and k_2, the range in which significant intensity occurs. For the calculation of phase from the short double-sided interferogram, triangular apodization is usually used.

This is, thus, an elegant method which works well [9,12,21] but which has been avoided by most chemical workers in favor of the multiplication method with its simpler computations. It is debatable whether the simpler computations alone remain a valid reason for this preference, now that computing efficiency has improved, and the convolution method should probably be re-examined as a possible cure for cases in which the precision of the multiplication method is not adequate.

Mertz's multiplication method [10] is related to the convolution method by the convolution theorem. Thus it is based on Eq. 36, that the calculated Fourier transform M(k) can be corrected by the phase factor $\exp(-2\pi ik\delta(k))$ to yield the spectrum. Recall that B(k) is real, so the spectrum is given by $\mathrm{Re}[M(k) \exp(-2\pi ik\delta(k))]$, where the phase angles $\delta(k)$ are found from a short two-sided interferogram. Mertz [10] and more recent authors [2,7,11] have given Eq. 36 (or its equivalent) in a form that is, to me, confusing, misleading, and suggestive of inefficient forms of computation. They have used

$$B(k) = |M(k)| \cos [\psi(k) - 2\pi k\delta(k)] \tag{37}$$

which implies that not only the modulus $|M(k)|$ but also the phase angles, $\psi(k)$, of the complex quantity $M(k)$ must be found. The calculation of the phase angles is a significant effort even for a computer. It can be seen that Mertz's equation expands to

$$\begin{aligned} B(k) &= |M(k)| \cos \psi(k) \cos 2\pi k\delta(k) \\ &+ |M(k)| \sin \psi(k) \sin 2\pi k\delta(k) \\ &= M_c(k) \cos 2\pi k\delta(k) + M_s(k) \sin 2\pi k\delta(k), \end{aligned} \tag{38}$$

where $M_c(k)$ and $M_s(k)$ are the cosine and sine transforms of the interferogram, i.e. to

$$B(k) = \mathrm{Re}[M(k)e^{-2\pi ik\delta(k)}]. \tag{38a}$$

The need to truncate and, if desired, apodize the interferogram, and the uncertainty in the location of the digitization positions and the zero-path-difference position (note that if $\delta(k)$ is a function of k the zero-path-difference position depends on k, the wavenumber of the radiation) all complicate the formal treatment of the recovery of spectra from one-sided interferograms. Sakai, Vanasse, and Forman [12], Sanderson and Bell [11] and R. J. Bell [1] have all discussed some aspects of these difficulties with Sakai et al.'s discussion [12] being the most extensive. To introduce the reader to these discussions, and to justify the use of Eq. 38a for one-sided transforms, the following clear treatment of the multiplication method is given. It was presented by C. R. Anderson and D. R. Mattson to the Nicolet Users Conference in October 1979, and I am grateful for their permission to reproduce it here in my notation.

Equation 25 (Sec. 4) indicates that if a symmetric interferogram $F(x)$ is multiplied by an apodization and/or truncation function $H(x)$ and Fourier transformed over infinite limits one obtains the approximation to the spectrum $I(k)$; i.e.

$$I(k) = \int_{-\infty}^{\infty} F(x)\, H(x) e^{-2\pi ikx}\, dx = B(k) * h(k)$$

where $B(k)$ and $h(k)$ are the infinite Fourier transforms of $F(x)$ and $H(x)$ (Eqs. 23b and 26). $F(x)$ is real and symmetric with respect to x so that $B(k)$ is real and given by

$$B(k) = 2 \int_{0}^{\infty} F(x) \cos 2\pi kx\, dx. \tag{39}$$

Although one could make $H(x)$ complex it is never done and $H(x)$ is always real. If $H(x)$ is also symmetric with respect to x, $h(k)$ is real. When $B(k)$ and $h(k)$ are real, $I(x)$ (Eq. 25) is the best approximation to the spectrum that it is possible to get from Fourier transform spectroscopy, and for later reference it is useful to note that Eq. 25 can in this case be written as:

$$I_B(k) = B(k) * 2 \int_0^\infty H(x) \cos 2\pi kx \cdot dx. \tag{40}$$

The Fourier transform of a typical, one-sided, asymmetric interferogram is given by

$$I(k) = \int_{-X_1}^{X_2} H(x)\ F(x)\ e^{-2\pi ikx}\ dx$$

where $X_2 > X_1$. This can be written as Eq. 25 if the finite integration limits are incorporated in the apodization function, $H(x)$, so we can write, using Eq. 24b,

$$\begin{aligned} I(k) &= [B(k)\ e^{i\phi(k)}] * h(k) \\ &= e^{i\phi(k)}\ [B(k) * h(k)]. \end{aligned}$$

The phase error is here written as $\phi(k)$ instead of $2\pi k\delta(k)$ for brevity. The phase exponential can be factored out of the convolution because it changes very little with frequency; R. J. Bell [22] has shown that this is valid provided that $d\phi(k)/dk$ is essentially constant over the range of k in which the instrument lineshape $h(k)$ has significant magnitude, and that this implies that $d\phi(k)/dk << 2\pi X_2$. Empirically this is usually the case. Thus we have

$$I(k)\ e^{-i\phi(k)} = B(k) * h(k). \tag{41}$$

We must now investigate $h(k)$. The apodization function $H(x)$ can be made the product of 2 parts,

$$H(x) = (\tfrac{1}{2} + H'(x))\ H''(x)$$

where $H''(x)$ is a symmetric apodization function, such as those described in Sec. 5, that is zero if $|x| > X_2$. $[\frac{1}{2} + H'(x)]$ is the so-called left-ramp function whose benefits Mertz [10] and Sanderson and Bell [11] have emphasized. $H'(x)$ is an odd function of x that is linear between $-X_1$ and $+X_1$. Its value is $-\frac{1}{2}$ at $x \lessapprox -X_1$ and $+\frac{1}{2}$ at $x \gtrapprox X_1$, so that $(\frac{1}{2} + H'(x))$ is linear between $-X_1$ and $+X_1$, zero at the left integration limit, $x = -X_1$, 0.5 at the zero path position, and +1 at $x = +X_1$. Further, $(\frac{1}{2} + H'(x))$ is zero at $x \lessapprox X_1$ and 1 at $x \gtrapprox X_1$.

Thus, using Eq. 26, and integrating only over the range where

$H(x)$ is non-zero,

$$h(k) = \int_{-X_1}^{+X_2} [\tfrac{1}{2} + H'(x)]\, H''(x)\, e^{-2\pi ikx}\, dx$$

$$= \int_{-X_2}^{+X_1} (\tfrac{1}{2} + H'(x))\, H''(x)\, e^{-2\pi ikx}\, dx$$

$$+ \int_{X_1}^{X_2} H''(x)\, e^{-2\pi ikx}\, dx.$$

$H'(x)$ is odd which yields, on rearranging

$$h(k) = \int_0^{X_2} H''(x) \cos 2\pi kx\, dx$$

$$+ 2i \int_0^{X_1} H'(x)\, H''(x) \sin 2\pi kx\, dx$$

$$+ i \int_{X_1}^{X_2} H''(x) \sin 2\pi kx\, dx. \qquad (42)$$

Thus, the real part of $h(k)$ is given by

$$\mathrm{Re}[h(k)] = \int_0^{X_2} H''(x) \cos 2\pi kx\, dx$$

$$= \int_0^{\infty} H''(x) \cos 2\pi kx\, dx$$

because $H''(x) = 0$ if $x > X_2$.

$H''(x)$ is a symmetric apodization function and $B(k)$ is real so, from Eq. 41 we have

$$\mathrm{Re}[I(k)\, e^{-i\phi(k)}] = B(k) * \int_0^{\infty} H''(x) \cos 2\pi kx\, dx$$

which is (Eq. 40) the best approximation to the spectrum that can be obtained from Fourier transform spectroscopy.

Empirically the multiplication method today gives spectra that appear free from distortion at wavenumbers up to 4000 cm^{-1}, if the interferogram is sampled at 0.6329 μm intervals, if X_1 ~0.3 mm, and if the phase spectrum calculation uses more than ~30 data points on each side of the zero-path position. But it has been reported that increasing the length of the short side of the interferogram does improve the quality of the spectra if extremely high precision (better than 1 in 10^4) is required [23].

I wish to close this chapter with an important practical point. It is clear from Eq. 5 that the phase angle $2\pi k\delta(k) = \phi(k)$ is the phase of the radiation of wavenumber k at $x = 0$, i.e. at the central peak of the interferogram. It is helpful to note that most digital-computer programs for Fourier transformation only yield this angle via Eq. 19 if the digitized point closest to (ideally at) $x = 0$ is the first point in the array to be transformed. If the first point in the array corresponds to one end of the interferogram, say $x = -X_1$, the angles given by Eq. 19 are $\phi(k) + 2\pi k(-X_1)$ and must be corrected for the second term. It is more convenient to rotate the points in all arrays that are to be transformed, after all apodization has been done, so that the first point is the zero-path-difference point, followed by the points on one side of the interferogram, followed by the zero-filling points, followed by the points on the second side of the interferogram starting at the end and working towards zero path. Then Eq. 19 yields $\phi(k)$ directly and ensures that the phases calculated from the low resolution transform are the phases that correct the high resolution transform.

REFERENCES

1. R. J. Bell, Introductory Fourier Transform Spectroscopy, Academic Press, New York (1972).

2. P. R. Griffiths, Chemical Infrared Fourier Transform Spectroscopy, John Wiley, New York (1975).

3. A. E. Martin, Infrared Instrumentation and Techniques, Elsevier, Amsterdam (1966), Chapter 4.

4. K. D. Möller and W. G. Rothschild, Far-Infrared Spectroscopy, Wiley-Interscience, New York (1971), Chapter 4.

5. D. C. Champeney, Fourier Transforms and their Physical Applications, Academic Press, New York (1973).

6. R. Bracewell, The Fourier Transform and its Applications, McGraw-Hill, New York (1965).

7. E. G. Codding and G. Horlick, Appl. Spectrosc., 27, 85 (1973).

8. Wo Yuen and G. Horlick, Anal. Chem., 49, 1446 (1977); Appl. Spectrosc., 32, 38 (1978).

9. M. Forman, W. H. Steel, and G. A. Vanasse, J. Opt. Soc. Am., 56, 59 (1966).

10. L. Mertz, Infrared Physics, 7, 17 (1967).

11. R. B. Sanderson and E. E. Bell, Appl. Optics, 12, 266 (1973).

12. H. Sakai, G. A. Vanasse, and M. L. Forman, J. Opt. Soc. Am., 58, 84 (1968).

13. Reference 5, page 9 & 10.

14. Reference 1, page 34.

15. Reference 5, page 22.

16. Wo Yuen, Ph.D. Thesis, University of Alberta, Edmonton, Canada (1978).

17. A. S. Filler, J. Opt. Soc. Am., 54, 762 (1964).

18. R. H. Norton and R. Beer, J. Opt. Soc. Am., 66, 259 (1976).

19. P. Jacquinot and B. Roizen-Dossier, Progress in Optics III, 29 (1964).

20. G. Horlick and G. M. Hieftje, Page 153 of 'Contemporary topics in Analytical and Clinical Chemistry', Volume 3, (Eds. D. M. Hercules, G. M. Hieftje, L. R. Snyder, and M. A. Evenson), Plenum Publishing Corporation, New York (1978).

21. J. Connes in Proceedings of the Aspen 1970 International Conference on Fourier Spectroscopy, U. S. Air Force Cambridge Research Laboratory, Bedford, Mass., (1971).

22. Reference 1, pages 159-164.

23. C. A. Anderson and D. R. Mattson, Reported to Nicolet Users' Meeting, Madison, Wisconsin, October, 1979.

NOTE ADDED IN PROOF

Our investigation of the instrument line shape of a truncated Gaussian apodization function has given the data in Table 2, and the curves of Fig. 9. It is clearly an attractive apodization function for some purposes, particularly since it yields pods of only 0.1% of the peak height with a full width at half height of 1.17 X^{-1}, and 0.02% pods with FWHH = 1.30 X^{-1}. With n = 1.75 the Gaussian yields much the same lineshape as the Happ-Genzel function and the ease of resolution of overlapping lines will be much the same for the two shapes.

The treatment of phase correction of one-sided transforms by Anderson and Mattson [23] has been published in Proc. Soc. Photo-Optical Instrumentation Engineers, 191, 101 (1979).

Table 2. Properties of Instrument Line Shapes from Gaussian Apodization Functions

Apodization Function[a]	Instrument Line Shape FWHH	Pod Height 1st	2nd	3rd	4th
n = 1	0.68 X^{-1}	-9%	+6%	-4.5%	+3.5%
n = 1.50	0.83 X^{-1}	-0.3%	+2%	-1.5%	+1.3%
n = 1.75	0.93 X^{-1}	-0.7%	+0.6%	-0.5%	+0.4%
n = 2.0	1.05 X^{-1}	-0.3%	+0.25%	-0.2%	+0.2%
n = 2.24	1.17 X^{-1}	-0.03%	+0.1%	-0.08%	+0.07%
n = 2.5	1.30 X^{-1}	-0.02%	+0.02%	-0.02%	+0.02%

[a] $H(x) = \exp(-n^2x^2/X^2)$ if $|x| < X$, and $H(x) = 0$ otherwise.

[b] $h(k) = \int_{-X}^{X} H(x)\, e^{-2\pi ikx}\, dx$, normalized to unity at the peak.

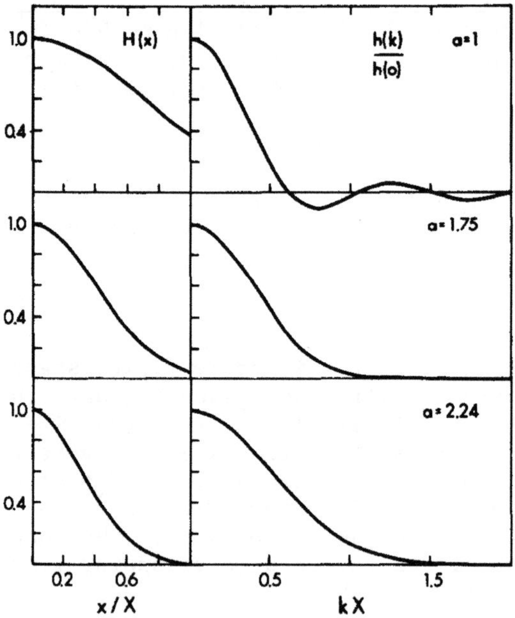

Fig. 9 Gaussian apodization functions and instrument lineshapes. $H(x) = \exp(-a^2x^2/X^2)$.

ENERGY-LIMITED SYSTEMS

John E. Bertie

Department of Chemistry
University of Alberta
Edmonton, Alberta, Canada

1. INTRODUCTION

A grating spectrometer is energy-limited if the diffraction-limited resolving power of the grating cannot be used because there is inadequate energy to yield an acceptable signal-to-noise ratio with the slits set to yield the diffraction-limited resolution. This is the case for most, if not all, infrared grating spectrometers, and traditionally the solution has been to open the slits to allow more energy to fall on the detector, so that an acceptable signal-to-noise ratio is obtained at the expense of decreased resolution. It is implicit in this practice that the detector noise is either independent of the signal or, less satisfactorily, proportional to a lower power of the signal than the first power, and that the signal-to-noise ratio of the radiation incident upon the detector is high. Infrared spectroscopists have usually had, however, little chance of ever approaching diffraction-limited resolution. Consequently the term 'energy-limited' has been used to describe the less well-defined situation where the sample cell or accessory transmits so little light that it is exceedingly difficult to obtain a spectrum of even the crudest sort.

I am not aware of a formal definition of the term 'energy-limited' as applied to a FTIR spectrometer based on a Michelson interferometer. An analogous definition to that for a grating spectrometer would be that an interferometer is energy limited if the signal-to-noise ratio is unacceptably high in a spectrum which results from a single scan of the moving mirror over the maximum available range of path-difference. As is now the case with modern computer-controlled grating-spectrometers, the

J. R. Durig (ed.), Analytical Applications of FT-IR to Molecular and Biological Systems, 51–76.

signal-to-noise ratio can be improved without loss of resolution by signal-averaging, i.e., by averaging a number of interferograms, or spectra, of the sample which were obtained under identical conditions.

Thus, although the term 'energy-limited' has a well-defined traditional meaning for grating spectrometers, its meaning for interferometers is less well defined and, in any case, its significance has been reduced by the ability to signal-average. For the purpose of this talk the term will be used to describe the situation where the sample, sample cell, or accessories transmit little light, so that one must take steps to improve the signal-to-noise ratio. Thus I will emphasize ways of improving the signal-to-noise ratio, with illustrations of what we have achieved and, without an exhaustive search of the literature, what others have achieved.

2. INSTRUMENT COMPONENTS WHICH AFFECT THE SIGNAL-TO-NOISE RATIO

It is obvious that the signal-to-noise ratio can be increased by decreasing the noise and/or by increasing the signal. Thus the detector, source, and beamsplitter of the interferometer, as well as any windows on the cells or accessories, should be chosen to maximize the signal in the wavenumber region of interest. Loewenstein [1] identifies four principal sources of noise in Fourier spectroscopy, the detector, the source, scintillation, and digitization. These sources of noise have been discussed in detail by Sakai [2, 3], who should be consulted for the many details which will be omitted from the rather qualitative discussion of this chapter. More recently, Mattson [4] has discussed the theoretically-expected signal-to-noise ratio in relation to that found experimentally.

For an FTIR spectrometer to yield the multiplex or Fellgett advantage [1-8] the detector noise must be essentially independent of the signal. This is the case for the detectors commonly used, namely the pyroelectric, triglycine sulphate detector and the liquid-nitrogen-cooled mercury cadmium telluride detector that are usually used in rapid-scanning interferometers, and the Golay cell and liquid-helium-cooled silicon or germanium bolometers that are used in slow-scanning far infrared interferometers. The cooled detectors yield much higher signal-to-noise ratios than the room-temperature detectors, and a cooled detector which yields a high signal in the wavenumber range of interest is the best choice for use with energy-limited systems--and for most other systems.

In rapid-scanning interferometers the signal-to-noise ratio can be enhanced by adjusting, if possible, the velocity, v, of the moving mirror, so that the electrical frequency, f, into which radiation of wavenumber k is converted [7] through the relation f = 2kv, is in the optimum range of the detector for all f.

These adjustments are made empirically, but for the wavenumber range 400 to 4000 cm^{-1} the optimum mirror velocity for a triglycine sulfate detector is about 0.35 cm sec^{-1} while that for an extended range (k_{min} ~ 400 cm^{-1}) HgCdTe detector is about 0.9 cm sec^{-1}. Thus one obtains two advantages by using the cooled HgCdTe detector, instead of the pyroelectric detector, a higher signal-to-noise ratio and a shorter time required for data collection.

The mid-infrared source of an FTIR spectrometer usually has a signal-to-noise ratio that is sufficiently high that no special attention need be paid to it. This is not necessarily the case for emission spectroscopy and, where possible, care must be taken to check and, if necessary, to enhance the signal-to-noise ratio of an emission source. In the case of atmospheric and astronomical sources, scintillation noise, which arises in the medium between the source and the detector and is directly proportional to the signal, is serious and must be overcome; methods have been described [1, 2, 9] for doing so.

For spectroscopy in the far infrared, the high-pressure mercury lamp is the most common source. Some of these lamps give a high signal-to-noise ratio when powered by basic circuits while others, sometimes of the same type as the 'quiet' lamps, are very noisy. Methods of stabilizing mercury lamps and, thus improving their signal-to-noise ratio have been described [10-12].

All of the above sources of noise are well known to infrared spectroscopists who have used grating spectrometers with analogue recording systems. The user of a commercial FTIR spectrometer has some control over these noise-sources by purchasing the best source and detector for the work envisaged and, if need be, by constructing a power supply for the source that is better than the one supplied. The user can also choose, from the wide range available, the beamsplitter to use for a particular experiment. Beamsplitters range in price from about \$1000 to \$7500, so a careful choice must be made, but the combination of mylar, germanium on potassium bromide, and silicon on calcium fluoride beamsplitters perform well over the range 10 to about 7000 cm^{-1}. The effect of the beamsplitter on the signal-to-noise ratio is shown in Fig. 1; the upper curve shows the intensity (increasing upwards) recorded by the HgCdTe detector with no sample in the spectrometer, a Ge on KBr beamsplitter, and a globar source with medium aperture (2 mm diameter); the lower curve shows the 100% line, which is the ratio of two independent recordings of the upper curve. The noise on the 100% line clearly increases markedly as the signal (upper

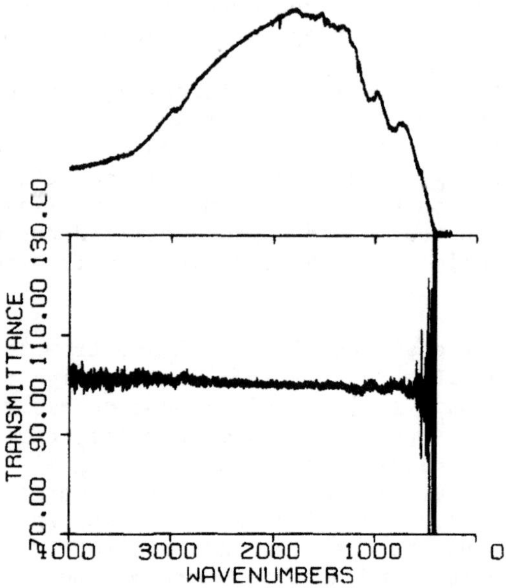

Fig. 1 Intensity (upper, increasing upward) and 100% line (lower) at 2 cm^{-1} resolution, no sample, no signal-averaging (NSS = NSB = 1) and APT = MD.

curve) decreases, because the noise on the intensity spectrum (upper curve) is independent of wavenumber. It should be noted that only one scan of the moving mirror was used for the upper curve and for each of the spectra whose ratio is the lower curve, so the noise is much higher than usual. The full specification and meaning of the operating parameters of the Nicolet 7199 FTIR instrument that were used to obtain the spectra given in this paper are in the appendix. Only deviations from these appended specifications will be noted with the later figures.

Digitization noise is not known to spectroscopists who are accustomed to analogue recording. As will be illustrated shortly, among its effects is the surprising one that increasing the gain can yield an increase in the signal-to-noise ratio under certain circumstances [2, 7].

Loewenstein [1] has identified two sources of digitization noise: (a) variation in the sampling interval, i.e., in the path-difference interval, Δx, at which the interferogram is digitized, and (b) inadequate dynamic range in the analogue-to-digital converter (ADC). Other authors [2, 7] reserve the term 'digitization noise' for noise from source (b), and treat source (a) separately. In any event the user of commercial instruments has little control over source (a) which depends on the quality of the device which measures the position of the moving mirror and on the quality of the mirror-driving mechanism. For far infrared work at less than 250 cm^{-1}, a stepping motor gives a sufficiently constant sampling interval for most studies. Alternatively, for wavenumbers up to 600-1000 cm^{-1} the Moiré fringes produced by a pair of diffraction gratings, one stationary and one mounted on the continuously moving mirror mount, give a sufficiently constant sampling interval. In rapid-scan mid-infrared interferometers, the sampling interval is determined by the zero-crossings of the interferogram from a He-Ne laser, i.e., by the laser wavelength, and is, thus, very precise. But noise due to source (a) can still be present if the mirror velocity is not constant, since the time constant of the electronics is finite so the signal at the ADC lags slightly behind the mirror [2]. As mentioned above, these are problems that the manufacturers must, and do, concern themselves with and the user can do little about.

The same is true of source (b) of digitization noise to some extent. The ADC must have an adequate dynamic range [7]. Thus a 4-bit ADC is not satisfactory [13] and a 15-bit ADC is usually used in mid-infrared instruments. But the user must use the instrument correctly to obtain the optimum signal-to-noise ratio, particularly if the detector has a low noise level. Digitization noise results if the noise amplitude is smaller than the voltage corresponding to the lowest bit of the ADC. This is a systematic error that cannot be overcome by signal averaging. The solution is to increase the gain of the amplifier so that the noise level is equivalent to the lowest two or three bits of the ADC. This is done in two ways: first to increase the gain used for the entire interferogram and, second, to use gain-ranging, i.e., to use a certain gain for the part of the interferogram near zero-path-difference where the AC signals are large, and to increase the gain at larger path-differences where the AC signals are small. This gain-ranging or gain-switching avoids saturating the ADC near zero-path-difference while ensuring a large signal (and sufficiently large noise) at larger path-differences.

On the Nicolet 7199 mid-infrared instrument in my laboratory we increase the gain by a factor of eight automatically at the 1024th data point for all samples, so that the first approximately 500 points to either side of the zero-path position have a lower gain than the remainder. For mid-infrared work to 4000 cm^{-1} we digitize at intervals of 0.6329914 μm times the refractive index of air, so the lower gain applies to path-differences of less than ~0.32 mm. The only additional adjustment is to increase the gain for the entire interferogram when the signal is small, and this increase in gain leads to a lower noise level in signal-averaged interferograms and the resultant spectra. This is illustrated in Fig. 2 by the comparison of the curves marked "X20" in the upper (GAIN = 8) and lower (GAIN = 1) boxes. For each box the sample was a 1 mm diameter aperture at the focus of the infrared beam in the sample compartment, and 128 interferograms, each containing 8192 data points, were averaged with GAIN = 1 or 8 (and the gain increased by a factor of eight at the 1024th point) and zero-filled to 16384 points before Fourier transformation to yield a sample (intensity) spectrum at 2 cm^{-1} resolution. Each sample spectrum was divided by a background spectrum that was collected in the same way but with a gain of one and without the 1 mm diameter aperture. The aperture transmits 8% of the incident light, as shown by the quiet flat line in each box. To show the noise, the ordinate scale was expanded by a factor of twenty so that for the curves marked "X20" the ordinate extends from 7 to 9% transmission. The noise with gain of one is about 50% greater than with gain of eight, because the signal-to-noise ratio in the sample interferograms with gain of one was quite low and the noise was too small, compared to the voltage corresponding to the first digitization bit, to be properly sampled, thus yielding a systematic error that was not removed by signal averaging. Increasing the gain increased the noise (and the signal) and allowed it to be properly digitized so that it became random noise in the digitized interferograms and was reduced by signal-averaging. The lower noise with a gain of eight is even clearer on the absorbance plots shown in Fig. 3, where the ordinate for the noisy curves is expanded ten times and extends from 1.06 to 1.12.

This feature of digitization noise is important when the signal is weak, i.e., for energy-limited systems.

3. OPERATING PARAMETERS WHICH AFFECT THE SIGNAL-TO-NOISE RATIO

Some operating parameters have been discussed in Sec. 2, namely the velocity of the moving mirror, the gain, and the switch-gain or gain-ranging parameters. By far the most important parameter for signal-to-noise enhancement is the number of scans (interferograms) that are averaged to produce the interferogram that is Fourier transformed. If successive interferograms are averaged coherently, that is if the nth data point in each digitized interferogram corresponds to exactly the same path-difference, the signal-to-noise ratio in the averaged interferogram and the resultant spectrum increases as the square root of the number of interferograms included in the average [7]. That this is the case with modern interferometers is shown in Fig. 4, which also shows again the beneficial effect of increasing the gain for small signals. The spectra are 100% lines (the ratio of two identical but independent spectra recorded with no sample) recorded with a 1 mm source aperture, gain of one or eight, and with thirty-two or

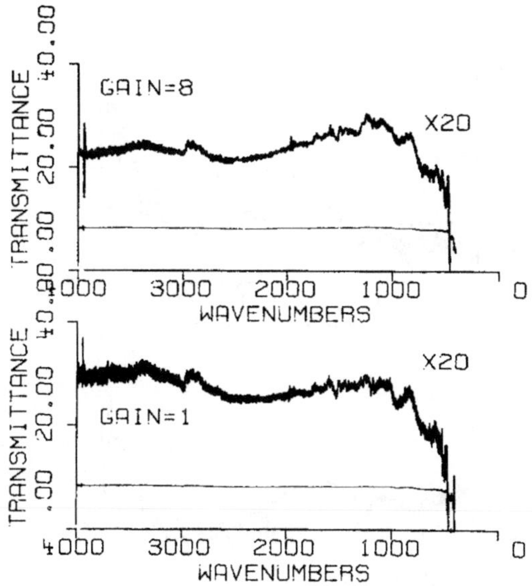

Fig. 2 Spectra of a 1 mm diameter aperture. The ordinate for the curves marked X20 extends from 7 to 9% T. GAIN = 1 or 8 for sample and 1 for background, NSS = 128.

two hundred fifty six interferograms averaged for each spectrum. The noise in the upper box is approximately $(32/256)^{\frac{1}{2}}$, i.e. $(2.8)^{-1}$ that in the lower box, and in each box the noise with a gain of one is about 50% larger than that with a gain of eight.

While signal averaging is by far the most commonly used and most valuable tool for signal-to-noise enhancement, there are other techniques that can be used. The spectra in Figs. 1-4 were recorded at 2 cm^{-1} resolution, although the frequency scale is too compressed to make use of this resolution. A resolution of 4 cm^{-1} or even 8 cm^{-1} is quite adequate for many systems, particularly liquid systems. If the time taken to collect an averaged interferogram at 2 cm^{-1} resolution is 2.5 minutes, as was the case for each sample and each background spectrum used for Figs. 2 and 3, a spectrum collected at 4 cm^{-1} resolution in 2.5 minutes would ideally have [7] double the signal-to-noise ratio shown in Figs. 2 and 3, and an 8 cm^{-1} resolution spectrum collected in 2.5 minutes would ideally have one-fourth of the noise shown in Figs. 2 and 3. Alternatively stated, since the signal-to-noise ratio increases as the square root of the number of scans, i.e., ideally as the square root of the collection time [7], an 8 cm^{-1} resolution spectrum and a 4 cm^{-1} resolution spectrum can ideally be collected with the same signal-to-noise ratio as a 2 cm^{-1} resolution

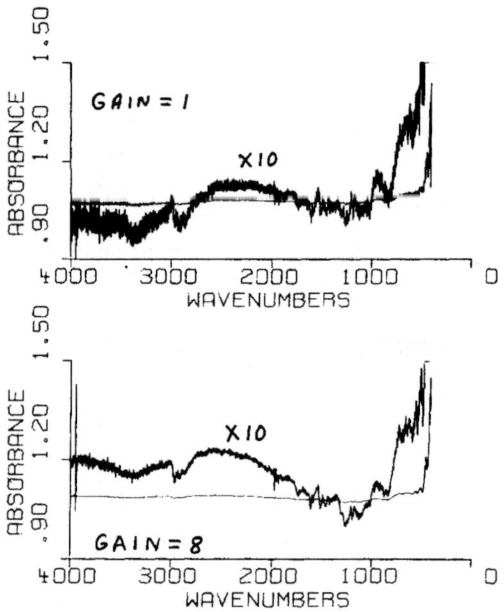

Fig. 3 Spectra of Fig. 2 plotted as absorbance. The ordinate for the curves marked X10 extends from 1.06 to 1.12.

spectrum in one-sixteenth and one-fourth, respectively, of the time required for the 2 cm^{-1} resolution spectrum. In fact the collection time is not proportional to the number of scans times the scan length, because of the times required to stop, start and return the mirror, so the numerical factors are not as advantageous as those given, but lowering the resolution is still an important option to consider when dealing with energy-limited systems, as it is when dispersive instruments are used.

The numerical factors in the last paragraph were calculated under the assumption that the source aperture is not changed when the resolution is changed. The source aperture limits the resolution achievable from a Michelson interferometer because the collimation is imperfect for a finite source. The highest resolution, Δk, (i.e., the smallest value of Δk) at wavenumber k that can be obtained with a circular source of diameter d and a collimating mirror of focal length F is [14, 15]

$$\Delta k = \frac{kd^2}{8F^2} .$$

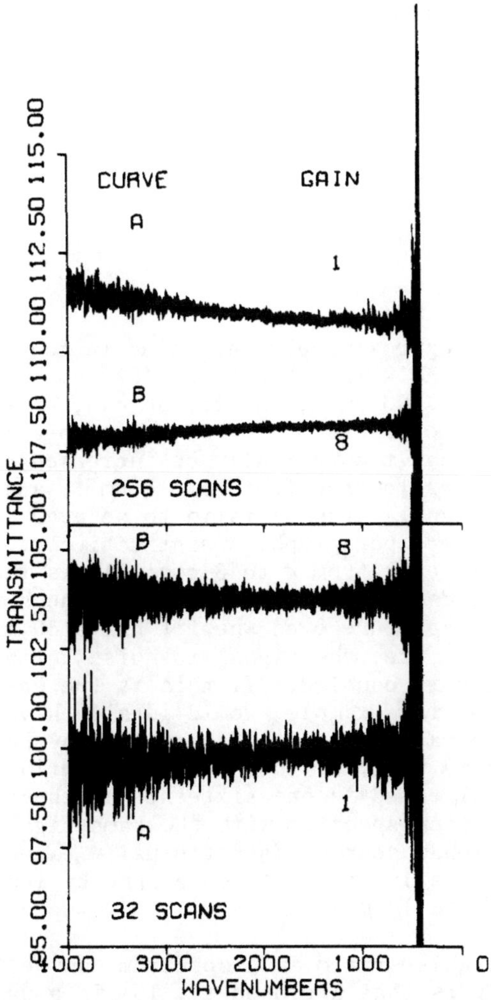

Fig. 4 100% lines. APT = SM, GAIN = 1 (curves A) or 8 (curves B), NSS = NSB = 32 or 256.

This expression is based on the relationship [14]

$$R\Omega = 2\pi,$$

where R is the resolving power at wavenumber k,

$$R = k/\Delta k,$$

and Ω is the solid angle subtended by the source aperture at the center of the collimating mirror. Thus

$$\Omega = \frac{\pi d^2}{4F^2}$$

and

$$R = \frac{8F^2}{d^2} .$$

If F is 24 cm, approximately the value in the Nicolet 7199 system for example, and k is taken to be 4000 cm^{-1}, a 6 mm source diameter yields $\Delta k = 0.03$ cm^{-1}, and for $\Delta k = 2$, 4 or 8 cm^{-1} the maximum source diameter is 15, 22 or 30 mm, respectively. Thus, the source diameter can theoretically be increased substantially when the resolution is lowered from 2 to 8 cm^{-1}, and this can cause the gain in signal-to-noise ratio to be even larger than indicated in the earlier paragraph. Quantitatively, halving the resolution from 2 to 4 or from 4 to 8 cm^{-1} allows the solid angle Ω of the beam to be doubled. If the image of the source completely fills the source aperture even when it has a 30 mm diameter, the signal and, therefore, the signal-to-noise ratio is doubled when the solid angle is doubled. If this is the case, a 4 cm^{-1} resolution spectrum, for example, would ideally have four times the signal-to-noise ratio of a 2 cm^{-1} resolution spectrum that is collected in the same time, or the 4 cm^{-1} resolution spectrum could ideally be collected in one-sixteenth of the time required for a 2 cm^{-1} resolution spectrum with the same signal-to-noise ratio. However, globar sources do not usually exceed 10 mm diameter so there is usually little to be gained by increasing the source aperture beyond 10 mm.

The signal-to-noise ratio in a spectrum can be affected by the electronic filters that are used [1, 16] in a rapid-scanning interferometer. The highest meaningful electrical frequency in the signal from the detector is $f_{max} = 2v\ k_{max}$, where v is the velocity of the moving mirror and k_{max} is the highest wavenumber of the radiation that reaches the detector. The signal from the detector may, however, contain noise frequencies that are higher than f_{max} and which, on Fourier transformation, yield noise at wavenumbers that are higher than k_{max}. If the interferogram is digitized at path intervals Δx given by $k_{max} = (2\Delta x)^{-1}$, wavenumbers from 0 to k_{max} are correctly sampled and appear as themselves in the spectrum. But wavenumbers from k_{max} to 2 k_{max}, which arise from electrical noise frequencies from f_{max} to 2 f_{max}, appear in the transformed spectrum folded about k_{max} into the k_{max}-to-zero range. Thus, k_{max} appears as itself and $k_{max} + k$

appears as k_{max}-k. Similarly, 2 k_{max} to 3 k_{max} appears in the spectrum as 0 ($\equiv$2 k_{max}) to k_{max} ($\equiv$3 k_{max}), and 3 k_{max} to 4 k_{max} is folded back so that 3 k_{max} + k appears as k_{max}-k, thus 4 k_{max} appears as zero. This phenomenon is known as aliasing. Its origin is described in detail elsewhere [1, 6-8] and it can be extremely useful for reducing the number of data points that need to be collected in certain circumstances [17].

To avoid noise appearing in the spectrum due to high-frequency electrical noise in the digitized interferogram, electronic filters are used to pass the desired frequencies but to stop the unwanted high frequencies from reaching the ADC. Unfortunately, electronic filters have rather gentle transmission edges, so that a filter that attenuates a frequency just over f_{max} could also attenuate many of the wanted frequencies below f_{max}. Thus, the electronic filter is usually chosen so that it blocks out frequencies above approximately 1.6 f_{max} to 2.5 f_{max}, and the sampling interval, Δx, is reduced so that frequencies above ~1.3 f_{max}, but usually not above 2 f_{max}, are correctly sampled [1,16]. The resultant spectrum is then calculated to wavenumbers above ~1.3 k_{max}, but usually not above 2 k_{max}, so that any high-frequency electrical noise is folded into the region above k_{max}, where it is known that no spectral information exists, and does not contribute to the wanted spectrum. Thus, the choice of the sampling interval and of the cut-off frequency of the low-pass electronic filter should be made with these ideas in mind, subject always to the more important consideration that the spectrum will be in error if radiation of wavenumber higher than $(2\Delta x)^{-1}$ reaches the detector. In fact, when the parameters given in the appendix ($(2\Delta x)^{-1}$ = 7899 cm^{-1}) are used on our Nicolet 7199 system to study the region 0 to 4000 cm^{-1}, the setting of the low-pass filter has little effect on the noise-level below 4000 cm^{-1}. The electrical frequency corresponding to 4000 cm^{-1} with V = 0.9 sec^{-1} is 7.2 kHz, and the frequencies at which the low-pass filter attenuates by 3db can be set to 50 kHz, 10 kHz, or even 5 kHz with no significant effect on the signal-to-noise ratio below 4000 cm^{-1}; although, of course, the 5 kHz setting badly attenuates the signal from 4000 to ~2000 cm^{-1}. On other systems, perhaps with other detectors, the low-pass filter setting apparently [1,16] is more important than it is on our system.

One method that appears to be a possible way of enhancing the signal-to-noise ratio of an energy-limited spectrum is to phase correct the Fourier transform of the noisy interferogram by using the phases calculated from a quiet interferogram instead of those calculated from the noisy interferogram. In fact, it is not a very useful technique, partly because the calculated phase spectrum is a very low-resolution spectrum, which can only introduce low-resolution noise, which is not usually a problem.

To end this section it should be noted that if a system is

sufficiently energy-limited the signal-to-noise ratio in the spectrum will be poor in spite of all attempts to improve it. In such cases digital filtering can aid the extraction of information from the spectrum, with the digital filter being chosen according to the specific information which is sought. A recent account of digital filtering for smoothing curves has been given by Horlick and Hieftje [18].

4. SOME SPECTRA

A rapid-scan mid-infrared Fourier transform spectrometer has much better sensitivity than even the best traditional analogue infrared grating spectrometers, and therefore yields much better spectra in energy-limited situations. For example, I have found the performance of the Nicolet 7199 Fourier transform spectrometer to be phenomenal compared to that of our previous mid-infrared spectrometer, the Beckman I.R. 12, which was among the best commercial spectrometers in 1967.

The performance of the interferometer has already been illustrated in Figs. 2 and 3 which show, on percent transmission and absorbance scales, the spectrum of a 1 mm diameter aperture placed at the beam focus in the sample compartment divided by the spectrum with no sample and the full beam diameter. The curves marked "X20" show the noise, which is totally due to the sample spectrum. The sample spectrum was calculated from the average of 128 interferograms, which took a total of 2.5 minutes to collect and ~0.5 minutes to transform. The maximum path-difference was ~0.5 cm yielding a resolution of ~2 cm^{-1}. The transmission of the aperture is ~8% and in 2.5 minutes with a gain of eight the system yields a spectrum with a signal-to-noise ratio of ~300 in the favorable region from 1000 to 2000 cm^{-1} and ~100 in the region near 4000 cm^{-1} where (Fig. 1, top curve) there is less energy available.

Thus, in contrast to the traditional situation with a grating spectrometer, a 1 mm diameter sample does not constitute a badly energy-limited situation with modern FTIR spectrometers. It is convenient to note here that, while the focus in the sample compartment is a 1:1 image of the source aperture, when the diameter of the source aperture is 6 mm an aperture of ~5 mm diameter in the sample compartment transmits over 95% of the radiation.

The spectrum of a 0.25 mm pinhole at the focus of the sample compartment, at 2 cm^{-1} resolution and collected and transformed in three minutes, is shown in Fig. 5 ratioed against a background spectrum (full beam, no sample). The pinhole transmits 0.65 to 0.7% of the incident light, and the signal-to-noise ratio obtained at 2 cm^{-1} resolution in three minutes is ~30 near 1500 cm^{-1}

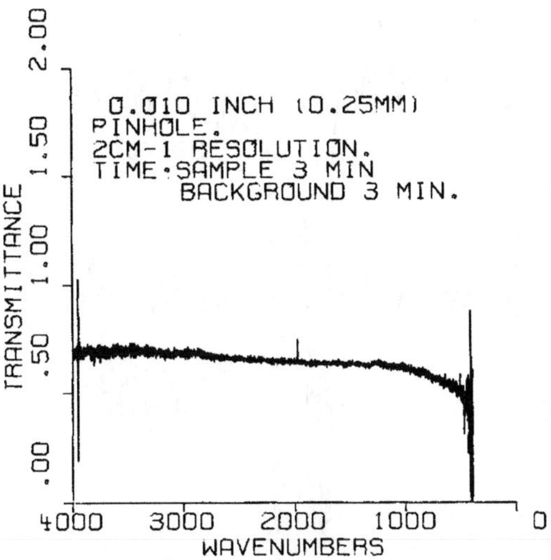

Fig. 5 Spectrum of a 0.25 mm pinhole. 2 cm^{-1} resolution, NSS = 128, GAIN = 64. 2.5 min collection time for each, sample and background, spectrum.

and ~14 near 4000 cm^{-1}, both of which are high enough to yield useful results. The signal-to-noise ratio can be improved by factors of 2, 4, etc. by signal-averaging for ten minutes, forty minutes, etc., instead of 2.5 minutes. Thus in approximately forty minutes our FTIR spectrometer (HgCdTe detector) can yield a 2 cm^{-1} resolution spectrum through a sample diameter of 0.25 mm with a signal-to-noise ratio (60 to 100) that compares favorably with that of perhaps the majority of spectra that chemists have recorded over the past twenty-five years using dispersive spectrometers.

The spectrum of polystyrene recorded through the 0.25 mm pinhole is shown as Curve B in Figs. 6 and 7, from 3950 to 1950 cm^{-1} and from 2000 to 400 cm^{-1}. The resolution is 2 cm^{-1}, the gain was 64, and 512 interferograms were averaged, requiring ten minutes to collect the data and thirty-five seconds to apodize, calculate the phases, Fourier transform, and phase correct the data. The background spectrum was taken with the parameters of the appendix and the full beam. The upper curves, A, in Figs. 6 and 7 show the spectrum of polystyrene recorded with the full beam, 32 scans averaged for both sample and background spectra, and the other parameters of the appendix. The ordinate scale for curves A in Figs. 6 and 7 is compressed by a factor of 125 and runs from 0 to 100% T with the curve offset by +25% T. It is clear that the spectrum collected in ten minutes through the 0.25 mm pinhole

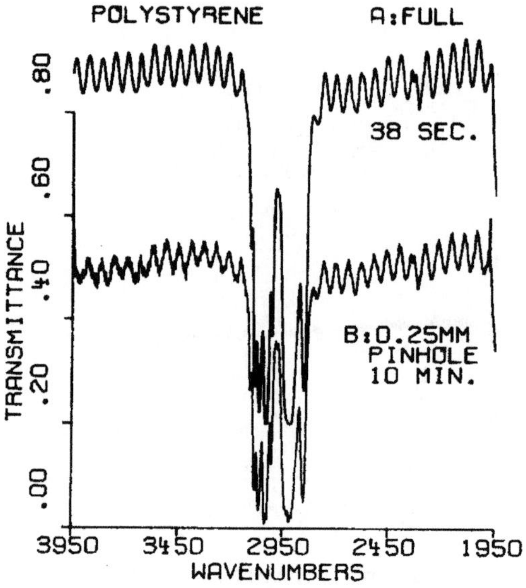

Fig. 6 2 cm^{-1} resolution spectra of polystyrene. Curve A: Full beam, NSS = NSB = 32. Curve B: Through 0.25 mm pinhole, GAIN = 64, NSS = 512.

is an excellent, faithful spectrum of polystyrene, with a noise level that is only significant below 450 cm^{-1} and above 3000 cm^{-1}, even though the spectrum is limited to the range 0 to 0.5% transmission.

Figure 8 shows the noisy region of the spectra of Figs. 6 and 7, above 3000 cm^{-1}, recorded under a variety of conditions to show the effect on the signal-to-noise ratio of the parameters discussed in Sec. 3. Curve A is spectrum B of Figs. 6 and 7 and curve E is spectrum A of Figs. 6 and 7. The transmittance scale applies to curves A to D and is 10% T per division for curve E. Curve B is the result of averaging 128 interferograms instead of the 512 of curve A and is, accordingly, noisier. Curve C is the result of 160 scans at 4 cm^{-1} resolution instead of the 128 scans at 2 cm^{-1} resolution of curve B, and is much quieter than curve B. Note that the data collection times of curves B, C and D were the same, 2.5 minutes. One can not collect twice as many scans at 4 cm^{-1} resolution as at 2 cm^{-1} resolution in the same time because of the 'unusable' time required to stop, start, and return the moving mirror. Curve D is, like curve C, the result of 160 scans at 4 cm^{-1} resolution but with a source aperture diameter of 12.5 mm instead of the 6 mm diameter (APT = F1) of curve C. Curve D is little better than curve C, indicating that our globar

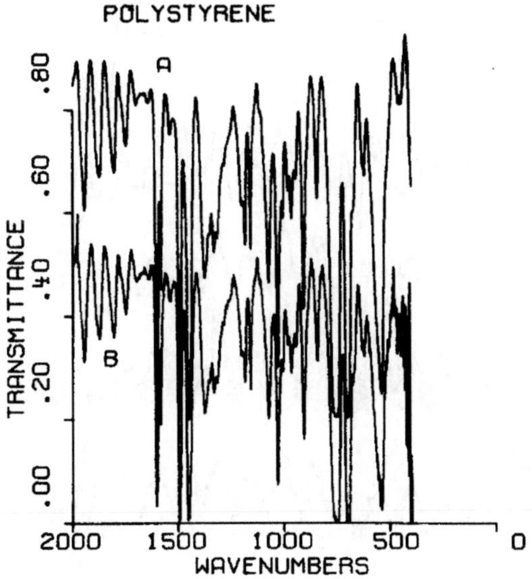

Fig. 7 2 cm^{-1} resolution spectra of polystyrene. Curve A: Full beam, NSS = NSB = 32. Curve B: Through 0.25 mm pinhole, GAIN = 64, NSS = 512.

source does not fill an aperture much larger than 6 mm diameter, so that we gain nothing by using a larger aperture. Curve E shows that the noise level in a spectrum from 400 to 4000 cm^{-1} collected with the full beam, and a gain of 1, in 38 seconds is undetectable on a scale of 10% T per division. Spectra that were recorded through the pinhole with 128 scans and the 3dB point of the low pass filter at 10 kHz, instead of the 50 kHz for curve B, had the same noise level as curve B.

Before I leave these spectra I should reiterate that Fig. 8 intentionally shows the noisiest parts of the spectra, and that 2.5 or 10 minutes are sufficient to yield good spectra at 2 cm^{-1} resolution through a 0.25 mm pinhole, which represents excellent instrument performance. It is also worth recalling (Appendix) that each spectrum contains information to 7899 cm^{-1}, although the signal decreases, so the signal-to-noise ratio decreases, steadily above 4000 cm^{-1} with the Ge on KBr beamsplitter.

It is clear from these examples that an interferometer can easily and quickly give spectra that have traditionally been more or less difficult to obtain with dispersive instruments - at least with the dispersive instruments of the pre-1968 vintage which are all that I have experience of. Spectra of cryogenic

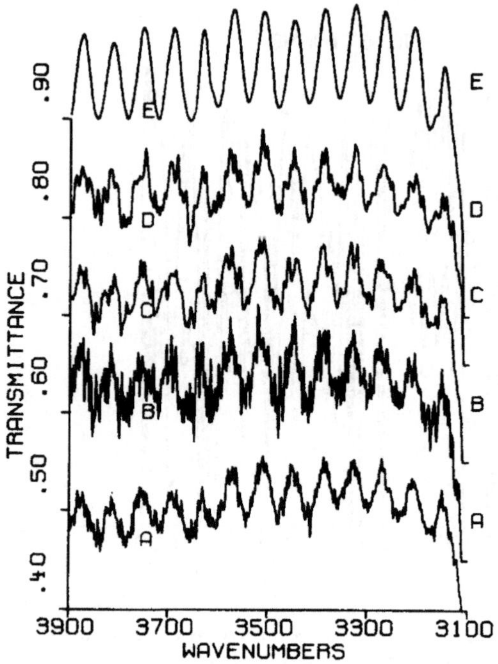

Fig. 8 Spectra of polystyrene film. Curves E and A are spectra A and B, respectively, of Fig. 6. B: as A but NSS = 128 not 512. C and D: 4 cm^{-1} resolution, 160 scans, source aperture 6(C) and 12.5 (D) mm.

samples have always presented some problems which, although not extreme, have normally required a reduction in the resolution to 2 cm^{-1} or worse and have required long scanning times, of the order of two hours or more for the region 4000 to 400 cm^{-1} and much longer than that if individual bands were studied in detail, and if a good ($\gtrsim$ 100) signal-to-noise ratio was desired. This arises from the three or four windows through which the light must pass in a cryostatic optical cell and because the aperture of the cell often attenuates some light with a dispersive instrument. The latter problem is alleviated by the small beam size in our interferometer (6 mm diameter at the sample focus), but the first problem of course remains, causing our cryostats to transmit only ~40% of the incident light whether they are for use with liquid-nitrogen (home-made cryostats) or liquid-helium (Air-Products, Helitran and Displex systems for the mid-infrared). But from the spectra presented above it is clear that 40% of the incident light is enough to obtain spectra of the highest quality. Three examples of such mid-infrared spectra of cryogenic systems are presented in Figs. 9 to 11.

Figure 9 shows the spectra that proved to our satisfaction, and we hope to the satisfaction of everyone else, that cyclobutadiene can not be square since it contains a C=C bond [19]. The band at 1523 cm^{-1} is not present in the starting compound, which is trapped in an argon matrix at 7°K, but it grows as the matrix is irradiated with ultraviolet light, which photolyzes the starting compound to cyclobutadiene and phthalan, and decays when the matrix is warmed to 35°K for ~1 hour and re-cooled, since molecules can diffuse in the matrix at 35°K and active species can react. However, the spectra contain many bands of different compounds, notably the starting material, the by-product phthalan, and the dimer of cyclobutadiene, and it is necessary to obtain sufficiently good spectra that the integrated absorbances of all of the bands that are believed to be due to cyclobutadiene can be shown to increase or decrease in the same proportion to their size during a period of photolysis or diffusion. To enhance the visibility of weak bands by making use of the fact that the bands of the matrix-isolated species are sharp, 0.5 cm^{-1} resolution was used and 500 scans were averaged for each spectrum (twenty minutes measuring time). The background spectrum was of a deposit of pure argon at 7°K, and the absorbance scale has been expanded to give 0.006 units per division, which corresponds to ~1% of the intensity in the background spectrum per division or ~0.4% of the intensity from the source per division. The signal-to-noise ratio is excellent and even though the band at 1523 cm^{-1} is extremely weak it is clear that it varies with treatment in exactly the same way as the band at 1240 cm^{-1}, which is about twenty times more intense.

Figure 10 shows the infrared spectrum at 1 cm^{-1} resolution that we have obtained for the solid that has been called ammonium oxide [20] but which is really ammonia hemihydrate, $2NH_3{\cdot}H_2O$, at 90°K. This solid and ammonia monohydrate (which has been called ammonium hydroxide [20]) melt at -79°C, their crystal structures are known [20,21] apart from the exact positions of the hydrogen atoms, and Waldron and Hornig [22] have reported their infrared spectra. We have grown these solids from stoichiometric mixtures of ammonia and water under equilibrium conditions and cooled them to 77°K. They were ground, X-rayed, and made into infrared samples while being kept below 100°K. Our results show clearly that Waldron and Hornig did not prepare these hydrates of ammonia, so the spectrum in Fig. 10 is the first authentic spectrum of $2NH_3{\cdot}H_2O$ to be reported [23]. The infrared samples were mulls of the powdered hydrate in propane, propylene and Freon-13 (CF_3Cl), and the absorption by the mulling agent has been substracted from the spectra shown. The straight lines in Fig. 10 near 2900 and 1450 cm^{-1} show where the propane mulling agent absorbed too strongly to yield usable information. The upper spectrum near 3000 cm^{-1} is of a mull in Freon-13 and shows the

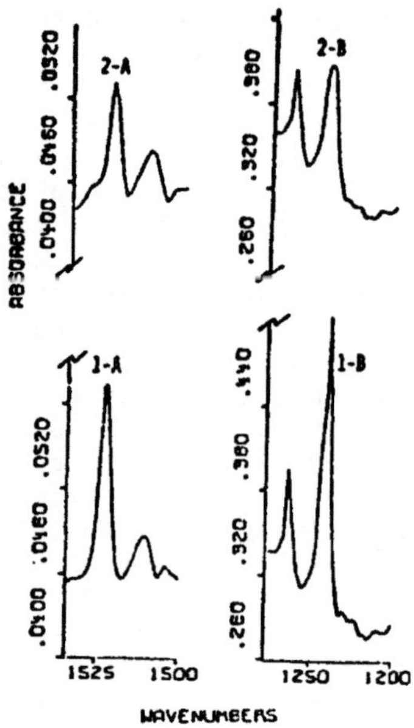

Fig. 9 The 1523 and 1240 cm^{-1} bands of cyclobutadiene in Ar matrix at 7°K, from Ref. [19]. Resolution 0.5 cm^{-1}, 20 min. collection time, unsmoothed. Lower: after irradiation, upper: after warming to 35°K for 1 hr. and recooling.

hemihydrate absorption near 2900 cm^{-1}, while other spectra have indicated clearly that no features exist where the line is straight near 1450 cm^{-1}. Figure 11 shows two of the original spectra from 3700 to 3100 cm^{-1}, to show the spectral quality. The sample is ammonia monohydrate mulled in propylene and the upper curve is the result of the average of 100 interferograms of length 1 cm (i.e., 1 cm^{-1} resolution, 2 2/3 minutes measuring time) while the lower curve is the result of averaging 500 interferograms (thirteen minutes measuring time). It is clear that even the weakest features on the lower curve are real because they are also present, with some noise, on the independent upper curve, and it should be recalled (Figs. 1 to 7) that this is the noisiest region of the spectrum. All of the data about this sample that one can want are available from the average of 500 interferograms, and we have found that the total instrument time required by such samples is about thirty minutes compared with frequently over twelve hours on our previous, dispersive, instrument. These samples require a lot of time to prepare and one can not always guarantee that a sample will be good. So it is particularly helpful to us that good spectra can be obtained on the interferometer from samples that would have yielded no useful

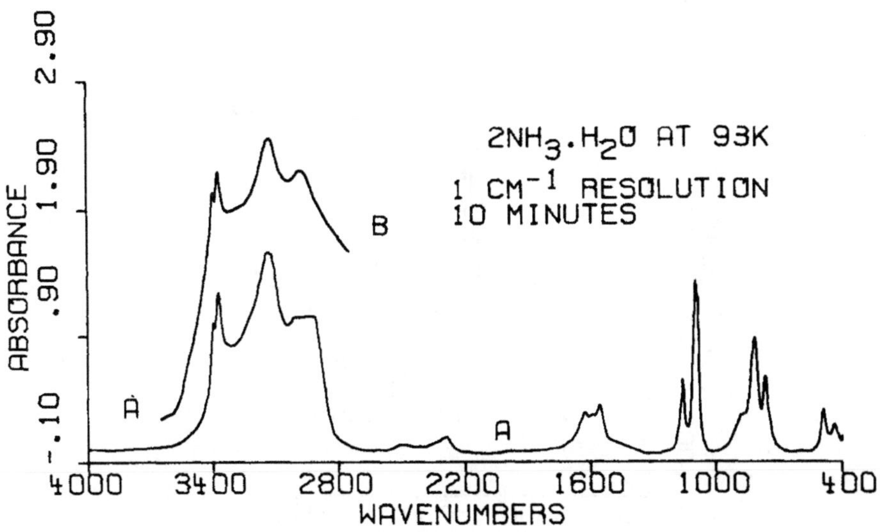

Fig. 10 Transmission spectrum of ammonia hemihydrate at 93°K. (A) from propane mull, after subtraction of propane peaks. The propane absorbed fully where the line is straight near 2900 and 1450 cm^{-1}. (B) mulled in CF_3Cl. Measurement time, ten minutes.

information on our earlier instrument, and that one can always return to the stored spectrum to extract further information.

The final examples of energy-limited mid-infrared spectra that I want to present are some preliminary reflection spectra [24]. Infrared reflection spectroscopy has not been extensively pursued, largely because it requires measurement of reflectances of the order of 5% and these have traditionally been difficult to measure with useful accuracy. It is clear from the spectra presented above that 5% of the incident intensity can be measured by FTIR spectroscopy with good precision, which makes reflection spectroscopy a straightforward matter nowadays. To illustrate this, Figs. 12 and 13 show reflection spectra of a glass microscope slide and a polished lithium fluoride window, both at 22°C.

Figure 12 shows the reflection by a glass microscope slide of p-polarized light at 57° angle of incidence, which is the Brewster angle of glass for near infrared and visible frequencies at which its refractive index is 1.54. The p-polarized light has the electric vector in the plane of incidence, which is perpendicular to the plane of the glass surface which reflects. The resolution was 4 cm^{-1}, the gain was 4, and only 32 scans were

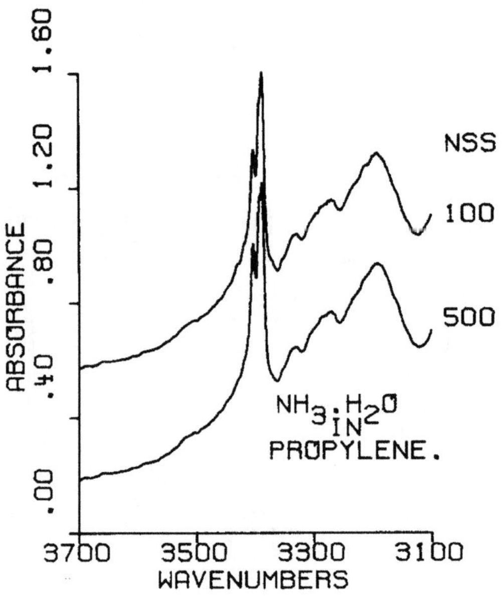

Fig. 11 The O-H and N-H stretching bands of $NH_3 \cdot H_2O$ in propylene at 93°K. Measuring times were 2.7 (upper) and 13 (lower) minutes.

averaged. This, plus water vapor in the unpurged sample compartment, explains the relatively high noise level. The reflectance above 1600 cm^{-1} is zero within the ±0.25% to ±1% noise level, indicating that the Brewster angle is very close to 57° in this range as well as for visible light. The peak reflectance at 1200 cm^{-1} of 28.5% was reproducible to ±0.25%, i.e., to 1 part in 114, as was the reflectance at the 1080 cm^{-1} shoulder and the zero reflectance above 1600 cm^{-1}.

The background spectrum for Fig. 12 was of the p-polarized light reflected at 57° by a front-surfaced aluminum mirror. Our reflectance accessory transmits only ~35% of the incident light and the polarizer transmits poorly and, in these experiments, attenuated the beam quite badly. Consequently the 28.5% reflection at the peak in Fig. 12 corresponds to only ~1.6% of the 1200 cm^{-1} radiation incident upon the accessory, and the total energy incident upon the detector at zero path-difference was ~1/700 of that with no sample or accessory in the beam.

Figure 13 shows the noise level and reproducibility of reflection spectra that were recorded more carefully than that of Fig. 12. Two spectra are shown of the reflection by a polished

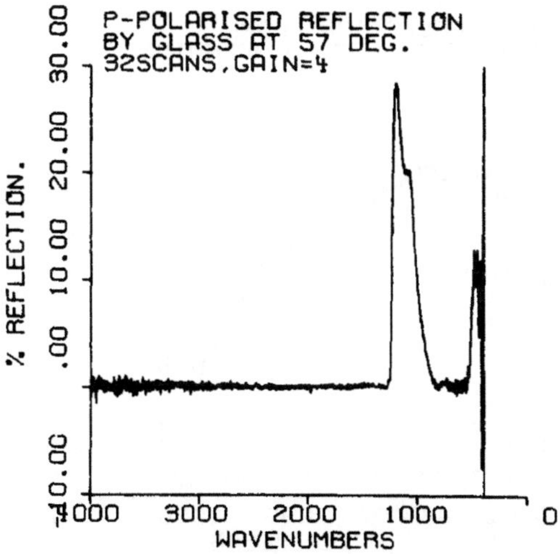

Fig. 12 p-polarized reflection by glass at 57° incidence. 4 cm^{-1} resolution (NDP = 4096), NPD = 60, NPT = 256, LPS = 5, SGH = 1, NSS = 32, GAIN = 4.

lithium fluoride window of unpolarized light at 45° incidence. For the background spectrum the LiF was replaced by an aluminium mirror. The two spectra and backgrounds were run on different days and the LiF surface was further polished between the two spectra, causing the peak reflectance at 630 cm^{-1} to be 3.5% greater in the second spectrum. Above 800 cm^{-1} the spectra coincide on the 20% per division reflection scale, and to show the excellent reproducibility, and the noise, the ordinate scale has been expanded by a factor of twenty for the curves marked X20, and runs from 0 to 4% reflection at 1% per division. In this case the peak reflectance of ~75% corresponds to ~25% of the intensity incident upon the accessory.

All of the examples of energy-limited spectra so far presented have been mid-infrared spectra, because this is the most common region of interest for analytical purposes. However, the far infrared region has traditionally been the really energy-limited part of the infrared spectrum. Further, the use of a liquid-helium cryostat in which the sample is surrounded by helium gas, to improve temperature control, makes this energy-limited region even more difficult, because the light must pass through four windows as well as the sample. Figure 14 shows spectra that we have obtained [25] using such a cell (Janis, model 8DT) in a Beckman-R.I.I.C. model FS-720 interferometer fitted with a Molectron, silicon-bolometer detector which operates at 1.3°K. The cryostat has two windows of TPX plastic at room

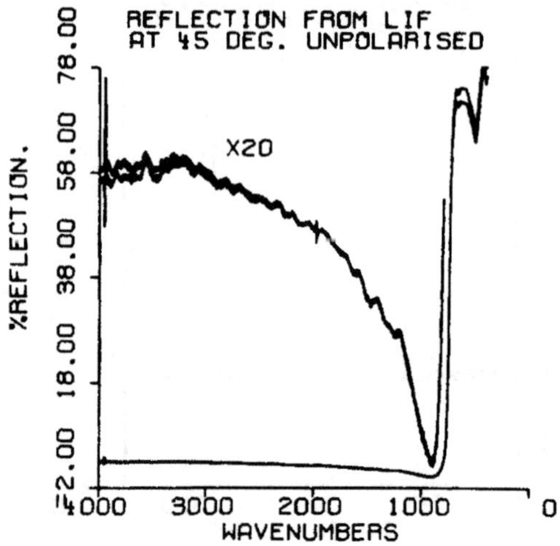

Fig. 13 Reflection by lithium fluoride of unpolarized light at 45° incidence. 4 cm^{-1} resolution (NDP = 4096), GAIN = 8, NSS = 500, SGH = 1. The ordinate for the curves marked X20 runs from 0 to 4% reflection.

temperature and two of quartz at the temperature of the sample. The sample was the structure II clathrate hydrate or deuterate of oxetane powdered and pressed into a pellet of adamantane. The instrument parameters are given in the caption and each curve shown is the average of four absorbance spectra, each of which was obtained from a different sample and background spectrum. The background spectra were of a pure adamantane pellet at 4.3°K. The appearance of analogous features in the spectra of the hydrate and deuterate show that the noise level really is as low as it appears to be, as is also the case down to 15 cm^{-1} (upper box). The detector is undoubtedly the most important factor in obtaining far-infrared spectra with a high signal-to-noise ratio, and Stone and Chantry [10] have discussed other factors.

There are many energy-limited systems that I have not discussed, some of which have been recently described by Griffiths, [26]. Emission spectra, both of earthbound and astronomical sources, and high resolution spectra are my most important omissions. A major driving force behind the development of interferometry was the desire to obtain high-resolution spectra of astronomical sources, and extraordinary spectra that would have been beyond belief fifteen to twenty years ago have been obtained by FT spectroscopy, notably by J. and P. Connes and their colleagues [27]. There are also commercial FTIR spectrometers that, when fitted with a cooled detector, yield in ~20 minutes measurement

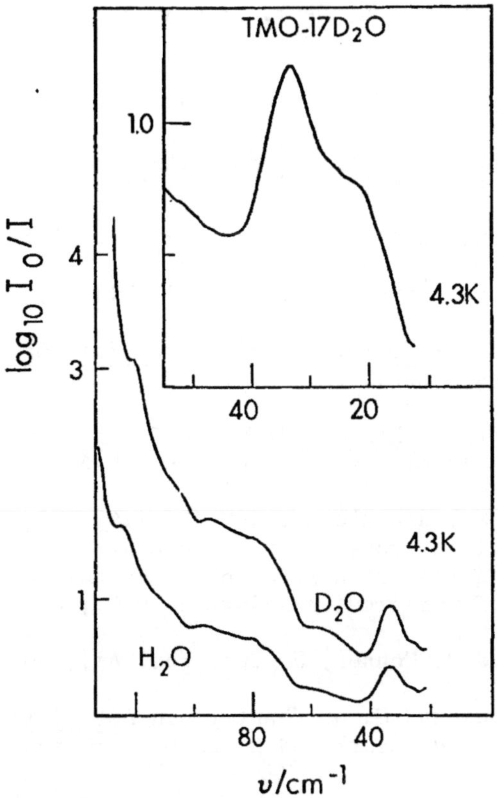

Fig. 14 Far-infrared transmission by the structure II clathrate hydrate of oxetane dispersed in an adamantane pellet. The deuterate curve is offset for clarity. 2 cm^{-1} resolution, sample interval 16 or 56 (inset) μm, beamsplitter 14 or 50 (inset) μm mylar, filter cutoff 200 or 80 (inset) cm^{-1}.

time high resolution ($\lesssim 0.1$ cm^{-1}) spectra below 100 cm^{-1} in the far infrared that have good signal-to-noise ratios. Thus we are today far less energy-limited in our spectroscopy than we were ten years ago, even though our society is far more energy-limited.

REFERENCES

1. E. V. Loewenstein. Page 3 in Proceedings of the Aspen International Conference on Fourier Spectroscopy, Aspen, Colorado, U.S.A., 1970. A.F.C.R.L. -71-0019, special report No. 114, Jan. 1971. Clearing house for Federal Scientific and

Technical Information, Dept. of Commerce, Washington, D.C., U.S.A.

2. H. Sakai. Page 19 of reference 1.

3. H. Sakai in Spectrometric Techniques, G. Vanasse, ed., Academic Press, New York, 1977, Volume 1, p. 1.

4. D. R. Mattson, Appl. Spectrosc., 32, 335 (1978).

5. P. Fellgett, J. Phys. Radium, 19, 187 (1958).

6. R. J. Bell, Introductory Fourier Transform Spectroscopy, Academic Press, New York, 1972.

7. P. R. Griffiths, Chemical Infrared Fourier Transform Spectroscopy, Wiley-Interscience, New York, 1975.

8. J. Connes, Rev. d'optiques, 40, 45, 116, 171, and 231 (1961). English translation: Document AD 409 869, Clearing house for Federal Scientific and Technical Information, Department of Commerce, Washington, D.C., U.S.A.

9. J. Connes and P. Connes, J. Opt. Soc. Am., 56, 896 (1966).

10. N. W. B. Stone and G. W. Chantry, in Adv. Infrared Raman Spect., (R. J. H. Clark and R. E. Hester, eds.), 3, 43 (1977).

11. D. G. Moss, J. Phys. E., 10, 1170 (1977).

12. R. E. Pulfrey, Appl. Optics, 15, 309 (1976).

13. G. Horlick and H. V. Malmstadt, Anal. Chem., 42, 1361 (1970).

14. R. J. Bell. Reference 6, page 149-150.

15. K. D. Möller and W. G. Rothschild, Far-Infrared Spectroscopy, Wiley-Interscience, New York, 1971, p. 145.

16. P. R. Griffiths, Reference 7, pages 62, 201.

17. G. Horlick and Wo Yuen, Anal. Chem., 47, 775 A (1975).

18. G. Horlick and G. M. Hieftje. 'Correlation Methods in Chemical Data Measurements' section 2.2, page 170 in Contemporary Topics in Analytical and Clinical Chemistry, (D. M. Hercules, G. M. Hieftje, L. R. Snyder and M. A. Evenson, eds.), Plenum Publishing Corporation, New York, 3 (1978).

19. Satoru Masamune, F. A. Souto-Bachiller, Takahisa Machiguchi and J. E. Bertie, J. Amer. Chem. Soc., 100, 4889 (1978).

20. W. J. Siemons and D. H. Templeton, Acta Cryst., 7, 194 (1954).

21. I. Olovsson and D. H. Templeton, Acta Cryst., 12, 827 (1959).

22. R. D. Waldron and D. F. Hornig, J. Amer. Chem. Soc., 75, 6079 (1953).

23. J. E. Bertie and Mary M. Morrison, Unpublished results.

24. J. E. Bertie and Joanne C. M. Zwinkels, Unpublished results.

25. J. E. Bertie and S. M. Jacobs, J. Chem. Phys., 69, 4105 (1978).

26. P. R. Griffiths. P. 153-169 in 'Transform Techniques in Chemistry', (P. R. Griffiths, ed.), Plenum Press, New York, 1978.

27. P. Connes and G. Michel. Appl. Opt., 14, 2067 (1975) and references therein. See also Reference 3 and R. J. Bell, page 107 of reference 3, and references therein.

APPENDIX

PARAMETERS OF NICOLET 7199 FTIR SPECTROMETER USED FOR FIGURES, EXCEPT AS NOTED IN FIGURE CAPTIONS.

Beamsplitter: Germanium on potassium bromide
Source : Globar coil
Detector : Mercury Cadmium Telluride type B (usable above 400 cm^{-1}).

PARAMETER	VALUE	MEANING
NDP =	8192	Number of data points collected per scan. At SSP = 2 this corresponds to 2 cm^{-1} resolution.
NTP =	16384	Number of points Fourier transformed. i.e. 8192 zero points added to the 8192 data points before Fourier transformation.

NSS =	as noted	on figures. Number of scans, i.e. interferograms, averaged before Fourier transformation to yield sample spectrum.
NSB =	128	As NSS, except for background spectrum.
GAN =	1	GAIN setting
SGL = SGH =	1 8	Gain given by GAN for the first 1024 data points and by SGH times GAN for the remainder.
AFN =	HG	Happ-Genzel apodization used for all spectra.
HPS = LPS =	1 6	High- and low-pass electronic filters set with 3 dB points at 100 Hz and 50 kHz.
VEL =	36	Mirror velocity equals 0.90 cm sec^{-1}.
NSK =	0	No data points skipped at beginning of scan. The zero path position is the (451 - NSK) data point.
SSP =	2	Path-difference increment between data points is (0.3164957 μm times n times SSP), where n is the refractive index of air at 632.9 nm; i.e., in our case Δx = 0.6329914.n μm, so k_{max} = 7899.00 cm^{-1}.
APT =	FL	Source aperture diameter. FL = 6.3 mm MD = 2.3 mm SM = 1.1 mm; intensity ratios 100:32:8.
NPD = NPT =	200 1024	To calculate the phase angles, 100 data points on each side of the zero-path-position were used and NPT-NPD = 824 zero points were added before Fourier transformation.
PHZ =	PH	The real part of the phase-corrected transform was calculated as the spectrum.
COR =	LO	Each interferogram was compared with those previously collected, and if necessary slid one data point in either direction to make the peaks match, before being included in the average.

TECHNIQUES

MODERN DATA PROCESSING TECHNIQUES FOR FOURIER TRANSFORM INFRARED SPECTROSCOPY

Jack L. Koenig

Department of Macromolecular Science
Case Western Reserve University
Cleveland, Ohio, U. S. A.

I. INTRODUCTION

Quantitative analysis of mixtures by infrared spectroscopy can be considerably improved using modern interferometric techniques compared to classical dispersion methods. The improvement arises from two different benefits. First, the signal-to-noise ratios for the Fourier transform spectra are considerably enhanced and secondly, the digitized nature of the spectra allows the introduction of computerized data processing techniques. This paper will discuss the nature of these new data processing methods as utilized for the quantitative analysis of the mixtures. The logic pattern to be followed in developing the spectral methods is shown in Fig. 1. The FT-IR spectrum of the mixture is obtained and recorded as a digitized absorbance spectrum. The first question to be answered concerns the number of spectroscopically distinguishable components in the mixture. A method termed factor group analysis will be described which answers this question [1]. The next logical step concerns a determination of the spectra of each of the components in the mixture. Two methods are available for making this spectral evaluation without chemical or physical separation of the components. If only two components are present, absorbance subtraction is simplest while the absorbance ratio method is more suitable for more complex mixtures [2, 3]. An analysis of the spectra of the components yields identification and structural information. A quantitative analysis of the mixture requires an evaluation of the spectra of the mixture in terms of the spectra of the components [4]. Classically, only a single analytic frequency was used for quantitative measurement, but with digitized spectra and a computer the entire frequency spectrum of each component can be fit to the mixture by

J. R. Durig (ed.), Analytical Applications of FT-IR to Molecular and Biological Systems, 79–88.

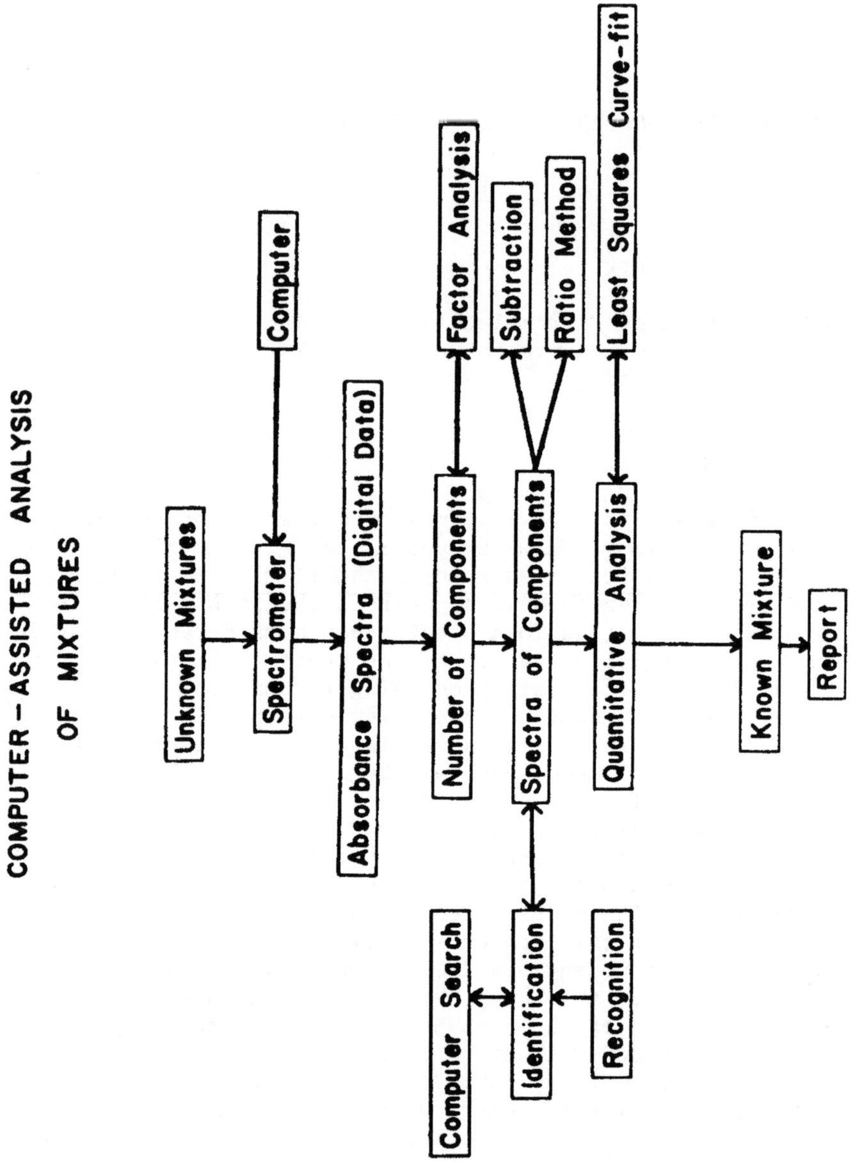

Fig. 1 Logic pattern of analysis of amorphous polymers.

curve-fitting techniques using various methods including least squares refinements which can yield an indication of the precision and the goodness of fit. In this fashion, quantitative analysis [5] of mixtures can be carried out.

We will illustrate the integrated use of the above techniques to study the amorphous and semicrystalline phases of poly (ethylene terephthalate) (PET). Factor analysis will be used to determine the number of spectrally identifiable components in a series of PET films annealed to different levels. The polymer has two rotational conformations assignable to glycol linkages; the most stable being the extended trans conformation (dihedral angle of 180°) and slightly higher in energy, the gauche conformation with the methylene units rotated out of the plane of the ring carbon atoms (dihedral angle of 60°). The glycol conformation has been shown by X-ray experiments to be exclusively trans in the crystalline phase of the polymer. The absorbance ratio method will be utilized to obtain the appropriate number of component spectra. A self-calibrating system will be used since external standard spectra are unobtainable for any real polymer system. With representative conformational spectra, a least-squares curve fitting procedure can be applied to analytically determine the concentration of the conformations in the PET films or arbitrary PET samples [5].

2. THEORETICAL BACKGROUND

Factor analysis is a mathematical procedure for determining the number of linearly independent components in a set of mixtures. This procedure has been applied to the study of the infrared spectra of mixtures [1].

Consider a matrix, A, dimensioned m by r where m represents a given mixture spectrum and r the absorbance value for the spectrum m for each spectral element (in cm^{-1}). Simply, it is a listing of each of the mixture spectra by column. We are interested in obtaining the rank of the A matrix. The rank will be equal to the number of linearly independent basis spectra which make up the series of mixture spectra. The task of determining the rank of A is accomplished by first multiplying A by A^T (A transpose) yielding a matrix C, termed the covariance matrix. The dimensions of C are condensed to m by m. The rank of C is equivalent to the rank of A. Diagonalizing C allows calculation of the eigenvalues and eigenvectors for the matrix. The rank of C is equal to the number of non-zero eigenvalues of C, the covariance matrix, which also indicates the number of component spectra in the initial set of mixture spectra. Experimentally, some degree of discernment is necessary in selecting non-zero eigenvalues since random spectral noise contributes small positive eigenvalues.

An absorbance ratio method has been applied to mixtures of simple organic molecules and also to polymers [2, 3] to determine the component spectra without separation or isolation. The method can also be applied to determine the relative concentrations of the components without external calibration [4].

For binary mixtures, two different mixture spectra have absorbances (assuming constant path length) which can be represented by Eqs. (1) and (2). A_m and A_m' represent the measured absorbances for the two components in mixtures m and m' where mixture spectra are composed of a concentration factor, c, times the pure component absorbance, a, in the most general sense.

$$A_m = a_1c_1 + a_2c_2 \tag{1}$$

$$A_m' = a_1c_1 + a_2c_2' \tag{2}$$

Equation (3) defines the ratio spectrum of A_m and A_m'.

$$\frac{A_m'}{A_m} = \frac{a_1c_1'a_2c_2'}{a_1c_1 + a_2c_2} \tag{3}$$

In regions of the ratio spectrum where the spectral contribution of a_2 is absent (frequency equal to r_1), the ratio is simplified to the ratio of the concentraction factors for the two mixtures as detailed in Eq. (4).

$$\left(\frac{A_m'}{A_m}\right)_{r_1} = \left(\frac{a_1c_1'}{a_1c_1}\right)_{r_1} = \frac{c_1'}{c_1} = R_{r_1} \tag{4}$$

The ratio coefficient, R_{r1}, is therefore defined. Similarly, in regions of the ratio spectrum where a_1 is absent, the ratio coefficient R_{r2} is experimentally obtained as shown in Eq. (5),

$$\left(\frac{A_m'}{A_m}\right)_{r_2} = \left(\frac{a_2c_2'}{a_2c_2}\right)_{r_2} = \frac{c_2'}{c_2} = R_{r_2} \ . \tag{5}$$

Since the concentration factors are defined as volume fractions they must add to unity

$$c_1 + c_2 = c_1' + c_2' = 1, \tag{6}$$

we can solve for the concentration coefficients as a function of

R_{r1} and R_{r2}:

$$c_1 = \frac{1 - R_{r_2}}{R_{r_1} - R_{r_2}} \qquad c_2 = \frac{R_{r_1} - 1}{R_{r_1} - R_{r_2}} \qquad c_1' = R_{r_1} c_1$$

$$c_2' = R_{r_2} c_2. \tag{7}$$

Since A_m and A_m' are measured, the unmixed pure component spectra A_1 and A_2 can be obtained:

$$A_1 = \left(\frac{1}{1 - R_{r_2}}\right) A_m' - \left(\frac{R_{r_2}}{1 - R_{r_2}}\right) A_m \tag{8}$$

$$A_2 = \left(\frac{1}{1 - R_{r_1}}\right) A_m' - \left(\frac{R_{r_2}}{1 - R_{r_1}}\right) A_m. \tag{9}$$

It is important to note that A_1 and A_2 are properly scaled to one another and as such can be used in least-squares fitting routines. This represents the internal calibration of the absorbance ratio technique.

Once the component spectra of a mixture have been determined and properly scaled, perhaps via the ratio method, other mixtures composed of identical components can be analyzed using the method of least-squares curve-fitting. A program to do so has been developed in our laboratory [5]. The fitting equation employed is shown in Eq. (10). N is the number

$$\sum_{i-1}^{N} R_{i,k} = \sum_{j=1}^{M} x \left(\sum_{i=1}^{N} W_i R_{i,j} R_{i,k} \right) X_j \tag{10}$$

$$S_i = \sum_{j=1}^{M} X_j R_{i,j} \tag{11}$$

of spectral elements (data points) in each spectrum and M the number of basis spectra used in the fitting procedure. $R_{i,k}$ represents the absorbance data for the i-th spectral element of the k-th basis spectrum. S_i is the data for the spectral range of the mixture spectrum. W_i is a weighting factor equal to $1/S_i$ (statistical weighting). X_i is the number which when multiplied by the appropriate basis spectrum and summed for all components

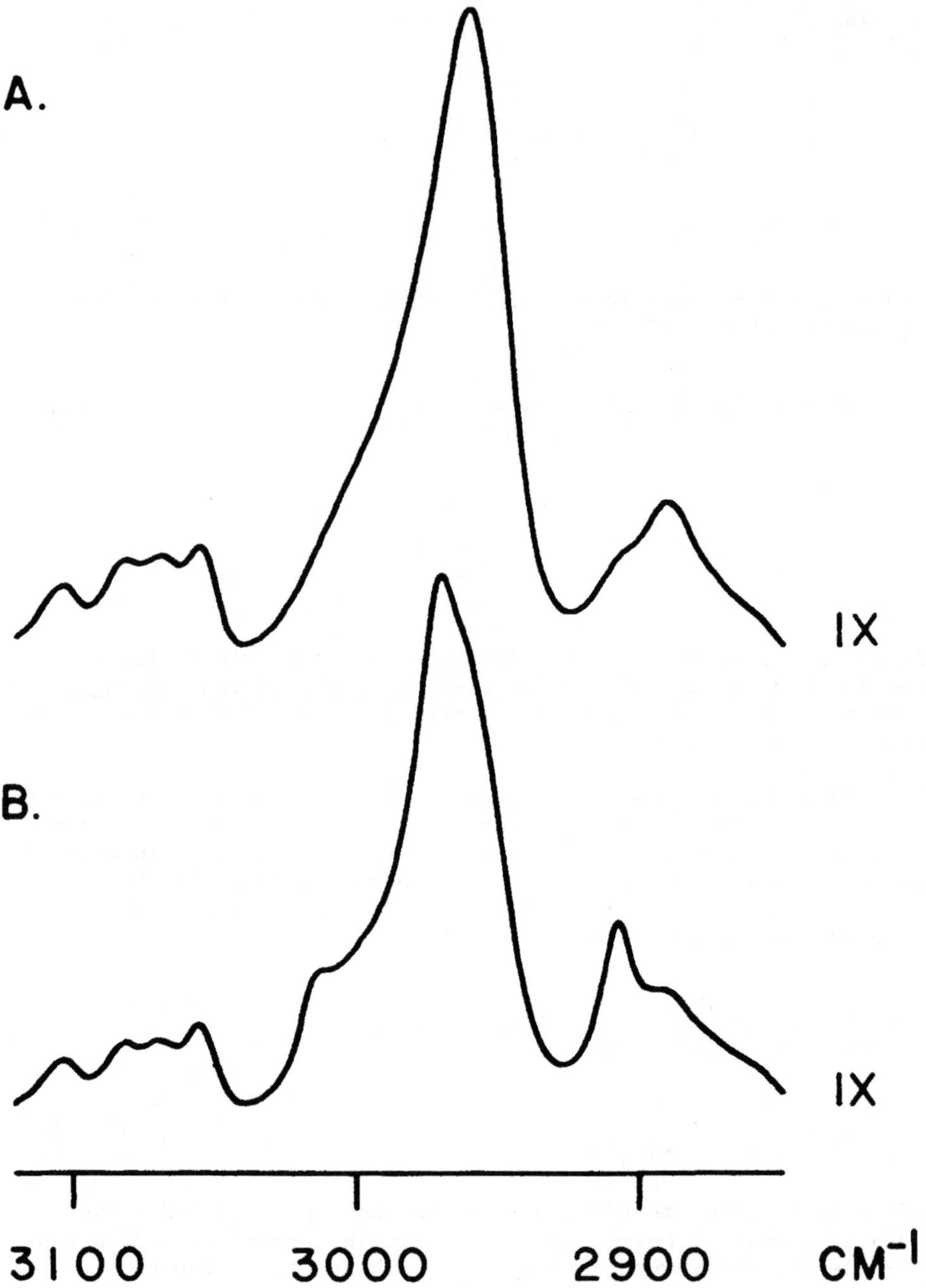

Fig. 2 PET C-H stretching region: (A) unannealed film; (B) film anneated at 160°C for two hours.

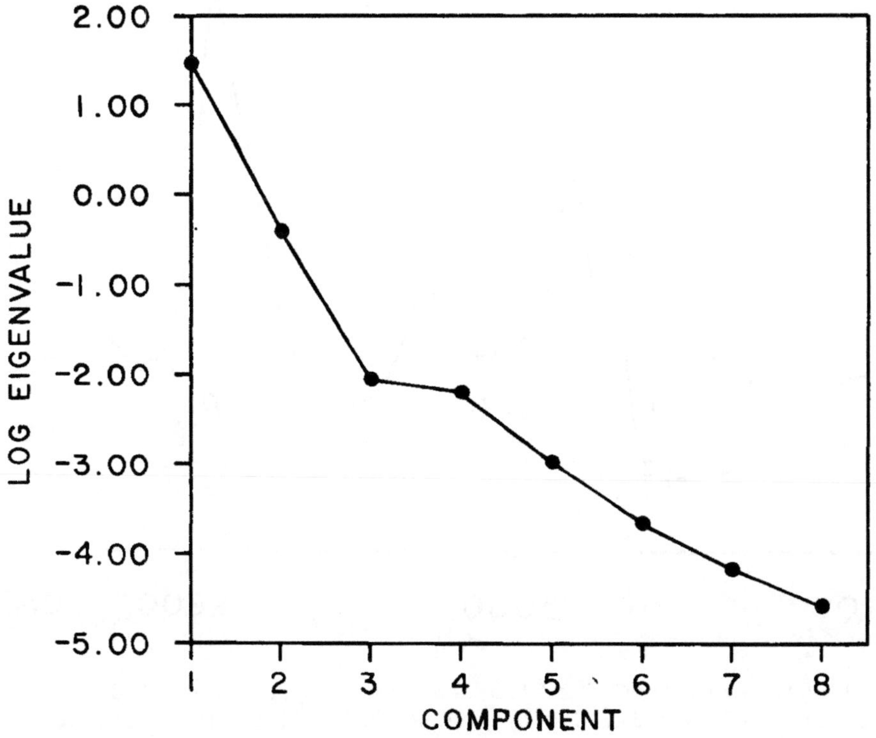

Fig. 3 Factor analysis of annealed PET films (3 mil) 3120-2850 cm^{-1}.

best fits the experimental mixture spectrum. When using calibrated component spectra, it is from the computer determined X_j values that volume fractions of the components in the mixture can be obtained.

3. RESULTS

Figure 2 illustrates the effect of annealing an amorphous PET film in the carbon-hydrogen stretching region. The spectra shown include the amorphous spectrum and the spectrum of the same sample after annealing at 160°C for two hours. A shoulder band at 2909 cm^{-1} becomes much more pronounced as does a band at 3015 cm^{-1}. Increases in the amount of the trans conformer were suggested due to either changes in the amorphous state and/ or crystallization.

Detailed in Fig. 3 is an application of factor analysis to

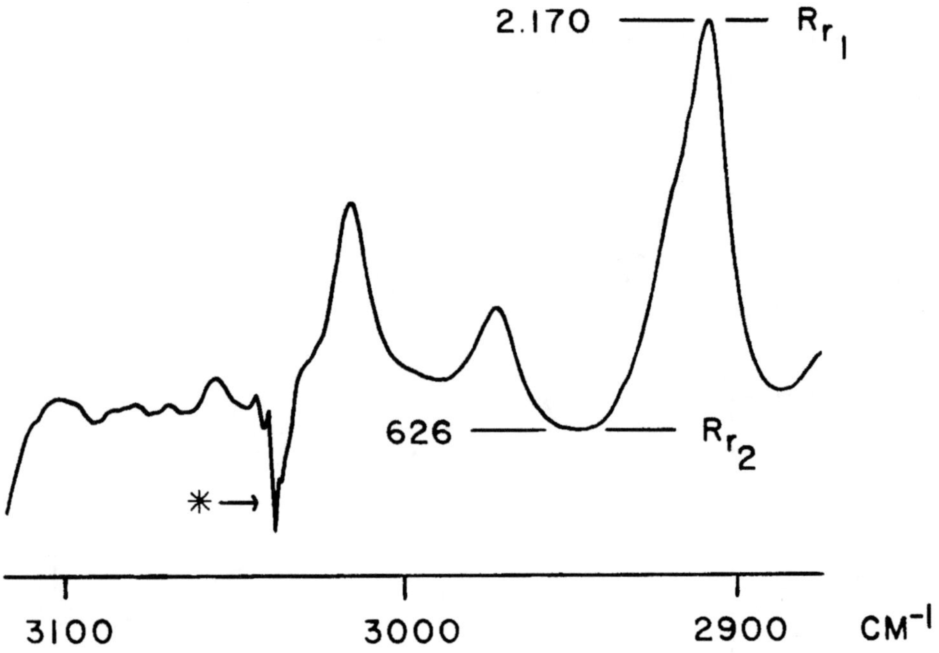

Fig. 4 Ratio spectrum for 3120-2850 cm^{-1} region; annealed at 160°C for two hours divided by unannealed. *Indicates artifact peak.

a set of eight spectra taken of annealed PET that covers the range of amorphous to annealed at 200°C for three hours. The figure shows a plot of log eigenvalue versus components. Two eigenvalues are clearly separated from the best of the values. By analogy to the work of Koenig and Antoon on factor analysis of infrared spectra [1], these two eigenvalues will be taken as non-zero and the others assumed to be zero due to experimental error. Assignment of the two components is made to the trans and gauche rotational isomers of the PET repeat. Since crystallization under the annealing conditions used was observed, the trans component must be present in both the amorphous and crystalline domains.

With the number of components determined, the ratio method can now be applied to the binary mixture of conformational isomers in PET. We seek information regarding the spectra of the two components as well as a measure of the amounts of each component in the samples investigated.

For the C-H stretching region, a ratio plot of an annealed film divided by an unannealed amorphous film is shown in Fig. 4.

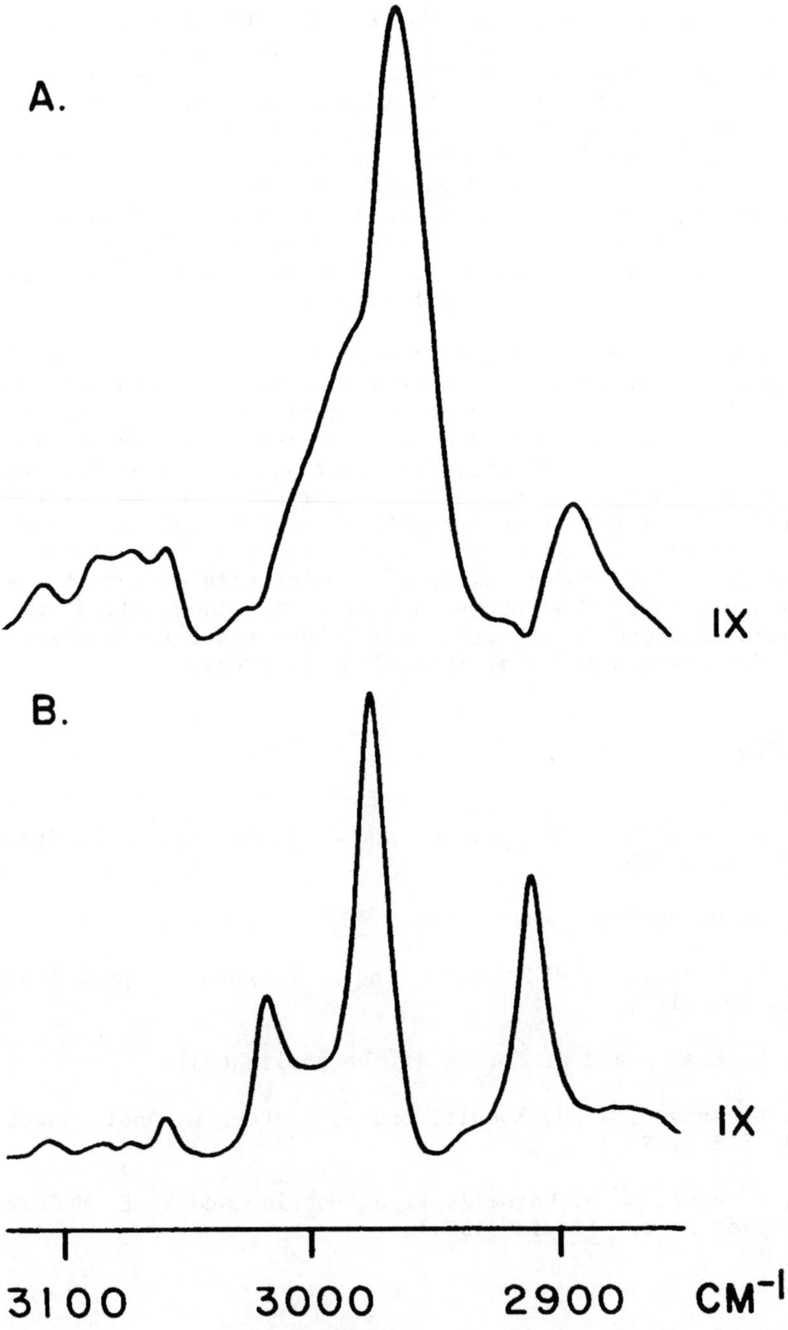

Fig. 5 PET: (A) Amorphous gauche C-H stretch; (B) crystalline and amorphous trans C-H stretch.

The ratio coefficients R_{r1} and R_{r2} are determined to be 2.170 and 0.626 respectively. Using this information and applying the ratio technique, the concentration of the trans isomer for the unannealed sample was found to be 47.4%, the remainder being gauche. The film annealed for two hours at 160°C was found to be 75.8% trans with 24.2% gauche. Figure 5 details the deconvoluted and internally scaled gauche and trans spectra. The trans spectrum shows characteristic bands at 2909, 2972, and 3015 cm^{-1} as well as C-H ring stretching modes at higher frequencies. The trans results agree with the theoretical vibrational analysis for crystalline PET by Boerio _et al._ [6].

We have, in conclusion, presented a systematic approach for investigating polymer chain conformations in the amorphous and semicrystalline materials. Fourier transform infrared spectroscopy has been utilized and combined with the techniques of factor analysis, absorbance ratioing, and least squares curve-fitting. The above approach was used to study amorphous and semicrystalline films of poly(ethylene terephthalate). Glycol trans and gauche conformational spectra were isolated and their volume fraction determined for a number of samples with different annealing histories. The success achieved with the application of the above techniques distinctly shows their value for characterization of amorphous and semicrystalline polymers.

REFERENCES

1. M. K. Antoon, L. D'Esposito, and J. L. Koenig, Appl. Spectrosc., _33_, 351 (1979).

2. T. B. Hirschfeld, Anal. Chem., _48_, 721 (1976).

3. J. L. Koenig, L. D'Esposito, and M. K. Antoon, Appl. Spectrosc., _31_, 292 (1977).

4. J. L. Koenig and D. Kormos (to be published).

5. M. K. Antoon, J. H. Koenig, and J. L. Koenig, Appl. Spectrosc., _31_, 518 (1977).

6. S. K. Bahl, D. D. Cornell, F. J. Boerio, and G. E. McGraw, Polymer Lett., _12_, 13 (1974).

HADAMARD TRANSFORM SPECTROSCOPY

C. S. Bartholomew, W. G. Fateley and N. P. Murarka

Kansas State University, Manhattan, Kansas, U. S. A.
and
IIT Research Institute, Chicago, Illinois, U. S. A.

1. INTRODUCTION

Fourier Transform Interferometry has become a useful tool in infrared spectroscopy. These instruments require very exacting optics and mechanical alignments which require micrometer accuracy in mirror alignments and the measurement of mirror displacement. Because of these required low tolerances, interferometers are very expensive and require constant calibration.

For the past fifteen years Hadamard Transform Spectroscopy (HTS) has offered us a very unique and promising technique for spectral observation. This HTS spectrometer shares the high throughput and multiplexing advantages of the Michelson interferometer. But, the Hadamard transform spectrometer has distinct advantages: (1) Mechanical tolerances on the order of millimeters are required, i.e. three orders of magnitude higher than an interferometer; (2) The higher tolerances mean lower cost in production and less maintenance; (3) Any conventional dispersive spectrometer can be converted to a Hadamard transform spectrometer quite easily if a minicomputer and a slot system are available. This type of conversion can reduce the cost of this instrumentation even more; (4) The transforms for creating spectral data from the encoded slot data are much simpler for HTS. Computer time is drastically reduced for HTS over that for Fourier transform spectrometry; thus simple computers can be used.

Yates, in 1935, [1] described how error levels could be reduced by weighing items which constituted a group rather than individually. It is amazing that even today with the use of

J. R. Durig (ed.), Analytical Applications of FT-IR to Molecular and Biological Systems, 89–118.

computers this simple form of multiplexing has not become a widely used tool for infrared spectroscopy.

To demonstrate HTS, Sloane and Harwit [2] have used systems of weighing as a good example. We shall use this, too. By weighing different combinations of several objects of unknown mass, the error in measuring an individual object mass can be reduced [2,3]. For example, suppose we were to weigh four objects with masses m_1, m_2, m_3 and m_4 on a double pan balance. (There are still one or two around in the archives of our laboratories.) For the first measurement we might place all four objects on the right pan and record the mass. Next we could place two objects on the right pan and two on the left, and so on in different combinations. By weighing four different combinations of the objects we could easily deduce the masses of the individual objects.

If the objects are weighed separately, the best estimates ($\hat{m}$) of the masses of any one object will be the actual mass of that object plus the error associated with the weighing of that object (the measured mass (M)).

$$M_1 = \hat{m}_1 = m_1 + e_1 \qquad M_3 = \hat{m}_3 = m_3 + e_3$$

$$M_2 = \hat{m}_2 = m_2 + e_2 \qquad M_4 = \hat{m}_4 = m_4 + e_4$$

If it is assumed that the error is random with a mean value of zero and variance σ^2, then the average variances of the above mentioned objects will be

$$(\hat{m}_1 - m_1)^2 = e_1{}^2 = \sigma^2, \ (\hat{m}_2 - m_2)^2 = e_2{}^2 = \sigma^2, \ \ldots$$

When all four objects are weighed simultaneously the average error due to the balance and weighing procedure is going to be the same as when the objects are weighed individually.

$$M_1 = m_1 + m_2 + m_3 + m_4 + e_1$$

$$M_2 = m_1 + m_2 - m_3 - m_4 + e_2$$

$$M_3 = m_1 - m_2 - m_3 + m_4 + e_3$$

$$M_4 = m_1 - m_2 + m_3 - m_4 + e_4$$

The masses of the objects can be determined by solving four simultaneous equations for each unknown.

$$M_1 + M_2 + M_3 + M_4 = 4m_1 + 0 + 0 + 0 + (e_1 + e_2 + e_3 + e_4)$$

The best estimate is:

$$\hat{m}_1 = \frac{1}{4}(M_1 + M_2 + M_3 + M_4) = \frac{1}{4}(e_1 + e_2 + e_3 + e_4) + m_1.$$

Since the variance of a sum of independent variables is the sum of the variances, the variances of the best estimate $\hat{m}_1$ (and of $\hat{m}_2$, $\hat{m}_3$, $\hat{m}_4$) is

$$\frac{1}{16}e_1^{\ 2} + \frac{1}{16}e_2^{\ 2} + \frac{1}{16}e_3^{\ 2} + \frac{1}{16}e_4^{\ 2} = \frac{1}{4}\sigma^2.$$

Using just a single pan balance, the following weighing scheme could be effectively used:

$$\begin{aligned} M_1 &= \phantom{m_1 + {}} m_2 + m_3 + m_4 + e_1 \\ M_2 &= m_1 + m_2 \phantom{{} + m_3 + m_4} + e_2 \\ M_3 &= m_1 \phantom{{} + m_2} + m_3 \phantom{{} + m_4} + e_3 \\ M_4 &= m_1 \phantom{{} + m_2 + m_3} + m_4 + e_4. \end{aligned}$$

Solving for m_1:

$$M_2 + M_3 + M_4 - M_1 = 3m_1 + e_2 + e_3 + e_4 - e_1 \text{ and}$$

$$\hat{m}_1 = \frac{1}{3}(M_2 + M_3 + M_4 - M_1) = m_1 + \frac{1}{3}(e_2 + e_3 + e_4 - e_1).$$

The variance for $\hat{m}_1$, then, is $(4/9)\sigma^2$. Similarly the variances of $\hat{m}_2$, $\hat{m}_3$ and $\hat{m}_4$ are $(7/9)\sigma^2$.

By weighing the objects in groups rather than individually, the errors for the estimated masses are reduced. Varying the combinations of objects being weighed or not weighed, as in the single pan balance, reduced the errors to $4\sigma^2/9$. Combinations of sums and differences of the objects' masses, as in the double pan balance, reduce the error even further to $(1/4)\sigma^2$ (from σ^2 for individual weighings of the objects).

The various position patterns (i.e. position on the right or left pan of the balance) used in the weighing of the objects are called weighing designs. These weighing designs can reduce the error in any group of measurements.

An absorption spectrophotometer measures the number of photons (i.e. intensity of radiation) of a certain wavelength light that are not absorbed by a particular sample. Since a quantity is being measured, the use of weighing designs becomes useful in reducing the error (which includes the noise) in spectrophotometric absorption measurements. The aforementioned "objects" correspond to wavelengths of light and the masses correspond to the number of photons passing through the sample and being counted by a detector.

It is apparent that by using weighing designs, the error (or noise level in spectroscopy) could be reduced drastically. Error reduction becomes very important in infrared spectroscopy for situations where the intensity of light is low. Subsequently, in the situation of limited intensity the sensitivity is low and the noise is proportionately larger than other spectroscopic regions. In these energy limited systems the most significant source of noise is the detector; thus, the error is independent of the signal applied.

The necessary conditions for applying weighing designs to infrared spectroscopy (and for that matter any spectroscopic technique) are: (1) The light must be separated into its component wavelengths, and (2) these components must be either transmitted or reflected and gathered as a negative contribution to the signal or, more simply, they must be blocked by some means. The first condition presents no problems. Dispersive gratings separate light into their various component wavelengths quite effectively. The technology in making dispersive spectrometers has been quite well developed and these spectrometers can be modified to do Hadamard Transform Spectroscopy.

The second criterion for HTS can be obtained by creating a series of masks (number masks = number slots) containing slots that can be either opaque or transparent. These masks would be placed in the light beam's path and thus allow certain wavelengths of light to enter while blocking other wavelengths. A weighing would thus be created quite analogous to the case of weighing objects on a single pan balance; the objects (or wavelengths of light) are either detected or not detected.

Another type of mask might be a mask that either lets light through or reflects it to a different detector; one detector's signal is then subtracted from the other's, creating a situation analogous to weighing objects on a double pan balance. The errors can be reduced much more drastically using a reflectance mask, but implementation of this kind of system might be extremely difficult.

2. INSTRUMENT

The Hadamard transform spectrometer is basically a dispersive spectrometer with an encoding mask at the entrance slot and/or exit slot of the instrument [2]. The basic components are a radiation source, an optical separator to separate the light into component wavelengths, the encoding mask, a detector, and a decoder to assimilate the data (see Fig. 1). The optical separator

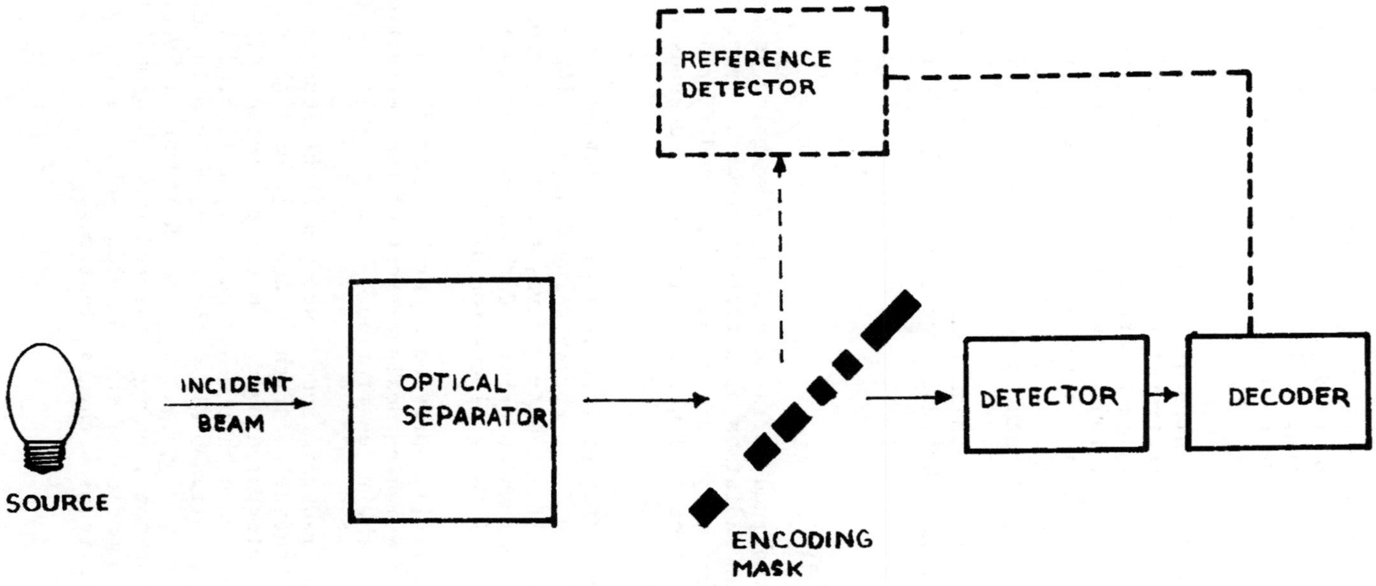

Fig. 1 Components used in Hadamard Transform Spectroscopy. The dotted lines indicate additional components necessary when an encoding mask utilizing reflecting and transparent slots is used instead of an encoding mask made of opaque and transparent slots.

could be any device such as a grating or prism that spatially separates the incident radiation into its optical components. The slotted mask encodes the signal according to some weighing design by either absorbing, reflecting, or transmitting the signal. The encoder need be nothing more complicated than a digital computer.

The basic parts of three different spectrometers are shown in Fig. 2. The first spectrometer in this figure is a classic dispersive spectrometer with a narrow slot at both the entrance and exit. The middle design in this figure is a Hadamard Transform spectrometer with one narrow entrance slot and a mask of some weighing design used as the exit slot. The bottom drawing in Fig. 2 has a mask at both the entrance and exit slots.

HTS requires as many measurements made as there are slots. For example, if a mask has 255 slots, then 255 measurements must be made.

3. WEIGHING DESIGNS

As can be imagined, a large number of weighing designs can be used to encode radiation of different spectral wavelengths. However, different weighing designs result in different error magnitudes. The choice of which design to use depends on which design gives the minimum error.

The Hadamard matrices (i.e., -1 or +1 elements) will give the minimum error in HTS. For weighing designs using +1 and 0 elements, the so-called S-matrix gives the minimum error for HTS. (Other weighing designs do exist and these will be indicated later.) See Appendix I for a discussion of these different matrices.

Sloane et al., [4], provide a very elegant discussion on how to evaluate a weighing design. Most of the derivation given below will follow their description.

In Hadamard Transform Spectroscopy a light source is spatially dispersed into N channels. Each channel, j, of light has an average energy, integrated over a specific time, of e_j. The channels of light are either measured (+1), reflected (-1) or blocked (0) out simultaneously according to a weighing design, W_i, where $W_i = (w_{ii}, w_{22}, \ldots. w_{in})$ which is the ith vector of weight. Thus, the total energy measured through a mask (weighing design) is the summation of the individual channel energies which have been allowed to pass through that mask,

$$\sum_{j=1}^{N} w_{ij} e_j .$$

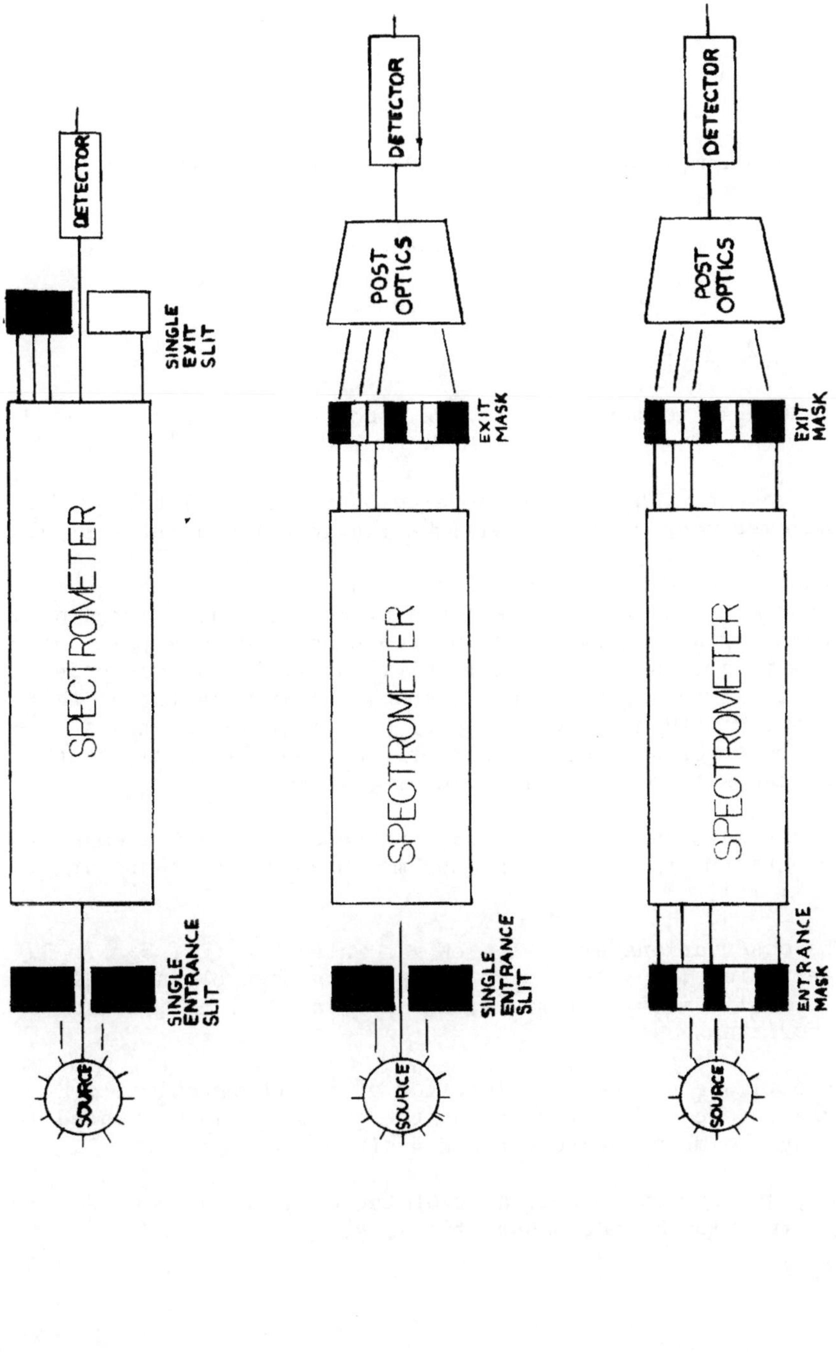

Fig. 2 Block diagrams of three spectrometers. The top spectrometer is a conventional monochromator. The middle spectrometer is a singly encoded Hadamard transform spectrometer and the bottom spectrometer is a doubly encoded Hadamard transform spectrometer.

In this summation, w_{ij} is the individual channel (j) weight for mask W_i. The mask weights are assumed to satisfy $|w_{ij}| \lesssim 1$ and correspond to attenuation. The photodetector adds a random noise, n_i, which gives a final measurement:

$$E_i = n_i + \sum_{j=1}^{N} w_{ij} e_j, \quad i = 1, 2, \ldots\ldots, M$$

A few assumptions must be made. (Note <> stands for ensemble averages.)

1) $<n_i> = 0$. The overall average for the noise is zero.

2) n_i is independent of the signal.

3) $<n_i^2> = \sigma^2$.

4) Successive measurements are uncorrelated ($<n_i n_j> = 0$ if $i \neq j$).

5) $<\hat{\vec{e}}> = \vec{e}$. The estimated value of the channel energy, if averaged over many readings, will be the actual channel energy.

Assumption number 3, above, says the ensemble mean square of the noise is the same for all measurements. This corresponds to integrating over equal time for those measurements [5]. If the noise is due to the detector only, and if it is independent of the size of the signal (this assumes linear behavior of the photodetector), then, when averaging over many measurements, the mean square error will depend directly on the time duration the detector takes to make the individual observation.

To estimate the energies, e_j, one needs at least as many measurements (M) (i.e. one mask per measurement) as there are unknowns (N) [4].

The observations are: $\vec{E} = \vec{e}W + \vec{n}$ where $\vec{E} = (E_1, E_2 \ldots E_m)$, the observations; $\vec{e} = (e_1, e_2, \ldots e_N)$, the channel energies; $(W_1^T, \ldots W_M^T)$, matrix of masks; and $n = (n_1, \ldots n_M)$, noise (T stands for transpose).

An estimate $\hat{\vec{e}}$ of $\vec{e}$ is a function of the observation, $\hat{\vec{e}}(E)$ lying close to $\vec{e}$. The accuracy of the estimate can be tested by minimizing the mean square error $\varepsilon = <(\hat{\vec{e}} - \vec{e})(\hat{\vec{e}} - \vec{e})^T>$.

Now, set the estimated value of the channel energies $\hat{\vec{e}}$, to be a linear function of the observations, E,

$$\hat{\vec{e}} = \vec{E}M$$

for some matrix M.

Minimizing $\varepsilon = \langle(\hat{\vec{e}} - \vec{e})(\hat{\vec{e}} - \vec{e})^T\rangle$

$$\langle\hat{\vec{e}}\rangle = \langle\vec{E}M\rangle = \langle\vec{E}\rangle M$$

$$\langle\vec{E}\rangle = \vec{e}W + \langle\vec{n}\rangle = \vec{e}W$$

$$\langle\hat{\vec{e}}\rangle = \vec{e}WM$$

Therefore according to assumption 4 above, $\langle\hat{\vec{e}}\rangle = \vec{e}$, WM must equal the identity matrix, I. If M = N, $M = W^{-1}$. If M > N (M cannot be smaller than N), $M = W^T(WW^T)^{-1}$. Substituting for $\hat{\vec{e}}$ and $\vec{e}$,

$$\hat{\vec{e}} - \vec{e} = \vec{E}M - e = \vec{e}WM + nM - \vec{e}$$

$$WM = I$$

so,

$$\hat{\vec{e}} - \vec{e} = \vec{n}M$$

$$= \vec{n}W^{-1}, \text{ if } M = N$$

$$= \vec{n}W^T(WW^T)^{-1} \text{ at } M > N.$$

From assumption 3,

$$\langle n_i^2\rangle = \sigma^2 \text{ and } \langle\vec{n}\vec{n}^T\rangle = \sigma^2 I;$$

thus,

$$\varepsilon/\sigma^2 = \text{trace } [W^{-1}(W^{-1})^T] \text{ if } M = N$$

$$= \text{trace } [W^T(WW^T)^{-1}(WW^T)^{-1}W] \text{ if } M > N.$$

Other methods for finding the best weighing design exist [3]. The A-optimal weighing design minimizes $\varepsilon = \sigma^2/4 \text{ trace } (W^TW)^{-1}$.

A D-optimal weighing design maximizes the determinant of W| det(W)|. D-optimality essentially minimizes the volume of the region in which the estimate, $\hat{\vec{e}}$, is expected to be found. E-optimality maximizes the smallest eigenvalue of W^TW.

These methods for finding the best weighing design do not always agree. A-optimality seems to be the best method assuming the individual mean square errors are about equal. The best weighing design can be found by applying one or more of the optimality criteria to that design. Weighing designs and their characteristics are given in Table 1.

Table 1[+]

W	Elements	N	Conditions	A-optimal $(\frac{\varepsilon}{\sigma^2})$	D-optimal	E-optimal
H_N	-1,1	4t	$H_N H_N^T = NI_N$	1/N	$N^{N/2}$	N
D_N	-1,1	4t+1 4t+2	(N,K,λ) sym block design*, change 0 to +1 and 1 to -1	$\frac{[N^2-N-4(N-2)n]}{4n[N^2-4(N-1)n]}$ for $n = k-\lambda$	$(4n)^{(N-1)/2}[N^2-4(N-1)n]^{1/2}$	min $[4n, N^2-4(N-1)n]$
P_N	-1,1	4t+1 (N=5, 13,25)	sym block design* with $N=(d^2+1)/2$, $k=(N+d)/2$, $\lambda=(N+2d+1)/4$	$\frac{2}{2N-1}$	$(N-1)^{(N-1)/2}(2N-1)^{1/2}$	N-1
C_N	1,-1,0	4t+2	$C_N C_N^T = (N-1)I_N$	$\frac{1}{N-1}$	$(N-1)^{N/2}$	N-1
E_N	-1,0,1	4t+2	$E_N^T E_N = [\begin{smallmatrix} Z & 0 \\ 0 & Z \end{smallmatrix}]$; $Z = (\frac{1}{2}N-2)I_{N/2} + 2J_{N/2}$#	$\frac{1}{N-1}$	$2(N-2)^{(N-2)/2}(N-1)$	N-2
G_N	-1,1	4t+3	Delete the first row and column of a normalized H_{N+1} matrix.	$\frac{2}{N+1}$	$(N+1)^{(N-1)/2}$	1

Table 1 (Continued)

W	Elements	N	Conditions	A-optimal $(\frac{\varepsilon}{\sigma^2})$	D-optimal	E-optimal
Q_N	-1,1	4t+3 (N=7,31)	Sym block design*; $N=(d^2+3)/4$, $k=(N+d)/2$, $\lambda=(N=2d+3)/4$	$\frac{4N-6}{(4N-3)(N-3)}$	$(N-3)^{(N-1)/2}(4N-3)^{1/2}$	N-3
A_N	0,1	4t 4t+2	Sym block design*	$\frac{k(N-2)+1}{k^2(N-k)}$	$k(k-\lambda)^{(N-1)/2}$	$k-\lambda$
E_N	0,1	4t+1	To the elements of E_N, change 1 to 0, -1 to 1	----	$2^{(-N+1)}N(N-1)^{(N-1)/2}$	---
S_N	0,1	4t+3	See Appendix I and Text	$\frac{4N}{(N+1)^2}$	$2^{-N}(N+1)^{(N+1)/2}$	$\frac{1}{4}(N+1)$

I is the identity matrix; J is the all ones matrix

* A symmetrical block design is a collection of subsets of size k acquired from a total N objects, such that any two blocks have λ common objects. Its incident matrix, A, forms the weighing design. The (N,k,λ) symmetrical block design converts to the incident matrix, A, by forming an N x N matrix, $A=(a_{ij})$, such that $a_{ij}=1$ if the ith block contains the jth object and $a_{ij}=0$ if the ith block does not contain the jth object.

+Note: A-optimal is the average mean square error. To get the total mean square error, multiply by N.

Hotelling [6] has shown that for any choice of mask W with $|w_{ij}| \leq 1$, the ε_i are bounded by $\varepsilon_i \geq (\sigma^2/N)$, and that it is possible to have $\varepsilon_i = (\sigma^2 N)$ for all i = 1, ..., N if and only if a Hadamard matrix H_N of order N exists.

The Hadamard matrix will provide the best weighing design for reduction of noise. This matrix and two other matrices will be considered in more detail. It is interesting to note that other matrices in Table 1 have not found much use in Hadamard Transform Spectroscopy, therefore only a mention of these matrices is given here.

A Hadamard matrix is an N x N matrix of +1's and -1's constructed in such a way that,

$$HH^T = NI \quad \text{(see Ref. [4])}.$$

H can be normalized by multiplying the proper rows and columns by -1 to obtain +1's in the first row and column ([3] and Appendix I of this chapter).

$$\begin{vmatrix} 1 & 1 & 1 & \cdot & \cdot & \cdot & 1 \\ 1 & & & & & & \\ 1 & & & & & & \\ 1 & & & \mathbf{G} & & & \\ \cdot & & & & & & \\ \cdot & & & & & & \\ \cdot & & & & & & \\ 1 & & & & & & \end{vmatrix}$$

If $N \geq 4$, and k and ℓ are two distinct rows of H then (excluding the first row),

row k has +1 and row ℓ has +1 in N/4 places

...+1	...	-1...
...-1	...	+1...
...-1	...	-1...

and N must be a multiple of 4. The dot product of any two rows of this matrix is zero and thus the matrix is orthogonal.

If one of the following conditions is satisfied, then G can be constructed as a cyclic matrix - a matrix where the (i + 1)th row can be obtained by shifting the ith row one space to the right or left. Here the G matrix has N - 1 rows and columns.

1) N = p + 1, p is a prime number

2) $N = p\,(p + 2) + 1$, p and $(p + 2)$ are prime

3) $N = 2^m$

For example:

$$G = \begin{vmatrix} 1 & -1 & -1 \\ -1 & 1 & -1 \\ -1 & -1 & 1 \end{vmatrix}$$

$$\text{row } k \cdot \text{row } \ell = \begin{cases} -1, & i \neq j \\ N - 1, & i = j \end{cases}$$

For more cyclic matrices, G, the reader is referred to the literature, Refs. [7,8 and 9].

The third choice for W is S, which can be generated from G by replacing +1's with 0's and -1's with +1's (see Appendix I).

$$\text{row } k \cdot \text{row } \ell = \begin{cases} N/4, & i \neq j \\ N/2, & i = j \end{cases}$$

By taking the trace $(W^{-1}(W^{-1})^T)$ it is found [4]:

Matrix W	trace (ε/σ^2)
H^T	1
G^T	$2 - 2N$
S^T	$4 - 8/N + 4/N^2$

4. EXPERIMENTAL

The S-weighing designed mask utilizing 0 or 1 cyclic elements is the most practical and easiest mask to design. Experimentally, zero denotes an opaque slot in the mask, +1 denotes a transparent slot, and -1 denotes a reflecting slot. The experiment is greatly simplified by the fact that only 2N - 1 slots are needed in one mask to record Hadamard Transform Spectroscopy if a cyclic weighing design is used. Thus, one can stepwise move a mask across an entrance or exit slot and record the intensity of radiation impinging upon the detector for each step of the mask. We refer the reader to Fig. 3 which demonstrates this stepping effect.

In practice the signal must be dispersed. Since noise increases by the square-root of the detector surface area, a small size detector is essential [5]. By refocusing the radiation back through the optics, the light can be focused to an area approxi-

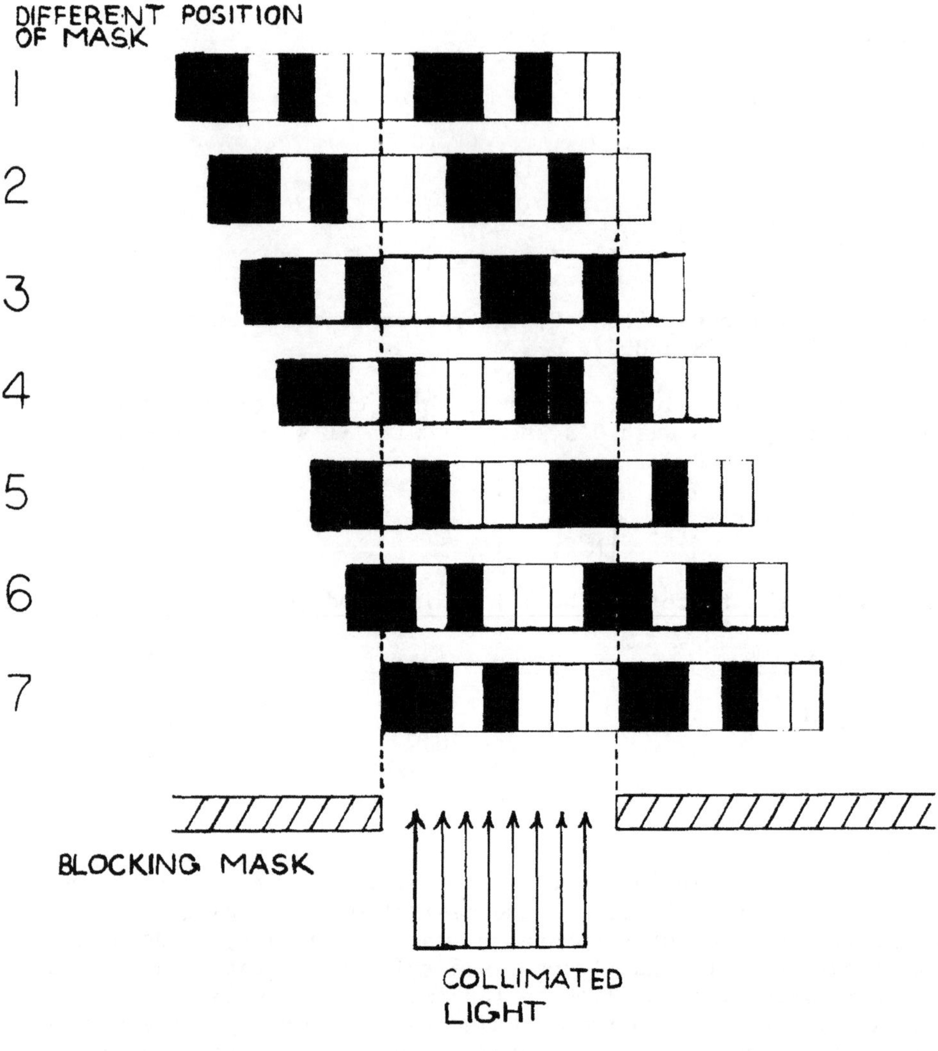

Fig. 3 A 2N - 1 = 13 slot mask is stepped across a blocking mask, allowing seven linearly independent seven channel weighing designs to be illuminated by the incident radiation. A measurement of the light intensity is made at each of the seven different mask positions. An S_7 weighing design matrix is thus created.

mately the same size as the entrance slot [10]. A typical arrangement used to accomplish this in Hadamard Transform Spectrometry is shown in Fig. 4 from Ref. [11]. Of course, if an entrance mask has the same number of slots as the exit mask, refocusing is not necessary since the entrance and exit slots will then be the same size.

In Fig. 4, the light is passed through the entrance aperature to the grating. The dispersed radiation from the grating is directed to a collimating mirror. The mirror reflects the radiation onto the mask. The encoded light beam is then returned by the same path through the dispersive optics and focused at the entrance focal plane [11]. In Decker's spectrometer, the entrance slot was slightly below the optical axis, so the returning exit signal was slightly above the optical plane to allow simple detection. The signal was detected by a 1.0 mm square lead sulfide detector. Of course the detector will vary depending on the region of the infrared spectrum used in the experiment.

Decker used an S_{255} matrix as a weighing design, a 509 slot (2N - 1) mask. The entering beam passes through a single narrow slot which is .185 mm high. Decker found the signal-to-noise (S/N) ratio to be 39.12. Using identical optics as the Hadamard Transform spectrometer, but using a monochromator instead of the weighing design mask, the S/N was only ~4.764. The Hadamard Transform spectrometer increased the S/N by a factor of 8.0612 ± 0.3110 over that of a conventional monochromator. The increase in S/N was very close to the theoretical increase of a factor of 8.01567.

An instrument utilizing an S_{2047} weighing design mask (4093 slots) created problems in calculating a S/N ratio. Decker reported just before termination of his experiments that he could not measure any detectable noise when any signal, no matter how small, was detectable in the spectrometer's output [12].

Harwit <u>et al</u>. [10] tested an instrument with an S_{255} exit mask (n) and an S_{15} entrance (m) mask. In this experiment Harwit used only (n + m - 1) measurements. The exit mask was used in n/m positions and the entrance mask was stepped through its m positions for each n/m exit mask position [10]. Theoretically, one can expect ε/σ^2 to equal 16/m + 1. Improvement over a single slot entrance, S_{255} mask exit spectrometer was shown to increase the S/N by 4.31 (computer simulation). The theoretical limit is 4.26. Assuming the RMS noise to be 0.2 times the peak to peak noise height, the signal-to-noise ratio was about 40 for the 1.7 μm mercury line measured [10,11]. An S_{509} mask entrance and S_{255} mask exit instrument performed with S/N of about 70. These results were not significantly better than Decker's single slot entrance, S_{255} mask exit instrument (S/N = 40), for the factor of

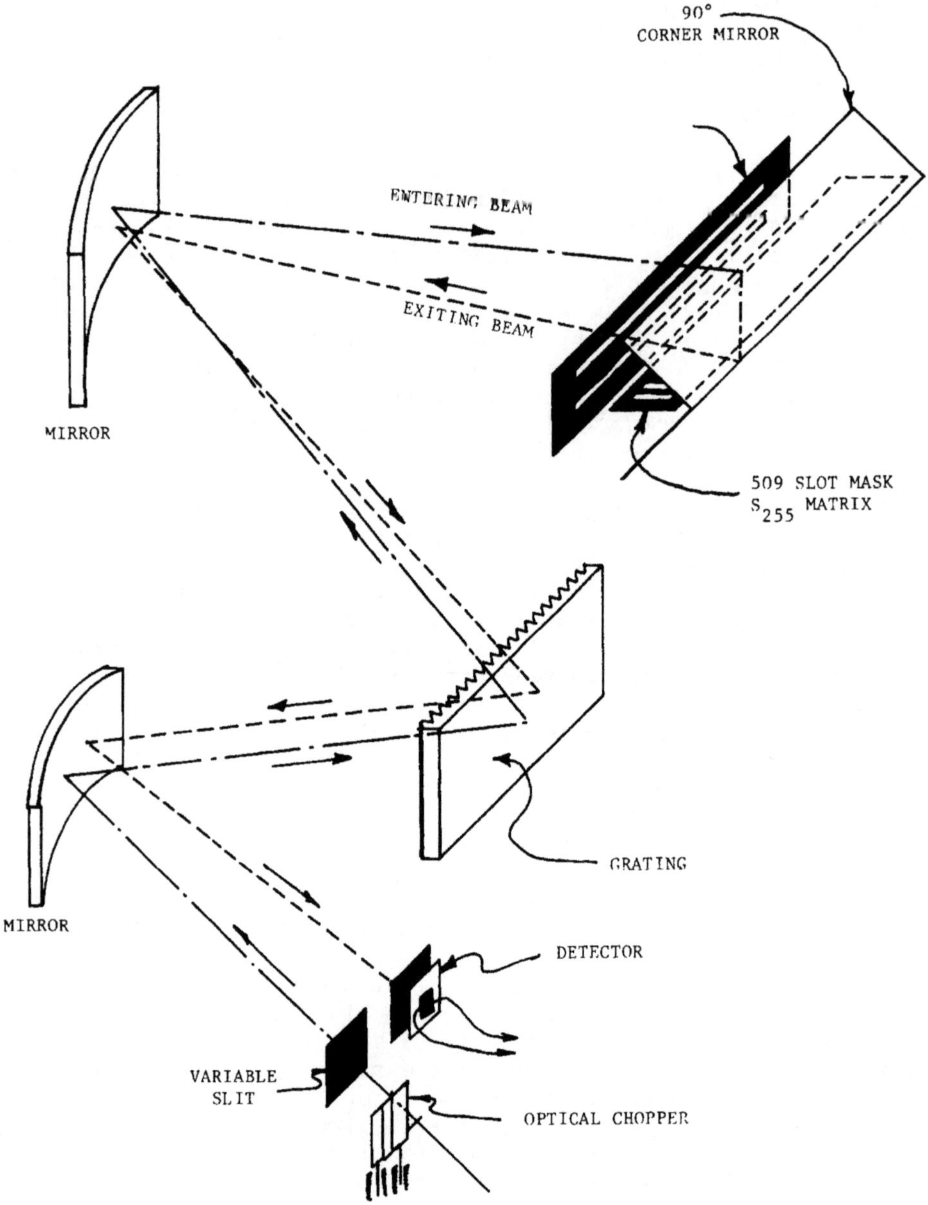

Fig. 4 Optics of a typical singly encoded Hadamard transform spectrometer.

4.26 was not realized in this experiment. Since the dispersive spectrometers used were not the same, differences in the design of the instruments could explain this lack of improvement.

If nm measurements are made, one can expect ε/σ^2 to equal $(22.2/n)(16/m + n)$ if $m >> n$. In an experiment utilizing nineteen slot exit and entrance masks, the mean square noise was found to be 4.2 times higher for $37(n + m - 1)$ measurements than for 361 (nm) measurements [13].

Spectral Imaging, Inc., using a configuration with an S_{255} exit and S_{15} entrance mask, claims "a figure of merit" for their instrument over 120 times that of an equivalent scanning monochromator spectrometer and up to about 30 times that of commercially-available interferometer spectrometers of equal resolution and spectral multiplexing [14].

"Figure of merit" was defined as:

$$E = RLN^{\frac{1}{2}}A_d^{-\frac{1}{2}} < \varepsilon_\lambda D_\lambda^R >_\lambda$$

where

$R = \lambda/\Delta\lambda = \nu/\Delta\nu$ is the resolution,

L is the "etendu" $\simeq nA_g/\ell R$ for HTS or grating spectrometer
$\simeq 2\pi A_M/R$ for Michelson FTS

N is the degree of spectral multiplexing = number of slots for HTS

n = number of entrance slots of HTS

A_d = detector area

A_g = grating area

A_M = Michelson's entrance aperature area

ℓ = focal length/slit height, for a grating instrument

$<\varepsilon_\lambda D_\lambda>_\lambda$ = average over the wavelength range for the product of detector detectivity times the net of filter, grating, beamsplitter, etc., efficiency

$E_{HTS}/E_{monochromator} \approx \frac{1}{2}nN^{-\frac{1}{2}}$

$E_{HTS}/E_{Michelson\ interferometer} \approx (n/2\pi\ell)(A_S/A_M)(e_g/e_{bs})$

ε_g = grating efficiency

ε_{bs} = beam splitter efficiency

It was somewhat disappointing that no experimental data was given to support the theoretical claims made by Spectral Imaging, Inc. Other experimental instruments utilized weight designs of S_{23} to S_{255} for entrance masks and S_{511} to S_{2047} exit masks [14]. It is interesting to note that the resolution was lower than 0.1 cm^{-1} in some of these instruments. Ranges of spectral bands were from 4 to 16 μ depending on optical configuration. Disappointing to us was the fact that no experimental data demonstrating S/N advantages and resolution of these spectrometers were given in these references.

5. DOUBLY MULTIPLEXED SPECTROMETER

The Michelson interferometer has demonstrated a unique advantage over conventional grating spectrometers. A large part of this advantage is due to the very high throughput of an interferometer (Jacquinot's advantage) and multiplexing (Fellgett's advantage) [16].

If high throughput can be added to the advantage of using polychromatic light (Fellgett's advantage), a Hadamard Transform spectrometer should show a similar increase in S/N ratios. To utilize Jacquinot's advantage in Hadamard Transform Spectroscopy, an entrance mask can be used in addition to an exit mask. For the S-matrix weighing design, HTS should increase the throughput to about one half that of an interferometer. Using Hadamard matrices the throughput of HTS should approximate that of a Michelson interferometer [16].

In practice an S mask is typically placed in the entrance of a dispersive grating spectrometer equipped with an S exit mask. For each position of the exit mask, the entrance mask is stepped through its m positions. The total number of measurements can be reduced from mn measurements to (n + m - 1) measurements.

If the experiments use (a) $n = m = N$, (b) N^2 observations, and (c) S_N-matrix weighing designs, then the mean square error, $\varepsilon \simeq (22.2/N)\sigma^2$, Ref. [16]. Compare this mean square error with the mean square error $N\sigma^2$ for a single slot entrance and exit spectrometer and the mean square error of $\{2 - (2/(N + 1)\}^2\sigma^2$ for a spectrometer with a single slot entrance slot which utilizes an S_N-matrix exit mask [16]. The results predict that, for large N, a doubly multiplexed instrument will theoretically reduce noise by a substantial factor.

6. ERROR

We have assumed that the only errors in the error analysis in our previous discussion are those associated with the detector. Of course, other errors can result from the optics and masks of the instruments. Tai _et al_. [17], using an S_{255} mask, found spurious negative bands appearing at higher and lower energies compared to the principal band. Tai and co-workers [17] were able to partially explain the effect when they simulated too wide and too narrow opaque mask slots. Their conclusion was that the opaque slots were too wide.

Tai _et al_. [17] analyzed the case where an opaque slot is too wide (i.e., when the transmitting slot is too narrow), and found the optical effect will give a spectrum of $(1 - \varepsilon, \varepsilon/2, 0...0, -\varepsilon/2, -\varepsilon/2, 0...0, \varepsilon/2)$; for slots too narrow the spectrum will be $(1 - \varepsilon, \varepsilon/2$, 228 0's, $\varepsilon/2, \varepsilon/2$, 22 0's, $\varepsilon/2)$. For any S-matrix mask, there will always be four equal intensity peaks in addition to the main peak. Along with the spurious bands, a spectrum will show a peak on either side of the main peak, causing a broadening appearance. Other errors may result from distortion, aberrations, and diffraction caused by the slots of an instrument, mask position, and mask movement. In a typical Hadamard Transform spectrometer, radiation that should pass through a certain slot will frequently spread onto other mask positions on the detector. Here T represents the Transfer matrix and it is the matrix describing the radiation distribution that results from light passing through a slot [18]. The matrix elements of T correspond to the proportional intensity impinging on a certain slot position of a specific detector measurement. W, as before, is the weighing design matrix.

The detector readings, $\vec{E}$, of measurments i from a Hadamard Transform spectrometer are

$$\vec{E} = TW\vec{e} + \vec{n}$$

where

$$\vec{E} = (e_o, E_1, \dots E_{n-1})^T$$

$$\vec{e} = (e_o, e_1 \dots e_{n-1})^T$$

$$\vec{n} = (n_o, n_1 \dots n_{n-1})^T \quad \text{(see Ref. [18])}.$$

In a conventional monochromator

$$\vec{E} = T\vec{e} + \vec{n}.$$

The results of all the incident radiation not falling in the same slot position after passing through the slot are artificially broadened spectral lines and increased noise.

If there were no distortion of light through a slot position then T would be the identity matrix. The mean square errors for the instruments would be those mentioned previously under weighing designs. For simplicity in our discussion we will assume that the signal appears only in the first slot position.

If the slot is sufficiently wide to eliminate diffraction, the intensity of radiation detected will be triangular shaped with the maximum at the mid point of the slot profile (slot function). The base will appear twice as wide as the slot and sym-

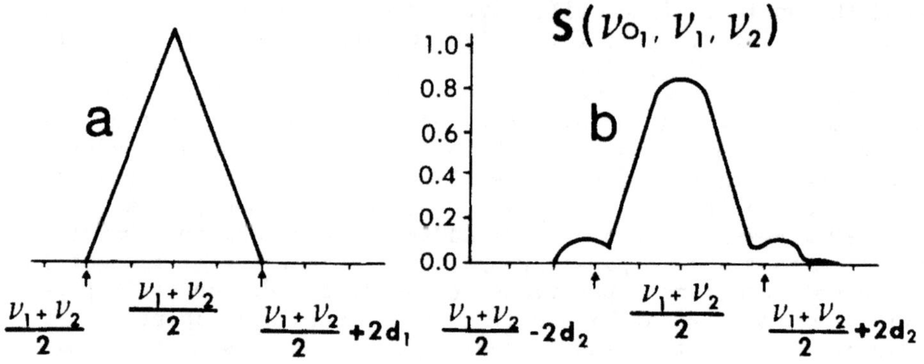

Fig. 5 Slot functions of (a) a sufficiently wide slot to eliminate all diffraction and (b) a narrow slot resulting in full diffraction.

metric about the slot position (Fig. 5a). The errors discussed in Ref. [18] arising from the above constraints are:

$$T = \frac{1}{6}\begin{vmatrix} 4 & 1 & 0 & . & . & . & 0 & 1 \\ 1 & 4 & 1 & . & . & . & 0 & 0 \\ . & . & . & . & . & . & . & . \\ 1 & 0 & 0 & . & . & . & 1 & 4 \end{vmatrix}$$

Conventional monochromator: $\varepsilon = 2\sqrt{3}\,N\sigma^2$
Hadamard Transform spectrometer (S exit mask) $\varepsilon \simeq 13.86\ \sigma^2$
HTS with quality matrix (S exit mask) $\varepsilon \simeq \{4 - 8/N + 4/N^2\}$, N is large
Multiplexing decreases the error by N/4.

A very narrow slot resulting in full diffraction gives the slot function shown in Fig. 5b [18].

The first row of T is (0.6667, 0.1482, 0.0080, 0.0031, ... 0.0017, 0.0031, 0.0080, 0.1482)(note T is a cyclic matrix).

The errors are:

Conventional monochromator: $\varepsilon = 2.93\ N\sigma^2$
Hadamard Transform spectrometer (S exit mask) $\varepsilon = 11.74\sigma^2$.

Again multiplexing decreases the error by N/4. This decrease in error is general for all T-matrices. Also, for any (0,1)-matrix W, $\varepsilon \geq 10.34\ \sigma^2$; N large [18].

The matrices T give an indication of the noise and line

broadening. For example, the larger in number and size of the off-diagonal terms, the larger the error.

If the slot mask is not aligned properly behind the blocking mask, broadening and increased noise result in a Hadamard Transform spectrophotometer.

T is an asymmetric circulant with first row:

$$1/6\ [4 - G\ \delta^2 + 3\delta^3,\ (1 - \delta)^3,\ 0 \ldots 0,\ \delta^3,\ 1 + 3\delta + 3\delta^2 - 3\delta^3].$$

A slight error in slot width compared to stepping size over a large number of step positions can result in broadening and increased noise [18].

The ith row of T is $1/6[0, \ldots 0, (i\Delta)^3, 1 + 3i\Delta)^3, 4 - 6 (i\Delta)^2 + 3(i\Delta)^3, (1 - i\Delta)^3, 0 \ldots 0; 4 - 6 (i\Delta)^2 + 3(i\Delta)^3$ is on the main diagonal, the initial position is assumed to be correct and the final position is off by $n\Delta$.

A continuously moving mask causes a pronounced increase in broadening over a stepped mask with the first row of T being:

$$\frac{1}{384}(230,\ 76,\ 1,\ 00,\ \ldots\ 0,\ 0,\ 1,\ 76)\ [18].$$

To reduce broadening, T must be incorporated into the solving of the spectral intensities. The best estimate of the e_i's are:

$$\hat{e} = W^{-1}\ T^{-1}\ E.$$

The transfer matrix T must be determined for the specific instrument being used.

7. IMAGERS

As mentioned earlier only (n + M - 1) measurements are needed to develop a spectrum from a doubly encoded Hadamard Transform spectrometer. But mn different measurements made from mn linear independent weighing designs could be taken (note the error analyzed for mn measurement (22.2/n) and that analyzed for n + M - 1 measurements (16/n + 1) described previously in this text). The extra measurements can provide one-dimensional spatial information (i.e. the infrared spectra at different distances across a source). However, this increase in spatial information can be obtained only at the cost of signal-to-noise ratios.

In general, each dimension added to a mask allows one to ob-

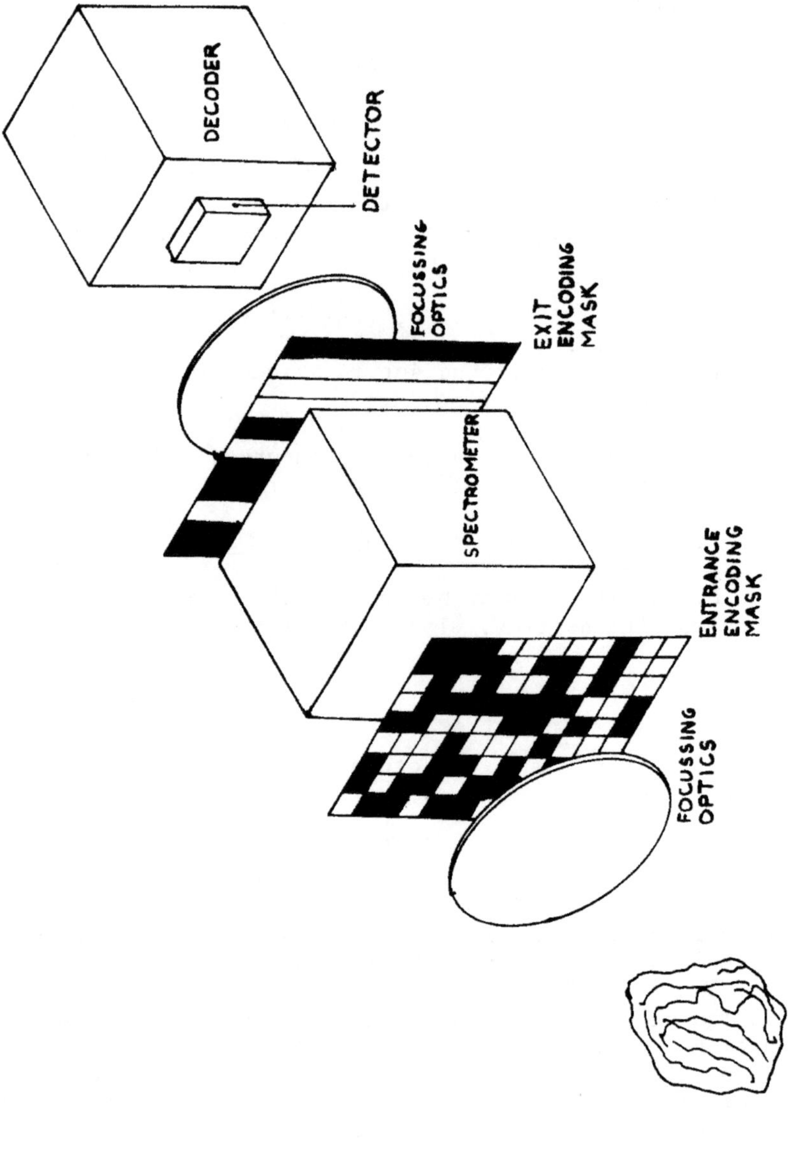

Fig. 6 Components of a doubly encoded Hadamard spectrometric imager. The two-dimensional entrance encoding mask and one-dimensional exit encoding mask allow spectral data in two-dimensions to be observed.

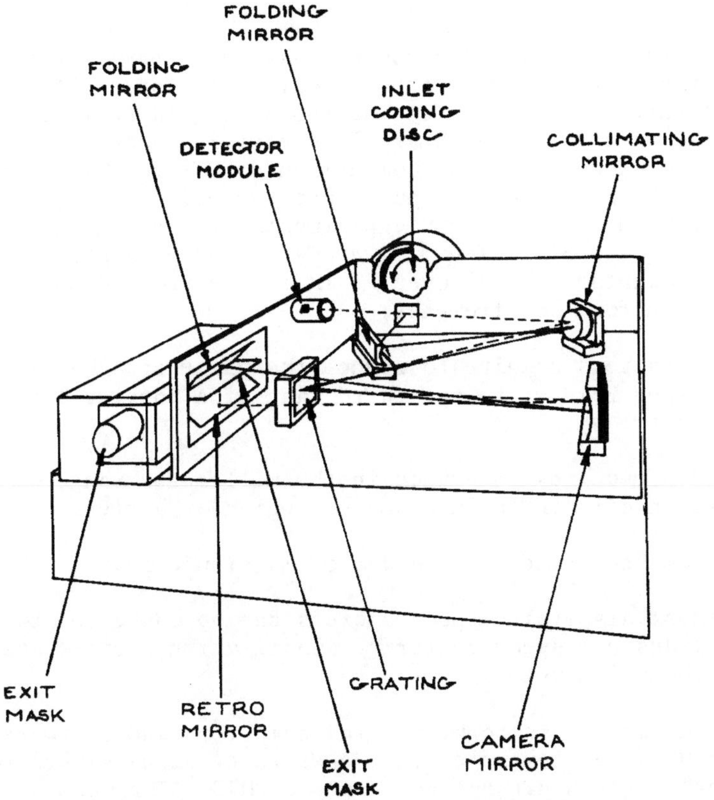

Fig. 7 Optics of a Hadamard Spectrometric Imager.

tain another parameter [19]. Thus, using a two-dimensional entrance mask and a one-dimensional exit mask (Fig. 6), spectral information in two dimensions is obtained. With two two-dimensional masks, an additional parameter could be resolved (e.g. polarization data).

A typical imager is shown in Fig. 7. Swift et al. [20] used an imaging spectrometer to obtain a cathode ray display of the 3.06 to 6.33 μm spectral region of the CO_2 emission of a flame. The source of the CO_2 emission was clearly displayed.

Hadamard Transform imaging spectrometers show great promise in collecting spectral data from surfaces. Spectral data from the earth's surface (such as infrared detection) in relation to its regional proximity is one obvious example for satellite applications.

8. CONCLUSION

Hadamard Transform Spectroscopy has some very nice applications in the detection of infrared radiation, especially for two dimension survey spectra. HTS incorporates both Fellgett's and Jacquinot's advantages which are so useful in today's Fourier Transform Spectroscopy. The comparison of HTS to FT-IR finds HTS within the same order of magnitude for performance, although much development is needed in experimental technology and weighting functions for HTS. The most dramatic and possibly the most important advantage of HTS over FT-IR is the reduced cost of a Hadamard Transform spectrometer.

Three factors combine to reduce the price of a Hadamard Transform spectrometer to one third that of a Fourier Transform interferometer:

(1) HTS requires tolerance in the millimeter range while FT-IR tolerance is in the micrometer region or better.

(2) Computer time is an order of magnitude less for HTS.

(3) Most dispersive spectrometers can be converted to a Hadamard Transform spectrometer by substituting slotted masks in place of the slit systems.

In addition, less sophisticated computers and software are needed in HTS. For example, the He-Ne laser signals used to track the mirror in FT-IR are not necessary in HTS. The Connes' effect is still realized in HTS since the mask will measure individual frequencies with great precision and allow subtraction and spectral stripping techniques which are now so important to spectroscopists applying FT-IR techniques. The minimization of moving parts makes HTS a viable instrument for situations where vibrations are present, i.e. air craft and satellite surveillance.

Hadamard Transform Spectroscopy does lack the full advantages realized by FT-IR in the far infrared region. Here interference from the mask can be a very serious problem.

Possibly the greatest potential for HTS will be the two dimensional spectral survey of a surface. Here specific chemical identification can be monitored in two-dimensional space. This two-dimensional monitoring can be very useful for systems as small as flames to pollution monitoring by aircraft surveillance over cities. Certainly of great interest would be the mapping of earth resources and the acquisition of meteorological data.

Hadamard Transform Spectroscopy represents a very unique and potentially useful tool for analytical chemistry.

ACKNOWLEDGMENT

This work is supported by AFOSR, Grant No. 78-3617. We would like to thank Mr. Ennis Scully and Mr. Kevin Fateley for drafting the figures contained in this article.

REFERENCES

1. F. Yates, J. Roy. Stat. Soc. Supp., 2, 181 (1935).

2. M. Harwit, "Hadamard Transform Analytical Systems," in Transform Techniques in Chemistry (P. R. Griffith, Ed.), Plenum Press, New York, 1978, pp. 173-197.

3. N. J. A. Sloane and Martin Harwit, Appl. Opt., 15, 107 (1976).

4. N. J. A. Sloane, T. Fine, P. G. Phillips, and M. Harwit, Appl. Opt., 8, 2103 (1969).

5. E. D. Nelson and M. L. Friedman, J. Opt. Soc. Amer., 60, 1664 (1970).

6. H. Hotelling, Ann. Math. Stat., 15, 297 (1944).

7. S. W. Golomb, Ed., Digital Communications with Space Applications, Prentice-Hall, Inc., Englewood Cliffs, New Jersey, 1964.

8. R. Thuene and S. W. Golomb, Calif. Inst. of Tech., Jet Propulsion Lab., Space Programs Summary No. 37-40, IV, 207 (1967).

9. L. D. Baumert, Calif. Inst. of Tech., Jet Propulsion Lab., Space Programs Summary No. 37-43, IV, 311 (1967).

10. M. Harwit, P. G. Philips, L. W. King, and D. A. Briatta, Jr., Appl. Opt., 13, 2669 (1974).

11. J. A. Decker, Jr., Appl. Opt., 10, 510 (1971).

12. J. A. Decker, Jr., Anal. Chem., 44, 127A (1972).

13. P. G. Phillips and M. Harwit, Appl. Opt., 10, 2780 (1971).

14. J. A. Decker, Jr., Analysis Instrumentation, 10 (1972).

15. Spectral Imaging, Inc., Product Briefs and Specifications for Hadamard Transform Spectrometers.

16. M. Harwit, P. G. Phillips, T. Fine, and N. J. A. Sloane, Appl. Opt., 9, 1149 (1970).

17. M. H. Tai, M. Harwit, and N. J. A. Sloane, Appl. Opt., 14, 2678 (1975).

18. N. J. A. Sloane, M. Harwit, and M. H. Tai, Appl. Opt. 17, 299 (1978).

19. M. Harwit, Appl. Opt., 10, 1415 (1971).

20. R. D. Swift, R. B. Wattson, J. A. Decker, Jr., R. Paganetti, and M. Harwit, Appl. Opt,, 15, 1595 (1976).

21. L. D. Baumert, in Digital Communications with Space Applications, edited by S. W. Golomb, Prentice-Hall, Englewood Cliffs, New Jersey, 1964, p. 51-55.

22. R. N. Ibbett, D. Aspinall, and J. F. Grainger, Appl. Opt., 7, 1089 (1968).

APPENDIX I. THE GENERATION OF THE H, G AND S MATRICES

There are two convenient methods for generating weighing design matrices. First, previous literature describes the construction of cyclic S-matrices which do not have length $N = 2^n - 1$ (e.g. $N = 4t - 1$, where N is a prime number; $N = p\ (p + 2)$ where p and p + 2 are both prime numbers) [4,21].

However, if the matrix does not have to be cyclic, construction of an H-matrix is quite straightforward. All H-matrices have the relation

$$H_{2i} = \begin{vmatrix} H_i & H_i \\ H_i & -H_i \end{vmatrix}$$

For example, Ref. [21]:

$$H_1 = \begin{vmatrix} 1 \end{vmatrix}$$

$$H_2 = \begin{vmatrix} H_1 & H_1 \\ H_1 & -H_1 \end{vmatrix} = \begin{vmatrix} 1 & 1 \\ 1 & -1 \end{vmatrix}$$

$$H_4 = \begin{vmatrix} H_2 & H_2 \\ H_2 & -H_2 \end{vmatrix} = \left|\begin{array}{cc|cc} 1 & 1 & 1 & 1 \\ 1 & -1 & 1 & -1 \\ \hline 1 & 1 & -1 & -1 \\ 1 & -1 & -1 & 1 \end{array}\right|$$

$$H_8 = \begin{vmatrix} H_4 & H_4 \\ H_4 & -H_4 \end{vmatrix} = \left|\begin{array}{cccc|cccc} 1 & 1 & 1 & 1 & 1 & 1 & 1 & 1 \\ 1 & -1 & 1 & -1 & 1 & -1 & 1 & -1 \\ 1 & 1 & -1 & -1 & 1 & 1 & -1 & -1 \\ 1 & -1 & -1 & 1 & 1 & -1 & -1 & 1 \\ \hline 1 & 1 & 1 & 1 & -1 & -1 & -1 & -1 \\ 1 & -1 & 1 & -1 & -1 & 1 & -1 & 1 \\ 1 & 1 & -1 & -1 & -1 & -1 & 1 & 1 \\ 1 & -1 & -1 & 1 & -1 & 1 & 1 & -1 \end{array}\right|$$

and so on.

A G-matrix can be constructed from this H-matrix by deleting the first row and first column, which in this case are positive one's. The S-matrix can be obtained from the G-matrix by the following substitution [22]. All elements which are +1 are replaced by 0, all elements that are -1 are replaced by +1. If this same substitution is made on the H-matrix, the resulting arrangement is the B-matrix.

A cyclic S-matrix can be constructed using primitive polynomials (modulo 2) for the $N = 2^n - 1$ order matrix [5]. To do this let $e_0, \ldots e_{N-1}$ be the ordered elements of the first row of S.

$P_n(X) = X^n + a_{n-1} X^{n-1} + a_0$ is a primitive polynomial of degree n. There is an excellent list of the polynomials of n degree in Table 2 of Ref. [4] and we shall not repeat these here. The first n elements of a row can be chosen arbitrarily, as long as the row is not all zero elements. The other elements of the row are generated from

$$e_{m+n} = \sum_{i=0}^{n-1} a_i e_{m+i} \quad [0 \leq m \leq (N - n - 1)].$$

The terms are added by modulo 2 ignoring carries, i.e., 1 + 1 = 0, 1 + 0 = 1, 0 + 1 = 1, 0 + 0 = 0.

For example, suppose $N = 2^3 - 1 = 7$. A primitive polynomial of degree 3 is $P_3(X) = X^3 + X + 1$ and $a_0 = 1$, $a_1 = 1$ and $a_2 = 0$ [5]. Choosing $e_0 = e_1 = e_2 = 1$, the first n = 3 terms, then

$$e_3 = a_0e_0 + a_1e_1 + a_2e_2 = 1 + 1 + 0 = 0$$

$$e_4 = a_0e_1 + a_1e_2 + a_2e_3 = 1 + 1 + 0 = 0$$

$$e_5 = a_0e_2 + a_1e_3 + a_2e_4 = 1 + 0 + 0 = 1$$

$$e_6 = a_0e_3 + a_1e_4 + a_2e_5 = 0 + 0 + 0 = 0$$

The elements of the first row of the S-matrix can also be produced using a shift register [17]. A shift register for $\rho(x) = 1 + x + x^4$ is shown in Fig. 8, taken from [17].

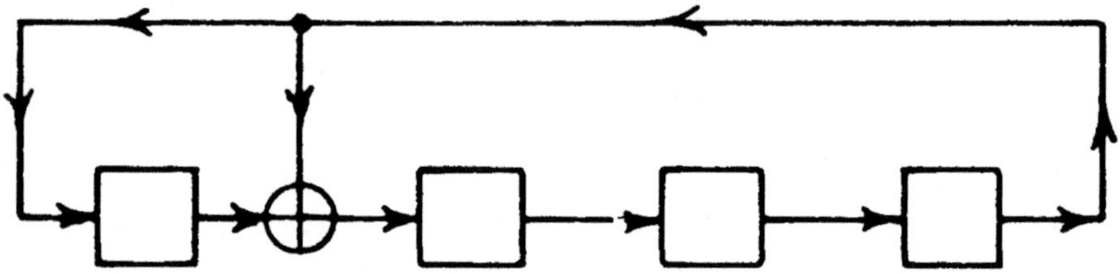

Fig. 8 Here the box, □, is the delay element which holds a zero or a positive one. The quartered circle, ⊕ is a modulo 2 addition. Used by permission [17].

The shift register will go through $2^n - 1$ states before repeating. A n x $(2^n - 1)$ matrix is created quite conveniently and the first row can be used as the first row of a cyclic S-matrix. Subsequent rows of the S-matrix are produced by cycling the previous row one space to the right or left. Reversing a previous process for generalizing an S-matrix from a G-matrix, and if the elements of the S-matrix which are zero are replaced by positive ones and the positive ones are replaced by negative ones, the G-matrix is created from the S-matrix. If this is followed by the addition of a first row and a first column of elements which are all +1, the H-matrix is generated.

APPENDIX II: FAST HADAMARD TRANSFORM

The individual spectral energies are found by $\hat{\vec{e}} = W^{-1} \vec{E}$. Therefore we need matrix W^{-1} and $\vec{E}$. $\vec{E}$ is the vector of measurements. The inverse of the weighing design, W^{-1}, can be found by the following steps. The inverse of an S-matrix can be determined by first taking the transpose of this S-matrix [5]. Second, the substitution of the fraction $2/(N + 1)$ is made for each +1 and $-2/(N + 1)$ for each zero in the transposed S-matrix. However, if S is a left cyclic matrix (i.e. the ith row is produced by shifting the previous row one space to the left), then the matrix S is its own transpose and the aforementioned transpose operation is not necessary. S^{-1} can be written as a product of (2/N + 1) and a matrix of +1's and -1's.

To find $\hat{\vec{e}}$, it follows that (N^2 - N) additions or subtractions are required by straightforward matrix multiplication.

Now, if N equals $2^n - 1$ and the weighing design, S, is generated using a primitive polynomial (mode 2) of degree n, the number of steps needed to determine $\hat{\vec{e}}$ can be reduced to (N + 1) $\log_2$(N + 1) [5]. This shorter method utilizes the fast Hadamard transform (FHT). Even though the Hadamard matrix multiplication performed by the FHT does not use the correct matrix, the FHT matrix is related to the correct matrix by two permutations.

The FHT algorithm, as presented by Nelson and Friedman [5], is a very useful technique. In FHT the following steps are used.

(1) The N Hadamard measurements, $\vec{E}$, are permuted by the permutation π_1. To the permuted measurement vector, $\vec{E}_\pi$, one need only add a zero to the front of the vector. This gives $\hat{\vec{E}}_\pi = (\vec{E}_\pi)$.

(2) The FHT is performed on $\hat{\vec{E}}_\pi$, giving a vector denoted as $\hat{\vec{S}}_\pi$.

(3) The first element of $\hat{\vec{S}}_\pi$ is deleted. The permutation, π_2, is then applied to the remaining vector forming $\vec{L}_\pi$.

(4) The vector, L_π, obtained in step 3 is then normalized (each element divided by -[(N + 1)/2]), resulting in the spectral intensity vector.

Before the algorithm mentioned above can be performed, the permutations, π_1 and π_2, must be determined. The permutations are discussed below.

First, the generation of the permutations of π_1 are described.

(1) Let

$$W_n^T = \begin{bmatrix} \alpha_{11} & \cdots \alpha_{1n} \\ \alpha_{n1} & \cdots \alpha_{nN} \end{bmatrix}$$

equal the first n rows of the transpose of matrix W = $\{w_{ij}\}$.

(2) Set $\pi_1(j) = \ell$ where ℓ is the integer expressed in binary notation by the sequence $\alpha_{nj}\alpha_{(n-1)j} \ldots \alpha_{1j}$. The value of the ℓth element of the permuted measurement vector is set equal to

the value of the jth element of the original measurement vector, $\vec{E}$.

The permutations of π_2 are generated in the following manner.

(1) Let W^T be the transpose of $W = \{w_{ij}\}$.

(2) Let R be the matrix whose ℓth column is the jth column of W^T, where $\pi_1(j) = \ell$.

(3) Let

$$P = \begin{bmatrix} p_{11} & p_{1n} \\ p_{N1} & p_{Nn} \end{bmatrix}$$

be the matrix where the jth column is the (2^{j-1})th column of R.

(4) Set $\pi_2(k) = \ell$, where k is the integer expressed in binary notation by the sequence $p_{\ell n} p_{\ell(n-1)} \cdots p_{\ell 1}$. The value of the ℓth element of the permuted vector, $\hat{\vec{L}}_\pi$, is set equal to the value of the kth element of the original vector.

Finally, the FHT algorithm which is compatible with the aforementioned permutations is presented.

(1) Let $[F(0), \ldots, F(N)]$ be the vector to be transformed. For all possible n-tuples $(\mu_{n-1}, \mu_{n-2}, \ldots, \mu_0)$, set

$$\hat{F}_0(\mu_{n-1}, \mu_{n-2}, \ldots \mu_0) = F(\sum_{i=0}^{n-1} 2^j \mu_j).$$

(2) For $1 < j < n$, set

$$\hat{F}_j(\mu_{n-1}, \ldots, \mu_j, \mu_{j-1}, \ldots, \mu_0) = \hat{F}_{j-1}(\mu_{n-1}, \ldots,$$
$$\mu_j, 0, \mu_{j-2}, \ldots, \mu_0) + (-1)^{\mu_{j-1}} F_{j-1}(\mu_{n-1}, \ldots,$$
$$\mu_j, 1, \mu_{j-2}, \ldots, \mu_0).$$

(3) Set

$$T(m) = F_n(\mu_{n-1}, \ldots, \mu_0), \text{ where } m = \sum_{j=0}^{n-1} 2^j \mu_j.$$

$[T(0), \ldots, T(N)]$ is then the Hadamard Transform of $[F(0), \ldots, F(N)]$.

TIME RESOLVED SPECTROSCOPY

D. E. Honigs, R. M. Hammaker and W. G. Fateley

Department of Chemistry
Kansas State University
Manhattan, Kansas, U. S. A.

1. INTRODUCTION

The basic procedures of Fourier Transform Infrared Time Resolved Spectroscopy (TRS) have been described previously by R. E. Murphy [1], A. Mantz [2], and G. Mamantov [3]. In this chapter we have chosen to present only a brief overview of this technique.

Time resolved spectroscopy appears to be a potentially useful method for the study of intermediates and products of gas phase reactions. The TRS technique involves the coordination of reaction stimulation, e.g. pulsed flash photolyses, laser excitation, or some other periodic reaction initiating techniques, and the recording of the interferogram at some unit time after stimulation. For versatility the reaction vessel may be designed to allow either emission or absorption studies. Time resolved spectroscopy provides a very attractive means for the elucidation of the structure of reaction intermediates and products because this technique may provide sufficient data for examination of the band shapes of materials and deduction of the energy distributions within the molecules.

The fundamental difference between TRS and normal FT-IR spectroscopy is the source characteristics. The source in this discussion may be the emission spectrum from the species created by the stimulation or a time independent source modulated by the absorption spectrum of the species created by the stimulation. In either case the source has been made time dependent rather than time independent as required for conventional FT-IR spectroscopy. Thus in either emission mode or absorption mode, the source in-

J. R. Durig (ed.), Analytical Applications of FT-IR to Molecular and Biological Systems, 119–124.

tensity has the time dependence of the spectrum of the species created by the stimulation.

2. INSTRUMENTATION

Figure 1 illustrates a possible energy build up in a molecular system after stimulation. One assumption is that the intensity at any time t_N after the stimulation is always reproducibly measured. To obtain an interferogram using this assumption for any time t_N, all that is necessary is an intensity file, I(X), composed of each mirror position at time t_N after stimulation. At the completion of the experiment the file will contain the total interferogram at time t_N. This file can be collected by the coordination of the mirror positions to a time - say t_5 - after the stimulation; here the He-Ne laser of the interferometer is used as a clock. The clock counts the number of zero crossings and after the correct number of crossings has occurred, the source is stimulated by a pulse sent from the controlling computer. The same clock is used to open the detector gate when the number of crossings is equal to time interval N which will be five for our example. After collection of these points in the interferogram, the flash start is shifted one zero crossing to the right and the above process is repeated until an intensity sample, I(X), is obtained for each mirror position, X. This is illustrated in Fig. 2.

Since TRS is most useful for kinetic studies, it would be advantageous to collect data for several time intervals after the stimulation event. This can be done by performing several of the above experiments; however, a more efficient method is available. Every zero crossing after the stimulation event represents a time related position where it is possible to open up the detector gate and record the intensity at some time, t_N, where N is the number of zero crossings after the stimulation event. With adequate storage capabilities, this can be done for as many times, t_N, as there are zero crossings between stimulation events. The final problem which remains when one uses TRS is the manipulation of the accumulated data into separate files for each t_N. The I(X) values for all X at each of the times t_N must be separated from those of the other times, t_j; and finally placed in order according to X value or increasing optical retardation in the file for each t_N. The computer software to do this is available; however, it can severely tax the memory and storage capacity of some laboratory computers.

Once enough data has been collected to adequately describe the interferogram of the transient species i.e., enough points to satisfy resolution and wavelength range requirements, the Fourier transform of the points can be taken using the equation

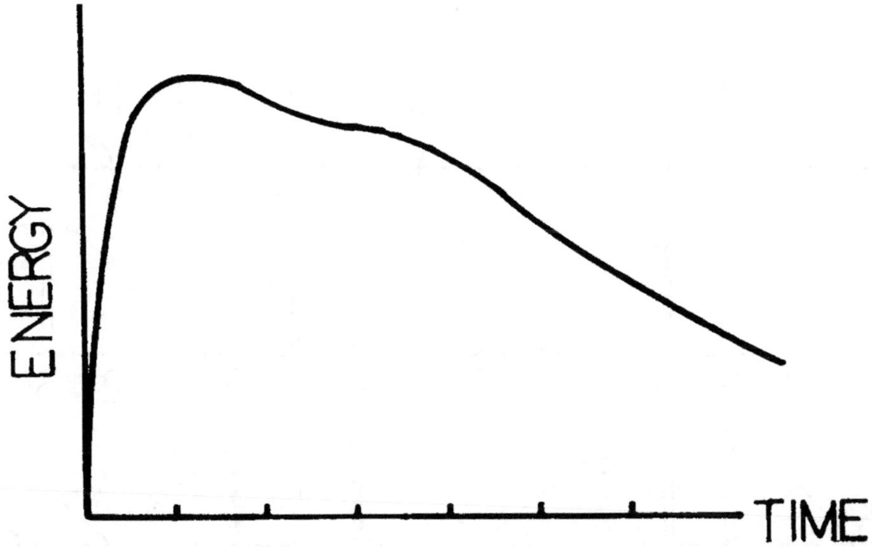

Fig. 1 Possible energy diagram of a molecular system after stimulation.

$$S(\nu) = 2 \int_0^X I(X_{t_N}) \cos 2\pi\nu \, XdX$$

where X is the maximum mirror displacement. The Fourier transform now provides the infrared spectrum of the reactive system at t_N after stimulation.

At this point it would appear that the goal has been reached but Mamantov et al. [3] have pointed out some serious practical limitations. They have shown that the interpretation of the data from a TRS experiment is, unfortunately, not as straightforward as it would appear. The problems occur when the theory makes certain assumptions which are not true in practice. The most obvious of these assumptions is that the source is exactly reproducible. In the conventional FT-IR experiment this isn't a bad assumption, but the close timing that makes the TRS experiment possible places abnormally high reproducibility and accuracy limits on the source.

One of the first problems with the source is its stimulation interval. The flash stimulation interval chosen must consider the reacting compounds since their products are the true source of radiation for operating the interferometer. The stimulation interval must be long enough for one and, hopefully, two things

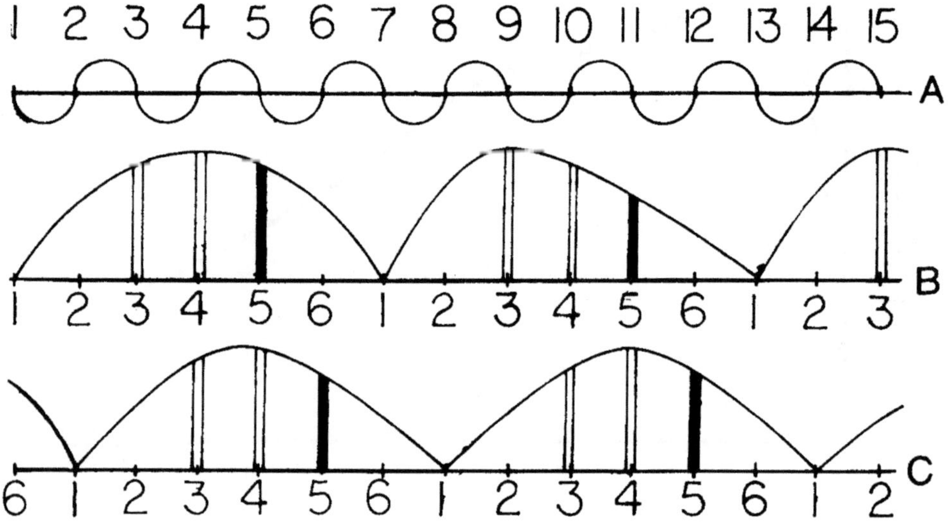

Fig. 2 Diagram of a TRS collection. This shows three files N = 3, N = 4, N = 5 being collected at a stimulation interval of six zero crossings. Collection file N = 5 is shaded. Each file will be collected until all mirror positions needed have been recorded with a pulse from each file. (A) is a plot of zero crossings by the He-Ne laser showing intensity as a function of optical retardation and time. (B) is a plot of two complete and one fractional stimulation intervals and the resulting source intensity as a function of optical retardation and time with time numbered with 1 corresponding to the start of a new stimulation interval. (C) is a plot of one fractional, two complete and one fractional stimulation intervals and the resulting source intensity as a function of optical retardation and time with time numbered with 1 corresponding to the start of a new stimulation interval. Note that the optical retardation scale is the same in both (B) and (C) but the stimulation interval is moved one zero crossing or 628 nm of optical retardation to the right along the optical retardation axis. Thus the N = 5 file (5 zero crossings after the start of a new stimulation interval) is moved from the optical retardation position corresponding to 5 in (A) for experiment (B) to the optical retardation position corresponding to 6 in (A) for experiment (C).

to happen. First, the stimulation interval must be long enough so that the excited species are swept out of the optical path by a flow system before the next stimulation occurs. If this is not done, the interferometer will detect the remnants of this previous stimulation along with the products from the next stimulation. The results of TRS analysis and spectral interpretation of this mixture from different stimulations can be disastrous, see Ref. [3]. If such a mixture does occur, it is not possible to interpret the spectra obtained.

A second problem involving the source is in the timing of the stimulation pulses. The electrical pulse or flash device must be stable enough that it will flash at the same time after the trigger pulse, with the same duration, and with the same intensity.

If the stimulation interval is not constant, the entire timing scheme collapses and the results are useless. In the TRS experiment the detector gate will still open at the same time period after the trigger pulse (400 μsec is a reasonable time after trigger) and the computer will make no compensation nor does it have any way to correct the signal for the change with respect to the stimulation event in the time period when the detector gate is open. The net effect is a broadening of the time period represented from 400 μsec to 400 μsec plus the detector gate to 400 μsec to 400 μsec plus the pulse accuracy.

Another problem occurs when the duration of the stimulation is variable, for the distribution curve of the species in their excited states will change. The same thing happens when the source intensity is variable, although the distribution curve will probably look different than from the one for stimulation duration. The net effect of both is the same. The light source appears variable to the interferometer which is exactly the reason we cannot run a normal FT-IR study of a time dependent system. The distortion in the spectrum from the changing intensity has been documented by Mamantov _et al_. Ref. [3] for systems of changing concentration. The distortion causes false absorption and emission peaks to be plotted so the spectrum shows both absorption and emission spectra components. This phenomenon is also what occurs for systems of changing intensity or flash stimulation duration.

A final problem lies in the physical design of the optical path. Simply because the detector gate is opened at some time after the stimulation, there is no assurance that the resulting data is from that stimulation interval only. For example, if the reaction occurs in a flowing system, it is very probable that the species representing the time period of interest has been swept down the tube and out of the optical path. Of course, this can be easily corrected by placing the observation point at the correct distance from the stimulation device. This correction applies only to some distinct time period, t_N, as determined by the flow rate. Thus any other information about another time period, t_j, will be inaccurate. In other words, all advantages from being able to monitor several different time periods are lost.

Additional complications are introduced by the interferometer if it is unstable. Much of the instability can be removed by the application of isothermal conditions within all components in the optical path.

On balance, TRS has many possible pitfalls. Some have been located; even more, probably lie ahead. The field of TRS holds a significant potential, but at this time any time resolved spectra must be interpreted with the utmost caution and skepticism.

ACKNOWLEDGEMENT

This work was sponsored by AFOSR Grant Number 78-3617.

REFERENCES

1. R. E. Murphy, F. Cook, and H. Sakai, J. Opt. Soc. Am., 65, 600 (1975).

2. A. W. Mantz, Appl. Spectrosc., 30, 459 (1976).

3. A. A. Garrison, R. A. Crocombe, G. Mamantov, and J. A. De Haseth, Appl. Spectrosc., July - Aug. issue 1980.

THE USE OF FOURIER TRANSFORM METHODS FOR THE MEASUREMENT OF INFRARED EMISSION SPECTRA

N. Sheppard

School of Chemical Sciences
University of East Anglia
Norwich, England, U. K.

1. SOME GENERAL PRINCIPLES

For a sample to be in thermal and radiative equilibrium with an environment at the same temperature, it is necessary for the amount of radiation absorbed from the environment to be equal to the amount emitted to the environment. This must be so in all directions and at all wavenumbers or wavelengths. This consideration is the basis of Kirchoff's Law [1] which states that at a given temperature the absorptance, $\underline{\alpha}$, of a sample is equal to its emissivity, $\underline{\varepsilon}$ [2]. The absorptance is the fractional absorption, and the emissivity the fractional emission, of the radiation that is characteristic of the temperature of the system. Such radiation is termed 'blackbody' radiation. It depends only on the temperature, and is independent of chemical composition etc. It follows from Kirchoff's Law that a perfect absorber is a perfect emitter, and a selective absorber is a selective emitter at the same positions in the spectrum.

In the presence only of the blackbody radiation which is appropriate to a uniform-temperature system it is not possible to distinguish one part of the system from another by radiation measurement. In order to measure the selective absorption or the selective emission of the sample it is necessary to introduce different temperatures to the system. The selective absorptance as a function of spectral position is normally measured as the fraction of energy absorbed by a near room-temperature sample from a strong radiation beam originating in a high-temperature source. Selective emission is measured from a hot body in a colder environment, so that the energy emitted is greater than the energy absorbed. The emissivity is calculated from the ab-

J. R. Durig (ed.), Analytical Applications of FT-IR to Molecular and Biological Systems, 125–140.

solute strength of emission by ratioing the measurements against those from a blackbody source at the same elevated temperature [3,4], see Fig. 1.

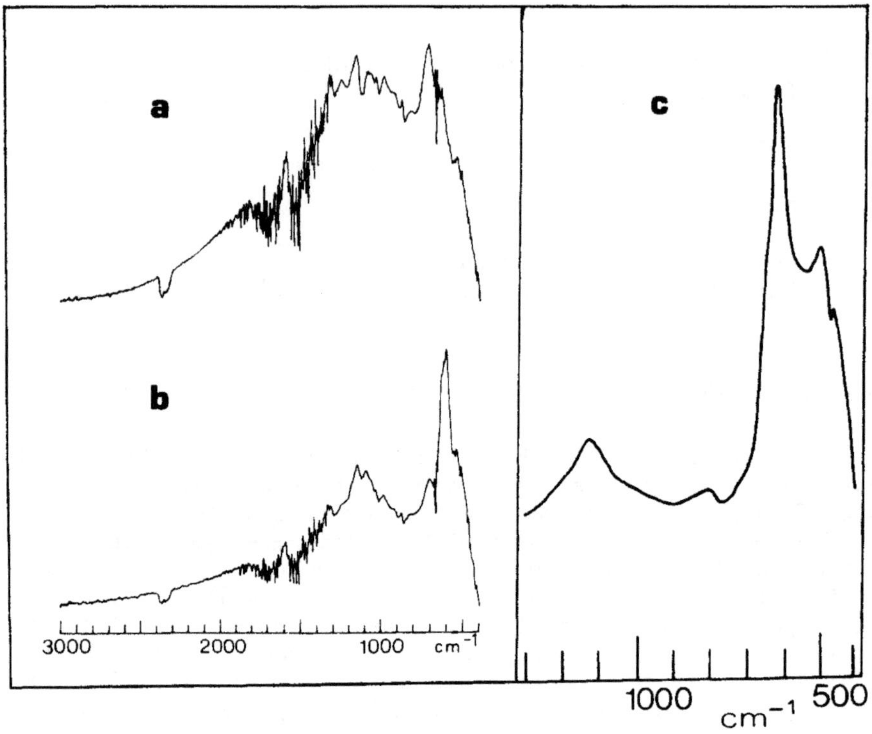

Fig. 1 (a) The infrared emission spectrum of a blackbody recorded at 135°C (b) the infrared emission spectrum from a copper plate which had been heated in air for one hour at 250°C to promote oxidation; the spectrum was recorded at 135°C; the ratio of the maximal values of (a) and (b) is approximately 0.75. (c) the ratio (b)/(a) to give the emissivity curve of the oxide layer. Reproduced with permission [4].

We note that it is always a net effect that is observed or measured. For example, when radiation from a high-temperature blackbody source is measured by a room-temperature detector, what is actually registered is the difference between the energy absorbed and the energy emitted by the detector. For this reason a room-temperature detector registers nothing in a room-temperature environment.

2. SIMILARITIES BETWEEN INFRARED TRANSMISSION SPECTRA AND INFRARED EMISSION SPECTRA

If, see later, no reflection or scattering of radiation occurs, the absorptance, $\underline{\alpha}$, of a sample with respect to an external light beam is given by

$$\underline{\alpha} = 1 - \underline{\tau}$$

where $\underline{\tau}$ is the transmittance (fractional transmission) of the sample [2]. Hence by Kirchoff's Law, the emissivity $\underline{\varepsilon}$ can be given by

$$\underline{\varepsilon} = \underline{\alpha} = 1 - \underline{\tau}.$$

In principle, therefore, the same information could be obtained by a transmission or by an emission experiment utilizing the same sample, provided the latter is not appreciably modified on being raised to a higher temperature in order to measure the emission spectrum.

In practice however the transmission experiment gives much better sensitivity in terms of signal-to-noise, because the high-temperature external source gives not only a much greater total amount of radiation, but also relatively greater amounts at higher wavenumbers [5]. This is illustrated in Fig. 2 which shows the blackbody emission appropriate to different temperatures. Except for refractory materials and gases it is rarely possible to heat a sample above a few hundred degrees centigrade without substantial chemical or physical change, and consequent change in the spectrum. The maximum possible emission at a particular position of the spectrum is at the blackbody level for that temperature and, see Fig. 2, the emission at for example 200°C (473 K) is greatly less than that of a normal infrared source at ca. 1500°C (1773 K). Note the logarithmic scale in Fig. 2.

In principle it would be possible to measure the selective emission of a sample even at room temperature, by use of a low-temperature detector and a low-temperature background, as illustrated experimentally by Low [6], and later by Brown and Young [41] but the signal/noise would be very poor compared with that from a transmission experiment (Fig. 2).

For the above reasons, when using dispersion spectrometers, transmission rather than emission experiments are usually carried out except when the sample itself has naturally a high temperature, or when transmission experiments are not feasible and perhaps temperature information is also sought from the measurements. Examples of the former type would be flames or molten salts; an example of the latter type would be emission from the atmosphere

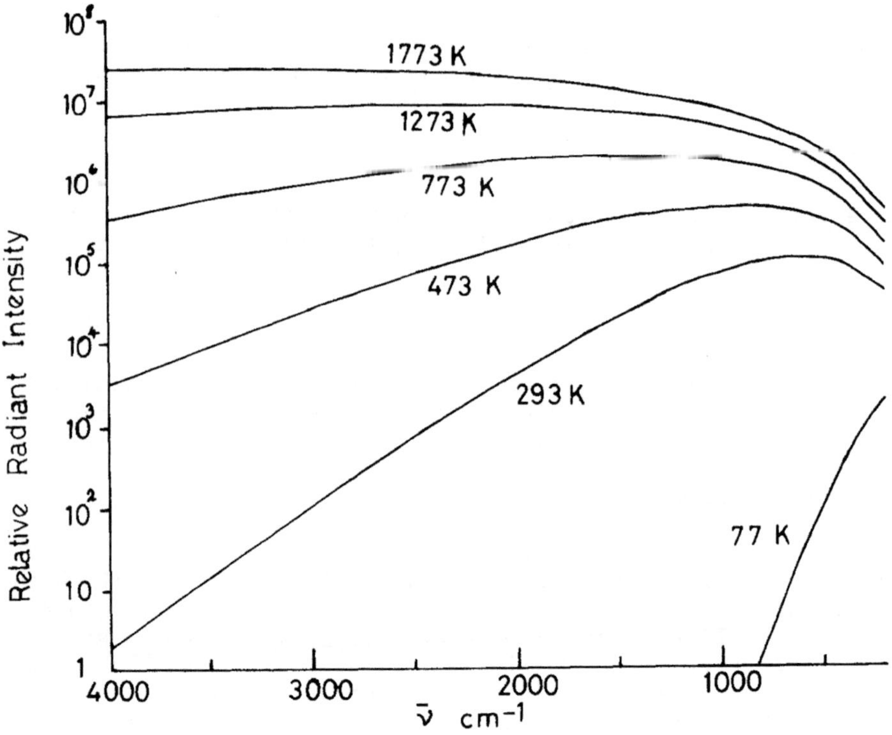

Fig. 2 Blackbody emission curves, log[E($\tilde{\nu}$,T)] at different temperatures as a function of wavenumber, $\tilde{\nu}$, at temperatures of 77 K (-190°C); 293 K (+20°C); 473 K (200°C); 793 K (500°C); 1273 K (1000°C).

or the surface of the earth as measured with instrumentation carried in satellites or rockets.

3. CONTRIBUTIONS TO EMISSION MEASUREMENTS FROM FOURIER TRANSFORM SPECTROSCOPY

One of the major advantages of Fourier Transform Spectroscopy based on interferometric measurements over conventional dispersion spectroscopic methods is the measurement of very weak signals. Under these circumstances the dynamic range of the analog-to-digital converter (ADC) is not exceeded and the full advantages of the FT method can be exploited. Further sensitivity can then be obtained by the use of low-temperature and low-noise detectors [7]. Thus, for example, a very weak absorption band is measured in a transmission experiment as the difference between two very large quantities and the dynamic range for the ADC may be exceeded. However, by the emission method only a very weak total signal, related directly to the strength of the weak absorption band (plus possibly a weak background), has to be measured and the dynamic range problem is no longer troublesome [8,9]. Overend has discussed the dynamic range problem in relation to dispersive spectrometers [5].

The earliest extensive applications of FT infrared spectroscopy were in far infrared spectroscopy where the sources of radiation are weak and even transmission experiments are very difficult [10]. In the mid infrared region Connes and his colleagues [11] carried out very early beautiful high-resolution measurements of the infrared spectra from the planets, or more distant stars, which are of course very weak sources. Also interferometers became increasingly used in rocket or satellite-based measurements of atmospheric or surface emission studies of the earth. The latter used mostly small compact interferometers of relatively low resolution [12].

M. J. D. Low [6,13-20] carried out a series of pioneering studies which showed that such equipment, then capable of a maximum resolution of about 20 cm^{-1}, could be used for chemical analytical problems. He and his colleagues carried out emission infrared measurements on a wide range of samples, including minerals [14], organic compounds [16] and biological specimens [18]. He also showed that such equipment could be used to identify toxic components in the hot gases emerging from chimneys on industrial plants [20]. This latter type of application has been pursued by others [21]. Similar work can doubtless be done with hot automobile exhaust gases. FT emission work using spectrometers capable of resolution down to 0.5 cm^{-1} have more recently been applied to studies of molten salts [22,23], hot gases [24], coatings on highly-reflecting metal surfaces [4,7,25], silicone grease [24], organic compounds [25], and the interaction of propylene with V_2O_5 in a catalytic study [26]. A comparison has also been made of emission spectra from organic compounds on a metal substrate or held between infrared transparent plates [25]. Recently a Michelson interferometer cooled down <u>in toto</u> to liquid N_2 temperatures has been used to measure chemiluminescence during gas-phase reactions [27,28].

Some earlier non-FT emission work of a chemical analytic nature has been reviewed by Low, McManus and Abrams [29], and several notable recent non-FT emission papers are listed in Refs. [30-33]. These include emission studies of inorganic salts [30], the oxidation of molybdenum metal [31] and MoO_2 to MoO_3 [32], and the interaction of formic acid with vanadium pentoxide [33].

Successful FT emission studies have also been made of the temperature of metal particles in a working Pd/SiO_2 catalyst [34] and of the compositions of 'lacquers' acquired on piston or bearing surfaces of internal combustion engines in the course of operation [35].

4. THE VARIATION OF EMISSION WITH THICKNESS; DIFFERENCES BETWEEN TRANSMISSION AND EMISSION SPECTRA

Low [16], Griffiths [24] and others [25,26,31] showed that features in selective emission spectra obtained from thin layers of materials corresponded very well to transmission spectra. An example is given in Fig. 3. However, thick samples were shown to give general and continuous emission spectra of little diagnostic value. Furthermore, sometimes these spectra showed minima close to where thin-layer spectra had shown maxima. These phenomena have since been elucidated, the appropriate explanation depending on the nature of the sample and the experimental conditions.

Let us first of all consider an emitting sample of variable thickness that is heated to a uniform temperature throughout. We shall compare the absorptance (or fractional absorption) spectra with emissivity spectra obtained by ratioing the observed emission against that of a blackbody at the same temperature.

An absorptance spectrum typically shows a number of sharp bands at small sample pathlengths. These increase in intensity with increasing thickness so as first to individually blackout and then to collectively do so. Provided the loss of energy in the transmission spectrum is entirely caused by absorption, Kirchoff's Law requires that the emissivity spectrum behaves in exactly the same way, i.e., that for a great thickness of sample the energy emitted corresponds to blackbody radiation, i.e. unit emissivity throughout, and gives no chemical information.

However, a different situation occurs if there are additional reflection or scattering effects which can contribute to the loss of transmission and to the form of the emission spectra. Reflection losses decrease transmission and hence enhance 'apparent absorption' when reflection effects are disregarded. On the other hand, reflected energy is not absorbed and Kirchoff's Law only applies to absorbed energy. The reflection effects therefore show up in emission experiments as reduced emission [4].

Metals can be very strong reflectors over wide spectral ranges; highly polar inorganic compounds such as minerals tend to have selective reflections in the vicinity of intrinsically strong absorption bands (i.e. bands which blackout at relatively low pathlengths); organic compounds tend to give weaker absorption

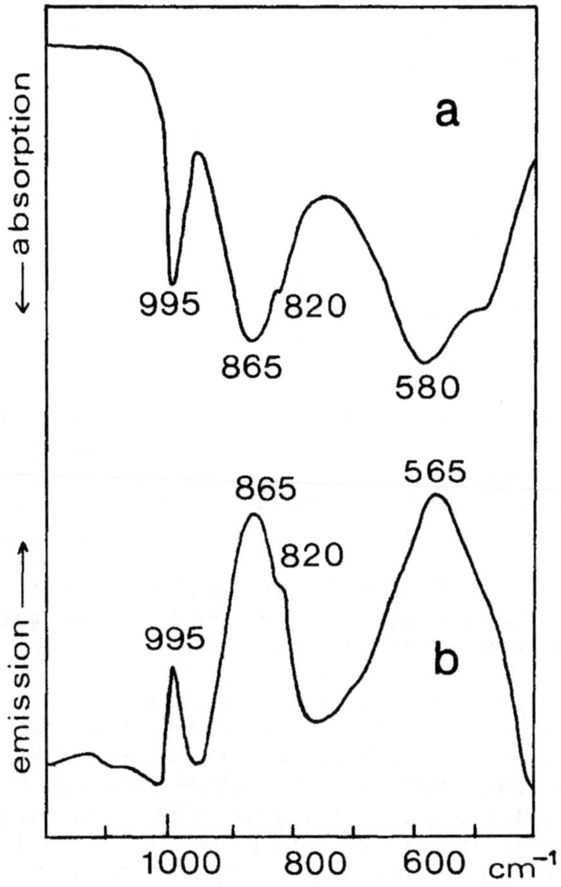

Fig. 3 Infrared spectra of molybdenum trioxide: (a) spectrum recorded in KBr (1.5 mg/g) by transmission method (b) emission spectrum of thin sample (0.1 mg cm^{-2}) recorded at 110°C under oxygen (redrawn, with permission, Primet _et al_. [26]).

spectra and only show appreciable reflectivity in the vicinity of the strongest bands. The selective reflections are related to high refractive indices in the vicinity of intrinsically strong absorptions but, beyond a certain sample thickness, are virtually independent of pathlength. A strong blackout absorption region which is caused by a product of small intrinsic absorption intensity but a large pathlength will not exhibit appreciable reflection; there would be a corresponding emissivity of unity. However where selective reflection occurs the emissivity will be reduced [3].

Taking reflectance [2] (fractional reflection, ρ) into account the energy impinging on a sample can be reflected, absorbed, or transmitted i.e.

$$\rho + \alpha + \tau = 1.$$

From Kirchoff's Law we see that

$$\varepsilon = \alpha = 1 - \tau - \rho.$$

Hence, only if $\rho = 0$ is α or ε equal to $(1 - \tau)$ as is frequently assumed to be the case. In the extreme case when $\tau = 0$ then

$$\varepsilon = \alpha = 1 - \rho.$$

This is illustrated in Fig. 4 where the emissivity of a thin layer of quartz is compared with the reflectance spectrum [4]. This phenomenon provides one possible explanation for the occurrence of a minimum in an emission spectrum associated with the position of an apparently strong absorption band [14,24].

In the absence of selective reflection, another possible cause of such a phenomenon occurs in emission spectra from samples of non uniform temperature, as was emphasized by Griffiths [24]. Many emitting samples are heated from the opposite side to the detector. This is particularly so when the sample to be investigated is backed by a highly-reflecting (and hence weakly-absorbing or emitting) metal surface. Under such circumstances the inner layers of the sample will be hotter than the outer layers that face the detector, and energy emitted by the inner layers may be selectively reduced by absorption in the cold outer layers. This self-absorption phenomenon is well known in the study of the emission spectra of gases, but can also be of importance for solid or liquid samples. Once again emission minima will be observed superimposed on the top of the strongest emission bands, or on a blackbody contour if the sample is thick.

The self-absorption situation can be avoided by careful design of the heating arrangements, and specifically if sample heating is by radiation from the front so that the outer layers are hot. Indeed this method can be used in reverse to measure a selective emission spectrum from a relatively thin outer layer of a thick sample which is opaque through blackout absorption. This has been demonstrated by Koga *et al.* [36] who obtained an emission spectrum from a thick opaque layer of a catalyst by front-heating with infrared radiation above 2000 cm^{-1}. With the help of a suitable filter the emission spectrum was observed in the complementary wavenumber range below 2000 cm^{-1}. This procedure is somewhat analogous to obtaining an infrared spectrum of the top layer of a sample by attenuated total reflection (ATR) [37].

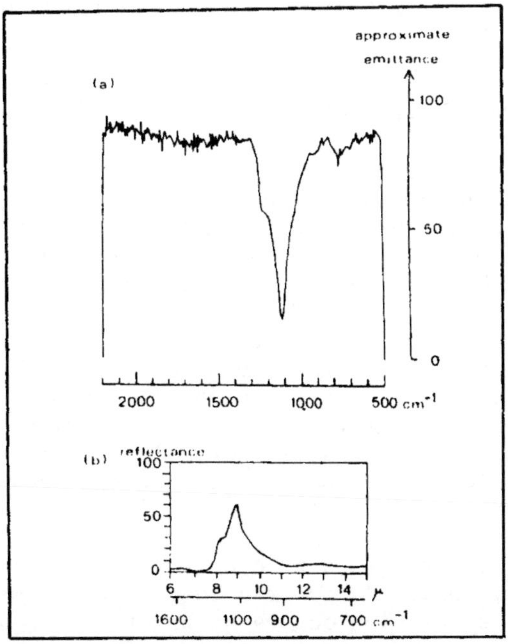

Fig. 4 (a) The emission spectrum of a 3 mm thick disc of fused silica heated to 118°C from below; (b) the reflectance spectrum of fused silica [O. Reinkober, Ann. Phys., Lpz. 34, 343 (1911)]. Reproduced with permission [4].

The emission method, while less sensitive, is flexible in that by varying the intensity of the heating radiation different thicknesses of material can be studied.

5. EMISSION SPECTRA FROM OPAQUE SAMPLES

It is sometimes implied that selective emission spectra can be obtained from opaque samples. However, if the opacity is caused by total absorption of light in a thick layer, then all that can be observed is blackbody emission which contains no structural information. If, however, selective reflection of radiation is also present then, as we have seen earlier, for a thick layer the blackbody emission will be diminished by the reflectance, ρ, (Fig. 4) and in this sense a characteristic 'reduced emission' spectrum can be obtained. It should however be borne in mind that the value of the reflectance depends critically on the quality of the surface. A flat surface will give a good quality 'reduced emission' spectrum, whereas a rough one may give an emission spectrum approximating to that of a blackbody once again.

6. EMISSION SPECTRA FROM OVERLAYERS ON REFLECTING SURFACES

A number of very good applications of emission spectroscopy have been made when the sample under investigation is a selectively emitting one on a highly reflecting metal layer [4,7,22-25,31]. In such cases in principle an alternative reflection/absorption experiment could be carried out by reflecting a light beam off the metal surface so that in effect it passes twice through the overlayer whose spectrum is sought. As in the transmission case, this would certainly have higher sensitivity than an emission spectrum, because of the use of a high temperature source. On the other hand the optical requirements are stringent and ideally the overlayer/metal sample should normally be optically flat and give specular reflection. The emission method does not have this strict requirement.

A perfect reflector is a zero emitter and good reflectors are very weak emitters. Hence if a thin layer of material which can give selective emission occurs as an overlayer on a metal mirror, the very weak continuum emission from the latter (perhaps 5% of a blackbody over most of the wavenumber range) interferes little with the emission from the overlayer. In a number of cases the weak continuum emission from a clean surface has been used as the reference spectrum [22,23,30]. For reasons already stated it is preferable, on practical and interpretational grounds, to use as the reference spectrum the emission from a blackbody at the same temperature [3,4] so as to give emissivity directly. For such work a spectrometer incorporating a minicomputer is most convenient. Slight temperature mismatch between sample and black-body reference can lead to approximately linearly sloping back-grounds.

Baumgarten [25] has pointed out that optical interference effects will cause the emission spectrum from an overlayer on a metal surface to differ from that from a sample of equivalent thickness on an infrared transparent plate [25]. In the extreme case of a very thin overlayer (thickness << wavelength) this leads to a situation of negligible emission in the direction normal to the metal surface, although highly oblique emission (i.e. emission at high angles to the normal) should be strong [5]. Greenler has worked out the quantitative theory associated with this phenomenon in the context of very thin overlayers such as those of monolayer thickness [38]. Greenler's conclusions have been verified qualitatively through measurements of the emission from stearate films on gold by Blanke, Vincent and Overend [5] and from a 60 A° layer of aluminium oxide on aluminium by Kember,

Chenery, Sheppard, and Fell [7,39]. The angular dependence of emission intensity is analogous to that pertaining to reflection/absorption experiments of overlayers on metals. In the same way, the emission viewed at highly oblique angles will be from vibrations of the overlayer that give dipole charges perpendicular to the metal surface [40].

When overlayers of adequate thickness on metals are measured by emission normal to the surface, it is common practice to subtract from the overall emission the background spectrum that is measured from the clean metal emitter alone. It should be noted, however, that by this means somewhat reduced intensities will be obtained in the stronger features of the emission spectrum of the overlayer. This is because some of the background metal emission will be absorbed by the overlayer, that is to say that subtraction of the full background amount from the composite spectrum will lead to overcompensation [7]. In such circumstances the precise evaluation of the emittance of the overlayer is difficult. However, when the mirror reflectivity is high, i.e. its emissivity is low, then a readily recognizable emission spectrum of the overlayer can be obtained which closely resembles that in transmittance. For weak emitters, when the thickness of the overlayer is such that interference effects are not important, the apparent emissivity of the overlayer will be approximately twice that from an overlayer on a transparent support because of the energy reflected back into the direction of the detector by the metallic mirror.

It should also be realized that a reduction in optical quality of the underlying metal surface, e.g. through corrosion, scratching, or pitting, will cause a stronger background emission from the metal, leading in the extreme case to a complete blackbody spectrum once again. Under these conditions it will be increasingly difficult to evaluate the emission spectrum of the overlayer. The situation is analogous to that in a transmission experiment when heavy background scattering of radiation by the sample leads to a major reduction in overall transmission.

Measurements of selective emission spectra of overlayers can, of course, be equally well made over more restricted wavenumber ranges and with reduced intensities, in the regions where the underlying material has strong selective reflections and hence 'reduced emission regions' as described earlier.

7. EMISSION SPECTRA FROM ADSORBED SPECIES

The emission spectrum from an adsorbed material on the surface of a sufficiently high area of adsorbent can, of course, be made without difficulty in regions where the adsorbent itself is non-absorbing and hence non-emitting [26,33]. However, this favorable situation rarely arises. More often the adsorbent itself has regions of strong absorption which will cause regions of blackbody emission. The emission situation is then little different from the transmission situation with respect to the ease or difficulty with which bands from the adsorbate can be measured.

If on the other hand part of the lack of transmission of the adsorbent is related to reflectance, then an emission 'window' will occur in part of the region of zero transmission. Because reflection of radiation does not occur literally at a surface but typically within a fraction of a wavelength of the surface, an absorption/reflection spectrum could be obtained from adsorbed molecules within that optical thickness. Similarly an emission spectrum of the adsorbate could be obtained. Even strong reflectors such as metals have appreciable penetration of radiation to a thickness of several hundred Angstroms; for less strong reflectors in the infrared region it could be that internal surfaces may be probed for adsorbates up to several micrometers by this method.

An alternative method of obtaining infrared spectra from adsorbed species when the overall sample is opaque due to adsorption is to heat the sample from above so as to get stronger emission from the top layer as described earlier and carried out by Koga et al. [36].

8. THE MEASUREMENT OF WEAK EMISSIONS AND THE USE OF LOW TEMPERATURE DETECTORS

When using an interferometer, the Ge/KBr or Ge/CsBr beamsplitter is often kept warm in order to prevent condensation of water vapor on the alkali halide plate which could lead to deterioration of the Ge film. Because the beamsplitter occurs between the emitting sample and the detector, additional emissions from the warm beamsplitter (mostly in practice from some germanium oxide on the germanium film which emits near 850 cm^{-1}) are superimposed on those from the sample. This can be very troublesome when the sample emission is very weak. Kember et al. [7] have shown that accurate compensation for the beamsplitter emission can be obtained by subtracting at the interferogram stage a spectrum from the beamsplitter alone. This method works because the signals originating from the beamsplitter have different phase relationships from those from the external emitting sample, as described by Brown and Young [41].

Profitable recourse can be made to the use of high-sensitivity, low-temperature infrared detectors, such as the mercury cadmium telluride photoconductive cell, when very weak emission spectra have to be measured. However, unlike room-temperature detectors, the low-temperature ones also respond to background room temperature radiation such as might be reflected onto the detector by the sample or intervening optics [7,41]. Kember *et al.* [7] and, in a somewhat different context, Brown and Young [41] have shown that this difficulty can effectively be overcome through measurement of the radiation reaching the detector when the emitting sample is at room temperature (any reflected radiation will still be present), and then - without moving the sample so that reflection effects are unchanged - when the sample is at the higher emission temperature. The emissivity, ε, can then be derived from the ratio:

$$\varepsilon = (E^S(T_2) - E^S(T_1))/(E^{BB}(T_2) - E^{BB}(T_1))$$

where T_1 and T_2 denote room temperature and the emission temperature, respectively, E^S denotes energy originating from the sample and E^{BB} denotes energy originating from the blackbody reference [7].

9. CONCLUSIONS

The FT-IR method is ideal for studying emission spectra from samples at moderate temperatures. This is because the overall emission signals are relatively weak such that the FT limiting factor of the dynamic range of the ADC is not brought into play. Full advantages of the interferometric method can then be utilized.

By ratioing an emission spectrum against that of a blackbody reference at the same temperature, the emissivity spectrum of the sample can be obtained directly. This, by Kirchoff's Law, is equal to the absorptance, i.e. fractional absorption, spectrum of the sample. In the absence of selective reflection it is also closely related to the conventional transmission spectrum.

In certain circumstances, e.g. the study of an overlayer on a highly reflecting metal surface, it can be more convenient to measure an emission spectrum than a transmission or reflection/absorption one. The emission method can often be applied, for example, to non-flat surfaces. For very thin layers (much less than the wavelength of radiation), such as result from adsorption monolayers, much greater emission is observed at highly oblique angles to the surface than when emissions are measured normal to the surface.

In some cases, as with molten salts or flames, the emission method is the natural one to use and FT-IR methods give enhanced sensitivity.

Since this article was written the author has become aware of another recent review of emission work involving FT techniques by J. B. Bates [42]. The article is somewhat complementary to the present one and gives much wider coverage to applications of infrared emission studies outside the area of chemical analysis.

ACKNOWLEDGEMENTS

I should like to thank Mr. D. H. Chenery for comments on the first draft of this article, and the Science Research Council, London, for its support of my group's researches in this area.

REFERENCES

1. J. Houghton and S. D. Smith, Infrared Physics, Oxford Univ. Press, Oxford, 1966.

2. The nomenclature and symbols used are those recommended by the International Union of Pure and Applied Chemistry (I.U. P.A.C.) [in "Manual of Symbols and Terminology for Physico-chemical Quantities and Units" (1979), I.U.P.A.C., Oxford, and Pure and Applied Chemistry, 51, 1 (1979)], and the International Union of Pure and Applied Physics, (I.U.P.A.P.), [in document U.I.P. 20, 'Symbols, Units and Nomenclature in Physics' (1978) and Physica, 93A, 1 (1978)].

3. E. Steger and R. Rasmus, Appl. Spectrosc., 28, 376 (1974).

4. D. Kember and N. Sheppard, Appl. Spectrosc., 29, 496 (1975).

5. J. E. Blanke, S. E. Vincent, and J. Overend, Spectrochim. Acta, 32A, 163 (1976).

6. M. J. D. Low, Appl. Spectrosc., 22, 463 (1968).

7. D. Kember, D. H. Chenery, N. Sheppard, and J. Fell, Spectrochim. Acta, 35A, 455 (1979).

8. P. R. Griffiths, Chemical Infrared Fourier Transform Spectroscopy, Wiley, New York, 1975.

9. P. R. Griffiths, C. T. Foskett, and R. Curbelo, Appl. Spectrosc. Rev., 6, 31 (1972).

10. H. A. Gebbie and R. Q. Twiss, Rep. Prog. Phys., 24, 729 (1966); K. D. Möller and W. G. Rothschild, Far-Infrared Spectroscopy, Wiley-Interscience, New York, 1971.

11. J. Connes and P. Connes, Infrared Astronomy, Gordon and Breach, New York, 1968, 193.

12. See for example: L. S. Bernstein, D. C. Robertson, J. A. Conant, and B. P. Sandford, Appl. Optics, 18, 2454 (1979), and other papers on atmospheric optics in the same Journal.

13. M. J. D. Low and H. Inoue, Anal. Chem., 36, 2397 (1964).

14. M. J. D. Low, Nature, 208, 1089 (1965).

15. M. J. D. Low and H. Inoue, Can. J. Chem., 43, 2047 (1965).

16. M. J. D. Low, L. Abrams, and I. Coleman, Chem. Commun., 389 (1965).

17. M. J. D. Low, J. Catal., 4, 719 (1965).

18. M. J. D. Low, Separatum Experientia, 22, 262 (1966).

19. M. J. D. Low and I. Coleman, Spectrochim. Acta, 22, 369 (1966).

20. M. J. D. Low and F. D. Clancy, Env. Sci. Technol., 1, 73 (1967).

21. H. W. Prengle, Jr., C. A. Morgan, C. S. Fang, L. K. Huang, P. Campani and W. W. Wu, Env. Sci. Technol., 1, 417 (1973).

22. J. B. Bates, A. S. Quist, and G. E. Boyd, Chem. Phys. Lett. 16, 473 (1972).

23. J. B. Bates and G. E. Boyd, Appl. Spectrosc., 27, 204 (1973).

24. P. R. Griffiths, Appl. Spectrosc., 26, 73 (1972).

25. E. Baumgarten, Spectrochim. Acta, A32, 865 (1976).

26. M. Primet, P. Fouilloux, and B. Imelik, Surface Science, 85, 457 (1979).

27. J. G. Moehlmann, J. T. Gleaves, J. W. Hudgens, and J. D. McDonald, J. Chem. Phys., 60, 4790 (1974).

28. J. G. Moehlmann and J. D. McDonald, J. Chem. Phys., 62, 3052, 3061 (1974).

29. M. J. D. Low, J. C. McManus, and L. Abrams, Appl. Spectrosc. Rev., 5, 171 (1972).

30. G. Fabbri and P. Baraldi, Appl. Spectrosc., 26, 593 (1972).

31. L. M. Gratton, S. Paglia, F. Scattaglia, and M. Cavallini, Appl. Spectrosc., 32, 310 (1978).

32. A. Veno and C. O. Bennett, Bull. Chem. Soc. Jap., 52, 2551 (1979).

33. M. Adachi, K. Kishi, T. Imanaka, and S. Teranishi, Bull. Chem. Soc. Jap., 40, 1290 (1967).

34. D. R. Kember, Ph.D. thesis, University of East Anglia, Norwich, 1976.

35. D. H. Chenery, J. Fell, and N. Sheppard, unpublished work.

36. O. Koga, T. Onishi, and K. Tamaru, Chem. Commun., 464 (1974).

37. R. G. J. Miller and B. C. Stace, Laboratory Methods in Infrared Spectroscopy (2nd Edition), Heyden, London, 1972.

38. R. G. Greenler, Surface Science, 69, 649 (1977).

39. D. H. Chenery and N. Sheppard, unpublished work.

40. R. G. Greenler, J. Chem. Phys., 44, 310 (1966); 50, 1963 (1969).

41. R. J. Brown and B. G. Young, Appl. Opt., 14, 2927 (1975).

42. J. B. Bates in Fourier Transform Infrared Spectroscopy; Applications to Chemical Systems, Vol. I, (J. R. Ferraro and L. J. Basile, eds.) Academic Press, New York, 1978.

ACCESSORIES AND SAMPLING TECHNIQUES FOR FT-IR SPECTROMETRY

Peter R. Griffiths

Department of Chemistry
Ohio University
Athens, Ohio, U. S. A.

1. SPECTROMETER OPTICS

The optics of most mid-infrared FT-IR spectrometers are designed so that absorption spectra of most samples may be measured using accessories which have been designed for grating spectrometers with relatively minor modification. The collimated beam from the interferometer is directed onto one of two off-axis paraboloidal mirrors which focus the beam at what would usually be considered as the "sample" and "reference" positions in the sample compartment of a conventional double-beam spectrometer. The beam is then refocused onto the detector. The beam can be passed through the front or back channel by simultaneously switching the position of two plane mirrors.

A third switchable plane mirror allows the operator the choice of either passing the (invisible) infrared beam from the interferometer through the sample compartment or passing a visible beam from a collimated white light source or a He-Ne laser through the sample compartment. This light beam is used for the rough alignment of sampling accessories such as beam condensers or ATR units. Once the mirrors in these accessories have been approximately aligned using this visible light source, the infrared beam from the interferometer is switched back to the sample compartment. The interferometer is then scanned rapidly and repetitively over the zero retardation point and a final alignment of the mirrors of the sampling accessory is made by maximizing the interferogram signal. Without the initial rough alignment using the visible light source, it can be quite difficult to obtain a detectable interferogram from an accessory whose mirrors are completely out of alignment. With the visible light source,

J. R. Durig (ed.), Analytical Applications of FT-IR to Molecular and Biological Systems, 141–148.

alignment of accessories is almost as rapid as it is on a grating spectrometer.

Some FT-IR spectrometers do not have two-beam sampling optics, but rather use a single channel with the sample "in" and "out" to obtain transmittance spectra. The 100% line obtained using a single channel with consecutive measurements of the sample and reference single-beam spectra is usually flatter than 100% lines obtained from a two-channel spectrometer, provided that the instrument is well-purged to eliminate absorption features due to atmospheric water vapor and carbon dioxide.

One word of warning should be given about the use of sampling accessories in FT-IR spectrometers. Although one can read in many places that the wavenumber accuracy of laser-referenced interferometers is very high, this is not necessarily true, and it is actually the repeatability or precision which is high. A single-beam spectrum of a vignetted or misaligned beam can exhibit a wavenumber shift up to one half of the resolution (but usually less than this) relative to a well-aligned unvignetted beam. If these spectra are ratioed, absorption bands appearing in both the sample and reference spectra will generate derivative-shaped features in the transmittance spectrum. Hence when sampling accessories are used in FT-IR spectrometers, the best results are obtained if a reference single-beam spectrum is taken through the accessory rather than through the empty reference channel.

2. CONVENTIONAL ACCESSORIES

The main differences between the beam in an FT-IR spectrometer and the beam in a grating spectrometer is that the beam is round and is usually larger at its focus in an FT-IR spectrometer because of the high throughput of an FT-IR spectrometer (Jacquinot's advantage). For large samples (e.g. 13 mm KBr discs, 10 cm gas cells, etc.), there is no difference in the transmittance of the accessory whether it is held in an FT-IR or grating spectrometer. Some cells designed for use on dispersive spectrometers have a rectangular shape, conforming to the shape of the slit, and a small amount of energy may be lost in an FT-IR spectrometer by vignetting. Similarly some accessories, in particular accessories for the measurement of ATR spectra, have rectangular mirrors and optical elements, so again some energy may be lost by vignetting. It is possible to purchase accessories from some companies (e.g. the Harrick Scientific Corp., Ossining, NY) which are designed to minimize light losses on an FT-IR spectrometer.

The sampling optics of most commercial Fourier spectrometers are designed so that the f/ number of the beam in the sample compartment is similar to that for conventional grating spectrometers.

It is therefore often possible to use accessories designed for grating spectrometers at quite high efficiency on a Fourier spectrometer. For example, this author has obtained 80% transmittance with a Perkin-Elmer 4x reflecting beam condenser on a Digilab FTS-14 FT-IR spectrometer.

Much of the published work using Fourier spectrometry has been performed using conventional sampling accessories. However one can make use of the increased sensitivity of FT-IR spectrometers to expand the range of sampling techniques available to Fourier spectroscopists. Two techniques which fall into this category are emission spectrometry and diffuse reflectance spectrometry. In addition two other techniques - photoacoustic spectrometry and circular dichroism spectrometry - are just starting to appear on the horizon. These four techniques should open up new fields for study by Fourier spectrometry.

3. EMISSION SPECTROMETRY

The energy emitted from isothermal samples is given by the product of the energy emitted if the sample were a blackbody and the <u>emissivity</u> of the sample. For optically thin samples, the emissivity is given by:

$$\varepsilon(\nu) = 1 - T(\nu)$$

where $T(\nu)$ is the transmittance of the sample. Thus for gases and samples deposited on infrared transparent or reflective substrates, the emission spectrum will appear like an inverted transmittance spectrum with the relative intensities of the bands across the spectrum varying as the blackbody background.

Measurement of the emission spectra of samples at temperatures less than 200°C with a grating spectrometer is difficult because the energy emitted is so low. However these measurements are often fairly simple on a Fourier spectrometer, and several reports of emission spectroscopy of gases and surface species on metal surfaces have appeared. For the latter types of samples, the emission spectrum of the background, which typically has an emissivity of 0.02 to 0.05, must be subtracted from that of the sample to leave the spectrum of the surface film.

In order to obtain an interpretable spectrum, it is most important that the sample be optically thin and isothermal. If temperature gradients exist in the sample, self-absorption by cooler molecules will lead to band inversions and quantitative measurements become extremely difficult, if not impossible.

Several commercial FT-IR spectrometers have provision for a

simple emission attachment, which is typically a plane mirror in the source optics which enables radiation from a heated sample held externally to the spectrometer to be correctly collimated and passed to the interferometer. Some emission measurements are made simply by holding a heated gas cell or a solid sample heated on a hotplate directly in front of the plane mirror.

Remote measurement of infrared emission spectra requires the use of a telescope. We have found the best type of telescope to use is a modified Cassegrainian in which the beam from the primary mirror is focused before being collimated by the secondary. This focus is at a common focus of the primary and secondary mirrors of the telescope. An aperture held at the focus is used to control the solid angle of the beam through the interferometer, and its diameter determines the spectral resolution (see "Theory and Instrumentation" section).

The lower the energy emitted by the sample, the greater the amount of interference by radiation emitted by the surroundings, including the interferometer itself. For extremely low sample emission, measured using a sensitive helium-cooled bolometer, the energy from ambient temperature surroundings can swamp the energy from the sample and even cause the interferogram to become photon-noise limited. In this case, the interferometer and sample chamber must be cooled with liquid nitrogen to avoid this effect. Exceptionally impressive chemiluminescence spectra of molecular beams have been made in this way.

An excellent discussion of infrared emission spectrometry has been given by Bates [1].

4. DIFFUSE REFLECTANCE SPECTROMETRY

Most sampling techniques for solids require tedious sample preparation. Pressing samples with KBr into pellets can cause ion-exchange and changes in crystal structure. The use of mineral oil mulls or solutions causes certain spectral regions to become obscured by the oil or solvent absorption bands. Ideally one would like to sample powdered solids directly. For this purpose, three sampling techniques - diffuse transmission, diffuse reflectance and emission spectrometry - may be considered. Sample thickness must be very carefully controlled for diffuse transmission measurements, and even then a severely sloping baseline is always observed, eventually eliminating measurements at short wavelength. The application of emission spectrometry requires that the sample is heated, and hence they must be thermally stable. In addition, the samples have to be completely isothermal, since even small temperature gradients make the interpretation of emission spectra very difficult. Therefore diffuse reflectance measurements would appear to present the best opportunity for characterizing powders, but before the 1970's, an efficient device for collecting diffusely reflected infrared radiation had not been developed.

In 1964, Kortum [2] speculated that diffuse reflectance infrared could not be measured with an adequate signal-to-noise ratio (SNR) using an integrating sphere in view of the very low efficiency of these devices for infrared radiation. However in 1976, Willey [3] reported interfacing a slow-scanning interferometer with a gold-coated integrating sphere for diffuse reflectance measurements. Although the photometric accuracy of this instrument is excellent, rather long scan times and low spectral resolution are needed to obtain spectra with an adequate SNR. In 1978, Fuller and Griffiths [4] described an ellipsoidal reflectometer which had a much higher collection efficiency than an integrating sphere, and this design formed the basis for the dramatically increased interest in diffuse reflectance now being manifested by FT-IR spectroscopists.

There are now two basic types of diffuse reflectometer being used extensively. One involves a modification of the original design of Fuller and Griffiths, but using a paraboloidal collecting mirror rather than an ellipsoid. The other type involves a modification of a typical specular reflectance attachment using a considerably larger collecting mirror. This design has the advantage of fitting in the sample compartment of commercial FT-IR spectrometers, and accessories based on this principle are now commercially available from Harrick Scientific and Digilab. The advantage of the paraboloidal reflectometer of Fuller and Griffiths is that it may be used for infrared microsampling and is the basis for an interface between an FT-IR spectrometer and a high performance liquid chromatograph (see "Chromatography and FT-IR" section).

Although not enough is currently known about diffuse reflectance infrared spectrometry, its benefits and drawbacks will become better understood in the next two or three years. This author believes that diffuse reflectance FT-IR measurements on such diverse samples as foodstuffs, catalysts, pharmaceuticals and coal will lead to real advances in the chemical analysis of bulk and surface species.

5. PHOTO-ACOUSTIC SPECTROMETRY

One of the biggest disadvantages to the measurement of the infrared spectra of bulk samples by diffuse reflectance is that the specular component of the reflected radiation (which does not penetrate into the sample) is measured along with the component which penetrates into, and subsequently re-emerges from, the sample. For most wavelengths, the refractive index of organic solids does not change dramatically, so that the specular (Fresnel) reflectance is approximately constant. However in the region of strong absorption bands, anomalous dispersion occurs leading to restrahlen bands in the specular reflectance spectrum. If the diffuse reflectance spectra of neat samples which exhibit restrahlen bands are measured, the absorption bands appear to be inverted at their center. The larger is the size of the particles being measured, the greater is the specular component of the total reflectance, and even for fairly weakly absorbing organic compounds with a particle diameter greater than 0.5 mm, the measured spectrum has the appearance of the specular reflectance spectrum rather than a diffuse reflectance spectrum. This effect makes quantitative measurements on neat samples with a large absorptivity or chunky samples very difficult, and an alternative technique would appear to be beneficial for such samples.

At the time of writing this section, no viable alternatives have appeared in print. However this author has seen some preliminary photoacoustic spectra of bulk solids measured by the personnel of the Nicolet Instrument Corporation which look quite promising. Although the SNR of photoacoustic spectra will probably never be as good as diffuse reflectance spectra of the same samples, it is very possible that photoacoustic FT-IR spectrometry could be a powerful alternative to diffuse reflectance when the specular reflectance component causes problems in the qualitative or quantitative interpretation of diffuse reflectance spectra.

6. VIBRATIONAL CIRCULAR DICHROISM

Ultraviolet circular dichroism (UV-CD) spectrometry is an important tool in the study of conformational changes in biopolymers. However, electronic spectra give no information on interactions occurring at sites which are not ultraviolet chromophores. Thus most UV-CD spectra are relatively simple with only two or three significant features. If vibrational circular dichroism (VCD) spectra of these samples could be easily measured, the information content of the spectrum should be much greater than that of the UV-CD spectrum. The main problem concerning the measurement of VCD spectra is the fact that the difference in the absorbance of samples to left and right circularly polarized infrared radiation per unit absorbance is about two orders of magnitude less than that for UV radiation. The measurement of VCD spectra below 2000 cm^{-1} is made even more difficult by the poor sensitivity of infrared detectors in this region.

In spite of these drawbacks, Nafie et al. [5] are developing what appears to be a viable method for the measurement of VCD spectra using a FT-IR spectrometer. The linearly polarized beam emerging from a continuously scanning interferometer with a low mirror velocity is passed through a photoelastic modulator which causes the radiation to alternate between left and right circular polarization at a rate of 50 kHz. The sample is held directly behind the modulator and the transmitted radiation is then passed onto either an InSb detector (for $\nu > 2000$ cm^{-1}) or an MCT detector (for $\nu > 2000$ cm^{-1}). The signal is amplified using a lock-in amplifier tuned to 50 kHz, so that only the VCD information is measured.

At this time, VCD spectra measured interferometrically using an InSb detector are only slightly better than VCD spectra measured using a grating spectrometer with the same type of detector. However there are probably several ways in which the FT-IR measurements may be improved, while much smaller improvements can be expected in the dispersive measurements. For VCD spectra below 2000^{-1}, it is probable that data will shortly be obtained using an FT-IR spectrometer in conjunction with an MCT detector, and VCD spectra are extremely difficult to obtain in this region using a grating spectrometer.

7. OPTICAL SUBTRACTION FT-IR SPECTROMETRY

One of the biggest problems with the measurement of weak absorption bands by FT-IR spectrometry is that detection limits cannot be decreased indefinitely by using progressively more sensitive detectors since the SNR becomes limited by the dynamic range of the analog-to-digital converter (ADC).

A technique has been described in which the sensitivity of an FT-IR spectrometer has been increased for weak absorption bands by using both the transmitted and reflected beams from the interferometer (see "Theory and Instrumentation" section). Since these two beams are 180° out-of-phase, the modulated components of each interferogram will cancel out if both beams are passed through equivalent paths to the same detector. When the measurement is made using a rapid-scanning interferometer with an a.c.-coupled amplifier, the d.c. component of the signal will never be observed and the a.c. component will be nulled out. If an absorbing sample is held in one beam and not in the other, the a.c. interferogram will only contain information on the sample absorption (which is weak) and not on the source emission (which is strong). Hence the SNR of the interferogram will be much lower than if the absorption spectrum were measured by conventional single-beam FT-IR techniques, and very sensitive detectors can be used without the SNR of the interferogram being limited by the dynamic range of the ADC [6].

This dual-beam (or optical subtraction) FT-IR technique has been applied by several workers to the measurement of GC-IR spectra and also for the characterization of surface species. Although dual-beam FT-IR spectrometry appears to give several advantages for the measurement of the spectra of weakly absorbing samples in theory, there is one major practical disadvantage to the technique in that it requires the use of special optics and cannot be performed using a simple accessory on a commercial FT-IR spectrometer.

8. SUMMARY

In conclusion, it can be said that most chemical measurements using an FT-IR spectrometer are still performed using conventional infrared sampling accessories and techniques. However several special techniques have been described for measurements which are difficult or impossible using conventional methods.

REFERENCES

1. J. B. Bates, in Fourier Transform Infrared Spectroscopy: Applications to Chemical Systems, (J. R. Ferraro and L. J. Basile, eds.), pp. 99-142, Academic Press, New York (1978).

2. G. Kortum and H. Delfs, Spectrochim. Acta, 20, 405 (1964).

3. R. R. Willey, Appl. Spectrosc., 30, 593 (1976).

4. M. P. Fuller and P. R. Griffiths, Anal. Chem., 50, 1906 (1978).

5. L. A. Nafie and M. Diem, Appl. Spectrosc., 33, 130 (1979).

6. D. Kuehl and P. R. Griffiths, Anal. Chem., 50, 418 (1978).

CHROMATOGRAPHY AND FT-IR SPECTROMETRY

Peter R. Griffiths

Department of Chemistry
Ohio University
Athens, Ohio, U. S. A.

1. GAS CHROMATOGRAPHY AND FT-IR SPECTROMETRY

The fact that is is possible to acquire an interferogram in less than one second makes FT-IR spectrometry an excellent technique for studying transient species. By far the most important application of this rapid-scanning property is the on-line identification of peaks eluting from a gas chromatograph (GC/FT-IR). GC/FT-IR measurements using low resolution (20 cm^{-1}) interferometers were described as early as 1967 [1], but only recently have GC/FT-IR measurements combining both adequate resolution (4 cm^{-1}), low detection limits (less than 1 μg of sample injected into chromatograph), and sophisticated data handling become routine.

In most GC/FT-IR measurements, the effluent from a GC column is continuously passed through a light-pipe gas-cell, i.e. a long, narrow gas cell with highly reflective walls. The dimensions of the light-pipe are critical for optimizing the sensitivity of GC/FT-IR measurements. The volume of the cell should be about equal to, or slightly less than, the volume of carrier gas containing each peak between its half-height points. A larger cell volume may permit two closely resolved GC peaks to be simultaneously present in the cell, while a smaller volume causes only a small fraction of the total sample to be present in the cell at any moment during the measurement. The optimum length and diameter then depend on the specific detectivity, D*, of the infrared detector. If a low sensitivity detector, such as a TGS pyroelectric bolometer, is used, the transmittance of the light-pipe should be as large as possible. On the other hand, if a short-range MCT photoconductive detector is used, the D* is so great

J. R. Durig (ed.), Analytical Applications of FT-IR to Molecular and Biological Systems, 149–155.

that the transmittance of the light-pipe should be kept below 10% to avoid limiting the sensitivity of the measurement by digitization noise. These calculations have been described in detail by Griffiths [2].

For chromatography carried out using 2 mm i.d. packed columns, the optimum cell volume is approximately 3 ml. Of the commercially available GC/FT-IR light-pipes, Digilab's is 2.2 ml in volume (2 mm in diameter, 70 cm long) while Nicolet's is slightly larger, wider, and shorter than this. No commercial GC/FT-IR system has yet been introduced for chromatography performed using wall-coated open-tubular capillary columns, for which a light-pipe approximately 100 μl in volume is required. Work to develop such a system is currently taking place in several laboratories, and it is almost certain that several systems for capillary GC/FT-IR will be described this year.

Several methods of making light-pipes have been investigated, but the most successful light-pipes are constructed by the method developed by Azarraga at the E.P.A. Environmental Research Laboratory at Athens, Georgia. In this method a quartz capillary tube is carefully cleaned and a layer of gold ceramic paint is deposited by pouring the diluted paint down the tube. The paint is then thermally decomposed leaving a smooth coherent layer of gold on the interior surface of the quartz tube. The reflectance of these tubes decreases as the temperature increases; for light-pipes with dimensions similar to those made by Digilab and Nicolet, the transmittance is approximately halved on raising the temperature from ambient to 200°C.

The main advances in GC/FT-IR over the past two years have been in the area of data handling. Until very recently, the time required to transform a 4K-point interferogram was much longer than the time needed to acquire each interferogram. Thus interferograms, rather than spectra, are always stored, and they are sequentially transformed at the end of the chromatogram. Several options are available for data acquisition in GC/FT-IR. For example, individual interferograms, or the average of several interferograms taken over a one or two second span, may be stored sequentially on either disc or magnetic tape. Since even a 4M-word disc can be filled very quickly at the rate of 4K-points every second, some way of sorting out these data to determine which interferograms should be stored and which discarded is necessary, and several means of achieving this end are feasible.

In one of these, low resolution (32 cm^{-1}) spectra are calculated for each individual interferogram or set of interferograms using only the first 512 points. The integrated absorption is then calculated over several specified wavenumber regions in which some functional group has a characteristic absorption [3,4].

These values are then plotted out in real-time to give "Chemigrams" (Nicolet) or "Infragrams" (Digilab). Only when the absorption in one of the specified wavenumber regions exceeds a certain threshold is the full interferogram stored. Of course it is possible that the sample may not have an absorption band in any of the specified regions, and a measure of the total absorption across the spectrum is needed to provide a better indication of which interferograms should be stored. Perhaps the best method involves the Gram-Schmidt vector othogonalization procedure on short (100 point) regions of the interferogram [5,6]. The chromatogram produced using this procedure is analogous to a total ion current chromatogram in GC/MS systems, and gives a very good measure of which interferograms should be stored during the chromatography for transformation at the end. When a peak is indicated as being present in the light-pipe, two methods of storing data are possible: either individual interferograms, or sets of interferograms, can be stored when the threshold absorption is exceeded or interferograms can be signal-averaged during this interval. The latter technique provides the most efficient usage of the available memory and is probably the best way of handling the data in high resolution chromatography. The former method is better for low resolution chromatography when peaks are less likely to be well resolved. In this case, two or more components may elute before the infrared absorption returns below the threshold level, so that spectral subtraction techniques may be required for the identification of each component.

A recent innovation in digital electronics will probably lead to another advance in the handling of GC/FT-IR data. Digilab has just introduced a microprogrammable processor which will enable a 4K-word double-precision interferogram to be transformed in less than one second. Thus instead of storing interferograms, it should be possible to compute GC/FT-IR spectra in real-time and obviate the need for transforming each interferogram at the end of the chromatogram. Indeed it is possible to speculate that data acquisition and the Fourier transform will all be done in background leaving the operator free to observe and manipulate any of the stored spectra in a similar manner to that now being used in advanced GC/MS data systems.

Most GC/FT-IR systems use a single-beam optical system in which the beam from the interferometer is focused by a fast aspherical mirror onto the entrance of the light-pipe, with the emerging beam being focused on the detector using at least one more off-axis aspherical mirror. Detection limits for moderate infrared absorbers are typically on the order of 100 ng of injected sample (although as little as 10 ng have been detected without trapping the sample in the light-pipe). It is possible that even lower detection limits for GC/FT-IR may be obtained using an optical subtraction technique (see "Accessories and Sampling

Techniques" Chapter 20). Detection limits about a factor of four lower than single-beam methods may be obtained in this way [7], but the optical system required for optical subtraction FT-IR is so much more complicated than the optics of conventional single-beam GC/FT-IR systems that it is unlikely that optical subtraction will be used by more than just a few people for GC/FT-IR.

2. LIQUID CHROMATOGRAPHY AND FT-IR SPECTROMETRY

Although it might be thought that the interface between a high performance liquid chromatograph (HPLC) and an FT-IR spectrometer would be similar to GC/FT-IR systems (but using a flow-cell rather than a light-pipe), there are some very significant differences. The most important differences are caused by the absorbing nature of the mobile phase. In order to maintain a sufficiently high transmittance to allow solute absorption bands to be observable at most wavelengths across the spectrum, the pathlength of the flow-cell must be kept short, typically about 100 μm. Since it is unnecessary to have a cell with a larger diameter than about 4 mm, it can be easily seen that no more than 2 μl of the eluting solvent is present in the flow-cell at any moment. It is rare that any HPLC peak eluting from conventional packed columns has a volume less than 250 μl, and it is therefore apparent that HPLC/FT-IR measurements will have detection limits about two orders of magnitude greater than those found in GC/FT-IR in those spectral regions where the solvent has a high transmittance, and even worse than this where the solvent absorbs more than half the incident radiation. If the pathlength of the flow-cell is increased, the detection limits may be increased in the spectral windows, but the proportion of the spectrum lost through absorption by the solvent is increased.

In spite of these disadvantages, several systems for this type of HPLC/FT-IR measurement have been described [8-10]. Undoubtly the most successful have not involved absorption or partition chromatography, but rather steric exclusion chromatography (SEC). There are several reasons for this. The column can be loaded with a greater quantity of solute in SEC than adsorption chromatography before the chromatographic resolution is degraded, so that up to 1 ml of the analyte solution has been injected onto SEC columns in some reported HPLC/FT-IR measurements. Also the choice of the solvent used in SEC is far less critical than in adsorption chromatography. To a great extent the choice of the mobile phase in SEC is based on the solubility of the sample rather than its effect on the retention of the sample. Hence it is often possible to choose a solvent for SEC/FT-IR whose infrared spectrum has large spectral windows. Conversely it is often rather difficult to use flow-cells for HPLC/FT-IR measurements of samples separated by adsorption or partition chromatography, where the choice of solvent is critical to the separation and gradient elution techniques often have to be applied. In the latter case, the composition of the mobile phase changes continuously throughout the run, and compensation for the infrared absorption of the solvent using scaled spectral subtraction routines is difficult (and usually impossible).

Ideally what we would like to do for HPLC/FT-IR is to remove the solvent automatically prior to the measurement of the infrared spectrum of the solute. The optimum conditions for infrared microsampling require that the sample be deposited in a small area (ideally about 3 mm in diameter for FT-IR). Thus the possibility of using a moving ribbon technique similar to the MacFadden HPLC/MS interface [11] is removed, since the solute would be deposited over a very wide area. We have investigated the feasibility of depositing the eluent from an HPLC column onto KCl powder, after preliminary concentration by partial evaporation. The diffuse reflectance spectrum of the deposited solute is then measured. This technique appears to be particularly promising in light of the high sensitivity of diffuse reflectance for microsampling compared with other conventional infrared microsampling techniques for solids [12].

Preliminary experiments [13,14] have demonstrated the feasibility of the technique. In our early work, a wheel with thirty-one sample cups on the perimeter was used. The KCl powder in each sample cup was supported on a fine mesh so that air can be drawn through the cup. After the preliminary concentration step, each peak indicated by the HPLC detector is deposited on a different sample cup, with the deposition being controlled using a microcomputer. When the first peak has been deposited, the wheel is rotated so that the cup on which this sample has been deposited is moved into a position where a slow stream of air is drawn through it, thereby eliminating any remaining solvent. The cup is then held in this position until the second peak has been deposited, at which time the wheel is rotated again and the process is repeated. After three more rotations of the wheel, the cup containing the first peak is positioned at the sample focus of a diffuse reflectance accessory of an FT-IR spectrometer, and interferograms are signal-averaged until the wheel again rotates to bring the next peak into the beam.

The detection limits obtained using this technique are at least two orders of magnitude better than those using a flow-cell for HPLC/FT-IR, with typical detection limits of approximately 100 ng of each injected component. However it should not be thought that this system represents anything more than a prototype HPLC/FT-IR system. The biggest drawback to this system is the fact each component separated by the chromatograph must give a response at the HPLC detector in order to be collected. For example, if a 254 nm UV detector is used, nonaromatic compounds will probably be missed. Therefore an analogous system allowing continuous deposition of every component must be developed. Hopefully such a system will be reported this year.

3. SUMMARY

GC/FT-IR is now at the point that it can be used routinely. The major manufacturers of FT-IR spectrometers are both developing spectral search systems for the automatic interpretation of GC/FT-IR spectra which will further enhance the power of this technique. The next advances should also be found in the development of fast processors for GC/FT-IR data reduction and in the construction of very small light-pipes for monitoring GC peaks separated using wall-coated open-tublar columns.

HPLC/FT-IR is by no means as advanced as GC/FT-IR, but methods for eliminating the solvent prior to the measurement of the spectrum of each peak are being developed. The next couple of years should see the introduction of a commercial HPLC/FT-IR system based on this technology.

REFERENCES

1. M. J. D. Low and S. K. Freeman, Anal. Chem., 39, 194 (1967).

2. P. R. Griffiths, Appl. Spectrosc., 31, 284 (1977).

3. P. Coffey, D. R. Mattson, and J. C. Wright, Amer. Lab., 10 (5), 126 (1978).

4. K. Krishman, R. Curbelo, P. Chiha, and R. C. Noonan, J. Chromatogr. Sci., 17, 413 (1979).

5. J. A. de Haseth and T. L. Isenhour, Anal. Chem., 47, 1977 (1977).

6. D. A. Hanna, G. Hangac, B. A. Hohne, G. W. Small, R. C. Wieboldt, and T. L. Isenhour, J. Chromatogr. Sci., 17, 423 (1979).

7. G. J. Kemeny, D. Kuehl, and P. R. Griffiths, Appl. Spectrosc., in press (1980).

8. K. L. Kizer, A. W. Mantz, and L. C. Bonar, Amer. Lab., 7, 85 (1975).

9. D. W. Vidrine and D. R. Mattson, Appl. Spectrosc., 32, 502 (1978).

10. D. W. Vidrine, J. Chromatogr. Sci., 17, 477 (1979).

11. W. H. McFadden, H. L. Schwartz, and S. Evans, J. Chromatogr., 122, 389 (1976).

12. M. P. Fuller and P. R. Griffiths, Appl. Spectrosc., submitted for publication (1980).

13. D. Kuehl and P. R. Griffiths, J. Chromatogr. Sci., 17, 471 (1979).

14. D. Kuehl and P. R. Griffiths, Anal. Chem., in press (1980).

FT-IR MATRIX ISOLATION STUDIES

A. J. Barnes and W. J. Orville-Thomas

Department of Chemistry
University of Salford
Salford, England, U. K.

1. INTRODUCTION

Matrix isolation is a technique for trapping isolated molecules of the species to be studied in a large excess of an inert material by rapid condensation at a low temperature, so that the diluent forms a rigid cage or matrix. At a sufficiently low temperature, diffusion of the solute species is prevented and thus reactive species may be stabilized for leisurely spectroscopic examination. The noble gases (primarily argon) and nitrogen are the most widely used matrix materials. Infrared, Raman, visible-ultraviolet, electron spin resonance, Mössbauer and magnetic circular dichroism spectroscopies have all been used in conjunction with the matrix isolation technique.

Infrared spectroscopy of matrix-isolated substances was first reported by Whittle, Dows, and Pimentel [1] in 1954. The early development of matrix isolation was directed principally to the study of unstable molecules and free radicals [2]. The ability to stabilize reactive species by trapping them in a rigid cage, thus inhibiting intermolecular interaction (apart from the low temperature itself, which effectively prevents the occurence of any process with an appreciable activation energy) is an important feature of matrix isolation. A logical extension to the range of reactive species studied in matrices was the application to high temperature vapors, first reported by Linevsky [3] in 1961. He condensed a molecular beam of lithium fluoride from a Knudsen furnace simultaneously with a large excess of noble gas or nitrogen diluent. Vapor phase studies of such materials have to be carried out at high temperatures, resulting in "hot bands" and the population of high rotational energy levels with consequent interpretive problems, apart from ex-

J. R. Durig (ed.), Analytical Applications of FT-IR to Molecular and Biological Systems, 157–170.

perimental difficulties such as corrosion of the cell. As a result of the obvious advantage of matrix isolation in the spectroscopic study of unstable species and vaporizing molecules, an enormous amount of work has been done in these fields.

Apart from the stabilization of reactive species, matrix isolation affords a number of other advantages over more conventional spectroscopic techniques. The isolation of monomeric solute molecules in an inert environment reduces intermolecular interactions, resulting in a sharpening of the solute absorptions compared with other condensed phases - the effect being particularly dramatic, of course, for hydrogen bonding substances. With the exception of a few small molecules (for example the hydrogen halides, water, ammonia, methane) rotation does not occur in matrices, thus giving much narrower bands than are obtained in the vapor phase. Also the low temperature itself results in small band widths. Consequently near-degenerate bands, which overlap completely even in the vapor phase or in dilute solution at room temperature, may often be resolved in matrix spectra. This has important applications in the spectroscopy of stable molecules, as well as unstable species. The resolution of near-degenerate fundamentals allows vibrational assignments to be made with greater confidence and the frequencies to be obtained more accurately. Matrix isolation has great potential in studying conformational isomerism, for example ethanol in an argon matrix shows resolved absorptions due to trans and gauche conformers for a number of its fundamental modes, whereas there is no definite evidence from the infrared spectrum of any other phase for the existence of the two conformational isomers [4].

Although the matrix isolation technique was developed to inhibit intermolecular interaction, two classic papers by Van Thiel, Becker, and Pimentel [5] on water and methanol demonstrated its value in studying hydrogen bonding interactions. Bands due to monomer, dimer, trimer, tetramer and higher multimers could be identified as the concentration of solute was increased. The matrix technique also provides a convenient means of studying the intermolecular interactions between pairs of unlike molecules in an inert environment.

The earliest reported application of Fourier Transform infrared (FT-IR) spectroscopy combined with matrix isolation was a far infrared investigation of the rotation of the hydrogen halides by Barnes et al. [6] in 1969. Despite the advances in technology which have made available FT-IR spectrometers covering the entire infrared range, to date only a small number of matrix studies have used FT-IR spectroscopy. This chapter will describe some of these studies and also mention other areas where the application of FT-IR may provide significant advantages over conventional infrared spectroscopy. Green and Reedy [7] have given a comprehensive review of matrix isolation studies with FT-IR to August 1976, and also consider in some detail the advantages and disadvantages of FT-IR spectroscopy in relation to matrix isolation work. The technique and application of matrix isolation infrared spectroscopy are covered in depth in the book by Hallam [8].

2. EXPERIMENTAL TECHNIQUES

A. Cryostat

The sample generally needs to be cooled to a temperature in the range 4 to 20 K for matrix isolation. The simplest method of attaining such temperatures is by the use of liquid hydrogen (20 K) or liquid helium (4 K) as the coolant in a double Dewar cryostat, the outer Dewar being filled with liquid nitrogen to minimize heat leakage. This type of cryostat was used for all early matrix isolation studies, and is still widely used where there is a readily available source of liquid refrigerant.

In the 1960's miniature open-cycle cryostats, operating on the Joule-Thomson principle, became commercially available. Hydrogen gas is cooled below its inversion temperature by liquid nitrogen and then expanded through a nozzle. The Joule-Thomson cooling effect leads to eventual liquefaction of the hydrogen giving a base temperature of 20 K. Addition of a further cooling stage using helium gas gives a base temperature of 4 K (pumping on the liquid lowers the base temperature slightly). Although the cooling power of those cryostats is much lower than the double Dewar type, the temperature may be more readily controlled over the range 4 to 300 K.

An alternative method of refrigeration is the closed cycle system, using helium as the working fluid in some variant of the Stirling refrigeration cycle. A two stage system can attain a base temperature of ca. 10 K, although the cooling power decreases with decreasing temperature. Again temperature control is readily achieved by balancing the refrigerator with a heater. Despite a somewhat higher capital cost, this type of cryostat has much lower running costs and is easy to set up and operate. The commercial availability of such miniature cryostats has led to widespread use of the matrix isolation technique in recent years.

B. Vacuum Shroud

The partial pressure of solute gas being deposited on the cold window in a matrix isolation experiment is often very low, thus it is vital to have an efficient pumping system to prevent impurities from influencing the spectra (as well as minimizing heat loss). An oil or mercury diffusion pump, backed by a rotary pump, and a liquid nitrogen vapor trap are normally used. The infrared beam

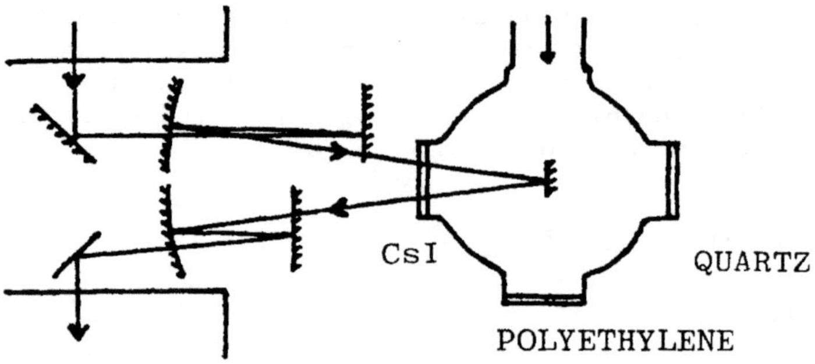

Fig. 1 Top view of beam projecting assembly and vacuum shroud for matrix spectroscopy using a cold mirror substrate (adapted from Ref. [9] with permission).

of the spectrometer has to pass through a pair of outer windows in the vacuum shroud and the cold window on which the sample is deposited. Alkali halide windows are used in the mid infrared, while in the far infrared region high density polyethylene outer windows and a silicon deposition window (because of the poor thermal conductivity of polyethylene) are required. The silicon window needs to be slightly wedged to avoid interference fringes.

It is readily apparent that the wide spectral range available with FT-IR is of limited value if it is necessary to change window materials to examine both the mid and far infrared spectra. Reedy and Gilby [9] describe a method of overcoming this problem using a mirror assembly to project the infrared beam of an FT-IR spectrometer to a focus outside the spectrometer. The matrix sample is deposited on a mirror located at the focus and the appropriate vacuum shroud window can be rotated into the beam according to the spectral range being examined (Fig.1).

C. Sample Deposition

The simplest form of the matrix isolation experiment is to mix the solute and the matrix gases in the desired proportions in a vacuum line using standard manometric techniques. The concentration is usually expressed as the matrix:solute ratio (M/S). The gas mixture is then sprayed, through a needle valve to control the flow rate, onto the deposition substrate which is turned at right angles to the gas inlet. This deposition technique is known as "slow spray on", flow rates of between 1 and 15 mmol h^{-1} of gas mixture being used. After a sufficient quantity of the gas mixture has been deposited, the needle valve is closed and the cryostat rotated in the vacuum shroud to bring the deposition substrate in line with the outer windows of the vacuum shroud for spectral scanning.

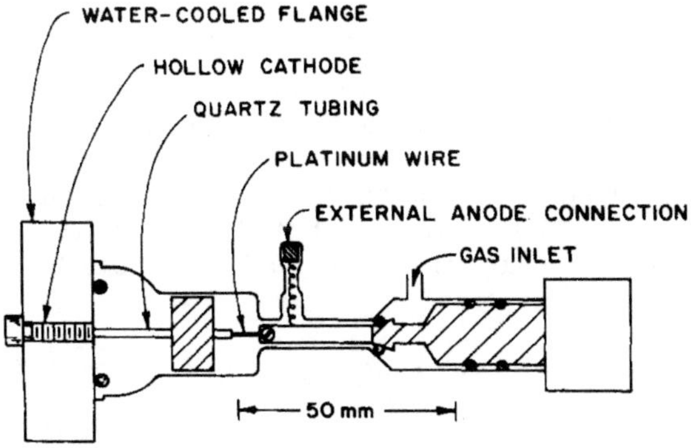

Fig. 2 Sputtering device for the production of high temperature atoms and molecules (reproduced with permission from Ref.[7]).

An alternative technique is pulsed deposition, in which a series of small quantities of the gas mixture is deposited in rapid succession. The heat released by each pulse anneals the matrix, giving samples which scatter light less than those deposited by the slow spray on technique. A comparative study [10] of the two methods showed that pulsed deposition does not lead to poorer isolation of the solute molecules.

Volatile liquids and solids (e.g. iodine) may be handled in the same way as gases, but involatile materials require a rather different technique. The solid solute is heated to an appropriate temperature in some type of furnace to obtain a beam of vapor directed at the deposition substrate, and simultaneously the matrix gas is sprayed on the cold surface. Organic materials may be evaporated in a simple glass tube wrapped with heating wire, while metals and other inorganic species required Knudsen cell effusion techniques. Moskovits and Ozin [11] have described a method of determining the solute in such a matrix by weighing the material effusing in the opposite direction to the deposition substrate (through a similar geometry of radiation shield collimators) on a quartz microbalance. Involatile materials have also been trapped in matrices from heated wires or filaments and by the use of sputtering techniques (Fig. 2).

Unstable molecules, free radicals, and ions may be generated by a wide variety of techniques, either in the gas phase immediately prior to deposition or in situ after deposition. Examples of a number of methods of generation are given in Table 1.

Table 1 Methods of Generating Unstable Species

1. In situ photolysis of a single precursor, e.g. CH_3 from the vacuum ultraviolet photolysis of CH_4 in argon.

2. Atom-molecule reactions, with the atom produced by in situ photolysis, e.g. HCO from the photolysis of HI + CO in argon or from the photolysis of HI in carbon monoxide (reactive matrix technique).

3. Metal atom-molecule reactions, e.g. CCl_3 from $CHCl_3$ + Li in argon.

4. In situ proton radiolysis, e.g. CCl_3^+ from proton radiolysis of CCl_4 in argon.

5. Gas phase pyrolysis prior to deposition, e.g. CH_3 from CH_3I.

6. Gas phase microwave discharge prior to deposition, e.g. O_3 from O_2 in argon.

3. MATRIX EFFECTS

An understanding of the various effects that the matrix can have on the infrared spectrum of the solute is vital, since the spectra of many of the species studied by the matrix isolation technique are previously unknown or markedly different from these in other phases (i.e. monomer instead of polymer). Neglect of the possibilities of matrix effects has led to a number of examples of misinterpretation of spectra in the matrix isolation literature.

The most obvious matrix effect is that the vibrational levels of the solute molecule will be perturbed by the matrix and thus the observed infrared frequencies will be shifted from their gas phase values. Additionally, multiplet band structure can arise from rotation of the solute in the matrix cage, alternative trapping sites in the matrix, aggregation of the solute or from lifting of the degeneracy of vibrational transitions. Inactive modes of the solute may be induced by the matrix environment and similarly inactive matrix lattice modes may be induced by the presence of the solute molecules.

A. Matrix Shift

Infrared bands of solutes trapped in matrices exhibit matrix shifts just as they exhibit solvent shifts in room temperature solution spectra. The same factors - electrostatic, inductive, dispersive, and specific interactions - contribute to both, but the rigidity of the cage means that repulsion forces also influence matrix shifts. Thus stretching modes usually shift to lower frequency, but bending modes often shift to higher frequency particularly in matrices (such as nitrogen) with symmetrical trapping sites. The shifts are normally quite small in inert matrices such as the noble gases or nitrogen, but a strongly interacting matrix (e.g. HCl in C_2H_4) or a highly polar solute (e.g. LiF) can give rise to large frequency shifts.

B. Rotation

A number of small molecules have been shown to rotate in noble gas matrices - hydrogen, hydrogen halides, water, ammonia, and methane. Restricted rotation has been postulated in a number of other cases e.g., sulphur dioxide or fluoromethane in krypton matrices, but these are less well established. Additionally several small free radical species, e.g. NH_2, exhibit rotational structure in argon matrices. In general rotation seems not to occur in matrices such as nitrogen with unsymmetrical trapping sites, although there are exceptions (e.g. hydrogen). Molecules trapped in low temperature matrices may show multiplet band structure for a variety of reasons. The characteristic feature of rotational structure on an infrared band is that it exhibits reversible temperature dependence, due to changes in the population of rotational energy bands as the temperature is raised or lowered. Confirmation that the small molecules listed above rotate in noble gas matrices has been provided by observation of the pure rotational transition in the far infrared or Raman spectra. In two such studies, FT-IR spectrometers were used: Barnes et al. [6] observed the $J = 1 \leftarrow 0$ pure rotational transitions of HCl and DCl in argon matrices at 18.6 and 10.9 cm^{-1} respectively, while Cugley and Pullin [12] observed the $J = 1 \leftarrow 0$ and $J = 2 \leftarrow 1$ transitions of NH_3 at 18.0 and 33.5 cm^{-1} respectively, and the $J = 2 \leftarrow 1$ transition of ND_3 at 19.75 cm^{-1}.

C. Multiple Trapping Sites

It is generally assumed that solute molecules occupy a substitutional site formed by removal of one or more matrix atoms or molecules. A very small solute, such as HF, could conceivably fit into an octahedral interstitial site, which has a diameter less than 2Å in argon or nitrogen. Other possible trapping sites are dislocation sites and grain boundaries, since a solid grown by rapid condensation from the vapor is unlikely to be ideally crystalline. However, the infrared bands of matrix isolated species are normally very sharp, so the trapping site must be well defined.

Where a solute molecule can be trapped in more than one type of site (for example, in substitutional sites formed by removal of different numbers of matrix atoms or molecules) each vibrational mode will generally exhibit more than one band since the interaction with the matrix environment will be different for each of the trapping sites. Multiplet structure can, however, arise from other causes and careful concentration dependence and temperature cycling studies are necessary positively to identify multiple trapping site effects. The relative intensity of the bands is usually unaffected by changes in solute concentration but changed irreversibly by annealing the matrix at higher temperatures for a few minutes or by varying the deposition conditions.

D. Aggregation

A group of two or more solute molecules will give rise to a frequency slightly different from the monomer even when the interaction is weak. Solutes which exhibit strong intermolecular interactions, for example hydrogen bonding substances, may give very large frequency shifts on aggregation. The effects of aggregation may be identified by varying the concentration of the solute in the matrix and eliminated by reducing the concentration sufficiently (unless appreciable dimerization occurs in the vapor phase, as for the carboxylic acids).

There are many examples in the matrix isolation literature of aggregate bands being incorrectly attributed to the monomeric solute, thus it is vital always to check for concentration dependence of matrix spectra. A related problem is the possibility that impurities may have influenced the observed spectrum - both nitrogen and water interact strongly with many solutes and great care should be taken to check for the possibility of leaks in the vacuum system.

4. APPLICATIONS

A. Unstable Molecules, Free Radicals and Ions

The matrix isolation technique has been extensively employed to study unstable species; recent reviews [13,14] show the range of such work. In principle, FT-IR should afford significant advantages in this area over conventional infrared spectroscopy: the better signal-to-noise ratio should enable transient species to be detected at lower concentrations, the wide spectral range should enable all vibrational modes of transient species to be observed, and computer subtraction of spectra should assist in identifying weak bands due to products of photolysis or reaction in the matrix. To date, however, few studies of unstable species have used FT-IR. Ault et al. [15] used an FT-IR spectrometer to search, unsuccessfully, for the vibrational fundamental of XeF, whose absorption and emission spectra in the visible-ultraviolet region had previously been observed in matrices [16]. Jacox and Milligan [17]

Table 2. Metal Nitrides and Oxides Studied by FT-IR Spectroscopy in Argon Matrices

Molecule	Mode	$\bar{\nu}$(cm^{-1})	Geometry	Ref.
FeO	ν	873.08	linear	[22]
CoO	ν	846.44	linear	[22]
NiO	ν	789.42	linear	[22]
CeO_2	ν_1	757.05	bent(146°)	[23]
	ν_3	736.80		
PrO	ν	817.10	linear	[23]
PrO_2	ν_3	730.45	linear	[23]
EuO	ν	668.00	linear	[23]
TbO	ν	824.25	linear	[23]
TbO_2	ν_1	758.75	bent(121°)	[23]
	ν_3	718.85		
ThN	ν	934.61	linear	[24]
ThO	ν	(880.00)[a]	linear	[23]
		(876.75)		
ThO_2	ν_1	787.40	bent(122.5°)	[23]
	ν_3	735.35		
UN	ν	<u>(1000.97)</u>[a]	linear	[25]
		(995.52)		
		(991.90)		
UN_2	ν_3	1050.9	linear	[25]
UO	ν	820.0	linear	[26]
UO_2	ν_3	776.10	linear	[26]
UO_3	ν_2	745.65	T-shaped	[26]
	ν_4	852.60		
PuN	ν	855.73	linear	[27]
PuN_2	ν_3	1029.74	linear	[27]
PuO	ν	(824.71)[a]	linear	[28]
		<u>(822.28)</u>		
		(818.81)		
		(816.04)		
PuO_2	ν_3	794.25	linear	[28]

[a]Multiple trapping sites (most stable site underlined).

used an FT-IR spectrometer to cover the far infrared region in a study of the vacuum ultraviolet photolysis of methylacetylene and allene; various C_3H_n species were found to be produced. FT-IR spectroscopy was used to obtain far infrared spectra of $GeBr_2$ in an argon matrix, produced both by photolysis of GeH_2Br_2 with a medium pressure mercury lamp and by trapping the vapors from solid $GeBr_2$ [18,19]. Green et al. [20] obtained spectra of $UO_2^+NO^-$ and $UO_2^+NO_2^-$ ion pairs produced by co-condensing UO or UO_2 with NO_2 in an argon matrix.

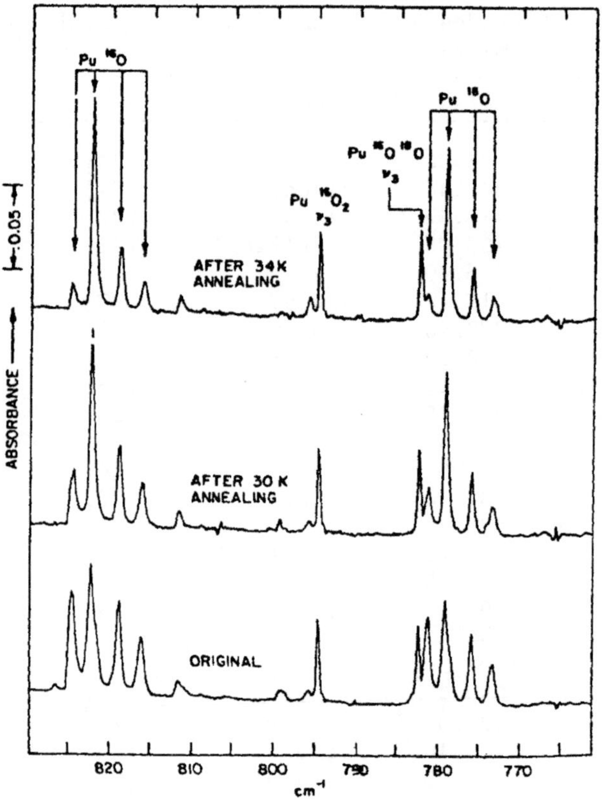

Fig. 3 Infrared spectra of an argon matrix containing PuO and PuO_2 recorded after deposition and after two successive annealing experiments (reproduced from Ref.[28] with permission).

B. High Temperature Species

A wide variety of high temperature species has been studied by the matrix isolation technique, in many cases with a view to evaluating thermodynamic properties of the molecules [21]. FT-IR spectroscopy has been used to study a number of transition metal oxides, lanthanide oxides and actinide nitrides and oxides trapped in argon matrices (Table 2).

It is instructive to consider one of these studies in some detail. Plutonium oxide species [28] were trapped in an argon matrix by passing an O_2/Ar gas mixture through a sputtering device (Fig. 2) with a plutonium-lined cathode. Identification of the species produced is achieved by using mixtures of $^{16}O_2$ and $^{18}O_2$. It can be seen (Fig. 3) that a group of four bands around 820 cm^{-1}

is shifted to ca. 780 cm^{-1} when ^{16}O is substituted by ^{18}O; this group of bands may be assigned to PuO. The PuO_2 species, on the other hand, gives a band due to the antisymmetric stretching mode of the mixed isotopic $Pu^{16}O^{18}O$ (782 cm^{-1}) in addition to those due to $Pu^{16}O_2$ (794 cm^{-1}) and $Pu^{18}O_2$ (754 cm^{-1}) - confirming the presence of two equivalent oxygen atoms. The symmetric stretching mode was observed as a weak band at 718 cm^{-1} for $Pu^{16}O^{18}O$, but was not found for $Pu^{16}O_2$ or $Pu^{18}O_2$, consistent with a linear geometry. Detailed analysis of the accurate isotope shifts obtained with the FT-IR spectrometer and the relative intensities of the bands confirmed that the molecule is linear PuO_2.

The appearance of four bands due to PuO, rather than the expected single band, could be attributed to multiple trapping site effects for the following reasons:

(a) All four bands were shifted similarly by ^{18}O substitution (Fig. 3).

(b) On annealing the matrix (Fig. 3), one of the bands increased in intensity at the expenses of the other three; this presumably is the most stable site.

(c) On changing the matrix to krypton, only a single band was observed.

C. Metal Complexes

Matrix isolation spectroscopy has been widely utilized to study transition metal complexes, particularly $M(CO)_n$, $M(N_2)_n$ and $M(O_2)_n$ as well as complexes with mixed ligands [29]. For example, MCO, $M(CO)_2$ and $M(CO)_3$ species have been produced by co-condensation of the metal and CO in an argon matrix or by photolysis of the stable $M(CO)_4$ precursor in an argon matrix. To date, the only such species reported by FT-IR spectroscopy is a "sideways bonded" Th...N_2 complex which has a N_2 stretching frequency of 1828.6 cm^{-1} in an argon matrix [24].

D. Stable Molecules

FT-IR spectroscopy has been used to obtain far infrared spectra of various stable molecules in low temperature matrices. Guillory and Smith [18] reported spectra of $GeCl_4$, $GeHCl_3$ and GeH_2Br_2 in argon matrices. Knözinger and Jacox [30] obtained a spectrum of methyl formate isolated in an argon matrix which enabled the COC bending mode and the torsion to be distinguished (in the gas phase spectrum a single broad band was observed). Barnes et al. [4,31,32] used an FT-IR spectrometer to observe the torsional modes of methanol, ethanol and trifluoroethanol isolated in argon matrices.

E. Molecular Complexes

The application of matrix isolation infrared spectroscopy to the study of molecular complexes has recently been reviewed [33]. Since detection of weak bands due to a complex against a background of the superimposed spectra of the uncomplexed donor and acceptor molecules is often required, FT-IR could give significant advantages over conventional infrared spectroscopy. Additionally, the wide spectral range of the FT-IR spectrometers would be a great advantage, since the low frequency hydrogen bond stretching and bending modes of molecular complexes have been little investigated. To date, few studies of molecular complexes have used FT-IR spectroscopy. Mann et al. [34] examined the far infrared spectrum of water at rather high concentrations in an attempt to identify intermolecular modes of the water dimer, and Barnes and Hallam [4, 31] carried out similar studies on methanol and ethanol. Knözinger and Jacox [30] used FT-IR spectroscopy to examine the intermolecular modes of methylcyanide dimers. They observed bands at 122, 105, 95, and 73 cm^{-1} for $(CH_3CN)_2$; and 113, 92, and 72 cm^{-1} for $(CD_3CN)_2$. The higher frequency bands were assigned to an antiparallel dimer and the lower frequency bands to a lower dimer. However, the bands displayed different dependence on concentration (the lower frequency bands are present even at M/S = 1000) and annealing, thus an alternative explanation is that the 95 and 73 cm^{-1} bands are phonon bands of the argon activated by the trapped species; similar bands were observed in a previous study of far infrared spectra of various solutes in argon matrices [6].

F. Analysis of Mixtures

Rochkind [35] applied the matrix isolation technique to the qualitative and quantitative analysis of multi-component gas mixtures. Overlapping rotational structure effectively prevents gas phase infrared spectroscopic analysis of mixtures, while intermolecular interactions lead to band broadening and spectral perturbations in the solid phase. Matrix isolation, on the other hand, provides narrow infrared bands at approximately the gas phase frequencies. Quantitative tests showed that Beer's law was obeyed for a variety of structurally dissimilar molecules. FT-IR spectroscopy has been applied in a similar manner to the analysis of mixtures of polycyclic aromatic hydrocarbons [36]. Using microsampling procedures [37] submicrogram quantities could be identified and their concentrations measured.

REFERENCES

1. E. Whittle, D. A. Dows and G. C. Pimentel, J. Chem. Phys., 22, 1943 (1954).

2. H. W. Brown and G. C. Pimentel, J. Chem. Phys., 29, 883 (1958); G. E. Ewing, W. E. Thompson and G. C. Pimentel, J. Chem. Phys., 32, 927 (1960); etc.

3. M. J. Linevsky, J. Chem. Phys., 34, 587 (1961).

4. A. J. Barnes and H. E. Hallam, Trans. Faraday Soc., 66, 1932 (1970).

5. M. Van Thiel, E. D. Becker and G. C. Pimentel, J. Chem. Phys., 27, 95, 486 (1957).

6. A. J. Barnes, J. B. Davies, H. E. Hallam, G. F. Scrimshaw, G. C. Hayward, and R. C. Milward, Chem. Comm., 1089 (1969).

7. D. W. Green and G. T. Reedy, Fourier Transform Infrared Spectrosc., 1, 1 (1978).

8. H. E. Hallam (Ed.), Vibrational Spectroscopy of Trapped Species, Wiley, London, 1973.

9. G. T. Reedy and A. C. Gilby, Appl. Spectrosc., 29, 532 (1976).

10. R. N. Perutz and J. J. Turner, J. Chem. Soc. Faraday II, 69, 452 (1973).

11. M. Moskovits and G. A. Ozin, Appl. Spectrosc., 26, 481 (1972).

12. J. A. Cugley and A. D. E. Pullin, Chem. Phys. Lett., 17, 406 (1972).

13. L. Andrews, Appl. Spectrosc., 33, 199 (1979).

14. M. E. Jacox, Rev. Chem. Intermed., 2, 1 (1978).

15. B. S. Ault, L. Andrews, D. W. Green, and G. T. Reedy, J. Chem. Phys., 66, 2786 (1977).

16. B. S. Ault and L. Andrews, J. Chem. Phys., 65, 4192 (1976).

17. M. E. Jacox and D. E. Milligan, Chem. Phys., 4, 45 (1974).

18. W. A. Guillory and G. R. Smith, Appl. Spectrosc., 27, 137 (1973).

19. R. J. Isabel, G. R. Smith, R. K. McGraw, and W. A. Guillory, J. Chem. Phys., 58, 818 (1973).

20. D. W. Green, S. D. Gabelnick, and G. T. Reedy, J. Chem. Phys., 64, 1697 (1976).

21. A. Snelson in M.A. Thermochemistry and Thermodynamics, (Skinner, Ed.), Butterworths, London, 1976, p. 81.

22. D. W. Green, G. T. Reedy, and J. G. Kay, J. Mol. Spectrosc., 78, 257 (1979).

23. S. D. Gabelnick, G. T. Reedy, and M. G. Chasanov, J. Chem. Phys., 60, 1167 (1974).

24. D. W. Green and G. T. Reedy, J. Mol. Spectrosc., 74, 423 (1979).

25. D. W. Green and G. T. Reedy, J. Chem. Phys., 65, 2921 (1976).

26. S. D. Gabelnick, G. T. Reedy, and M. G. Chasanov, Chem. Phys. Lett., 19, 90 (1973); J. Chem. Phys., 58, 4468 (1973); J. Chem. Phys., 59, 6397 (1973).

27. D. W. Green and G. T. Reedy, J. Chem. Phys., 69, 552 (1978).

28. D. W. Green and G. T. Reedy, J. Chem. Phys., 69, 544 (1978).

29. G. A. Ozin and M. Moskovits (Eds.), Cryochemistry, Wiley, New York (1976).

30. E. Knözinger and M. E. Jacox, Ber. Bunsenges. Phys. Chem., 82, 57 (1978).

31. A. J. Barnes and H. E. Hallam, Trans. Faraday Soc., 66, 1920 (1970).

32. A. J. Barnes, H. E. Hallam, and D. Jones, Proc. Roy. Soc. London A, 335, 97 (1973).

33. A. J. Barnes in Molecular Interactions, (H. Ratajczak and W. J. Orville Thomas (Ed.)), Vol. I, Wiley, London, 1980, p. 273.

34. B. Mann, T. Neikes, E. Schmidt, and W. A. P. Luck, Ber. Bursenges. Phys. Chem., 78, 1236 (1974).

35. M. M. Rochkind, Anal. Chem., 39, 567 (1967); Anal. Chem., 40, 762 (1967); Spectrochim. Acta A, 27, 547 (1971).

36. G. Mamantov, E. L. Wehry, R. R. Kemmerer, and E. R. Hinton, Anal. Chem., 49, 86 (1977).

37. D. M. Hembree, E. R. Hinton, Jr., R. R. Kemmerer, G. Mamantov, and E. L. Wehry, Appl. Spectrosc., 33, 477 (1979).

SOME REMARKS ON FAR INFRARED TECHNIQUES FOR MEASURING ABSORPTION SPECTRA

A. Hadni

Laboratoire Infrarouge Lointain
Universite de Nancy I
Nancy, France

1. INTRODUCTION

This is not a review on Far Infrared Fourier Transform techniques: several lectures in this Institute have been devoted to the problem. Assuming we have at our disposal one of the best far infrared interferometer or spectrometer we shall consider the problem of how to get the absorption spectrum of a sample?

2. TRANSMISSION OF A PLATE WITH PLANE PARALLEL FACES

In Fig. 1 we see a plate of thickness d; SB_1 is one incident ray striking with angle of incidence i; r is the angle of reflection. The transmission T is a function of the complex refractive index $\bar{n} = n - jk$:

$$T(n,k) = \frac{(1 - R^*)^2 (1 + \frac{k^2}{n^2}) T^*}{(1 - R^* T^*)^2 + 4R^*T^*\sin^2(\phi + \psi)} , \qquad (1)$$

with

$$R^* = \frac{(n - 1)^2 + k^2}{(n + 1)^2 + k^2} , \qquad (2)$$

the reflectivity for $d \to \infty$.

J. R. Durig (ed.), Analytical Applications of FT-IR to Molecular and Biological Systems, 171–199.

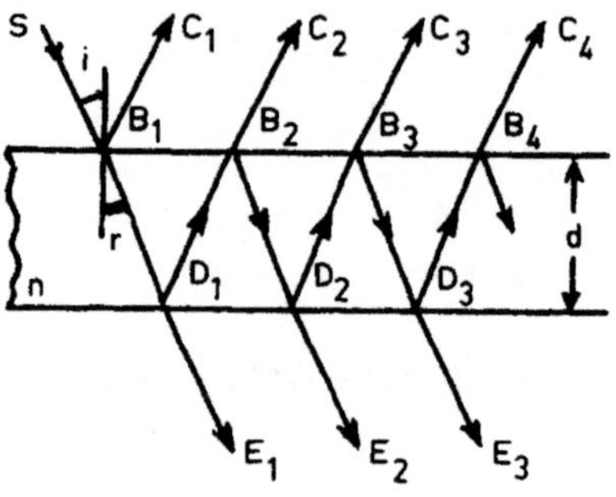

Fig. 1 Multiple internal reflexion in a plate with parallel faces.

$$T^* = e^{-Kd} \tag{3}$$

d being the thickness of the plate, and K the absorption coefficient:

$$K = \frac{2k\omega}{c} \tag{4}$$

ψ is the phase retardation at each internal reflection:

$$tg\psi = \frac{2k}{n^2 + k^2 - 1} \tag{5}$$

ψ can be either zero or π for $k = 0$ (glassy reflection) according to Eq. (5), but it is shown that it is zero for the internal reflection here considered (it should be π for the reflection from empty space to the plate). ϕ is the phase retardation at each traversal of the plate:

$$\phi = 2\pi \frac{nd \cos r}{\lambda}, \tag{6}$$

A. Case of a Small Absorption Index ($k \ll 1$)

$$\psi \simeq 0; \ \sin^2(\phi + \psi) \simeq \sin^2 2\pi n\mathring{\nu}d$$

i) T is a periodic function of $\mathring{\nu}$, with a period $\Delta\mathring{\nu} = 1/2nd$ from which n is obtained if the resolution is high enough to let us see the fringes.

ii) $$T(n,k) \simeq \frac{(1 - R^*)^2 T^*}{(1 - R^*T^*)^2 + 4R^*T^*\sin^2\phi} , \qquad (7)$$

T is oscillating between

$$T_{max.} = \frac{(1 - R^*)^2 T^*}{(1 - R^*T^*)^2} \qquad (8)$$

and

$$T_{min.} = \frac{(1 - R^*)^2 T^*}{(1 + R^*T^*)^2} . \qquad (9)$$

These two values can be measured and then (8) and (9) give two equations with two unknowns, R* and T*. It is assumed the plate is thick enough to give $\Delta\tilde{\nu}$ sufficiently small for being able to neglect the variations of T* and R* between a maximum and the nearest neighbor minimum

iii) In the case where the spectral resolution is not large enough to give fringes, an average transmission is observed:

$$T_{average}(n,k) \simeq \frac{(1 - R^*)^2 T^*}{1 + T^{*2} R^{*2}} \qquad (10)$$

Now R* may be known from a different experiment; then T* (and K) can be obtained.

B. Case of a High Absorption Index

We assume an absorption index k high enough to allow the use of a plate thickness $d \ll \lambda$ to get significant transmission spectra.

Because of the high absorption, multiple interferences are negligible and

$$T = e^{-Kd}(1 - R^*)^2 \qquad (11)$$

The knowledge of both T and R* for each frequency leads to $K(\omega)$.

C. General Case

From the apparent transmission spectrum $T[n(\omega),k(\omega)]$ given by Eq. (1), it is possible to get $n(\omega)$ and $k(\omega)$ from Kramers-Kronig relations. However T has to be known on the entire spectrum and the method has not been much employed. The apparent transmission T given by Eq. (1) may be compared with the apparent reflectivity R measured in the same conditions (i.e. normal incidence). Now R is given by Eq. (12) as shown later, and we have for each frequency two equations with two unknowns n and k.

3. REFLECTIVITY OF A PLATE WITH PLANE PARALLEL FACES

It is shown that for normal incidence:

$$R(n,k) = R^* \frac{(1 - T^*)^2 + 4T^*\sin^2\phi}{(1 - T^*R^*)^2 + 4T^*R^*\sin^2(\phi + \psi)} \tag{12}$$

Spectral range of negligible transmission. Assuming $T^* << 1$, we have $R(n,k) \simeq R^*$, and we are referred to paragraph 3.

Spectral range of small absorption. Assuming $k^2/n^2 << 1$, we know that $\phi \simeq \pi$ and

$$R(n,k) \simeq R^* \frac{(1 - T^*)^2 + 4T^*\sin^2\phi}{(1 - T^*R^*)^2 + 4T^*R^*\sin^2\phi} \tag{13}$$

i) R is periodic functions of either $\dot{\nu}$ or d (Fig. 2), with

$$R_{min.} = R^* \frac{(1 - T^*)^2}{(1 - T^*R^*)^2} \tag{14}$$

$$R_{max.} = R^* \frac{(1 + T^*)^2}{(1 + T^*R^*)^2} \tag{15}$$

It is seen on Eq. (14) that a small variation of T around 1 will affect drastically $R_{min.}$. There is thus a possibility of getting T^* with some accuracy from the measurement of $R_{min.}$, when R^* is approximately known.

ii) Average value of R.
From Eq. (13), the average value of R is obtained

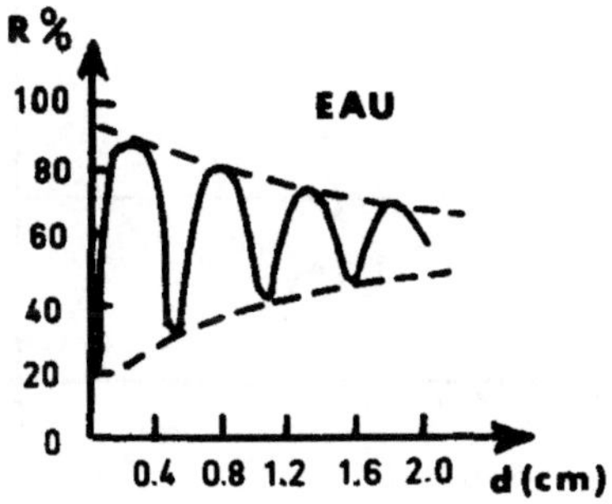

Fig. 2 Reflectivity of an absorbing plate with parallel faces, versus thickness d for λ = 9.35 cm (adapted from Ref.[14]).

$$R_{average} = R^*[1 + \frac{T^{*2}(1 - R^*)^2}{1 - R^{*2}T^{*2}}], \tag{16}$$

It can be compared to $T_{average}$ (Eq. 10) to get two relations with unknowns R* and T*.

4. REFLECTIVITY OF A THIN DIELECTRIC PLATE DEPOSITED ON A CONDUCTIVE HOLDER

Figure 3 gives the scheme of the experiment, SI is the incident ray, and the dielectric film has a thickness e.

Let us write the incident s-polarized wave as:

$$E_s^i = E_o\ e^{j\omega(t + \frac{r}{v})}$$

and the reflected one:

$$E_s^r = -E_o\ e^{j\omega(t - \frac{r}{v})}$$

assuming in both cases a positive direction parallel to OY to measure the electric field, and r being the distance to I' measured either along I'I or I'J. So that

$$E_y = E_s^i + E_s^r,\ E_y = E_o\ e^{j\omega t}(e^{j\omega \frac{r}{v}} - e^{-j\omega \frac{r}{v}})$$

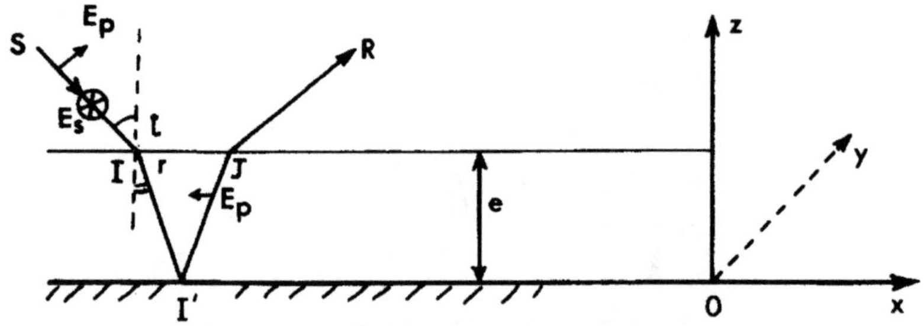

Fig. 3 Double crossing of a metal coated dielectric plate.

$$E_y = 2E_o\ e^{j\omega t}\ \sin\frac{\omega r}{v};\ \text{and}\ E_y = 0\ \text{for}\ r = 0.$$

We have also for the incident p-polarized wave:

$$E_p^i = E_o\ e^{j\omega(t + \frac{r}{v})},$$

and for the reflected one:

$$E_p^r = E_o\ e^{j\omega(t - \frac{r}{v})}$$

assuming in both cases that the positive direction is parallel to $\hat{P} \wedge \hat{E}_S$, $\hat{P}$ being the Poynting vector. We thus have:

$$E_x^i = E_o\ \cos r\ e^{j\omega(t + \frac{r}{v})};\ E_x^r = -E_o\ \cos r\ e^{j\omega(t - \frac{r}{v})};$$

$$E_x = E_x^i + E_x^r;\ E_x = 2\ E_o\ \cos r\ e^{j\omega t}\ \sin\frac{\omega r}{v};\ \text{and}$$

$$E_x = 0\ \text{for}\ r = 0.$$

We have also

$$E_z = E_z^i + E_z^r;\ E_z = 2\ E_o\ \sin r\ e^{j\omega t}\ \cos\frac{\omega r}{v},$$

and

$$E_z = 2\,E_o \sin r\, e^{j\omega t} \text{ for } z = 0.$$

It is thus concluded that for $e << -\lambda/4$, we have, anywhere in the film $\bar{E}_x \simeq 0$; $\bar{E}_y \simeq 0$ and $E_z \simeq 2\,E_o$ Sin $r\,e^{j\omega t}$ for thin dielectric plates deposited on a conductive holder and illuminated in a non perpendicular incidence, we have electro-magnetic standing waves with the electric field practically perpendicular to the surface. The field intensity is increasing with r. This situation is comparable to the one of a longitudinal electric field (Fig. 4) which has been shown possible for $\varepsilon_R = 0$, value only achievable for the frequency of the LO branch at $q = 0$. The standing electro-magnetic waves will excite forced vibration of the lattice, with special enhancement for ω_{LO} where there is resonance with a lattice vibration at the same frequency and same wave vector.

This enhanced absorption can be seen in another way. The electric field E_N of the standing wave is accompanied by a polarization $\hat{P} = \varepsilon_o\,(\varepsilon_R - 1)\,\hat{E}$ which is also perpendicular to the conductor, with $\hat{E} = \hat{E}_N - \hat{P}/\varepsilon_o$, $\hat{P}/\varepsilon_o$ being the depolarizing field. Hence

$$\hat{P} = \varepsilon_o\,(\varepsilon_R - 1)(\hat{E}_N - \frac{\hat{P}}{\varepsilon_o});\ \hat{P}(1 + \varepsilon_R - 1) = \varepsilon_o\,(\varepsilon_R - 1)\hat{E}_N,$$

and for $\varepsilon_R \neq 0$,

$$\frac{\hat{P}}{\hat{E}_R} = \frac{\varepsilon_o\,(\varepsilon_R - 1)}{\varepsilon_R};\ \frac{\hat{P}}{\hat{E}_R} = \varepsilon_o(1 - \frac{1}{\varepsilon_R}).$$

It is seen that for $\varepsilon_R = 1$, $\hat{P}$ is null for $\varepsilon_R < 1$, $\hat{P}$ is opposite to $\hat{E}_N$, and the smaller ε_R is, the larger $\hat{P}$ will be and hence the absorption.

Now for $\varepsilon_R = 0$, $\hat{P} = \varepsilon_o(\varepsilon_R - 1)\,\hat{E}$, which gives $\hat{P} = -\varepsilon_o\hat{E}$ as we have seen it.

Figure 5 shows the experimental result. There is a large absorption at ω_{LO} but also a sharp peak at ω_{TO} since there is a small component of $\hat{E}$ on either ox or oy as soon as e is a finite part of λ.

5. REFLECTIVITY OF A PRISM SHAPED SAMPLE

Multiple reflections are avoided and the true reflectivity R* is obtained. For normal incidence:

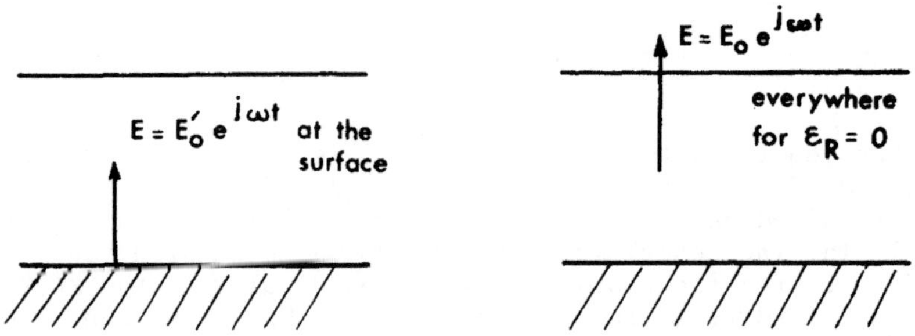

Fig. 4 Comparison of a standing electro-magnetic wave and a longitudinal electro-magnetic wave.

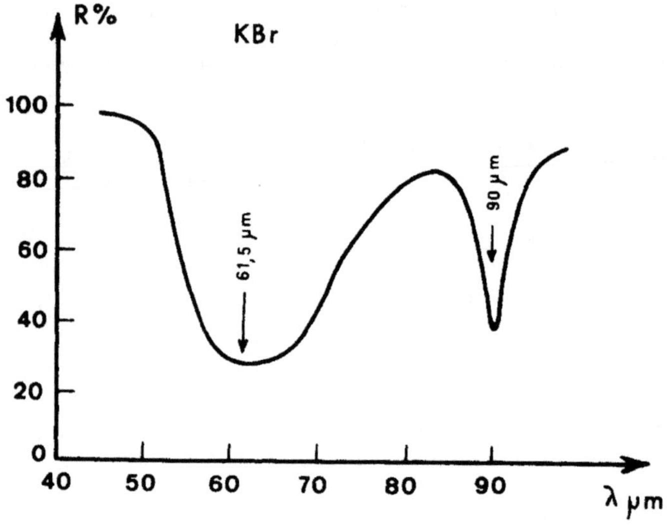

Fig. 5 A KBr thin film evaporated on a brass plate presents two minima in its reflection spectrum corresponding to LO and TO frequencies for q = 0.

$$R^* = \frac{(n + 1)^2 + k^2}{(n + 1)^2 + k^2} \qquad (17)$$

Now $n = n(\omega)$ and $k = k(\omega)$ so that for each frequency, the measured value of R^* gives one equation with two unknowns. In fact, from the general Kramers-Kronig relations, if $k(\omega)$ is known, $n(\omega)$ can be computed and vice-versa:

$$n(\omega) - n(\infty) = \frac{2}{\pi} \int_0^\infty \frac{n\, k(u)}{u^2 - \omega^2}\, du$$

$$k(\omega) = \frac{2\omega}{\pi} \int_0^\infty \frac{n(u) - n(\infty)}{u^2 - \omega^2}\, du. \tag{18}$$

It is thus possible to get $n(\omega)$ and $k(\omega)$ from the measurement of $R^*(u)$. There are some experimental difficulties. The most important is that $R^*(\omega)$ goes very close to zero for any resonance in the crystal, and a small error for R^* at these frequencies will affect the whole spectrum of either $n(\omega)$ or $k(\omega)$.

6. ATTENUATED TOTAL REFLECTION

A. Total Reflection

Let us consider a transparent prism of high refractive index n_1, the large face of it is supported by a transparent sample of refractive index $n_2 < n_1$ (Fig. 6).

It is well known that total reflection is observed for $i > i_1$ with:

$$\sin i_1 = \frac{n_2}{n_1}.$$

B. Attenuated Total Reflection

Even in the case of total reflection, the electro-magnetic wave goes through the sample at some depth d which is a function of i, n_2/n_1, and λ: $d = d(i, n_2/n_1, \lambda)$ there is an evanescent wave. It is shown that d decreases when either i increases, or n_2 decreases. Now when the sample shows some absorption (n_2 is now complex), there is still reflection but there is also attenuation because of some absorption of the evanescent wave. It is a case of Attenuated Total Reflection (ATR). The reflectivity spectrum is then very often similar to a transmission spectrum Fig. 7a shows the dispersion of the index of refraction n_2 around the resonant frequency ω_o. For a prism of high refractive index n'_1, the relation $n'_1 > \overset{o}{n}_2$ is kept all through the resonance and there is total reflection (attenuated) for all frequencies. In

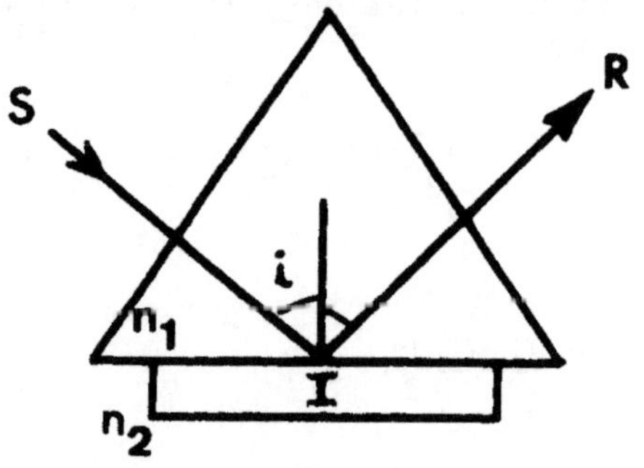

Fig. 6 There is total reflection for $n_1 > n_2$ and $i > i_1$.

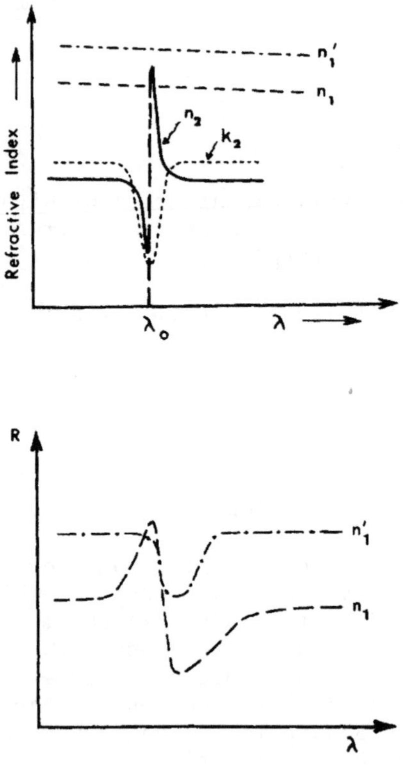

Fig. 7 Abnormal dispersion of the refractive index n_2 leading to attenuated total reflection (adapted from Ref.[15]).

Fig. 7b is shown the corresponding reflectivity. Now let us call n_1 the refractive index in a case where for a small frequency range the relation $n_1 > n_2$ is not fulfilled. The total reflection is not possible and the electro-magnetic wave goes deeply into the sample giving a large distortion seen in Fig. 7b. To avoid this distortion, a high refractive index prism has to be used. Table 1 gives the characteristics of some materials. It is useful to remember that the highest value of n for KBr, which shows a strong anomalous dispersion, is only four. The angle of incidence i is often much larger than i_1 in order to decrease d. The absorption at each reflection is small. To get a significant change of reflectivity, a number of reflections has to be used. Fig. 8 gives the scheme of a device allowing multiple ATR, it is called Multiple Internal Reflexion (MIR). A good contact between prism and sample is a need. It is easy for liquid samples. A good polishing is necessary for solids. In both cases it is possible to get quantitative data and one advantage is to avoid interference fringes. It sometimes happens that sample thickness is smaller than depth penetration d. It is then shown that i_1 is not dependent on the refractive index of the sample ($\sin i_1 = 1/n_1$). The ATR or the MIR method are then specially attractive since the critical angle is a constant. For low absorption samples, MIR spectroscopy is a powerful method for micro-samples that can be spread over a broad area of an MIR plate. As shown in Fig. 9 all of the sample which lies beyond the depth penetration is unable to affect the radiation beam. A quantitative treatment has been given by several authors Hansen [1], Vincent [2] and others. It is shown that $R = 1 - 2B\chi + 2B^2\chi^2 - 2C\chi^3 + \ldots$ with $\chi = k_2/n_1$, and $B(i, n_2, k_2, n_1)$.

Now the transmission of a plate of thickness d is written $T \simeq 1 - Kd + K^2/2\, d^2$, with $K = 2k_2\, \omega/c$ and up to the second order, the ATR spectrum is identical to the transmission spectrum of the sample with a thickness d which is a function of B. This thickness d is 2 or 3 times smaller in the case ($d_\perp$) when the electric field of the electro-magnetic wave is perpendicular to the plane of incidence. It is shown in Fig. 10 that $d_\perp/\lambda$ can be very small. The use of ATR could thus be a means of avoiding to grind and polish very thin crystal plates to study a high intensity transition (the intensity considered here being however limited by the assumption $\chi^3 \ll 1$). In that case ATR must be used in such a

TABLE 1

	n (IR)	i_1
Ge	4	15°
Si	3,5	17°
KRS5	2,4	29°
AgCl	2	30°

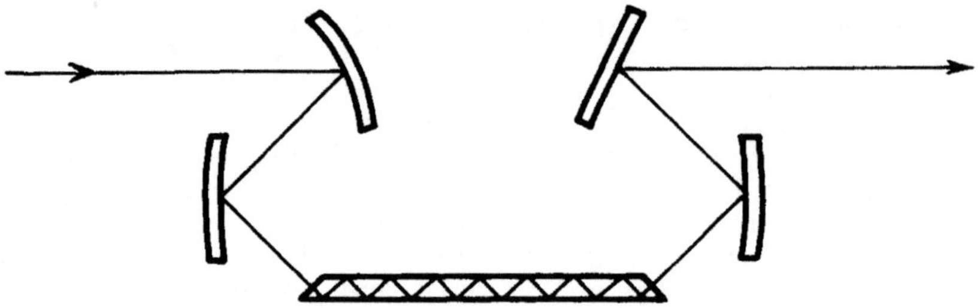

Fig. 8 Multiple internal reflection.

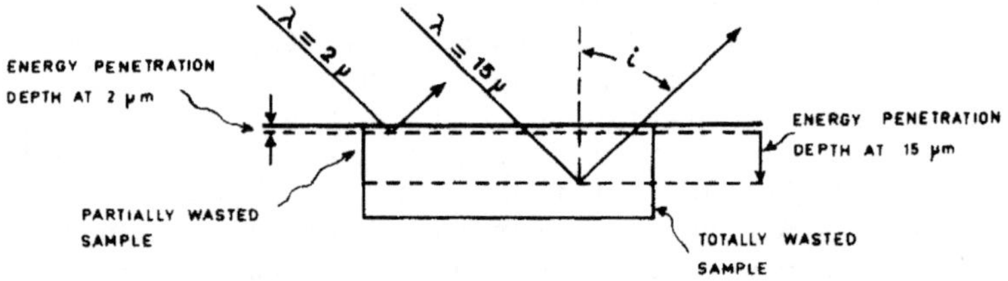

Fig. 9 For long wavelengths the electro-magnetic wave goes deeply into the sample (adapted from Ref.[15]).

way that $d_\perp$ is very small: i must be much larger than i_1, n small (hence the need of a prism with a high refractive index n_1 since $n = n_2/n_1$). Values of $d_\perp/\lambda$ around 0.1 are obtainable in the whole infrared. This gives $d_\perp \simeq 1$ μm at 1000 cm^{-1}; but $d_\perp = 10$ μm at 100 cm^{-1}. Now experiments show that the highest values of the absorption coefficient have the same order of magnitude in any part of the spectrum ($K_{max.} \simeq 10^4$ cm^{-1}), hence $d_{optimum}$, (giving for instance a transmission T = 1/e = 37% at the bottom of the absorption bands) has also the same value (for instance $d_{optimum} \simeq 1$ μm). It is then understood why ATR has been widely used in infrared up to 16 μm but not farther at our knowledge.

The above quantitative treatment is not applicable to high intensity transitions because χ^3 is not negligible and also because it is assumed that n is a constant all through the absorption band. More elaborated theories have shown that ATR, is still usable in that case. A single reflection experiment like that represented by Fig. 6, with $n_1 = 4$ (germanium) and $i \simeq 80°$ (the prism angle A is then very small), can involve a depth penetration limited to 0.01 λ, i.e. $d_\perp \simeq 0.1$ μm at 1000 cm^{-1} and $d_\perp = 1$ μm at 100 cm^{-1}.

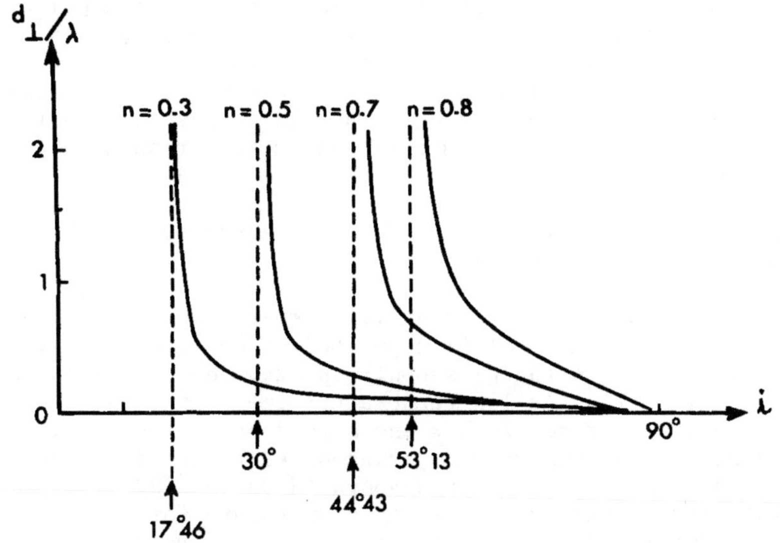

Fig. 10 $d_\perp/\lambda$ can be very small for high values of i and small values of $n = n_2/n_1$ i.e. high values of prism index n_1 (from Ref.[2]).

7. ABSORPTION

A. Bolometer With Direct Absorption

A plate made with the absorbing material is illuminated with monochromatic light from a spectrometer. There is absorption for a discrete series of wavelengths. This means very often, that electro-magnetic energy is converted into phonons having these characteristic frequencies. Finally they decay into other phonons until a new thermal equilibrium is obtained with some increase ΔT of temperature. This increase ΔT is proportional to the flux of absorbed energy. Any physical property sensitive to temperature will give a signal proportional to ΔT in a first approximation, and hence to the absorbed energy.

The temperature sensitive properties which can be used are the electrical resistivity, the dielectric constant, the spontaneous polarization etc.... We have, respectively, a resistance bolometer, a dielectric bolometer, a pyroelectric one etc. Recently the far infrared absorption coefficient of a TTF-TCNQ single crystal, an organic semi-conductor, has been obtained from bolometric measurements (Eldridge [3]). The single crystal was too small to allow reflectivity measurements, and too thick to give usable transmission data. However it has led to a sensitive bolometer with a response proportional to the absorption coefficient. Let us notice that the total response in this case is the sum of two parts. One is the bolometric response considered above. The other one is a photo-conductivity signal. By the use of a high frequency chopping, it is possible to isolate this contribution and to get the bolometric response by substraction.

B. Bolometer With Heat Transfer From the Absorber to the Temperature Sensitive Element

Such a device is efficient at low temperature. The absorber (any material to be studied) is deposited on one side of a very thin (e = 10 μm for instance) sapphire plate, the sensor on the other side (Fig. 11). At low temperature the heat capacity of the sapphire plate is negligible and a good percentage of the heat produced in the absorber is transported to the detector. This one can be made of silicon evaporated on the sapphire plate. Its resistivity is very sensitive to temperature for $T < 3$ K. The heat conductivity is specially high in sapphire around 2 K and the response time is not too much increased.

The method has been successful for low absorption materials (i.e. absorbed mono-layers). It has not been used for crystals. It is often a case of high reflectivity and thus it is also a case of low absorption, and the method might be feasible.

C. Spectrophone, Spectro-Acoustic Cell

The sample is put in a gas filled cell equipped with an infrared window. The chopped radiation is absorbed by the sample at some discrete wavelengths. For this radiation the pressure in the cell is modulated. This pressure modulation is detected with a choice of devices. One of the most used is a piezoelectric one. There has been a recent meeting on photoacoustic spectroscopy at Iowa State University in 1979. Several liquids and solids have been studied in the infrared with acoustic spectroscopy. There is a problem of sensitivity. However, with the number of lasers transitions now available in the infrared, it is less critical and Patel has shown the possibility of measuring absorption coefficients as small as 10^{-6} cm^{-1}. The method is specially recommended when there is a large scattering of radiation by the sample (The case of inhomogeneous materials which are difficult to shape onto a well polished slab or the case of samples with any complicated geometry which has to be kept).

8. SPONTANEOUS AND STIMULATED RAMAN SCATTERING, NON LINEAR EFFECTS, FAR INFRARED EMISSION

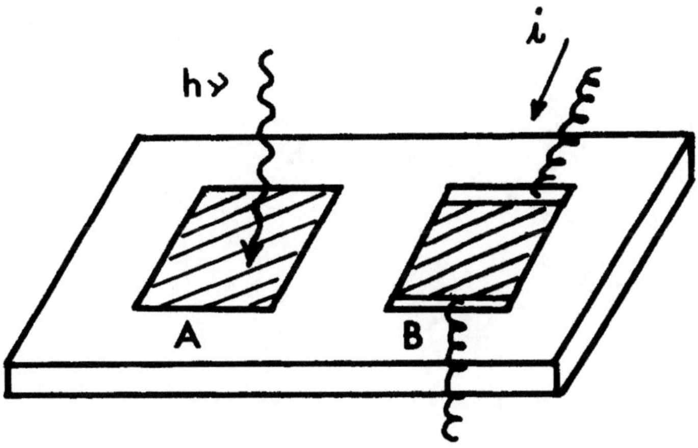

Fig. 11 Bolometer with separated absorber and sensor mounted on a thin (10 μm) sapphire plate.

Let us consider the possibilities of first order Raman scattering in the crystal excited by radiation ω_1. In Fig. 12a is the scheme for the diffusion of a Stokes line $\omega_2 = \omega_1 - \omega_3$, ω_3 being the frequency of a crystal phonon of wave vector $\vec{q}_3$, with $\vec{k}_2 = \vec{k}_1 - \vec{q}_3$ and in Fig. 12b is the scheme for the diffusion of an anti-Stoke line $\omega_2 = \omega_1 + \omega_3$. In both cases the phonon frequency is the difference of the two electro-magnetic frequencies ($\omega_3 = \omega_1 - \omega_2$ in the first case, $\omega_3 = \omega_2 - \omega_1$ in the second one). We have

$$q_3^2 = k_1^2 + k_2^2 - 2k_1k_2 \cos\theta. \tag{19}$$

Now

$$k_1 \simeq k_2 \simeq k; \text{ and thus } q_3 \simeq 2k \sin\frac{\theta}{2}$$

θ can be varied from zero to 180°; $\sin\theta/2$ from 0 to 1 and q_3 from 0 to 2k; i.e. very small values compared to $q_{Max} \simeq 2\pi/a$ at the limit of the Brillouin zone, with $a \simeq 10^{-10}$m. We thus have two extreme possibilities.

a) $\theta = 0$: $\hat{k}_1$ and $\hat{k}_2$ parallel (Fig. 13a), $q3 \simeq 0$
b) $\theta = \pi$: $\hat{k}_1$ and $\hat{k}_2$ antiparallel (Fig. 13b), $q3 \simeq 2k$.

Now the probability of the process in the crystal is proportional to $n_1\ (n_2 + 1)(n_3 + 1)$, n_1, n_2, n_3 being the density of bosons of frequency ω_1, ω_2 and ω_3, respectively. $n_2 = 0$ is the case of spontaneous Raman diffusion. $n_2 > 1$ is the case of stimulated

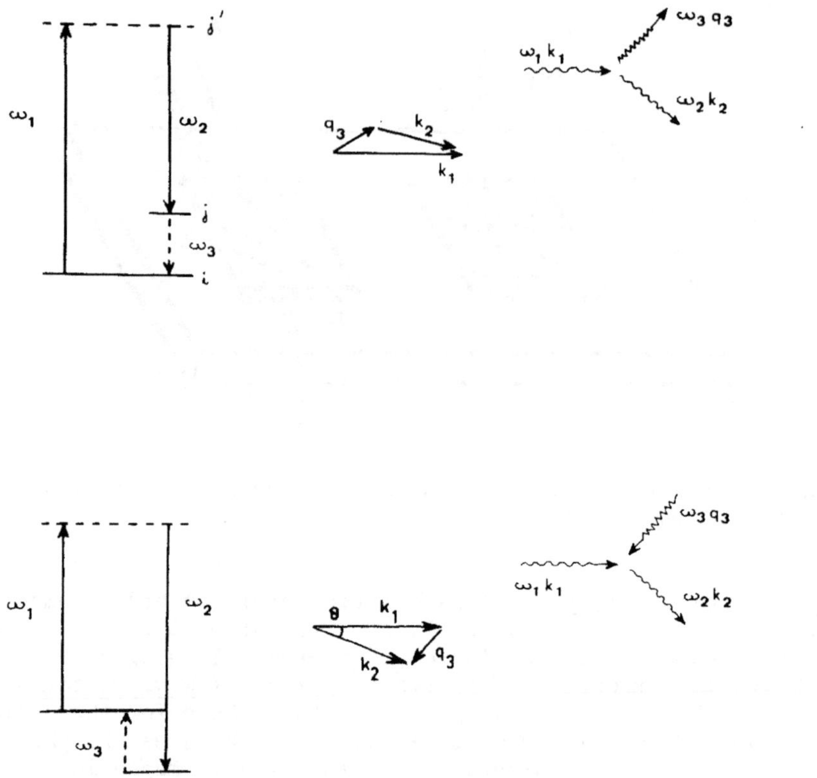

Fig. 12 (a) Scattering of a Stokes line. (b) Scattering of an anti-Stokes line.

Fig. 13 (a) $\hat{k}_1$ and $\hat{k}_2$ parallel. (b) $\hat{k}_1$ and $\hat{k}_2$ antiparallel.

Raman diffusion.

A. Spontaneous Raman Scattering

The observation of polariton dispersion effects requires

very small values of q_3 and hence forward scattering (Fig. 13a) with values of θ limited to some 10°.

Experiments on a number of cubic and uniaxial crystals have shown that for a given excitation frequency ω_1, the scattered frequency ω_2 depends on θ, and hence on the phonon wave-vector q_3 responsible for the scattering. The value of q_3 is given by Eq. (19); and $\omega_3 = \omega_1 - \omega_2$. It is thus possible to get $\omega_3 = \omega_3 (q_3)$, i.e. the polariton dispersion curve. Figure 14 shows such a curve in the case of lithium niobate. For each value of θ, two values of ω_3 are observed since there are two branches, one going to zero for $q_3 \to 0$, the other one tending to a finite frequency which corresponds to a longitudinal mode.

B. Stimulated Raman Scattering

It has been shown that the use of a second beam at frequency ω_2 for instance will increase considerably the intensity of the Raman scattering when the intensity of the excitation beam is over some large threshold. We shall describe as an example the experiments reported in Miss Paraire's dissertation [4]. Ruby laser pulses of high intensity are used to pump two different dye-lasers C_1 and C_2 (Fig. 15). The frequencies of ω_1 and ω_2 are tunable. Their polarizations are parallel to the ruby radiation. The instantaneous powers (4 MW) are larger than the threshold for non linear effects. We shall assume for instance $\omega_2 < \omega_1$ and we are limited to the scheme of Fig. 12a since $\omega_3 = \omega_1 - \omega_2 > 0$.

For a given geometry described for instance in Fig. 13a, the wave-vector q_3 of the phonons used in the process is determined by Eq. (19)(nearly zero for Fig. 13a; slightly larger for Fig. 13 b).

Now if we look at the TO phonon dispersion curve for each value of q_3 there are two or more polariton frequencies available. The frequency ω_1 being given, ω_2 may be adjusted to get $\omega_1 - \omega_2$ equal to one of these resonant frequencies. There will be then a detectable decrease for the ω_1 beam, an increase for the ω_2 beam, and there will be a creation of phonons ω_3. They are confined in the material for the frequencies $\omega_3 > 100\ cm^{-1}$ (pure phonons). They can emit far infrared radiations (i.e. exit from the crystal) for $\omega_3 < 100\ cm^{-1}$ (case of polaritons). It is possible by modifying the geometry of the beams to modify q_3 from 0 to some low value and to measure the polaritons frequency vs. q_3 by either measuring the decrease of intensity of the ω_1 beam or the far infrared emission. Figure 16 shows results in good accordance with those given by spontaneous Raman scattering.

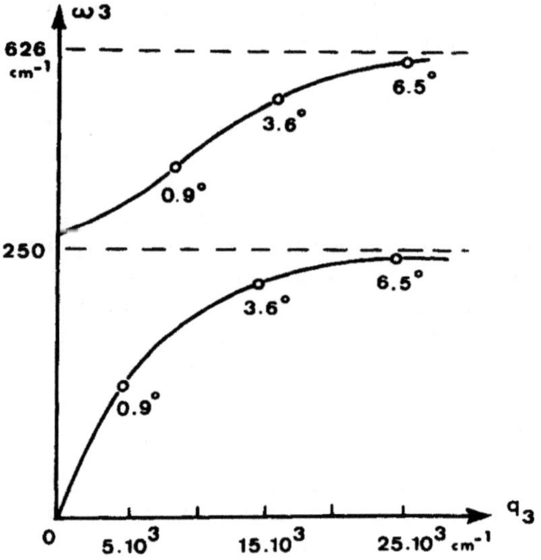

Fig. 14 Dispersion curve of polaritons in the case of lithium Niobate obtained by spontaneous Raman scattering (from Ref.[16]).

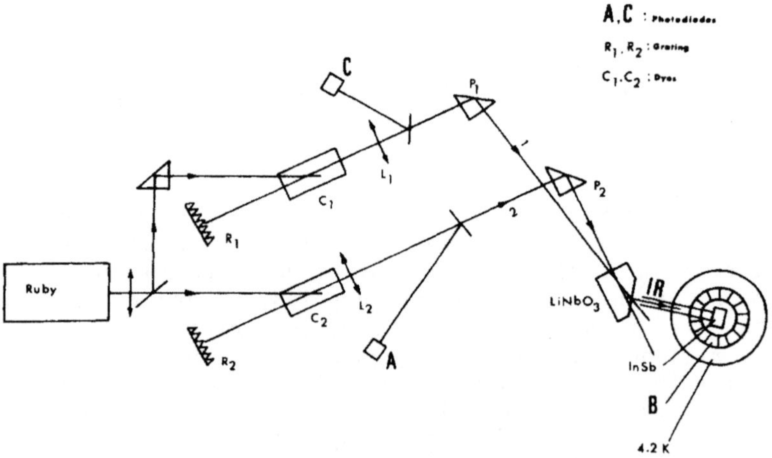

Fig. 15 Scheme for generating polariton in $LiNbO_3$ with two powerful dye laser pulses (from Ref.[4]).

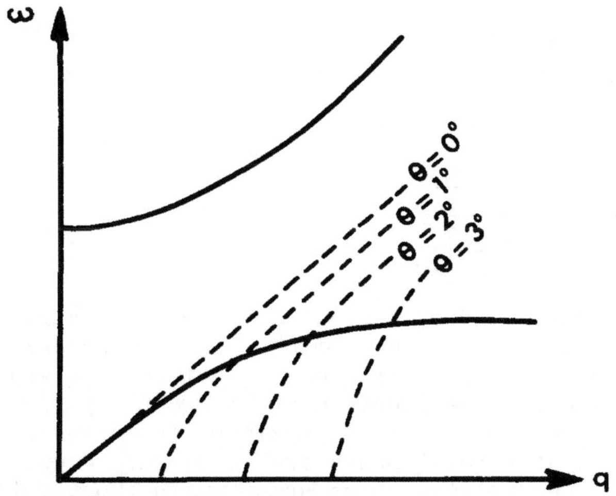

Fig. 16 Dispersion curve of polaritons in Lithium Niobate obtained by the use of stimulated Raman scattering (from Ref. [4], used by permission).

9. INTERFACE ELECTRO-MAGNETIC MODES

Surface phonons are a consequence of the finite extent of the crystal. Let us define surface modes, i.e. electro-magnetic waves which propagate in a wave-like manner along the interface (i.e. in the x direction) but decay exponentially to zero as $y \to \pm\infty$. They are revealed in two types of experiments: far infrared spectra of dielectric powders and the two-prisms technique which allows generation of surface waves in one direction with measurement of their wave-vector and the detection of these waves after propagation at a finite length on the surface (a few centimeters).

A. Far Infrared Spectra of Dielectric Powders

(1) Spheric particles. A long time ago, Fröhlich [5] demonstrated that the fundamental infrared absorption of a small diameter alkali-halide sphere will occur at a frequency ω_S given by:

$$\omega_S^2 = \frac{\varepsilon_o + 2\varepsilon}{\varepsilon_\infty + 2\varepsilon}\,\omega_{TO}^2 \tag{20}$$

ε being the dielectric constant of the medium in which the crystal is embedded. We have:

$$\frac{\omega_S^2}{\omega_{TO}^2} = \frac{\varepsilon_0/\varepsilon_\infty + 2\varepsilon/\varepsilon_\infty}{1 + \frac{2\varepsilon}{\varepsilon_\infty}}$$

$$\frac{\omega_S^2}{\omega_{TO}^2} = \frac{\omega_{LO}^2}{\omega_{TO}^2} \cdot \frac{1 + 2\frac{\varepsilon}{\varepsilon_0}}{1 + 2\frac{\varepsilon}{\varepsilon_\infty}},$$

and it is seen that $\omega_{TO} < \omega_S < \omega_{LO}$. More recently Englman et al. [6] have shown that besides this mode of uniform polarization, a series of surface modes with frequencies between ω_{TO} and ω_{LO} should absorb light. Experiments have been done by T. P. Martin [7] (Fig. 17). They show that the maximum absorption is shifted beyond ω_{TO}, and that several peaks are observed. However their location is not in quantitative agreement with those calculated. It has also been shown that when translational symmetry is destroyed by making one or more dimensions of the crystal finite, the low frequency acoustical modes acquire a dipole moment and absorb light also (Fig. 18).

(2) Other particles. In the general case of an ellipsoidal particle with one principal axis parallel to the electro-magnetic wave field E_{ext}, we assume dimensions small enough to permit a uniform polarization, hence:

$$P = \varepsilon_o(\varepsilon_R - 1)(E_{ext} + E_{dep}).$$

In this general case, $E_{dep} = -NP/\varepsilon_o$, with $N = 1/3$ for the upper considered case of a spherical particle, $N = 1$ for a disc perpendicular to the field, $N = 0$ for a disc parallel to the field etc.. Hence

$$\frac{\overline{P}}{E_{ext}} = \frac{\varepsilon_o}{\frac{1}{\varepsilon_R - 1} + N}$$

and P should go to infinity for $1/\varepsilon_R - 1 + N = 0$, or $\varepsilon_R = 1 - 1/N$

a - spherical particles, $N = 1/3$, $\varepsilon_R = -2$ and the maximum of absorption is located between ω_{TO} and ω_{LO} as was discussed in (1).

b - discs irradiated at a high angle of incidence, E_{ext} is nearly perpendicular to the disc, $N = 1$, $\varepsilon_R = 0$, and the maximum of absorption is located at ω_{LO}.

c - discs irradiated perpendicularly, E_{ext} is parallel to the disc, $N = 0$, $\varepsilon_R = \infty$ and the maximum of absorption is located at ω_{TO}.

We see there the need of small single platelet crystals to

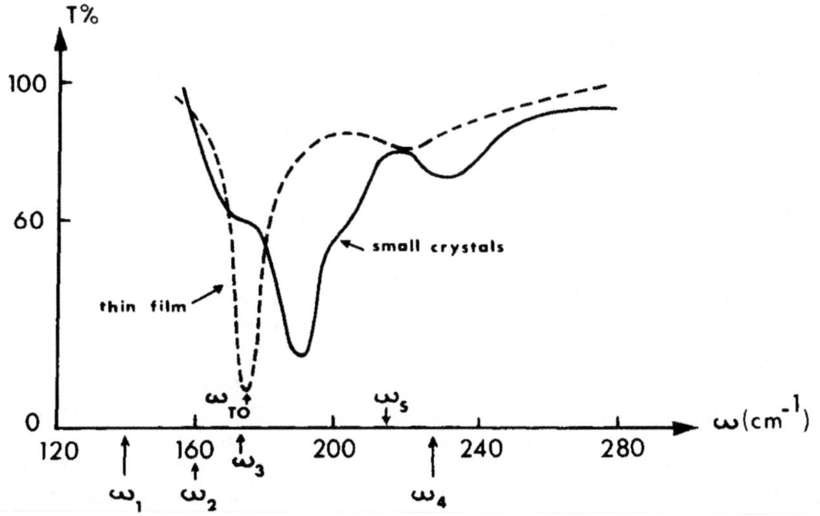

Fig. 17 Comparison of the far infrared transmission of a thin KBr film and of powders (from Ref.[7], used by permission).

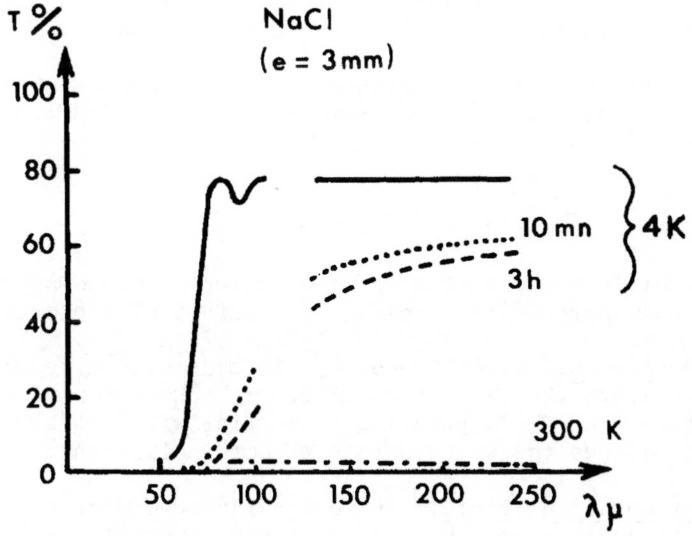

Fig. 18 Comparison of the far infrared transmission of a NaCl plate, and of NaCl powders (from Ref.[17], used by permission).

get ω_{TO}. When there is dichroism these platelets must have the same orientation. This is technically difficult when we want very thin platelets. It is possible to grind and polish a single crystal down to 5 µm, but to get a smaller thickness, special epitaxial methods have to be used. Figure 19 from Gerbaux et al. [8] gives their principle.

B. Generation and Detection of Surface-Waves by the Two-Prism Method

(1) Generation. The method has been first described by Otto [9]. We shall consider it in a well definite case, for instance Begley et al. experiments [19]. A CH_3OH laser line at 84 cm^{-1} is introduced in a TPX (methyl-pentine polymer) prism at an angle θ (Fig. 20) and is reflected. When the prism is at an optimum distance (about 190 µm) from the AsGa surface, for some values of θ, there is a dip in the reflectivity. This is explained by a conversion of p-polarized bulk polaritons from the prism into surface polaritons propagating in the x direction of the AsGa surface. The wave-vector of the bulk polaritons in the prism is written $k = 2\pi\nu n/c$, and its component on the surface to be tested, $k_x = 2\pi\nu n/c \sin\theta$, can be varied with θ. For a given ν, if $k(\nu) > q(\nu)$, $q(\nu)$ being the electro-magnetic surface wave (ESW) wave-vector at the AsGa surface, it is possible, by adjusting θ, to get $k_x(\nu) = q(\nu)$. For this value of θ there is conversion of p-polarized bulk polaritons into ESW propagating along x, and there is a dip in the reflectivity curve versus θ.

The dispersion curve of surface polaritons can be obtained in this way. Figure 21 is an example in the case of $SrTiO_3$. It is shown that

$$q(\omega) = \frac{\omega}{c}\left[\frac{\varepsilon'(\omega)\varepsilon}{\varepsilon'(\omega) + \varepsilon}\right]^{\frac{1}{2}} \tag{21}$$

ε being the dielectric constant of the medium above the surface, and ε' the real part of the dielectric constant of the crystal.

(2) Detection. Detection of ESW is made with a second prism identical to the first one. There is conversion of surface polaritons into bulk polaritons emerging with the same θ angle which matches the wave-vectors of both polaritons.

(3) Attenuation length. By moving the second prism along the x direction it is possible to get the attenuation length L_x, distance at which the ESW intensity is divided by e. For a doped AsGa sample ($N = 3.10^{18}$ carries per cm^3), for $\dot{\nu} = 84$ cm^{-1}, L_x = 1.17 cm is in good accordance with calculations. For a platinum surface, the theory gives L_x = 0.46 cm for $\dot{\nu} = 1000$ cm^{-1} and L_x = 137 cm for $\dot{\nu} = 84$ cm^{-1}. The first length is well verified, the

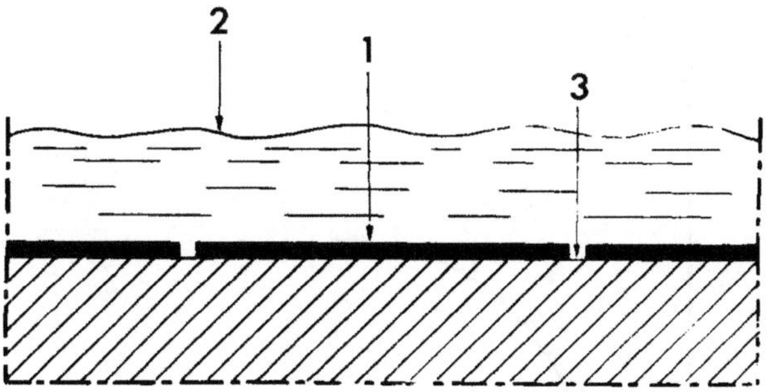

Fig. 19 The TGS solution (2) is kept in contact with the single crystal holder through the holes (3) in the metal layer (1).

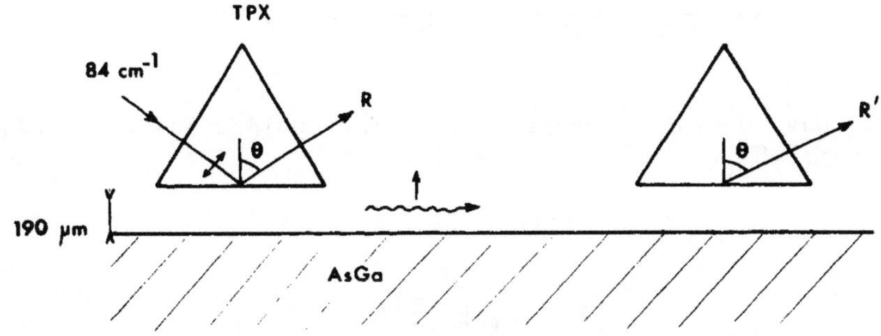

Fig. 20 Generation and detection of surface polaritons by the double prism method (adapted from Ref. [10]).

measurements of the second one gives a much shorter distance (1.07 cm). The data on the optical constants of platinum in the far infrared are probably erroneous.

(4) Conclusion. Surface waves can be generated and detected. They play an important role not only in solid state physics but also in optics. For instance, the Wood's anomalies of reflection grating are explained by ESW.

More generally, Burstein [11] has defined interface electromagnetic modes as those which involve a single electro-magnetic wave in each medium. Thus the Brewster mode is an interface mode (Fig. 22); Fig. 23 gives also the definition of Fano modes and Zennick modes.

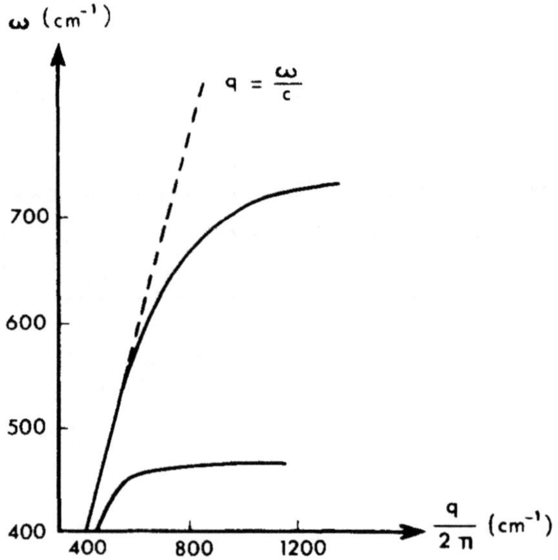

Fig. 21 Wave vector dispersion of surface polaritons in $SrTiO_3$ (after Ref.[18]).

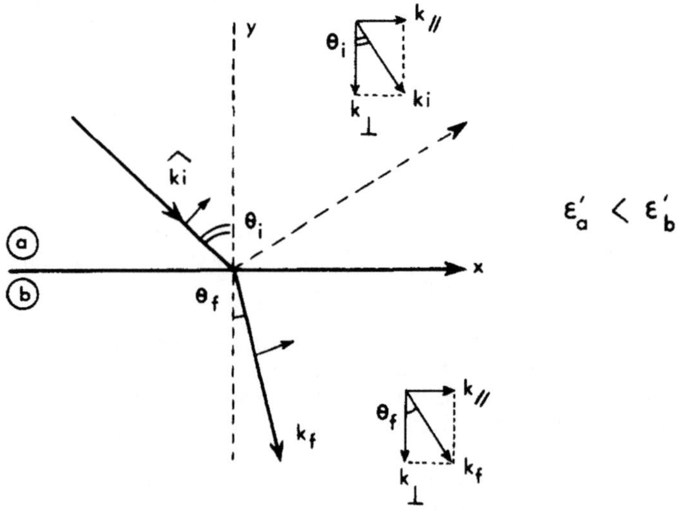

Fig. 22 Brewster mode.

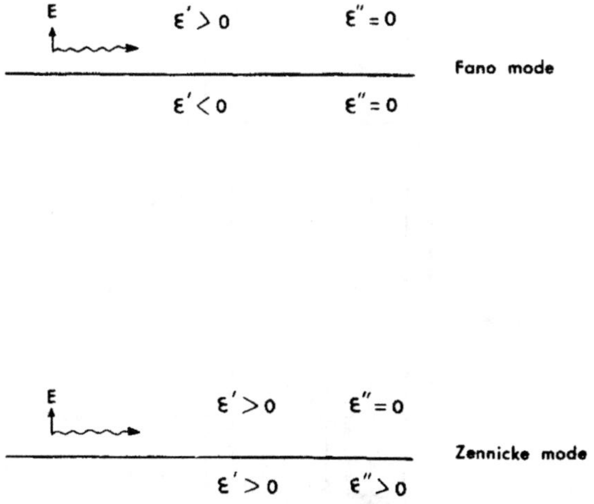

Fig. 23 Fano and Zennicke modes (from Ref. [11]).

10. LOW TEMPERATURES AND FTIR SPECTROSCOPY

The use of low temperature is very important for Far Infrared Spectroscopy for different purposes:

1 - To use high detectivity bolometers or quantum detectors.
2 - To decrease the width of Far Infrared lines.
3 - To cancel phonon difference processes which give a considerable background absorption, and let appear:
- one phonon transitions
- electronic transitions
- phonon addition processes

4 - To allow water to be transparent enough for its use as a solvent of biological materials.
5 - To look at low temperature phase transitions.
6 - To operate super-conductive coils for Zeeman effect studies.

A. High Detectivity Bolometer

A gallium doped germanium bolometer working at 1.8 K (liquid He^4 pumped at 20 Torr) has a detectivity $D^* \simeq 10^{12}$ W^{-1} cm $Hz^{\frac{1}{2}}$. Similar bolometers operated at 0.4 K (liq. He^3 pumped at 0.02 Torr) achieve a $D^* \simeq 10^{13}$ W^{-1} cm $Hz^{\frac{1}{2}}$. Fig. 24 gives the scheme of a He^3 cryostat.

B. Sharp Lines

Figure 25 gives the absorption spectra of TGF B at 290 K and

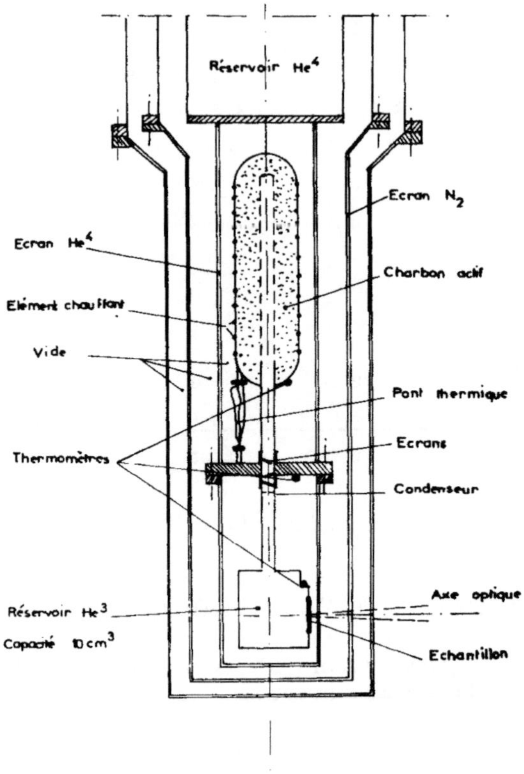

Fig. 24 Scheme of a He^3 cryostat.

80 K. Sometimes lower temperatures can still sharpen the lines. It is difficult to get lines narrower than 0.5 cm^{-1} even at 1 K.

C. Supertransmission

Figure 26 gives a typical case where the transmission of a 1 mm thick CsI plate goes from zero at 300 K to 100% at 4 K for $\lambda > 200$ μm.

D. Water As a Far Infrared Solvent?

Water is not so opaque in the far infrared as it is in the mid and near infrared. Below 20 cm^{-1} the absorption coefficient is about 150 cm^{-1}. Heavy water is slightly more transparent. However this absorption limits to 0.1 mm the thickness of the cell which might be operated in the far infrared, which is generally too small.

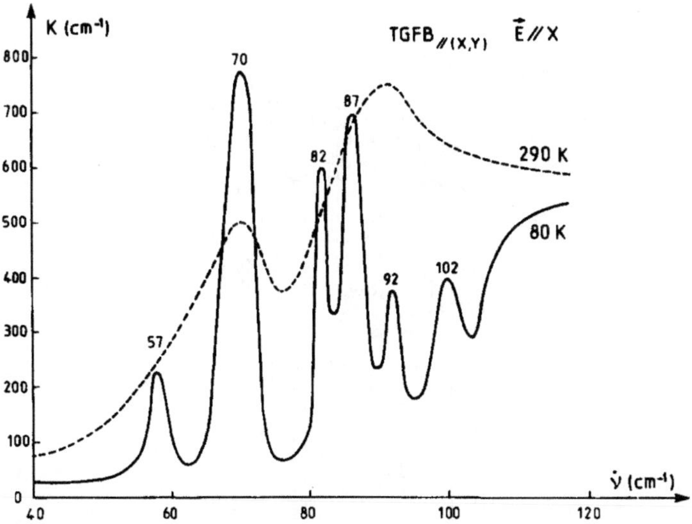

Fig. 25 TGFB single crystal absorption spectra in the far infrared at 80 K (——) and 290 K (---) E parallel to x axis.

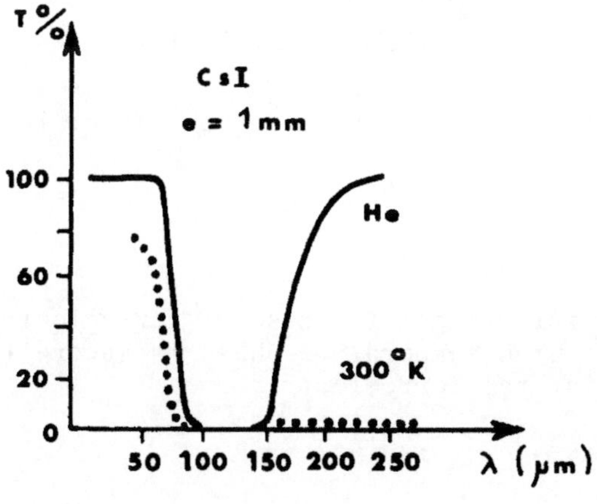

Fig. 26 At liquid helium temperature a 1 mm thick Cesium iodide plate is quite transparent for $\lambda > 250\ \mu m$.

However ice is much more transparent. At 100 K the absorption coefficient is 0.5 cm^{-1} for $\tilde{\nu}$ = 25 cm^{-1} [12]. Recently Champion and Sievers [13] have used a 6 cm long sample where hemoglobin was dissolved.

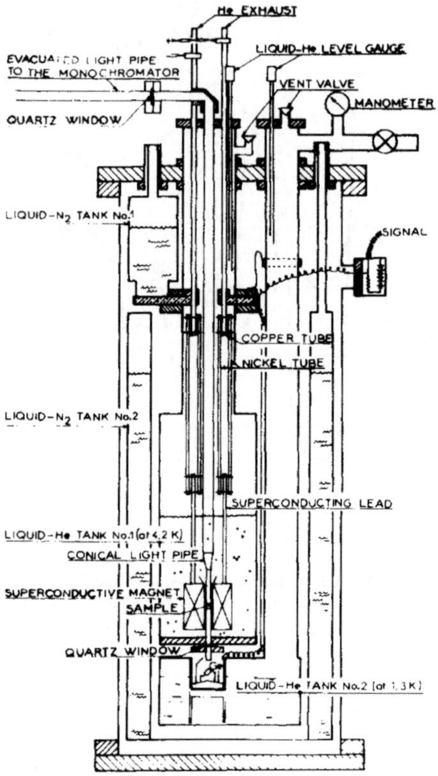

Fig. 27 Cryostat for the study of Zeeman effect in the far infrared.

E. Low Temperature Phase Transitions

Let us cite for example the case of ferroelectric phase transitions. The high temperature phase is generally paraelectric, the lower one being the ferroelectric one.

F. Zeeman Studies

Figure 27 gives the scheme of a cryostat with two vessels. One operating at 4 K contains a super-conductive coil. The other one at 1 K is used to cool a germanium bolometer.

REFERENCES

1. W. N. Hansen, Spectrochim. Acta, 21, 815 (1965).

2. J. Dayet and J. Vincent-Geisse, Journal de Physique 29, 1005 (1968).

3. J. E. Eldridge and F. E. Bates, Solid State Comm., 30, 195 (1979).

4. N. Paraire, Production de Rayonnement Infrarouge par Excitation Cohérente de Polaritons dans le Niobate de Lithium. Thèse de Doctorat d'Etat, Paris, 1979.

5. H. Frohlich, Theory of Dielectrics, Oxford University Press, Oxford, 1958.

6. R. Englman and R. Ruppin, J. Phys., C1, 614 (1968).

7. T. P. Martin, Phys. Rev., 8, 3480 (1970).

8. X. Gerbaux, J. M. Waldschmidt, and A. Hadni, Appl. Opt., 17, 1616 (1978).

9. A. Otto, Z. Phyzik, 216, 398 (1968).

10. D. L. Begley, R. W. Alexander, C. A. Ward, R. Miller, and R. J. Bell, Surface Science, 81, 245 (1979).

11. E. Burstein, A. Hartstein, J. Schoenwald, A. A. Maradudin, D. L. Mills, and R. F. Wallis, Surface Polaritons-Electromagnetic Waves at Interfaces, 1973.

12. E. Whalley, Physics and Chemistry of Ice Symposium, (E. Whalley Ed.), Ottawa, Canada, 1972.

13. P. M. Champion and A. J. Sievers, J. Chem. Phys., 1980.

14. J. A Stratton, Electromagnetic Theory, McGraw Hill, New York, 1942.

15. P. A. Wilks, "A Practical Approach to Internal Reflection Spectroscopy" in Laboratory Methods In Infrared Spectroscopy, Heyden, London, 1972, p. 205.

16. B. N. Mavrin, T. E. Abramovitch, and K. E. Sterin, Sov. Phys. Solid State, 14, 1562 (1974).

17. A. Hadni, G. Morlot, X. Gerbaux, D. Chanal, F. Brehat, and P. Strimer, C. R. A. S., 260, 4973 (1965).

18. B. Fisher, D. Bauerle, and W. J. Buckel, Comm. Sol. State. Phys., 14, 291 (1974).

APPLICATIONS

QUANTITATIVE ASPECTS OF FT-IR SPECTROSCOPY IN INDUSTRIAL APPLICATIONS

P. H. G. van Kasteren

Koninklijke/Shell-Laboratorium,
(Shell Research B.V.)
Amsterdam, Netherlands

1. INTRODUCTION

Infrared spectroscopy is an old technique that is widely used for the analysis and identification of materials. It is an optical method based on the absorption of radiation due to vibrations of molecular groups of which the frequencies fall in the infrared region of the spectrum. Up to some ten years ago this kind of spectroscopy was mainly carried out with wavelength - or frequency - dispersive instruments to record e.g. the energy transmitted through a sample as a function of the wavelength - i.e. the IR spectrum. The limited sensitivity of this technique, however, particularly in comparison with that of fast growing other spectroscopic methods such as mass spectroscopy and nuclear magnetic resonance, caused IR spectroscopy to lose a great deal of its importance.

Although the Michelson interferometer, an optical device for spectroscopy providing an enhanced sensitivity, had already been known for some one hundred years, its wide application was hampered by the fact that the primary output of an interferometer is an interference signal, called the interferogram, instead of a spectrum. The translation of the interferogram, the so-called Fourier transformation, into a spectrum, necessary in order to be of practical use for spectroscopists, requires a digital computer to perform this complex mathematical operation rapidly with high accuracy. Up to some ten years ago this had to be done off-line and the resultant time gap between experiment and result introduced by this remote processing of the data was the major reason why this technique found only incidental and limited application in industry. With the advent of the minicomputer, however,

J. R. Durig (ed.), Analytical Applications of FT-IR to Molecular and Biological Systems, 203–228.

it became feasible to build an interferometer and a computer together in one instrument at a reasonable price, thus providing with the so-called FT-IR spectrometer the analytical tool for industrial spectroscopists. Since then infrared spectroscopy has again become a powerful technique for spectroscopic analyses.

The experience gained so far at universities and research laboratories has clearly demonstrated that this technique is useful not only in science, but also - and may be even predominantly in the near future - in industry, where it can be used for quick checks on the composition of process streams or purity determinations of final products. Another important field of industrial application of FT-IR spectroscopy concerns the characterization and development of catalysts used in petrochemical conversions. Valuable information on intermediate species and hence on catalytic reaction mechanisms can be obtained with relative ease from the high-quality spectra provided by the FT-IR spectrometers. As this subject will be covered by the lectures of Prof. N. Sheppard, it will not be discussed any further here. In addition, structural information about the catalyst itself can be obtained from adsorption experiments using test gases such as carbon monoxyde. The minor shifts in vibrational frequencies of the adsorbed molecules of such gases are characteristic of the interactions with the active sites and therefore provide information about the distribution of the active sites over the surface of the supporting material. Thus the IR spectra provide valuable information about the initial distribution of the active sites and activity of the catalyst, about poisoning during operation, and about the redistribution of the active sites upon regeneration.

The major aspects of a FT-IR spectrometer are its high speed of measurement and its high wavelength repeatability. These features can be used either to record a spectrum of reasonable quality in a very short time, typically one second, or to record a spectrum of extremely high quality, i.e. with a high signal-to-noise ratio (S/N), in a reasonable period of time. These aspects have not only promoted the application of IR spectroscopy to dynamic systems but have also rendered subtraction of such high-quality spectra meaningful, as is necessary for the resolution of minor differences which are normally hidden in the overwhelming spectroscopic information from the bulk of the sample. Such difference spectra, obtained from spectra of a sample and a representative reference, however, show not only the differences in chemical composition but also those originating from physical deviations between them and from imperfections of the spectrometer. One should therefore carefully examine the difference spectra for these three sources in order to discriminate between facts and artifacts.

2. MAIN PRINCIPLES OF THE SPECTROMETER

For the evaluation of the limitations imposed by the spectrometer, we will briefly discuss the major components of the FT-IR instrument (Fig. 1). The IR beam, covering the spectral region of interest, is divided into a transmitted and a reflected beam by a semitransparent mirror, the beamsplitter. The returning beams, reflected from a fixed (FM) and a movable mirror (MM), interfere again at the beamsplitter, pass through the sample, and are subsequently detected. With a monochromatic-light source, the intensity of this interference pattern shows, as a function of the optical path difference between the two arms, maxima and minima due to alternating constructive and destructive interferences to the two beams. Upon moving the mirror at a constant velocity, the detector signal will resemble a cosine function.

As for all optical frequencies, constructive interference occurs when the path difference is zero, all their intensities add up at this position of the movable mirror, whereas at all other positions the cosines add up to an only slightly varying average value. Therefore, the coherent superposition of all the cosine signals, the interferogram, exhibits a center burst at zero path difference between the two arms. The center burst therefore reflects the energy content of the whole spectrum, rather than yielding specific spectroscopic information, which is mainly contained in the other parts of the interferogram outside this center region. Digitization of the interferogram, in order to perform the translation into a spectrum, is carried out with the help of the interference signal of an auxiliary interferometer equipped with a He-Ne laser (mono-signal). The main interferogram is sampled at each zero-crossing of the mono-signal or multiple thereof. Therefore the path difference between two successive data points, in the digitized IR interferogram, is always a multiple of half wavelengths of the laser, or 0.316 micron, and thus an internal calibration is obtained. One stroke, or scan, of the moving mirror produces one interferogram. Coherent co-addition of interferograms, point by point addition, can be accomplished when somehow the starting points or the center burst of consecutively produced interferograms can be made to coincide. This can be done using the interferogram of another auxiliary interferometer equipped with a visible-light source (WL). The center burst of this interferogram, which can be physically made to occur earlier than in the IR interferogram, marks the position where the sampling of the main interferogram is started, notably at the next zero-crossing of the mono-signal.

Such a co-addition of scans results in a higher signal-to-noise ratio as the signal increases in a manner directly proportional to the number of scans co-added, while the noise - because of its statistical nature - only increases with the square root

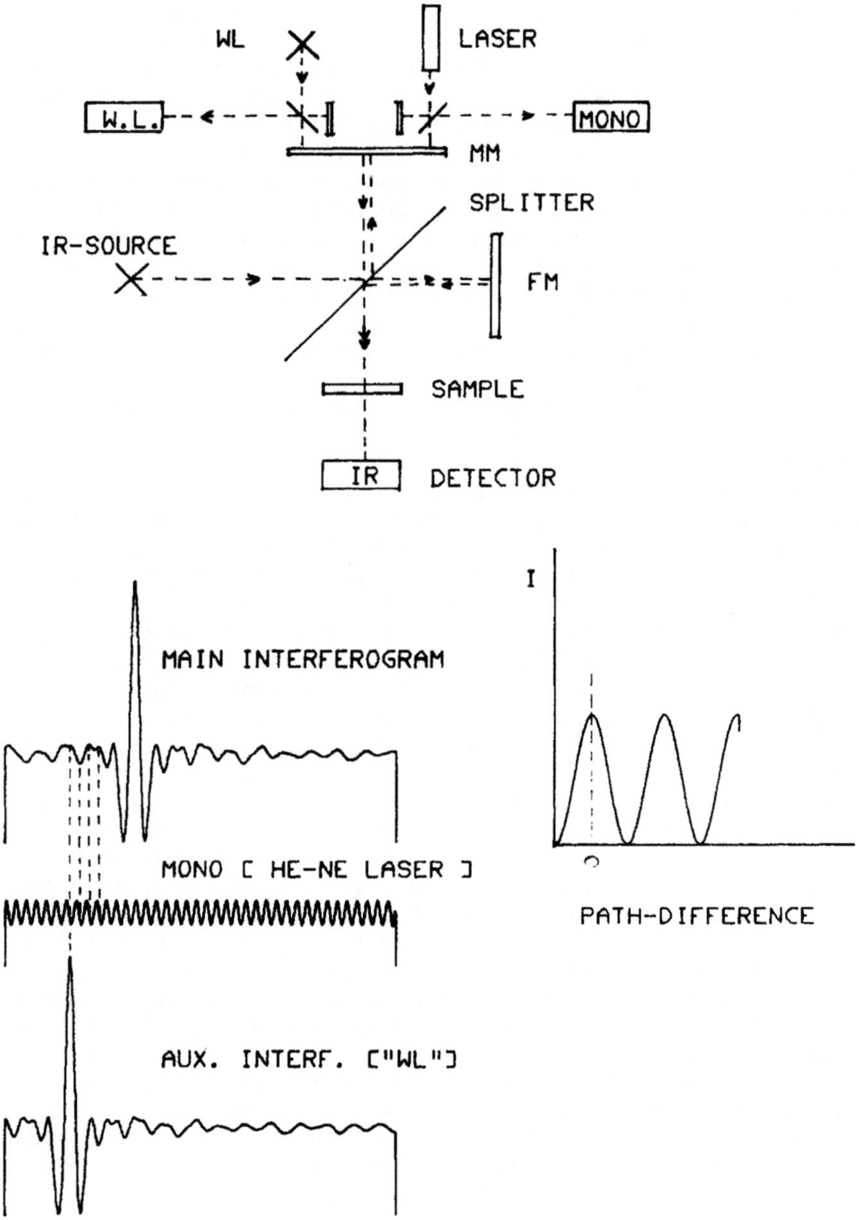

Fig. 1 Schematic signal-representation.

of it. As this holds for statistical noise at any level, this procedure can in principle be repeated indefinitely for ever increasing S/N, but instrumental restrictions or imperfections will gradually become dominant and limit the efficiency of co-addition. Factors which might limit the S/N ultimately obtainable are:

a. the computer word length and the dynamic range,
b. the stability or the repeatability of the spectrum.

3. COMPUTER WORD LENGTH AND DYNAMIC RANGE

The maximum intensity in an interferogram, at full throughput of the sample, should match as closely as possible the maximum input voltage of the A/D converter for highest accuracy in the digitized signal, while simultaneously the noise should be represented by at least four or five units of the least significant digit in the word representing the intensity. The ratio between the maximum intensity in the interferogram and the noise is called the dynamic range of the interferogram. Usually this dynamic range is artificially enlarged, e.g. by recording the center burst with a lower amplification than the other parts of the interferogram, thus compressing it before it is fed into the A/D converter. After digitization the center-burst region is reconstructed by computer multiplication.

In commercially available FT-IR spectrometers, word lengths of 16- (or 32- in double precision), 20- and 24-bits are used. Co-addition of scans is effective until the noise in the averaged, or normalized to one scan, interferogram becomes less than <u>one</u> bit. Thus when the noise is represented by five in a 16-bit word, in the least significant digit, co-addition is no longer effective above $5^2 = 25$ scans and above $25 \times 2^4 = 400$ scans for a 20-bit word length, etc.

The maximum S/N ratio in the interferogram is therefore limited by the word length as it becomes bit-noise limited. Further improvement of the S/N ratio in the corresponding spectra can be achieved by signal averaging in the frequency domain rather than in the time domain (interferograms) by co-addition of spectra. This is possible because the dynamic range requirements are, for a given S/N ratio in the spectra, much larger for the time domain than for the frequency domain as the intensities of all the frequency domains (distance between two data-points ≡ resolution elements) in the spectrum add up in the interferogram at the center burst, zero-path-difference, constituting the total energy content of the spectrum at a single point. This is expressed by the relation $(S/N)_I = (S/N)_S * \sqrt{n}$, where $(S/N)_I$ and $(S/N)_S$ are the signal to noise ratios for one scan in the interferogram and the spectrum, respectively, and n is the number of resolution elements.

The computer requirements and the procedure of measurements can be derived for a practical case. A practical value for $(S/N)_S$ at a resolution of 8 cm^{-1} is 400. Since the whole IR range is divided into 500 resolution elements, the $(S/N)_I$ should have a value of 400 * $\sqrt{500}$ ≈ 9000. If co-addition of 5000 scans, with one hour measurement time, is necessary for the noise level desired in the final spectrum, the required computer word length will be 5000 * 9000 ($\approx 2^{25}$) or 25-bits. It will be clear that in using a computer with a smaller word length, co-addition in the frequency domain (of spectra) instead of in the time domain (interferograms) has to be performed to prevent the spectra from becoming bit-noise limited at an early stage of the measurements.

4. STABILITY OF THE INTERFEROMETER

When spectra are not bit-noise limited, by proper signal averaging in either the time or the frequency domain, the difference of two such spectra should be ideally dominated by statistical noise. After some time, however, upon signal averaging, the noise will become gradually so small that instrumental imperfections become noticeable. Ultimately, the resultant rest absorptions or non-specific absorptions will dominate the difference spectra, being detrimental to the detection limit and the quantitative analysis of minor components in the sample. These imperfections become most pronounced in difference spectra of samples with many narrow absorption bands, since then the wavelength repeatability of the instrument is strongly involved. For a band with an absorption of 2.00 absorbance units (A.U.) and a band width, at half height, of 2 cm^{-1}, one may want to claim that the rest absorption in the difference spectrum is less than 0.1%, which is comparable with the absorption level of e.g. additives in polymers. Such a rest absorption of 0.002 A.U. could result - apart from photometric inaccuracy, which can be removed by proper compensation - from mutual frequency shifts between the two spectra of only 0.004 cm^{-1}. Such shifts between spectra originate from non-coherence of co-added interferograms, due to instabilities of the interferometer, and can be arbitrarily divided into three classes:

a. short-time irregularities, within a single scan causing deformation of the interferogram and thus partial non-coherence of co-added interferograms;
b. medium-term variations, with a time scale of several minutes or the time scale of a single experiment;
c. long-term variations, with a time scale of several hours; two spectra consecutively recorded within several minutes are identical, but two spectra recorded after an interval of several hours are not.

As the resulting features in the averaged interferograms, or in the spectrum, have different time scales, particularly compared with that of statistical noise, they invalidate the trading rule between the number of scans and the "noise" in different ways.

As pointed out previously, these effects will become noticeable only when the spectra are not bit-noise limited and the level of the statistical noise is fairly low. Although it sounds paradoxical, it is therefore true that the better the instrument, i.e. the lower the noise level in a single scan, the sooner the above-mentioned phenomena will be observed upon co-addition. Tests carried out to evaluate the stability of an interferometer should therefore be done at the poorest resolution, say 8 cm^{-1}, of the instrument as the noise level is then lowest. The results of such experiments are given in Fig. 2, showing besides a detail of the solvent spectrum of 1,2,4-trichlorobenzene, the difference spectra of two successive measurements with four different numbers of scans co-added. The absorbance scales have been expanded proportionally to the square root of the numbers of scans. Thus, as long as the co-addition is fully effective, the difference spectra should be identical. For a proper judgement of the wavelength repeatability of the interferometer from the residual absorptions of the solvent bands in the difference spectra, one should realize that the noise, expressed in absorbance units, is much higher at the positions where band maxima occur than outside this region. A noise level of 0.2% T (transmission) represents at the base line of e.g. 90% T an absorbance of 10^{-3} A.U., while the same noise corresponds at a band maximum, of say 10% T, to 10^{-2} A.U., this being about an order of magnitude larger. Thus the larger excursions in the difference spectra at the positions where absorption bands occur in the parent spectra should not be attributed to frequency shifts before the contributions of the noise have been taken into account. With these considerations in mind it appears from Fig. 2 that the co-addition is fully effective in going from 1 to 100 scans, but upon further co-addition up to 1000 scans, the statistical noise becomes so small that other features become preponderant. It is clear that in this case the detection limit and the quantitative accuracy are determined not by the noise but by the residual absorptions. These spectral artifacts are most likely due either to variations in the mirror velocity during a scan or to changes in the optical paths of the arms of the main or the reference (mono) interferometer.

5. EFFECT OF VARIATIONS IN THE VELOCITY OF THE MIRROR

Variations in the mirror velocity during a stroke, i.e. short-term instabilities, do not in principle affect the sampling

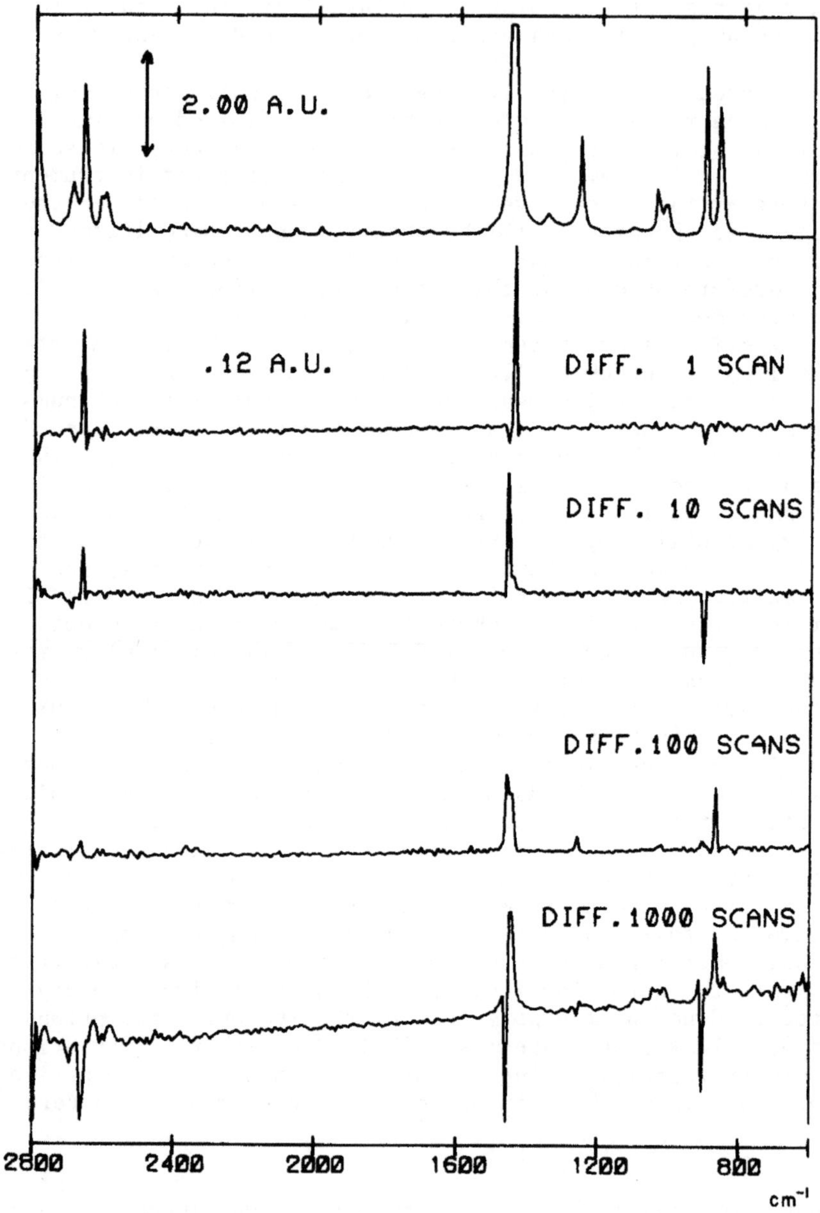

Fig. 2 Difference spectra with scale expansion proportional with $\sqrt{NSS^*}$. *Number of scans

intervals as the mono-signal becomes frequency-modulated to the same extent. However, deviations do occur as a result of mutual phase shifts of the IR and the mono-signals due to different time constants of the IR and mono-signal channels. Zachor [1] has shown for a number of electrical filters that a periodic velocity variation of 1% already causes significant ghost amplitudes, or artifacts in the final spectra. Since the IR-signal channel comprises electrical band-pass filters to remove energy and the noise occurring at higher optical frequencies than the range of interest from the detected signal, the time constants of this channel are different from that for the transfer of the mono-signal. In particular around the cut-off frequencies of the filters, significant phase shifts will occur upon frequency modulations due to variations in the velocity of the mirror. Such phase shifts cause intra- and interscan incoherences of the interferograms, which can be regarded as short- and medium-term instabilities of the interferometer. Variations in mirror velocity may originate directly from oscillations of the servo-control loop due to improper adjustment or indirectly from variable frictions of the moving mirror over the path length involved. The sensitivity of the interferometer to velocity variations, in combination with the filter actions, can be evaluated by deliberately off-setting the timing. After completion of a scan and the subsequent fast fly-back, it takes some time for the mirror to be at the proper speed again. If now the timing for the data collection is set, by physical misadjustment, to start in the acceleration period of the mirror, the effect will be clearly noticeable in the difference spectrum of two successively recorded spectra. The contribution of the electrical filter can be subsequently examined by bypassing it, however, at the expense of the noise level, which should be taken into account.

6. CHANGES IN OPTICAL PATHS

The long-term stability of the interferometer plays a role in noise-level measurements in which co-addition of scans over several hours is required and in compensations of spectra which have been recorded after several hours' intervals. The stability can be evaluated from 1:1 subtraction of sequentially recorded interferograms. The differences at the center bursts between two interferograms as a function of time over a period of about fifteen hours are given in Fig. 3. One notices that these differences are not randomly distributed but show a gradual variation. As the difference interferograms show a resemblance with either the reference interferogram itself or its reflected image, the deviations have been taken to be positive and negative, respectively. These differences may result from intensity variations, which are not harmful to the spectra, or from mutual shifts.

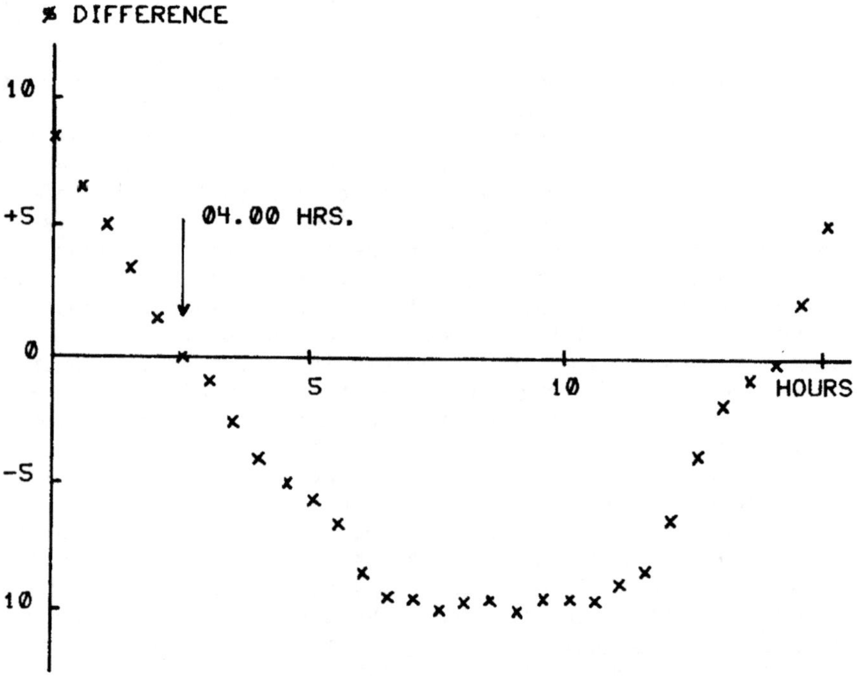

Fig. 3 Stability of interferogram over fifteen hours. Interferograms of 500 scans recorded at intervals of about twenty-five minutes. Differences with respect to the interferogram recorded at 04.00 hours.

As it appears that two interferograms, arbitrarily selected from this series, could not be matched by changing the multiplication factor (≠ 1.0), it is clear that the interferograms are shifted mutually. Figure 4 shows three interferograms out of the series, of which two are about equal (A and C) while the third (B) shows a difference of about 9%, as is evident from the difference spectra A - B and A - C. The mutual shifts, however, are far less than one data point. We have attempted to obtain co-incidence of the interferograms by a mutual shift, after interpolating the interferograms, over distances of tenths of a data point. From the results, Fig. 4, it appears that the actual shift, which can also be estimated from the slope of the center burst, amounts to 0.13 data points, while the amplitude of the interferograms is the same. Such a long-term drift of about 0.1 data point over five hours, is most likely caused by temperature or density gradients, or differences, between the two arms of the main

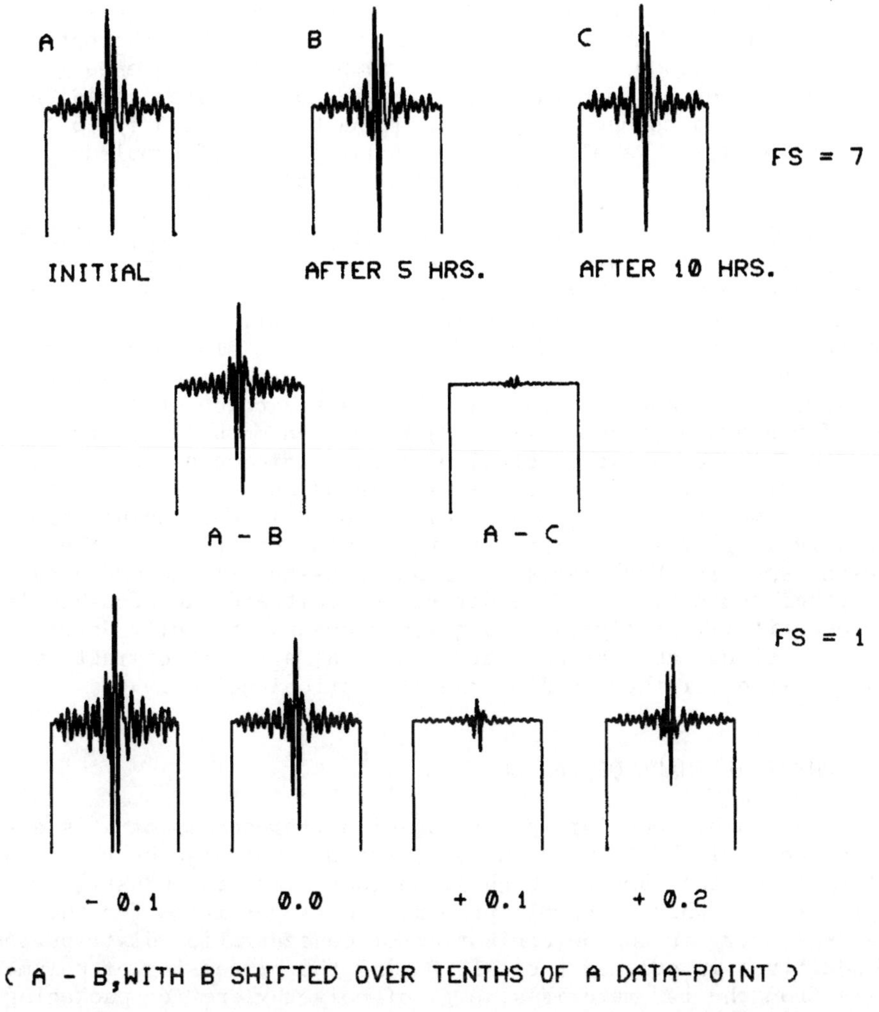

Fig. 4 Interferograms of TCB in 1 mm cell over ten hours.

interferometer or between the main and the auxiliary interferometer. For an estimate of the magnitudes of the path length variations required for the above-mentioned drift, one has to consider the construction of the interferometer, since the total distances between the beamsplitter and the mirrors are involved and not only the stroke of the moving mirror.

For a distance of 5 cm between beamsplitter and mirror, the

optical path, defined as the physical path multiplied by the refractive index, in one arm is given by 10 x refractive index of air (or nitrogen). With a temperature coefficient of this refractive index of about -1.0×10^{-6}, it follows that the optical path changes with 0.1 μm per degree temperature difference (or gradient). Thus a temperature, or equivalent density, gradient of 1°, causes a change in the optical path of about $0.1/0.6 \approx 0.15$ of the laser wavelength or 0.15 datapoints if sampled at every second zerocrossing of the mono-signal.

In view of the construction of the interferometer, which has different gas flow patterns in the various arms, among other things due to an air bearing of the mirror and a temperature difference between the interferometer and its surroundings, it is not unlikely that the observed drift is due to slowly varying density or temperature gradients. The impact of such a drift on the spectra is shown in Fig. 5. The spectra calculated from the interferograms A, B and C (Fig. 4) have been mutually subtracted, A - B and A - C, while a detail of these difference spectra has been plotted with a four-fold scale expansion. These results clearly demonstrate that minor shifts, of 0.13 data points cause not only large rest absorptions at the band positions in the present spectra, but also significant non-specific absorbances in the other regions. Thus the detection limit and the accuracy in the quantitative analyses of a given compound are fully determined by the residual absorptions, artifacts, as a result of instrumental instabilities, rather than by the statistical noise level.

7. PHYSICAL SHAPE OF SAMPLES

The unambiguous interpretation of difference spectra is affected not only by the instrumental factors but also by the physical shape of the sample. Such situations arise in industrial applications, when no sample preparation is permitted for the analyses, e.g. in the determination of concentration distributions of additives in polymer sheets. During the processing of polymer films from the raw materials, e.g. of polyethylene for packaging applications, often small amounts of compounds are blended in, in the extruder, to improve the surface properties of the final films. As such an additive, at an overall concentration of less than 0.1%, is supposed to migrate to the surface, the analysis of commercial films usually implies the determination of the overall content and the magnitude of the surface enrichment. The overall concentration can be determined, though with desctrucion of the film, by dissolving the film in, or extracting it with, a suitable solvent. As this solvent is usually not suited to infrared analysis, the additive has to be isolated by precipitation of the other components and evaporation of this solvent, followed by redissolution in a suitable medium. This method has the disadvantage

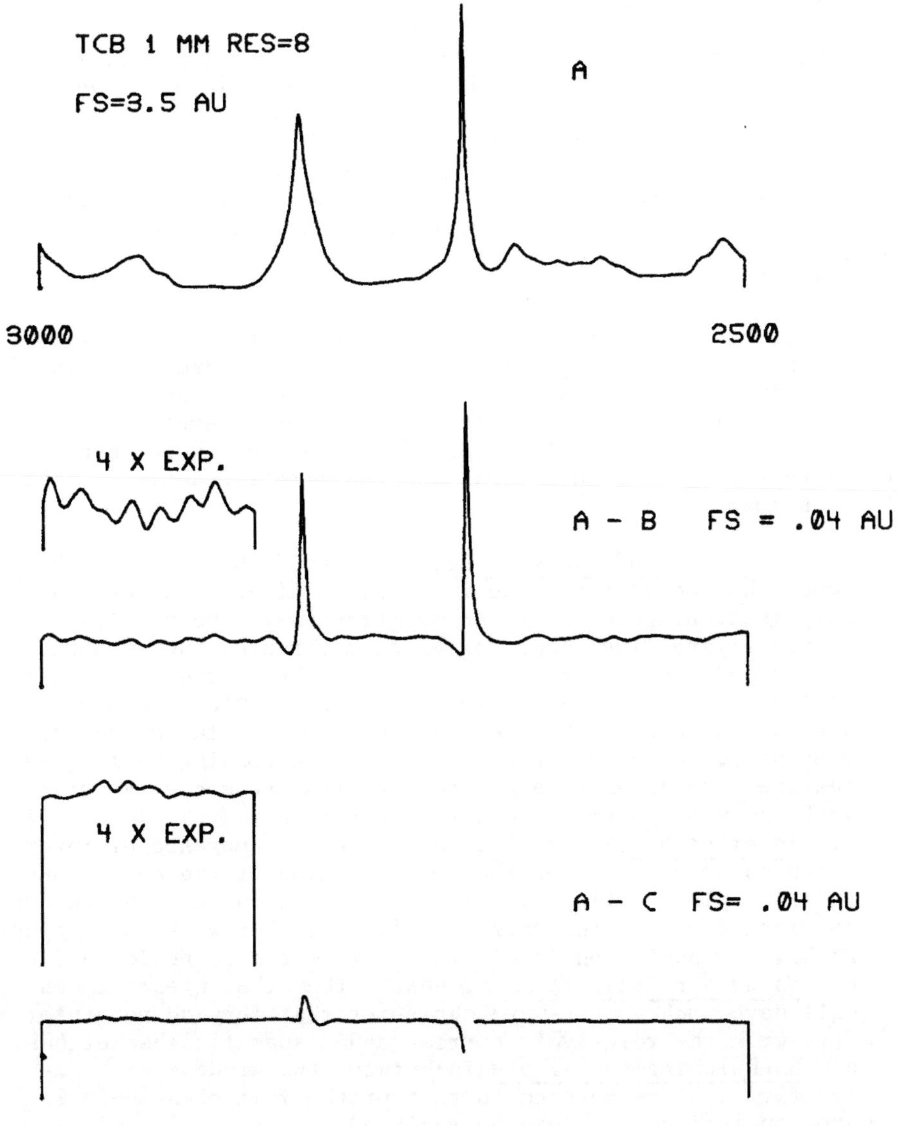

Fig. 5 A = initial spectrum, B = five hours later, C = ten hours later.

of being laborious and is hampered by the uncertainty of stability and recovery of the product in the various steps. Therefore combination of transmission - and attenuated total reflection (ATR) - results is to be preferred for the analysis.

The transmission spectra of films, however, usually show signs of multiple reflections, or interferences, in the film. The amplitude of these is for a non-absorbing plane-parallel layer in a parallel beam on perpendicular incidence in air given by:

$$A_F = 4 * (n - 1)^2/(n + 1)^2, \quad (1)$$

where n is the refractive index of the film material. The distance between two successive maxima of these oscillations, expressed in wavenumbers, is given by: $F = 1/(2 * n * d)$, where d is the thickness of the film.

The specific absorptions of low concentrations of additives may, in the case of strong interference, even be hardly distinguishable in the spectra as they are buried in the interferences. Various ways of getting around this problem of intereferences have been proposed in the literature. As many of these have serious restrictions or drawbacks for quantitative analyses, they will be summarized here.

a. Roughening of the surface, to destroy the coherence between the incident and the internally reflected beams, has the disadvantage of introducing stresses in the sample. In partly crystalline and/or oriented materials, the structure is changed, leading to deformation of the crystalline bands in the spectra. Moreover reflection/absorption phenomena are perceptible in the spectra, dependent on the degree of roughness. Upon compensation, the corresponding "derivative" features in the difference spectra might severely interfere with the weak polymer-additive absorptions.
b. Tapering of the sample, to eliminate coherence of interferences occurring over the cross-section of the beam, causes in principle non-linearities in the relation between measured absorbance versus concentration [2, 3]. For weak absorptions at high transmission levels this effect can be neglected.
c. Using very thin films to ensure that the interferences will have such a long-wave character that they do not interfere with the relatively narrow-banded specific absorptions.
d. Sandwiching the sample in between two windows with refractive indices matched to that of the film might help if good optical contact can be realized.
e. Using polarized light incident on the film surface at the Brewster angle. This method is effective, like the other experimental techniques, in that it prevents the occurrence of interferences, in this case by decreasing the reflection coefficient at the interfaces. The reflection coefficient for polarized light, with its electrical vector parallel to the plane of incidence, passes through a minimum when the angle of incidence is increased.
f. Computer-assisted removal of the interfercences. The

oscillations in the frequency spectrum, when originating from a single-thickness, perfectly parallel layer and a parallel beam, are theoretically described by a sine wave. Therefore, a sine wave can be generated with a computer which, when matched both in amplitude and in phase to that occurring in the spectrum, eliminates the disturbing interferences in the non-absorbing regions of the spectrum. In Fourier transform spectroscopy another approach can be applied as the sine wave in the spectrum corresponds with a spike in the parent interferogram. Removal of this spike, to eliminate the fringes in the spectrum, is possible by zeroing or replacement either by a straight line at the average value of the interferogram or by the same region of a similar interferogram without a spike at that position. The principal drawback of these techniques is - apart from the problems associated with matching, i.e. finding the spike among the other spectral contributions and proper replacement of it - that only the effects perceptible in the spectra are removed and not the physical cause. This is particularly important in quantitative analyses, since the periodicity of the interferences changes underneath an absorption band. This is the result of the changes in the reflection coefficient due to the variations in the optical constants, e.g. the refractive index, at these positions. The problems involved in computer elimination of the fringes will be discussed along with relevant spectra.

As the analyses of polyethylene films, with 600 ppm oleamide mainly at the surface, do not allow any of the sample preparation techniques, we have applied the method of using parallel polarized light, incident at the Brewster angle, to avoid interferences. The selection of the angle of incidence for the minimum reflection coefficient was not too critical, as the curve is known to be rather flat around the Brewster angle. Since compensation of the spectra was necessary to resolve the specific absorbances of the additive, additional spectra have been recorded of polyethylene films without this additive. The interferograms of the blank films (stacks of eight sheets of 0.04 mm) at various angles of incidence and the differences between them are shown in Fig. 6. It appears that the spike is barely perceptible in the interferogram recorded at perpendicular incidence (0°). The difference spectra, with ten-fold scale expansion, show the features related to the interferences. Although in the interferograms recorded at various angles of incidence, the spikes should appear at different positions, this is not very clear in the differences because they are all so broad that they overlap to a great extent. Therefore replacement of the spike in one interferogram by the same part without a spike of one of the others is hardly possible, if at all. This problem of recognition and subsequent replacement is even greater in the interferograms of the

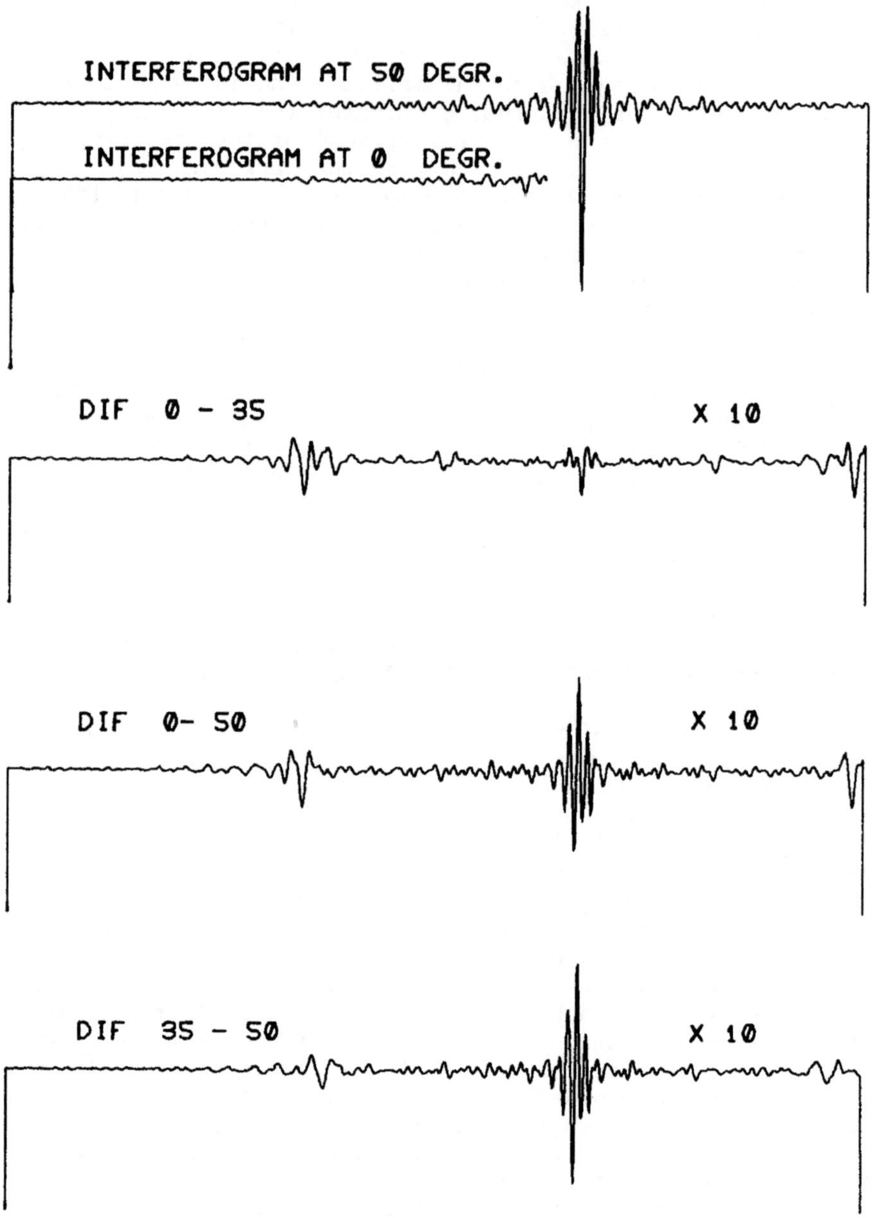

Fig. 6 Interferograms and differences of PE-film with polarized light (//) at various angles of incidence.

film with additive (Fig. 7). The broadness of the spike is due to a distribution of thicknesses in the sample and a distribution of angles of incidence originating from the converging beam of the interferometer. The difference spectra obtained from sample minus reference both at the same angle of incidence are shown in Fig. 8. From the top graph of Fig. 8, giving the spectrum recorded at 50°, it appears that spectrum compensation is required to resolve the specific bands of the oleamide groups of the additive, notably at 1660 cm^{-1} for the C=O stretch and at 1635 cm^{-1} for the NH_2 bending. The difference spectra show that the interferences, or in fact the difference of two interferences, decrease with increasing angle of incidence, as expected. The intensity ratio of the two specific bands, however, changes with the angle of incidence, even when the same oscillations outside this region are assumed to be present under these bands. This does not happen to the two, much broader, N-H stretch absorptions at the higher frequencies, where the interferences appear to be much weaker probably due to a decreased coherence of the reflections at these higher frequencies. The observed anomaly for the other two bands is therefore most likely due to a different effect of the angle of incidence on the interference phenomena for the two bands, as a result of different optical constants of the C=O and the NH_2 groups. Another possibility is that it is due to preferential orientation of these groups. This has been examined by comparing the transmission spectra with two ATR spectra of the same film. The ATR spectra have been recorded with the sample parallel and perpendicular to the stretch direction of the film, thus mutually rotated over 90° (Fig. 9). Figure 9 shows, besides the specific absorptions of the oleamide group, the CH_2-rocking vibrations of the PE matrix at 730 and 720 cm^{-1}. Long-chain CH_2 units in PE exhibit in the crystalline state a splitting up of the CH_2 rocking into an in-phase and out-of-phase vibration due to interactions between adjacent chains. It appears from the literature that the dipolar momenta of these two vibrations, at 730 and 720 cm^{-1}, are perpendicular to each other. In stretched PE films, the crystallites or the chains are lined up to some extent and this orientation becomes evident in the intensity ratio of the two vibrations. Now in the ATR spectra, these rocking vibrations having preferential orientation in the plane of the films clearly show the effect of sample rotation, while the oleamide absorptions do not. Thus if the oleamide groups do have a preferential orientation, it is not in the plane of the film. They cannot be oriented perpendicular to the surface either since the absorption in transmission, at perpendicular incidence, would have been absent as there is no coupling between the electrical vector and the dipolar momentum. Moreover, the intensity would be strongly dependent on the angle of incidence, which appeared not to be the case. All the evidence therefore indicates that the anomalous behavior of the intensities of the oleamide bands at perpendicular incidence originates from artifacts due to interferences. After all, this is not so surprising since

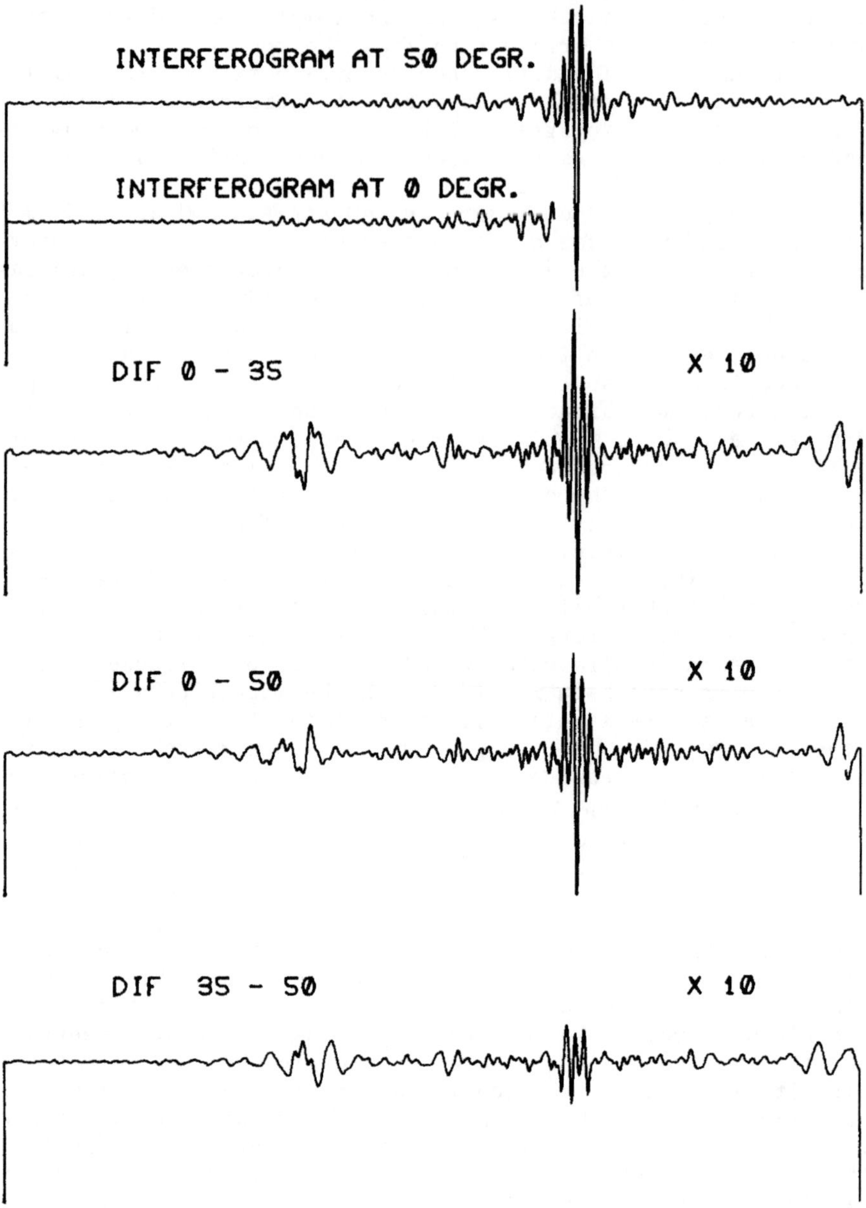

Fig. 7 Interferograms and differences of PE-film with 600 PPM oleamide, with polarized light (//), at various angles of incidence.

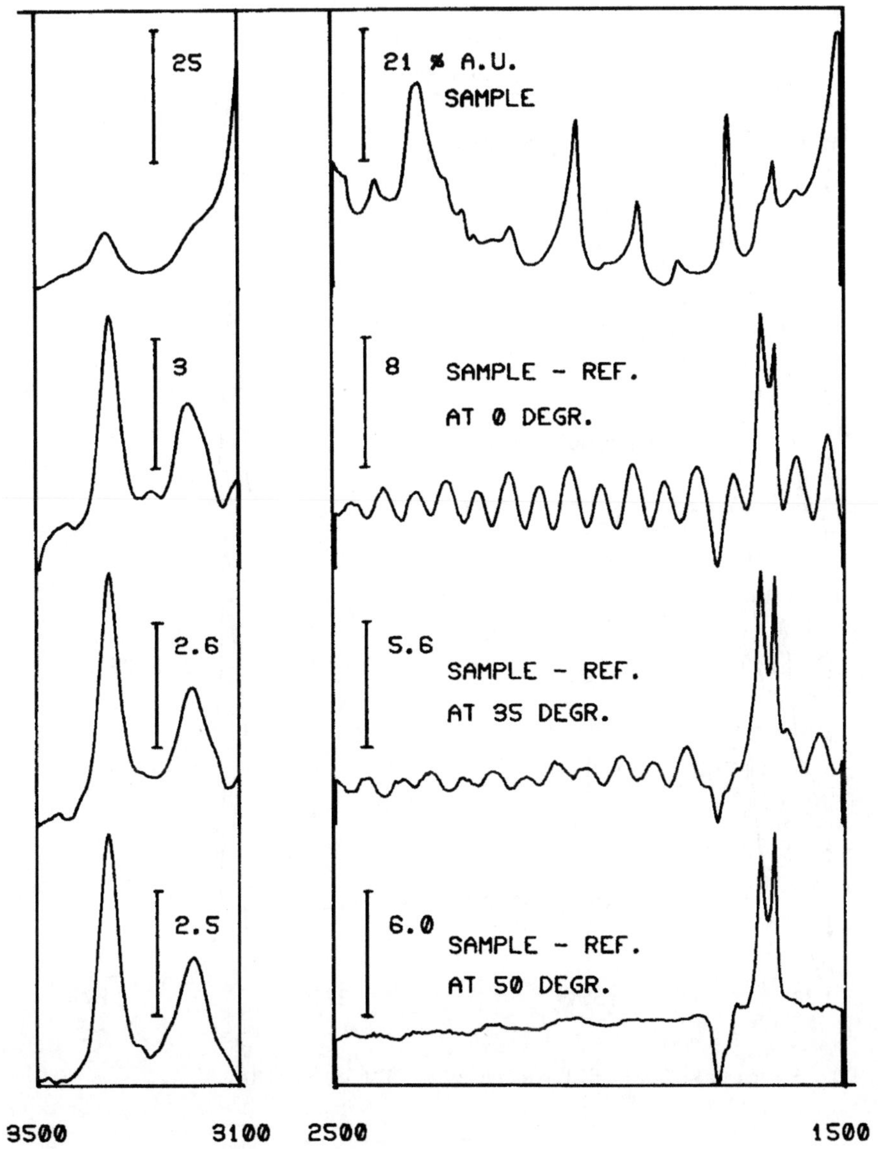

Fig. 8 Difference spectra, parallel-polarized light at three angles of incidence, of PE-film with 600 PPM oleamide.

from the ATR and the transmission spectra it appeared that the oleamide at the surface, over a penetration depth of about 2 µm is 50 times enriched, while ESCA experiments indicated a 1000-fold

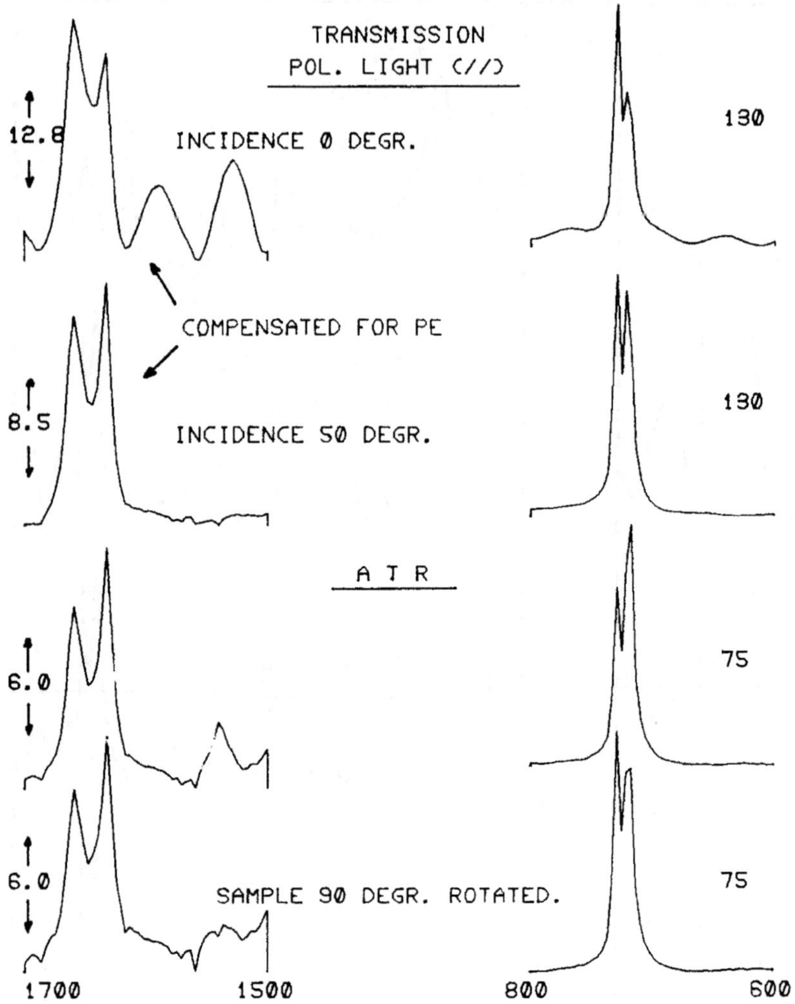

Fig. 9 Spectra of PE-film with 600 PPM oleamide.

enrichment over a depth of 2 nm. Therefore this film with additive can, with some simplification, be regarded as a three-layer (oleamide/PE/oleamide) system, having a complex transmission behavior particularly upon varying the angle of incidence. One could in this simplified model even imagine that the number of reflections in the oleamide layers is different from that in the PE film, thus off-setting the basis for a quantitative analysis.

The analyses of the spectra demonstrate that removal of the interferences from the spectra by computer techniques is inadequate for quantitative analyses of this type of samples, but that it is necessary to avoid the interferences.

8. ON-LINE DETECTION AND ANALYSES

The potentials of FT-IR for quantitative on-line analyses provide valuable possibilities for industrial use. An interesting application appeared to be the determination of the CH_2 and the CH_3 group content in a sample peak of a polymer in the eluent of a gel permeation chromatograph (GPC/FT-IR). The concentrations of alkanes to be expected in the solvent, 1,2,4-trichlorobenzene (TCB), are usually in the range of 10 to 100 ppm. A preliminary approach was, as a simulation, to analyze the absorptions of n-C_7 into CH_2 and CH_3 content in a liquid stream using n-C_5 and n-C_{15} as references.

A simple flow-through system, comprising a solvent reservoir, a pump, a sample-injection device, a column, and an IR liquid flow-through cell, were the means of realization. The samples, n-C_5, n-C_7, and n-C_{15}, were introduced into the solvent stream by injecting 10 µl of a solution containing 0.42 mg sample. The column was introduced to cause a peak broadening in order to be able to split up the band spectroscopically into a number of spectra with different concentrations of the sample. The IR flow cell was made of two KBr disks clamped together with a spacer in between to give a path length of 0.4 mm. The supply and return lines were cemented in holes drilled in the KBr disks. With this simple construction a cell volume was obtained of only 0.1 ml at an optical cross section of 1 cm^2.

Averaged interferograms of the solution flowing through the IR cell, each obtained by co-addition of sixty scans at a resolution of 8 cm^{-1}, were stored in separate data files. The time of measurements for such an interferogram amounts to 42.6 seconds, while the time interval between data collections for two such interferograms is only 0.7 s (≡ 1 scan). The resultant loss of information is small because this time interval is short compared to the residence time of the sample in the IR cell, being about six seconds. The transformations of the interferograms can be carried out simultaneously with the continuing data collection. The spectra obtained should be compensated for the solvent (1:1) and only the sequence showing sample absorptions is stored (Fig. 10). The CH_2 and CH_3 content of n-C_7 can in principle be calculated from their specific, though overlapping, group absorptions. The specific absorbances of these groups have been derived from the spectra of n-C_5 and n-C_{15} recorded in the same set-up.

<u>The isolation of the CH_2 and CH_3 group absorptions from spectra of n-pentane and n-pentadecane</u>

Spectra of n-C_5, n-C_7 and n-C_{15} can be described in terms of the contributions of CH_2 and CH_3 groups as follows:

$$A_{15} = C_{15} \times 0.8585 \times \alpha_2 + C_{15} \times 0.1415 \times \alpha_3 \quad (2)$$

$$A_7 = C_7 \times 0.7000 \times \alpha_2 + C_7 \times 0.2000 \times \alpha_3 \quad (3)$$

$$A_5 = C_5 \times 0.5833 \times \alpha_2 + C_5 \times 0.4167 \times \alpha_3 \quad (4)$$

where: $A_{(\ )}$ = absorbance of the compound
$C_{(\ )}$ = amount of compound in the beam (mg/cm^2),
$\alpha_{(\ )}$ = absorption coefficient (cm^2/mg) for CH_2 and CH_3.

Thus the absorption spectrum of n-C_7 can be expressed in those of n-C_5 and n-C_{15} with the relation:

$$A_7 = f_1 \times A_{15} + f_2 \times A_5. \quad (5)$$

The fractions f_1 and f_2, or solution vectors, can be determined by a multi- linear-regression analysis. Combination of the above relations gives:

$$0.7000 \times C_7 = 0.8585 \times f_1 \times C_{15} + 0.5833 \times f_2 \times C_5 \quad (6)$$

$$0.300 \times C_7 = 0.1415 \times f_1 \times C_{15} + 0.4167 \times f_2 \times C_5. \quad (7)$$

As to the ratio between the concentrations of n-C_5 and n-C_{15} it follows from (6) and (7) that

$$C_5 = 1.358 \times \frac{f1}{f2} \times C_{15}. \quad (8)$$

Subsequently, expressions for the absorption of isolated CH_2 and CH_3 groups, derived from relations (2), (4) and (8), are given by:

$$A_{CH_2} = C_{15} \times \alpha_2 = 1.514 \times A_{15} - 0.3786 \times \frac{f2}{f1} \times A_5$$

and

$$A_{CH_3} = C_{15} \times \alpha_3 = 2.297 \times \frac{f2}{f1} \times A_{15} - 2.120 \times A_5.$$

These calculations can be carried out accordingly with the experimental values of 0.322 and 0.706 for f1 and f2, respectively. The results are shown in Fig. 11. Subsequently these group absorption spectra are used in a curve-fit routine to analyze the absorptions of n-C_7 in the various selected spectra. The solution

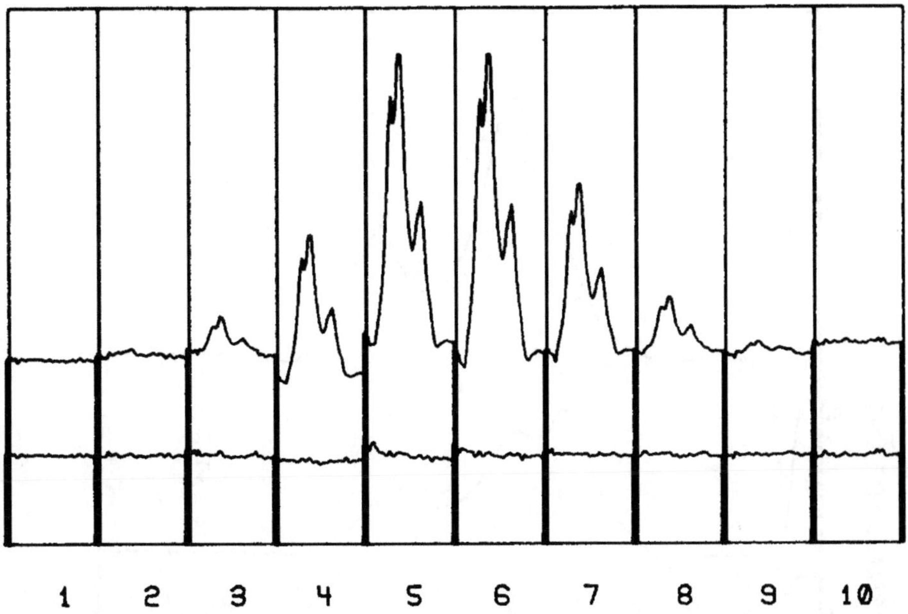

Fig. 10 IR-fractions of on-line detection of GPC-peak: solvent: 1,2,4-trichloro-benzene; flowrate: 1 ml/min.; sample injected: 0.42 milligram of n-heptane; spectra: res=8, 60 scans (43 sec.); bottom lines: restabsorptions after comp. for CH_2 and CH_3.

vectors, or compensation factors, with the corresponding uncertainties and the standard errors in the base line (the rest absorption) are given in Table 1. The CH_2 content, in each spectrum, follows directly from the values in the first two columns.

The average concentration of n-C_7 in trichlorobenzene, during the time of recording, can also be calculated as follows. The sum of the solution vectors, e.g. 2.057 for CH_2, corresponds to the total amount of n-C_7, 0.42 mg, injected. With a flow rate of 1 ml/minute of TCB (ρ = 1.454 g/ml) and a time of measurement of 43.3 s one finds that a value of 1.0 for the solution vector of CH_2 groups corresponds to 194.6 ppm and for the CH_3 groups to 458.1 ppm n-C_7. It appears from Table I that the concentrations calculated from the CH_2 and the CH_3 groups agree very well with each other. The close fit of the deconvolution appears from the rest absorptions, shown in Fig. 10 bottom lines, calculated from the n-C_7 absorption diminished with the group absorptions using the solution vectors as compensation factors. The last column

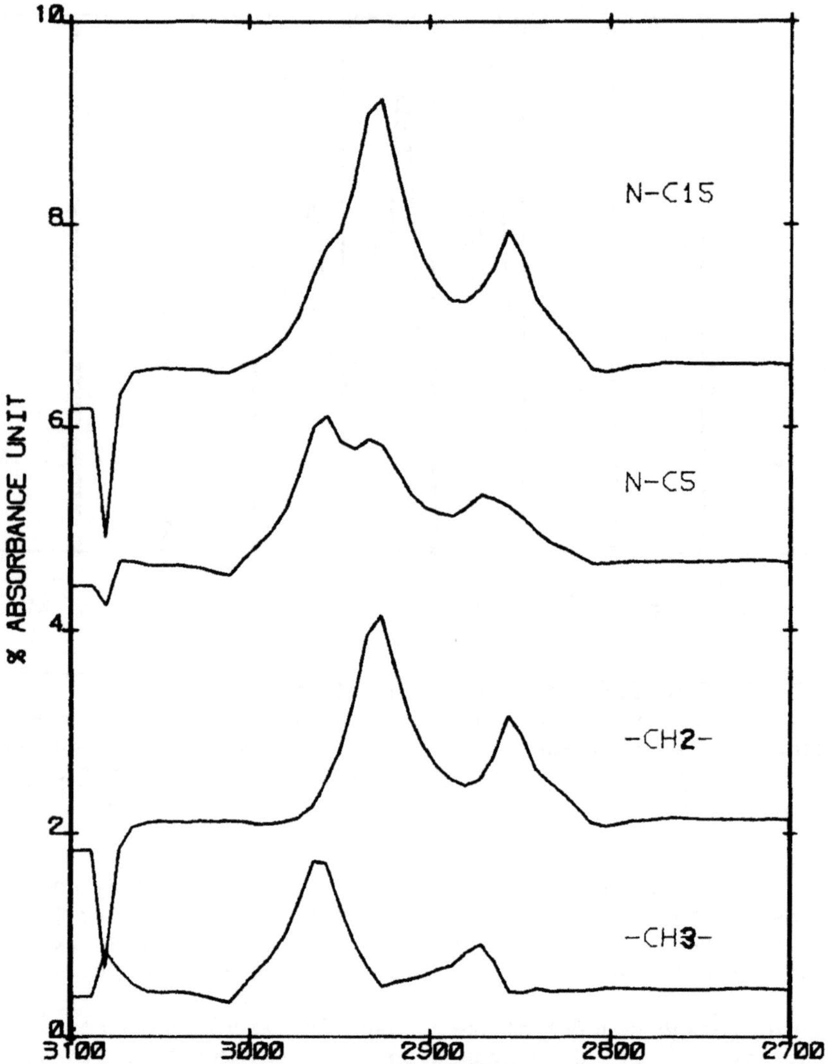

Fig. 11 CH_2 and CH_3 group-absorptions isolated from spectra of n-pentadecane and n-pentane (both from on-line measurements).

of Table 1 shows that even at a n-C_7 concentration as low as 2 ppm a reasonable value for the CH_2/CH_3 ratio is obtained (64.2% compared with the real value of 70.0%). The detection limit for n-C_7 is estimated at about 1.5 ppm. From these experiments it

TABLE 1

RESULTS OF THE ANALYSES OF SPECTRA OF n-C_7

Spectrum no.	Solution vectors (region 3000-2750 cm^{-1}) x 10^{-2}		Uncertainty in solution vector x 10^{-2}		Standard error in base line %Aux 10^{-2}	Concentration n-C_7 p.p.m. from		% CH_2 in n-C_7
	CH_2	CH_3	CH_2	CH_3		CH_2	CH_3	
1	0.00	0.15	0.17	0.28	0.06	0.0	0.7	-
2	1.04	0.58	0.25	0.41	0.87	2.0	2.7	64.2
3	7.01	2.88	0.37	0.61	1.30	13.6	13.2	70.9
4	28.77	12.09	0.25	0.41	0.87	56.0	55.4	70.4
5	58.92	25.23	0.76	1.25	2.63	114.7	115.6	70.0
6	61.89	26.49	0.77	0.47	1.65	120.4	121.4	70.0
7	34.45	14.62	0.24	0.40	0.85	67.0	67.0	70.1
8	10.57	4.49	0.29	0.47	1.00	20.6	20.6	70.2
9	2.46	0.85	0.22	0.36	0.77	4.8	3.9	74.3
10	0.62	0.00	0.29	0.47	1.00	1.2	0	-
Total	205.73	87.38				400.3	400.5	

appears that quantitative analysis of the composition of an alkane directly from the solution spectra, recorded in less than one minute, is possible at concentrations down to 2 ppm, provided representative reference spectra are available.

9. CONCLUSIONS

The introduction of FT-IR instruments, with their high speed and accuracy, has opened up those fields of industrial application for IR-spectroscopy, which have been inaccessible so far. It should be realized, however, that in exploiting the high sensitivity of these instruments (high-signal-to-noise ratio), the final results should be closely examined for artifacts, because these results consist of a difference between two large quantities. It is therefore the task of spectroscopists to ensure that the results are indeed significant.

REFERENCES

1. A. S. Zachor and S. M. Aaronson: Delay Compensation; Its Effects in Reducing Sampling Errors in Fourier Spectroscopy. Appl. Optics, 18, 68 (1979).

2. J. L. Koenig: Effect on Non-uniform Distribution of Absorbing Material on the Quantitative Measurement of Infrared-band Intensities. Anal. Chemistry, 36, 1045 (1964).

3. T. Hirschfield: Diagnosis and Correction of Wedging Errors in Absorbance-subtract Fourier Transform Infrared Spectrometry. Anal. Chemistry, 51, 495 (1979).

APPLICATIONS OF FOURIER TRANSFORM INFRARED TO POLYMERS

Jack L. Koenig

Department of Macromolecular Science
Case Western Reserve University
Cleveland, Ohio, U. S. A.

1. INTRODUCTION

Fourier transform infrared spectroscopy is having a major impact on the study of polymer systems. The advantages of FT-IR arise primarily from the higher signal-to-noise ratio obtainable from the higher energy throughput of the system and the capability of signal averaging of the spectra to achieve a further increase in the signal-to-noise ratio. Additionally, the full benefit of the higher signal-to-noise ratio can be realized through the use of the data processing potential arising from the availability of the minicomputer.

The experimental difficulties arising from the study of polymer spectra result from the fact that the systems are all complex mixtures of components and structures. Therefore the observed bands are broad and weak for polymers. The structure elucidation is considerably improved by using absorbance subtraction to remove interfering absorbances and absorbance magnification to enhance the observed signals. Curve fitting techniques allow the observation of very small differences in absorbance and the removal of spectral features which do not contribute to the problem at hand. Several reviews have been written on the application of FT-IR to polymers [1, 2, 3] and it is unnecessary at this time to reiterate the various techniques which are used as some of these newer methods are covered in the adjoining papers. It will rather be our aim to acquaint the reader with some of the newer applications using FT-IR for the study of polymers from our laboratory.

J. R. Durig (ed.), Analytical Applications of FT-IR to Molecular and Biological Systems, 229–240.

2. MULTICOMPONENT POLYMER SYSTEMS

Nearly all practical polymer systems contain more than one component either in the form of additives to improve stability, processibility or performance or as other polymers in the form of blends or as grafted systems. In this manner, the ultimate properties are improved and better utilization of the system is realized. However, nearly all of these multicomponent systems have difficulty in achieving homogeneity or in maintaining it since most systems are incompatible with polymers. FT-IR is a potential tool for the investigation of the mutual compatibility of the various components. The very small spectral effects which are introduced as a result of the interactions required for the polymers to be compatible can be detected using FT-IR. As a result FT-IR has been used to investigate the interactions of additives such as antioxidents [4] as well as for blends of different polymers in a search for the structural basis of compatibility [5, 6].

The conceptual basis of the application of FT-IR to the study of compatibility is easy to understand. If two polymers are completely imcompatible, which implies phase separation, one can synthesize the spectrum of the blend by coadding the spectra of the two pure components in proportion to the composition of the blend. If proper care is taken, a comparable spectrum will be obtained which will compare with the observed spectrum of the blend. On the other hand, when the polymers or additives are compatible, it implies that there is only one phase and that a distinct chemical interaction occurred between the components of the mixture. In this case, there should be observable spectral differences between the coadded spectra of the pure components and the spectra of the mixture. The spectral changes most easily observed will be spectral frequency shifts arising from chemical interactions since such shifts will not exist in the spectra of the components and therefore cannot be subtracted from the spectra of the mixture. Absorbance subtraction of the coadded spectra and the experimental spectra will highlight these spectral frequency shifts and make them easily observable. Chemical interactions may not be sufficiently strong to induce frequency shifts but changes in the specific absorbances can occur since the intensities are more sensitive to the environment than are the frequencies. Changes in intensities as a result of chemical interaction are more difficult to recognize but in the case of antioxidents in polybutadiene these effects have been observed [4].

Weak interaction effects are observed between the phenolic hydroxyl and the arylamine NH groups of antioxidants with the rubber unsaturation. The FT-IR spectra of 2,6-di-tert-butyl-4-methyl-phenol, (CA-1) in the cis-1,4-polybutadiene (BR) reveal these interactions. Fig. 1 shows the spectra of purified BR (spectrum A) and BR stabilized with CA-1, 6.4% by weight (spectrum B). Efforts were taken to minimize oxidation by purification and infrared sampling was taken under a N_2 atmosphere. At this concentration some of the CA-1 bands are hidden in the spectrum of BR. When a spectral subtraction of BR + CA-1 minus BR is performed (B - A), the difference spectrum resembles that of pure CA-1 in KBr (spectrum C). Since A and B are the spectra of two different samples, subtraction was undertaken relative to a specified BR band to eliminate all of the BR bands from the spectrum. Scaling of the subtraction to the band at 1655 cm^{-1} (#1) produces negative bands at 740 and in the 3100-2800 cm^{-1} region. If scaled to the band at 740 cm^{-1} (#2), the effect is reversed and some of the BR absorbances are now positive. In #1 the BR spectrum is oversubtracted and, in #2, it is undersubtracted. Apparently, none of the BR bands can be relied upon as a proper internal thickness since they all change in band shape and intensity with addition of CA-1. The deviations in band shape and intensity are not due to oxidation because the experiments are performed in N_2 and none of the spectral changes which accompany oxidation are present such as formation of trans-methine (975 cm^{-1}) C-O groups (1090-1060 cm^{-1}), and carbonyl groups (1740-1700 cm^{-1}). Furthermore, the effects appear to be reversible; removing the CA-1 from BR by acetone leaching reduces the spectral changes that are proposed to arise from molecular interaction between the BR and the antioxidant. Fig. 2 shows the FT-IR spectra in the range 700-2000 cm^{-1} of pure polymethyl methacrylate (PMMA), pure poly(vinylidene fluoride) (PVDF) and a blend of PVDF/PMMA containing 75:25 pars by weight. It is apparent that the synthesized spectrum is almost identical to that of the experimental spectrum of the blend. Hence, this blend is incompatible. Figure 3 shows the results of an identical experiment using PVDF/PMMA blend of 39:61 parts by weight (equivalent to a 1:1 mole ratio). It is not possible to approximate the spectrum of the blend by absorbance addition of the two polymer spectra. In particular, bands at 1180, 1072, and 840 cm^{-1} are not matched. Adjusting the weighting parameters does not produce a better spectrum. Hence, the blend of this composition is compatible. The spectral changes probably arise from the chemical interactions introducing conformational changes in the polymer chains in the compatible blend [6].

It was proposed very early that spectral "separations" could be made using FT-IR without physical and chemical separations which are extremely difficult for many polymer systems. Recently [7] a particularly interesting example has been presented where the spectral separation was used to identify a pigment which could not be easily separated or identified by procedures other than FT-IR. Two films were cast from different solvents and the spectra indicated that an acetone cast film had more blue pigment

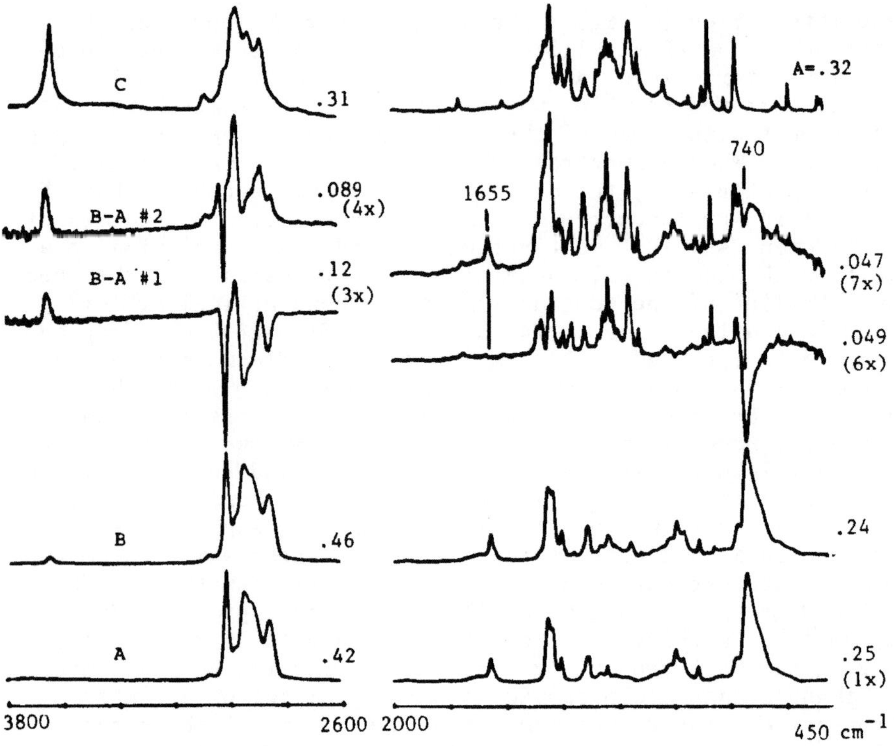

Fig. 1 FT-IR spectra showing interaction effects in BR + CA-1: (A) BR, unoxidized and without additives; (B) BR + CA-1, unoxidized; (B - A) #1, difference spectrum of (B) minus (A) normalized to band at 1655 cm^{-1}; (B - A) #2, difference spectrum of (B) minus (A) normalized to band at 740 cm^{-1}; (C) spectrum of CA-1 in KBr.

than the one cast from chloroform. The major spectral features attributed to the pigment were at 1588 and 804 cm^{-1}, but these two bands were not sufficient to characterize the pigment completely. Compensation of the bands due to the major components such as the resin allowed the complete infrared spectrum to be obtained and identified as methyl violet.

3. MULTIPHASE POLYMERS

Polymer systems often exhibit multiple phases and the absorbance subtraction technique allows, for the first time, the determination of the spectra of each of the pure phases. This application, although of practical importance, has historical

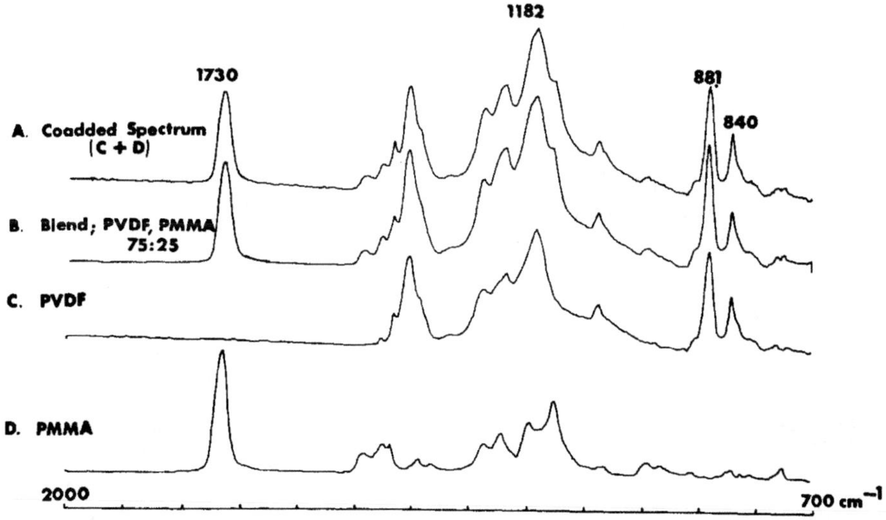

Fig. 2 FT-IR spectra of pure polymethyl methacrylate (PMMA), pure poly(vinylidene fluoride) (PVDF) and a blend of PVDF/PMMA.

significance as this application was the first described in the literature for the use of absorbance subtraction [8]. A recent application involves the study of poly(tetramethylene terephthalate) (PTMT) which undergoes a stress-induced reversible crystal-crystal phase transition [9]. Spectral assignments were made using phase differences generated by stretching and solvent casting. Cast films were made with three different solvent/nonsolvent ratios producing films with different amounts of each phase. Absorbance subtraction allowed significant improvement to be made in the band assignments. The bands at 1409 and 1504 cm^{-1} are due to benzene ring modes. The relatively weak bands at 1322 and 1350 cm^{-1} appear in the amorphous state, increase in the α sample, and almost disappear in the β sample. The low absorbance of these bands and the frequency range suggest they are assignable to the methylene wagging mode vibrations of the α phase. The three remaining bands characterize the methylene bending region for the other phases. In the amorphous phase, two bands are observed at 1388 and 1394 cm^{-1}. Upon annealing, the 1388 cm^{-1} increases in intensity. Upon stretching, the 1388 cm^{-1} peak decreases in intensity relative to the 1394 cm^{-1} band. The 1394 cm^{-1} band is assigned to the CH_2 bending mode of the structure with the two methylene groups adjacent to the central trans bond in the methylene chain. It is possible by using the above assignments and others to estimate the relative contribution of each of the phases to the infrared spectrum: (i) the bands at 917 and 1456 cm^{-1} indicate α crystallinity; (ii) bands at 960 and 1470

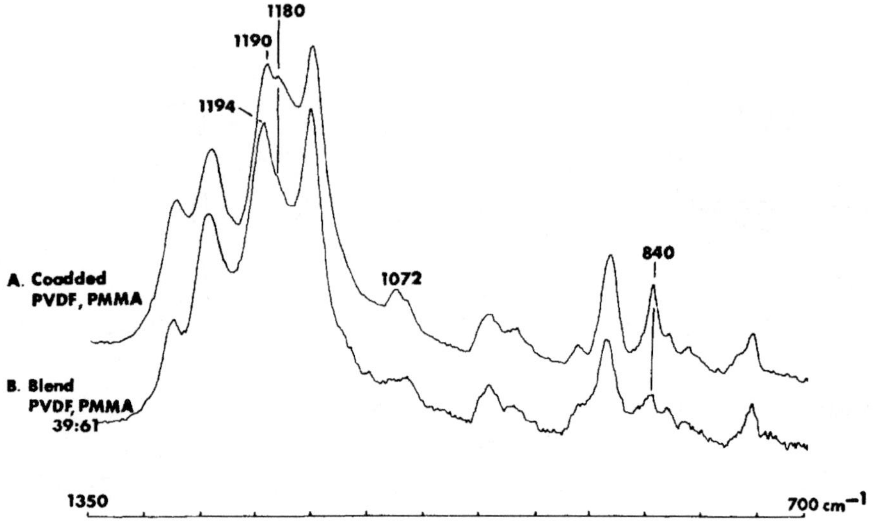

Fig. 3 PVDF/PMMA blend of 39:61 parts by weight.

cm^{-1} indicate β crystallinity; (iii) the 935 and 1388 cm^{-1} bands contribute from both the α and β phases.

Another multiphase polymer, poly(vinylidene fluoride) (PVF_2), has received considerable study because it exhibits piezoelectric and pyroelectric properties. Two types of molecular conformations are found to exist with one form (phase I) having a planar (all trans) zig zag structure and phase II having a TGTG' chain structure. A third form of PVF_2 is obtained when the polymer is crystallized under certain conditions. For the infrared studies [10], phase I was prepared by casting films from a 9% (by volume) solution in N-N-dimethylacetamide (DMA). The cast films were then heated to 200°C and the air quenched to room temperature. This resulted in films of unoriented phase II which was transformed to phase I by biaxially drawing the film at room temperature. Samples of phase II were prepared by casting 0.2 mil films from a 6.3% (by volume) solution of DMA at 60°C. The cast films were then heated to 200°C and then air quenched to room temperature. Samples of phase III were prepared by casting the films from 6.3% solution in DMA at 60°C. Each of the films was studied under conditions which reveal the crystalline phases of each of the forms and that of the amorphous phase. In order to obtain the crystalline spectra, two samples with different degrees of crystallinity are required. The less-crystalline spectrum is then subtracted from the more-crystalline spectrum. The criteria for the amount of subtraction is to remove the contribution from the amorphous regions of the film. These spectra permit accurate determination of band location of the ordered material unshifted by the amorphous material. The crystalline spectrum of phase III yields approximately 45 bands in the 3800-400 cm^{-1} range which was found to be inconsistent with the originally proposed structure for phase II. Several possible phase III chain conformations are acceptable from the spectral point of view, but the conformation TTTGTTTG' is consistent with the X-ray results and the theoretical calculations.

4. CHEMICAL TRANSFORMATIONS IN POLYMERS

Since the infrared bands of polymers are inherently broad and weak, there is considerable difficulty in detecting minor chemical changes occurring on the polymer chain. However, using absorbance subtraction, it is possible to remove the interfering component, that is the polymer, and just examine the reacting portions of the chain and the reaction products. This is illustrated in Fig. 4 where no discernable changes are apparent between the spectra of an unoxidized and oxidized BR [11]. However, the difference spectrum quite clearly shows an oxidative change. The increases in absorbance above the baseline are due to increases in absorbing species (oxidation products), and any decreases are due to losses of absorbing species (point of attack on polymer chain). The ordinate scale is expanded about a hundred fold, as indicated. This spectrum, obtained after only one hour at 25°C, shows the initial oxidative changes. The 1080-1110 cm^{-1} region is characteristic of C-O groups. Loss of cis-methine groups is shown by the negative 3007 and 740 cm^{-1} bands in the difference spectrum. As the oxidation proceeds, the number of absorbing species increases and the complexity as well as the total absorbtivity. Using absorbance subtraction techniques the spectral changes can be followed and the concentration of the various species sorted out [11]. The difference spectrum after three hours shows formation of trans-methine at 975 cm^{-1} and specific carbonyl groups at 1727 and 1700 cm^{-1}. The thirty-four hour difference spectrum shows a broadening of the C-O band. After six hundred-forty hours, only 27% of the cis-methine remains. Similar work has been reported on thermal degradation of polyacrylonitrile [12].

5. POLYMERS AT INTERFACES

Polymers are often used at surfaces to enhance adhesion, to lubricate, to reduce wear, etc. One of the most important industrial applications is the role of polysiloxanes in improving the performance of fiber glass reinforced composites. It has been found that when silane coupling agents are applied on the surface

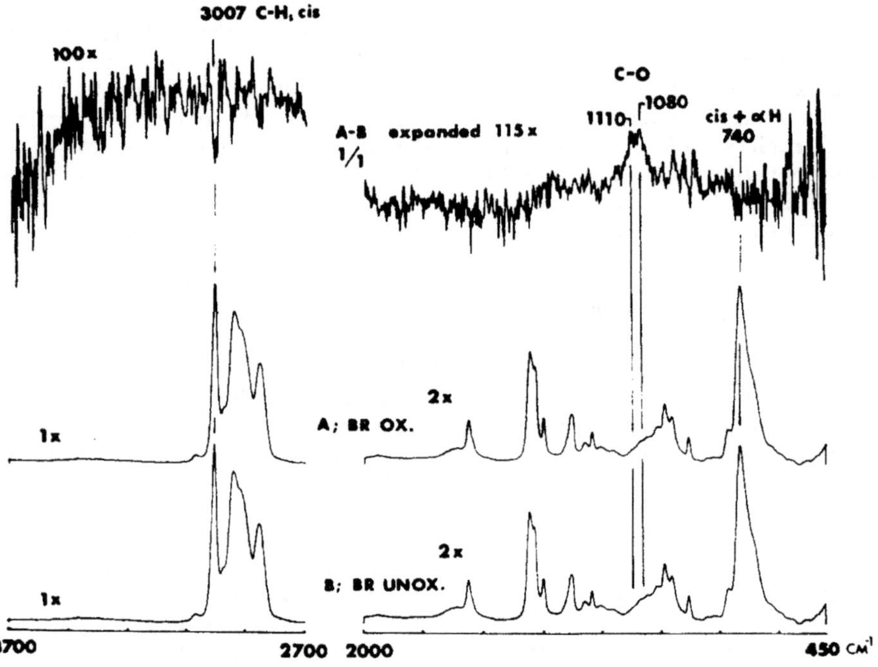

Fig. 4 Cis-1,4-polybutadiene oxidation, 1 hour at 25°C. Bottom, unoxidized; center, oxidized; top, difference spectrum.

of the fiber glass, the composite has improved mechanical properties particularly shear modulus and improved lifetimes especially under humid conditions. Although for the past thirty years application of coupling agents has been standard industrial practice, the mechanism of the reinforcement has not been understood. One of the most useful theories of the reinforcement is based on the presence of chemical bonding of the coupling agent to the glass surface, but no experimental evidence has been obtained to verify such a chemical bonding theory. Application of FT-IR techniques to this problem have led to a better understanding of this process [13, 14, 15]. The first problem was to detect the presence of the coupling agent on the surface of the strongly absorbing glass. This was accomplished by dispersing the samples in KBr pellets and using absorbance subtraction to remove the spectrum of the glass. This process is further complicated by the fact that the coupling agent is present at very low concentrations and the inorganic portion of the spectrum is very similar to that of the interfering glass. For the initial study [13], Cab-o-sil was selected as this silica has a very high surface area and so would require more of the coupling agent. The

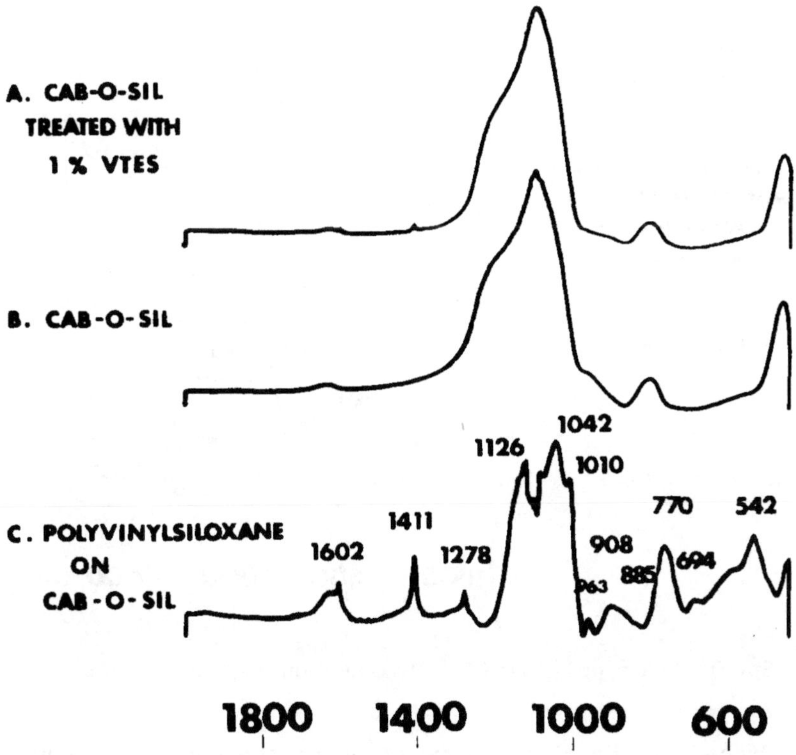

Fig. 5 The spectrum of Cab-o-sil, Cab-o-sil treated with an aqueous solution containing 1% by weight VTES, and the spectrum of the treated fiber after the glass has been subtracted from it.

coupling agent was vinyl (triethoxy silane) (VETS). Figure 5 shows the spectrum of Cab-o-sil, Cab-o-sil treated with an aqueous solution containing 1% by weight VTES, and the spectrum of the treated fiber after the glass has been subtracted from it. This difference spectrum corresponds to the spectrum of poly (vinyl siloxane) indicating that the coupling agent on the glass surface exists in polymeric form. Similar studies were carried out on E-glass fibers where the surface area of the fibers is 500 times less than that of the porous silica. Figure 6 shows the result of this experiment. The interesting aspect of this work is that the amount of coupling agent on the fiber surface is the same as for the silica where it is known that only a monolayer exists. Consequently, the spectral results demand that the coupling agent not only be polymeric but also exist as multilayers, perhaps as many as 100-200 layers.

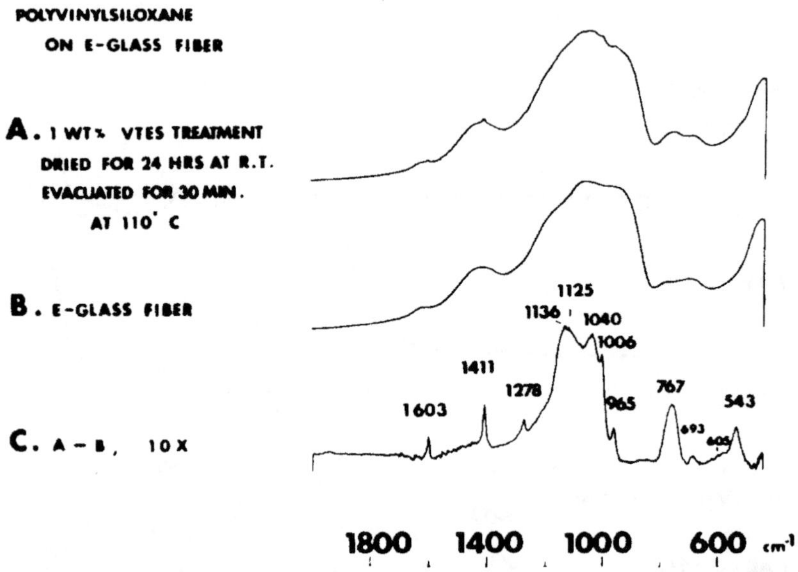

Fig. 6 Spectrum of poly(vinyl siloxane) on E-glass fiber.

Evidence for the chemical reaction of the coupling agent with the surface is shown in Fig. 7. The spectra were obtained by absorbing an oligomeric polyvinylsiloxanol on the surface of the fiber and subjecting the sample to a heat treatment to induce reaction. After subtraction, new bands appear at 1235, 1170, 1080 and 1045 cm^{-1}, while the bands at 970 cm^{-1} and 893 cm^{-1} decrease in intensity. The band at 970 cm^{-1} is assigned to the SiO stretching mode of the glass hydroxyls. The band at 893 cm^{-1} is attributed to the SiO stretching mode of polyvinylsiloxanol. This indicates that the surface hydroxyls and siloxanols of the coupling agent have reacted. Possibly the substantial reduction of intensity of the 970 cm^{-1} band must be due to the siloxane formation through the interface. Bond formation at the interface leads to the appearance of new bands at 1170 and 1080 cm^{-1}. This experimental evidence for the oligomeric silane treated Cab-o-sil system supports the contention of siloxane bonds through the interface.

6. CONCLUSION

The application of FT-IR to the study of polymers has proven to be very useful through the elimination of interfering

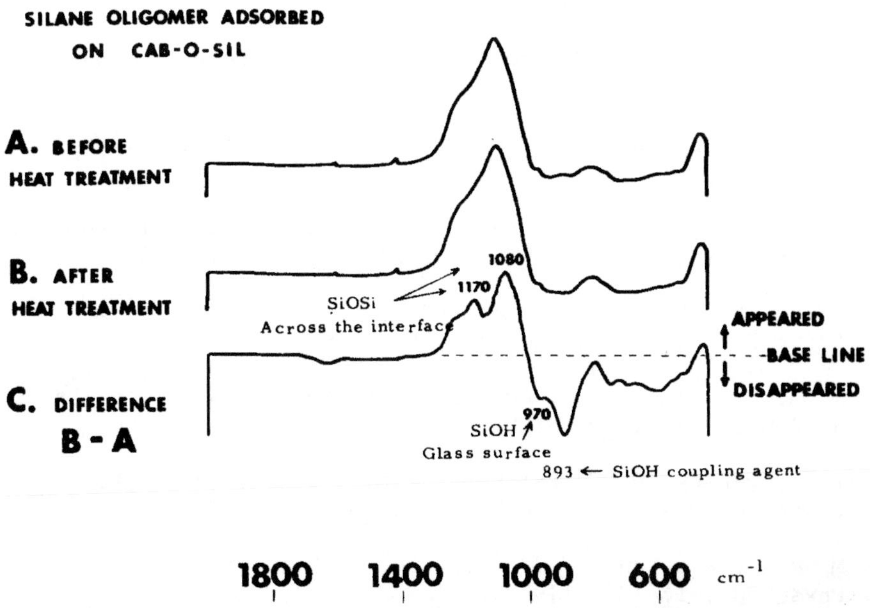

Fig. 7 Spectrum of silane oligomer adsorbed on Cab-o-sil.

absorbances and the detection of small spectral changes associated with low levels of interaction between substances and chemical reactions occurring at the early stages of reactions. Further applications will be found and the continued development of FT-IR in the characterization is expected.

REFERENCES

1. J. L. Koenig, Appl. Spectrosc., 29, 293 (1975).

2. L. D'Esposito and J. L. Koenig, Fourier Transform Infrared Spectroscopy, Vol. 1, Academic Press, 1978, p. 61.

3. M. M. Coleman and P. C. Painter, Rev. Macromol. Sci., C16, 197 (1977).

4. R. L. Pecsok, J. R. Shelton and J. L. Koenig, Rubber Chem. Technol., in press.

5. S. Wellinghoff, E. Baer and J. L. Koenig, J. Polymer Sci.: Phys., 15, 1913 (1977).

6. M. M. Coleman, J. Zarian, D. F. Varnell, and P. C. Painter, Polymer Letters, 15:745 (1977).

7. A. S. Manocha and R. M. Montgomery, Appl. Spectrosc., 32, 344 (1978).

8. M. M. Coleman, P. C. Painter, D. L. Tabb, and J. L. Koenig, J. Polymer Sci.: Poly. Letters Ed., 12, 577 (1974).

9. B. Stambaugh, J. B. Lando, and J. L. Koenig, J. Polymer Sci.: Polymer Phys. Ed., 17, 1063 (1979).

10. M. A. Bachmann, W. L. Gordon, J. L. Koenig, and J. B. Lando, J. Appl. Phys., 50, 6106 (1979).

11. R. L. Pecsok, P. C. Painter, J. R. Shelton, and J. L. Koenig, Rubber Chemistry and Technology, 49, 1010 (1976).

12. M. M. Coleman and R. J. Petavich, J. Polymer Sci.: Polymer Phys. Ed., 16, 821 (1978).

13. H. Ishida and J. L. Koenig, J. Colloid Interface Sci., 64, 555 (1978).

14. H. Ishida and J. L. Koenig, J. Colloid Interface Sci., 64, 565 (1978).

15. H. Ishida and J. L. Koenig, J. Polymer Sci., Polymer Phys. Ed., 17, 615 (1979).

INFRARED SPECTRA OF GLOBULAR PROTEINS IN AQUEOUS SOLUTION

Jack L. Koenig and D. L. Tabb

Department of Macromolecular Science
Case Western Reserve University
Cleveland, Ohio, U. S. A.

1. INTRODUCTION

Globular proteins exist in their native state in aqueous environments. However, the most detailed structural models for protein conformation are obtained in the solid state by X-ray diffraction of single crystals [1]. These solid state structures cannot be assumed to remain invariant upon dissolution in aqueous media [2]. Vibrational spectroscopy is sensitive to conformational changes in biopolymers, and it can be used to obtain structural information in both the solid and solution states. Vibrational spectra are most commonly obtained by infrared absorption and Raman scattering techniques. The information obtained from these methods are highly complementary [3]. Raman spectroscopy has been used to study globular proteins in both the solid and aqueous phases [4], while infrared spectroscopy has been fundamental in determining the secondary conformation of fibrous proteins in their native solid state. Infrared absorption studies of aqueous solutions of globular proteins have been severely limited due to the strong interferring H_2O absorption [5]. The majority of infrared solution studies have been concerned with the amide I and amide II bands in D_2O solutions [5]. However, the amide frequencies observed in D_2O solution are affected by hydrogen-deuterium exchange in the peptide groups [5]. Infrared studies of globular proteins in H_2O has been limited to the amide I region of myoglobin, β-lactoglobulin, and α-casein, and the spectra were obtained by extremely difficult and tedious differential techniques [6].

In this chapter we present the infrared absorption spectra of several globular proteins in H_2O solutions in the frequency

J. R. Durig (ed.), Analytical Applications of FT-IR to Molecular and Biological Systems, 241–255.

range of 2000 to 800 cm^{-1}. The spectra were obtained with a Fourier transform infrared spectrophotometer. The absorption contribution from the H_2O was digitally subtracted from the solution spectra. Spectra obtained in native solution and solid phases are presented for hemoglobin, bovine serum albumin, ribonuclease, β-lactoglobulin, and α-casein. The effect of dissolution in H_2O on the secondary structures of these globular proteins will be examined.

These proteins have different secondary structures. Hemoglobin has been found to be highly alpha helical (70-80%) in both solid state and aqueous solution [7]. Optical rotary dispersion (ORD) studies indicate that native bovine serum albumin in approximately 55% alpha helix and 45% disordered [8]. Crystal structural analysis of ribonuclease by X-ray diffraction [9-11] has shown a small (15%) helical content and a larger extended chain contribution. From the circular dichroism (CD) studies of aqueous solution of ribonuclease, Greenfield and Fasman [12] have calculated 9-12% alpha helix, 33-43% anti-parallel beta sheet, and 38-44% disordered for the secondary structure. From the previous studies, the secondary structure of beta lactoglobulin appears similar to ribonuclease. ORD [13, 14] and CD [15] studies indicate an alpha helical content of 10% with the remainder approximately equally divided between the beta sheet and disordered conformations for β-lactoglobulin. Alpha casein has been showed by CD studies of cast films to exist in the disordered conformation with little ordered secondary structure [16].

2. EXPERIMENTAL

Bovine albumin, β-lactoglobulin B, and α-casein were purchased from Miles Laboratory. Hemoglobin and ribonuclease were obtained from Sigma Chemical Co. and Worthington Biochemical Corp., respectively. All solutions were at a concentration of 10%. The proteins were dissolved in distilled H_2O except for β-lactoglobulin where 0.1 M NaCl solution was used. For solid state samples, films were cast from these solutions on the AgCl plates. They were then dried in vacuo at room temperature.

The infrared spectra were obtained interferometrically on a Digilab FTS-14 Fourier Transform Spectrophotometer at a resolution of 2 cm^{-1}. The solution samples were run in 5-10 μ thick AgBr liquid cells. The procedure for obtaining the spectra of proteins in aqueous solutions is as follows. First, the absorption spectra of H_2O and the protein solution are individually recorded and stored in digitized form in the unit's data system. Then, the absorbance contribution of the H_2O in the solution spectrum is digitally subtracted from the solution spectrum in the absorbance mode. This method is shown for hemoglobin in

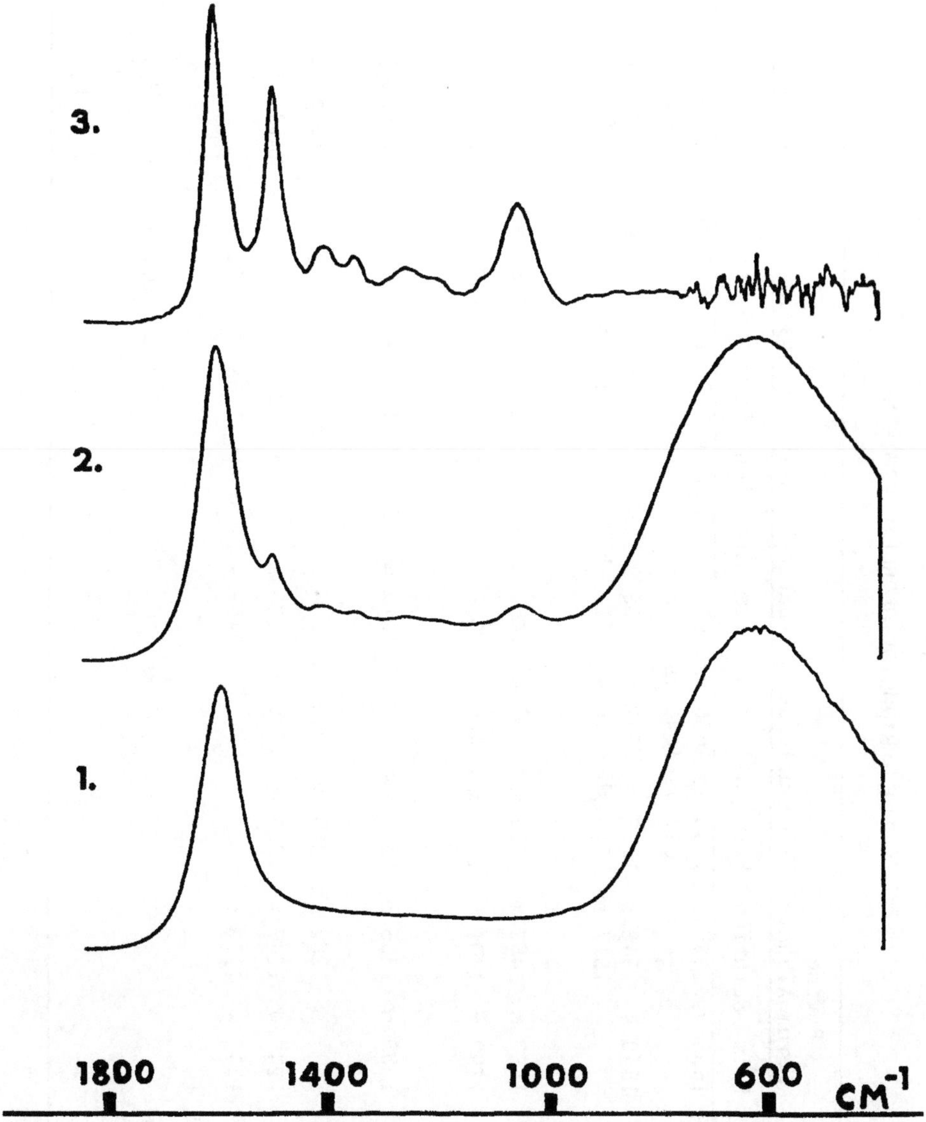

Fig. 1 Aqueous solution infrared spectrum of hemoglobin obtained with Fourier transform spectrophotometer: 1) absorbance spectrum of H_2O; 2) absorbance spectrum of aqueous hemoglobin solution; 3) absorbance spectrum of hemoglobin in aqueous phase (spectrum 2-spectrum 1).

TABLE I

INFRARED BANDS OF GLOBULAR PROTEINS[a]

Hemoglobin		Bovine Serum Albumin		Ribonuclease		β-Lactoglobulin		α-Casein		Assignments
Film	Solution	Film	Solution	Film	Solution	Film	Solution	Film	Solution	
1657 vs	1656 vs	1656 vs	1655 vs	1653 vs	1656 vs	1633 vs	1632 vs	1656 vs	1655 vs	Amide I
				1647 vs	1646 vs					Amide I
1541 s	1547 s	1540 s	1548 s	1538 s	1548 s	1538 s	1551 s	1538 s	1551 s	Amide II
			1517 sh	1518 sh	1520 sh	1517 sh		1519 sh		Try[b]
1467 w	1469 w									
1455 m	1455 m	1452 m	1452 m	1450 m	1453 m	1451 m	1450 m	1448 m	1453 m	$\delta(CH_2)$, $\delta(CH_3)$
1390 m	1400 m	1390 m	1401 m	1394 m	1405 m	1398 m	1401 m	1399 m	1402 m	ν_s (COO^-)
1367 w	1367 w									
1306 m	1308 m	1305 m	1305 m	1309 vw		1313 w	1315 w	1312 w	1319 w	ν_t (CH_2), ν_w (CH_2)
			1282 sh	1260 sh	1260 sh					Amide III
1245 w	1255 w	1248 w	1247 w	1237 m	1240 m	1238 m	1247 m	1242 m	1245	Amide III
1170 w	1170 w	1172 w	1173 w			1168 w	1165 w	1158 w	1162 w	
1124 sh	1130 sh									
1098 ms	1106 ms			1089 ms	1101 ms			1101 m	1098 m	ν_t (NH_2), ν_r (NH_3^+)[c]
1050 sh										ν (C-N)

TABLE 1 (Continued)

Hemoglobin		Bovine Serum Albumin		Ribonuclease		β-Lactoglobulin		α-Casein		Assignments
Film	Solution	Film	Solution	Film	Solution	Film	Solution	Film	Solution	
970 w								980 m	979 m	

Key: s-strong; m-medium; w-weak; v-very; sh-shoulder.

[a] Frequencies in cm^{-1}.
[b] Reference [29].
[c] Reference [30].

Fig. 1. Spectra one and two are the absorbance spectra of H_2O and of hemoglobin in H_2O, respectively. Spectrum three is the resultant solution spectrum of hemoglobin after the absorbance contribution of H_2O has been digitally subtracted.

3. RESULTS AND DISCUSSION

The infrared spectra in aqueous solution and solid state of hemoglobin, bovine serum albumin, ribonuclease, β-lactoglobulin, and α-casein are shown in Figs. 2-6, respectively. The infrared frequencies are listed in Table 1. Only the more salient features of the spectra will be discussed.

A. Amide I Frequencies

For each individual protein the amide I frequency is essentially the same in solid and aqueous phases. Susi et al. [6] observed the same result with the amide I frequencies of myoglobin and β-lactoglobulin.

The amide I bands of hemoglobin and bovine serum albumin, both highly α-helical, occur at 1656 ± 1 cm^{-1}. The amide I frequency of α-helical myoglobin has previously been observed at 1652 cm^{-1} [6]. α-Casein, which is highly disordered, also has an amide I band at 1655 cm^{-1}; previous workers [6] reported the same value. For β-lactoglobulin the amide I frequency is at 1632 cm^{-1}, as was found by Susi et al. [6]. From polypeptide and fibrous protein studies [5, 17], this particular amide I frequency is indicative of the anti-parallel beta sheet conformation. The amide I band of beta lactoglobulin is highly unsymmetric. The skewness of this band towards higher frequencies is indicative of the high disordered content present in the protein. Ribonuclease in both solution and solid phases shows two amide I frequencies: 1653 and 1647 cm^{-1} in the solid and 1656 and 1646 cm^{-1} in solution. The stronger band is at 1646 cm^{-1}. With the consideration of the 1632 cm^{-1} value for the amide I of β-lactoglobulin, the 1646 cm^{-1} amide I value does not indicate a well formed anti-parallel beta sheet structure in ribonuclease. Even though there is extensive extended chain structure present, X-ray diffraction results [9-11] have indicated that the β-structure is somewhat irregular. The 1646 cm^{-1} band seems tobe indicative of irregularly formed β sheet together with disordered structure in ribonuclease. The shoulder at 1656 cm^{-1} can be assigned to the smaller α-helical content of ribonuclease.

B. Amide II Region

The amide II frequency of these five globular proteins does not allow a distinction to be made between the α helix, β sheet,

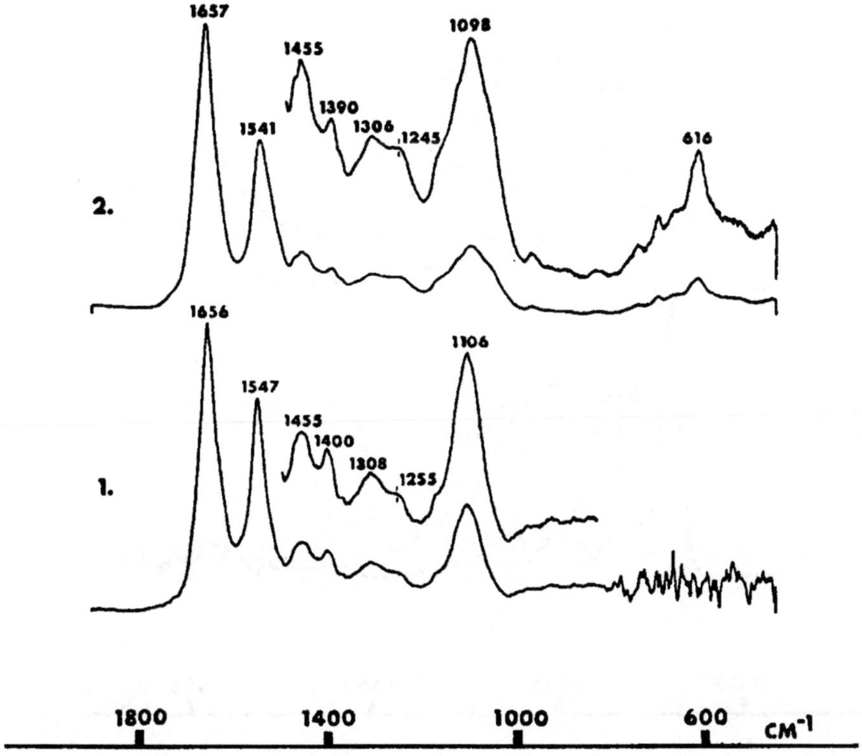

Fig. 2 Infrared absorbance spectra of hemoglobin; 1) aqueous solution (pH = 4.8); 2) cast films.

and disordered conformations that are present in these proteins. This point is in contrast to the case of fibrous proteins where the amide II frequency has been used to distinguish between the α helix and β sheet conformations [5]. Upon dissolution of the proteins in aqueous media, the amide II frequencies increase. This increase indicates an increase in the degree of hydrogen bonding in the protein [5], probably resulting from an increase in solvent penetration into the protein. As was previously stated, there was essentially no change in the amide I frequency. The amide II vibration which is mostly -NH bending should be affected to a greater extent by hydrogen bonding changes than the amide I vibration which is essentially C=O stretch [5].

The magnitude of the shifts of the amide II frequency that

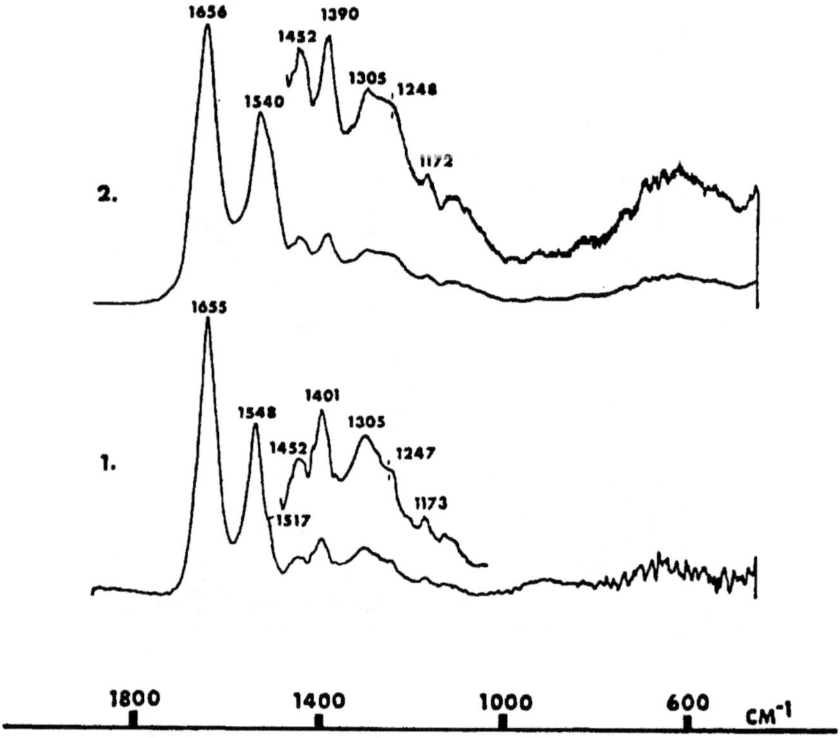

Fig. 3 Infrared absorbance spectra of bovine serum albumin: 1) aqueous solution (pH = 5.6); 2) cast films.

occurs upon dissolution of the proteins in H_2O can be seen from Table 2. It is noted that the more ordered proteins (hemoglobin, especially, but also bovine serum albumin) experience smaller shifts than the more disordered proteins (ribonuclease, β-lactoglobulin and α-casein). Thus, upon dissolution in H_2O the more ordered proteins require less of an increase in hydrogen bonding to acquire the maximum state of hydrogen bonding that is present in the aqueous protein solution. If one assumes the proteins are completely hydrogen bonded in aqueous solution, the above result is consistent with the fact that in the solid state the hydrogen bonding is more complete for the more ordered globular proteins.

C. Amide III Region

The absorption in the amide III region for hemoglobin is

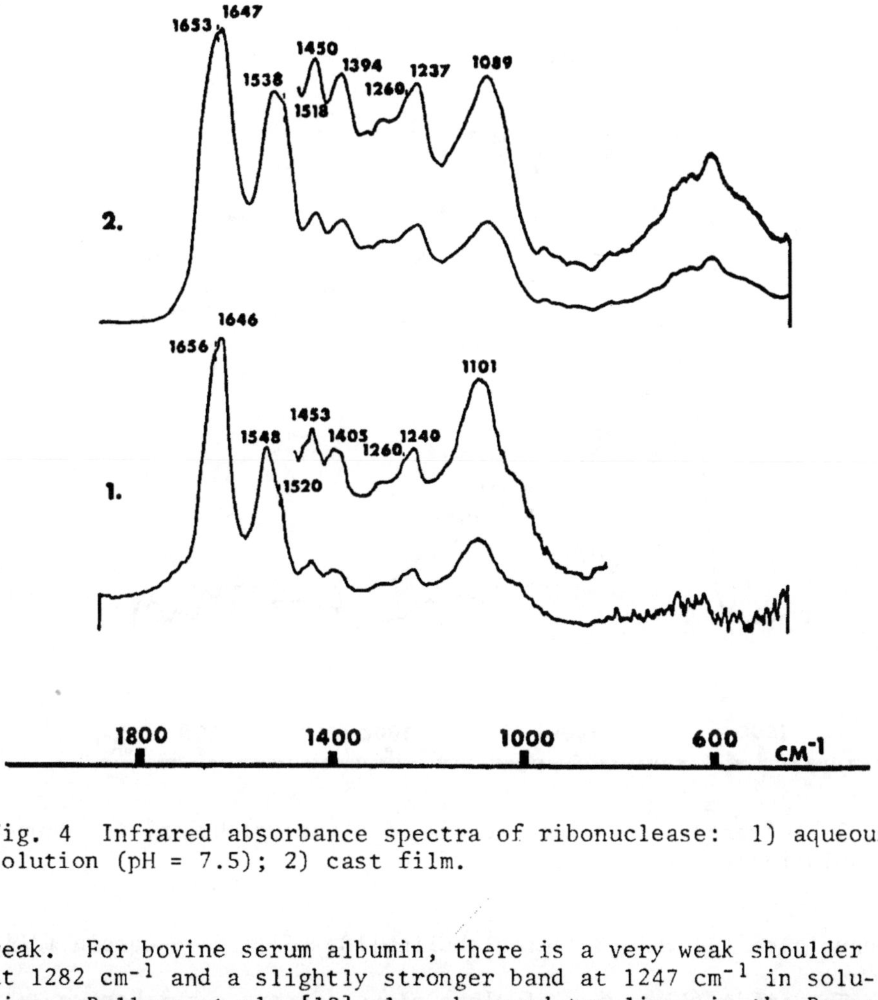

Fig. 4 Infrared absorbance spectra of ribonuclease: 1) aqueous solution (pH = 7.5); 2) cast film.

weak. For bovine serum albumin, there is a very weak shoulder at 1282 cm^{-1} and a slightly stronger band at 1247 cm^{-1} in solution. Belloco et al. [18] also observed two lines in the Raman spectrum of native bovine serum albumin at 1280 and 1250 cm^{-1}. The two frequencies can be assigned to the presence of the α-helical and disordered conformations in bovine serum albumin with the weaker 1280 cm^{-1} band being indicative of the α helix. Weak amide III bands for α-helical proteins have previously been noted in the Raman spectra of α-helical model polypeptides [19, 20] and the globular protein tropomyosin [21], which is reported to be highly α-helical in both solid [22] and native aqueous solution [23]. In the amide III region of ribonuclease, β-lactoglobulin, and α-casein there are stronger bands of medium intensity. Raman spectroscopy has also shown amide III lines of medium intensity for these globular proteins of small helical content:

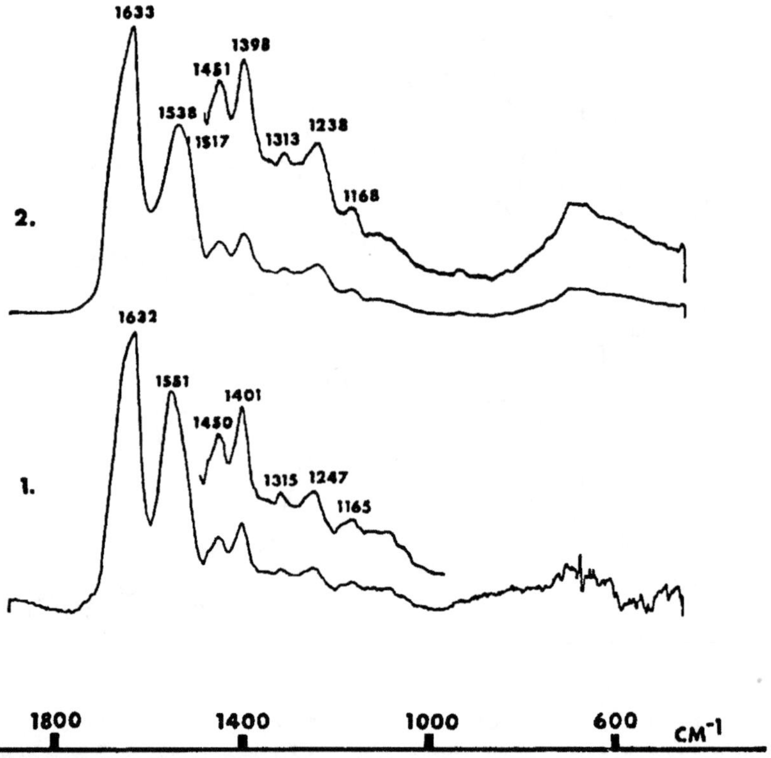

Fig. 5 Infrared absorbance spectra of β-lactoglobulin: 1) aqueous solution (pH = 6.0); 2) cast film.

ribonuclease [24-26], β-lactoglobulin [18, 27], and α-casein [21].

The amide III region of ribonuclease deserves further comment. In addition to the main amide III band, there is a much weaker band occurring as a shoulder at 1260 cm^{-1} in both the solid state and aqueous solution spectra. In the Raman spectra of ribonuclease [24-26], there are also two lines in the amide III region, occurring at 1261 and 1239 cm^{-1} in the solution spectra. However, in the Raman spectra the 1261 cm^{-1} line is much stronger than in the infrared. Yu et al. [26] has suggested that the 1261 cm^{-1} line should be assigned to the α-helical conformation in ribonuclease based on Yu and Liu's study [28] of the hormone glucagon where they concluded that the amide III line for the α helix should occur with strong intensity near 1260 cm^{-1}. However, based on their conclusions that α-helical structures

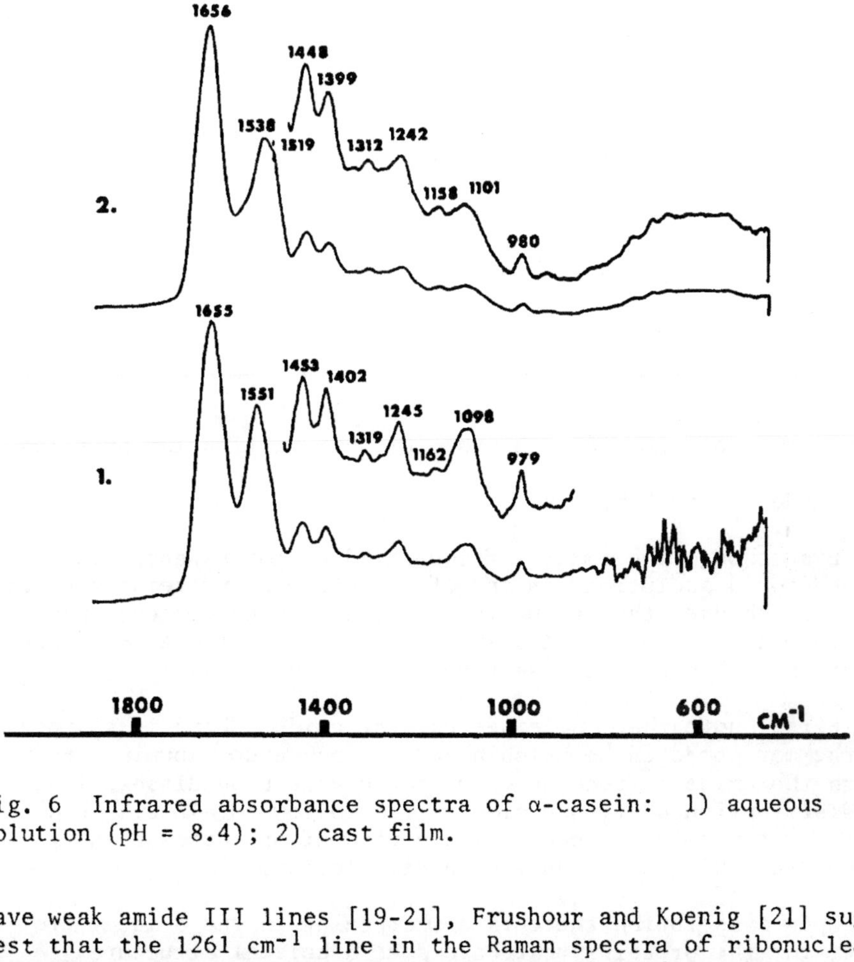

Fig. 6 Infrared absorbance spectra of α-casein: 1) aqueous solution (pH = 8.4); 2) cast film.

have weak amide III lines [19-21], Frushour and Koenig [21] suggest that the 1261 cm^{-1} line in the Raman spectra of ribonuclease should be assigned instead to the disordered conformation. In this current infrared study the weakness of the 1260 cm^{-1} shoulder in the amide III region would seem to indicate assignment to the α-helical conformation based on the infrared results for hemoglobin and bovine serum albumin and the Raman results for α-helical structures [19-21]. The other amide III band of medium intensity can be assigned to both the irregular β sheet and other disordered conformations. These assignments of the two amide III frequencies are similar to those made earlier for the two bands observed in the amide I region. The discrepancy between the infrared and Raman intensity of the two amide III frequencies is unexplained at this time. Clearly, further work needs to be done in order that the bands in the amide III region can be adequately assigned

TABLE 2

AMIDE II FREQUENCY SHIFT UPON PROTEIN DISSOLUTION IN H_2O

Protein	$\Delta\nu(cm^{-1})$
Hemoglobin	6
Bovine Serum Albumin	8
Ribonuclease	10
β-Lactoglobulin	13
α-Casein	13

to specific conformations that are present in globular proteins.

D. Amide I Band Contours

Even though the amide I frequency does not essentially change upon dissolution for any of the globular proteins studies, there are changes that occur in the amide I band contour upon dissolution. The bandwidths at half-height for the amide I bands are given in Table 3 for the five globular proteins.

First, note that, as expected, the amide I band is sharper for the more ordered hemoglobin and bovine serum albumin. Even though the amide I frequencies do not change upon dissolution, the distribution of frequencies shifts to lower values. Such a shift for the amide I mode is indicative of an increase in hydrogen bonding [5] which occurs upon dissolution.

Upon dissolution there is a sharpening in the amide I frequency for the proteins that have some α-helical structure (hemoglobin, bovine serum albumin, and ribonuclease). Proteins are not in their native state in the solid phase. It appears that upon dissolution the α-helical components become more regularized into the specific α helix conformation. This ordering of the α-helical components upon dissolution is greater than for the other protein conformations.

4. CONCLUSIONS

Infrared absorption spectra of the globular proteins, hemoglobin, bovine serum albumin, ribonuclease, β-lactoglobulin, and α-casein, have been obtained in H_2O solution. The absorbance

TABLE 3

AMIDE I BANDWIDTHS

	Amide I Bandwidth[a] at Half-Height in cm^{-1}	
Protein	Film	Film
Hemoglobin	45(24/21)	38(17/21)
Bovine Serum Albumin	50(27/23)	41(17/24)
Ribonuclease	66(44/22)[b]	58(36/22)[b]
β-Lactoglobulin	54(44/10)	53(39/14)
α-Casein	59(31/28)	58(27/31)

[a]For each protein the total bandwidth is given first. The values in the parenthesis are the contribution of the bandwidth from the high and low frequency sides of the bands, respectively.

[b]Center of band is taken to be 1647 and 1646 cm^{-1} for film and solution, respectively.

contribution of the water was digitally subtracted from that of the protein solution spectra. These solution spectra were compared to spectra obtained in the solid state from cast films.

The secondary structure of globular proteins can be distinguished by infrared spectroscopy by utilization of a combination of amide I frequency and the comparative intensity in the amide III region. Such a procedure is illustrated in Table 4.

Changes occur in the infrared spectra upon dissolution of the proteins in H_2O. The increase of the amide II frequency and the shifting of the amide I frequency distribution to lower frequencies which occur upon dissolution reflect the increase in the hydrogen bonding pattern of the globular proteins. The results indicate that such an increase by hydrogen bonding occurs especially for the more disordered proteins. This result is indicative of the more complete hydrogen bonding pattern that is present in the solid state for α-helical globular proteins as opposed to the more disordered proteins.

TABLE 4

DETERMINATION OF SECONDARY STRUCTURE IN GLOBULAR PROTEINS BY INFRARED SPECTROSCOPY

	Secondary Structure		
	α helix	Anti-parallel β sheet	Disordered
Amide I Frequency (cm^{-1})	1652-56	1632	1655
Amide III Intensity	weak	strong	strong

REFERENCES

1. R. E. Dickerson and I. Geis, The Structure and Action of Proteins, Harper and Row, New York, 1969, Chapter 2.
2. C. Tanford, Advan. Protein Chem., 23, 122 (1968).
3. E. B. Wilson, Jr., J. C. Decius, and P. C. Cross, Molecular Vibrations, McGraw-Hill Book Co., New York, 1959, Chapter 3.
4. J. L. Koenig, Raman Spectroscopy of Biological Macromolecules: A Review, J. Polym. Sci., Part D, 1972.
5. H. Susi, Structure and Stability of Biological Macromolecules, Marcel Dekker, New York, 1969, Chapter 7.
6. H. Susi, S. N. Timasheff, and L. Stevens, J. Biol. Chem., 242, 5460 (1967).
7. P. Urnes and P. Doty, Advan. Protein Chem., 16, 401 (1961).
8. E. Shechter and E. R. Blout, Proc. Natl. Acad. Sci. U.S., 51, 695 (1964).
9. G. Kartha, J. Bella, and D. Harher, Nature, 213, 862 (1967).
10. H. W. Wyckoff, K. D. Hardman, N. M. Allewell, T. Inagami, L. N. Johnson, and F. M. Richards, J. Biol. Chem., 242, 3984 (1967).

11. H. W. Wyckoff, D. Tsernoglou, A. W. Hanson, J. R. Knox, B. Lee, and F. M. Richards, J. Biol. Chem., 245, 305 (1970).

12. N. Greenfield and G. Fasman, Biochemistry, 8, 169 (1969).

13. T. T. Herskovits, R. Townend, and S. N. Timasheff, J. Am. Chem. Soc., 86, 4445 (1964).

14. S. N. Timasheff, R. Townend, and L. Mescanti, J. Biol. Chem., 241, 1863 (1966).

15. R. Townend, T. F. Kumosinski, and S. N. Timasheff, J. Biol. Chem., 242, 4538 (1967).

16. G. D. Fasman, H. Hoving, and S. N. Timasheff, Biochemistry, 9, 3316 (1970).

17. S. Krimm, J. Mol. Biol., 4, 528 (1962).

18. A. M. Belloco, R. C. Lord, and R. Mendelsohn, Biochim. Biophys. Acta, 257, 280 (1972).

19. J. L. Koenig and B. G. Frushour, Biopolymers, 11, 1871 (1972).

20. B. G. Frushour and J. L. Koenig, Biopolymers (submitted).

21. B. G. Frushour and J. L. Koenig, Biopolymers (submitted).

22. D. L. D. Caspar, C. Cohen, and W. Longley, J. Mol. Biol., 41, 87 (1969).

23. S. Lowey, J. Biol. Chem., 240, 2421 (1965).

24. R. C. Lord and N. Yu, J. Mol. Biol., 50, 509 (1970).

25. J. L. Koenig and B. G. Frushour, Biopolymers, 11, 2505 (1972).

26. N. T. Yu, B. H. Jo, C. S. Liu, J. Am. Chem. Soc., 94, 7572 (1972).

27. B. G. Frushour and J. L. Koenig, Biopolymers (submitted).

28. N. T. Yu and C. S. Liu, J. Am. Chem. Soc., 94, 5127 (1972).

29. E. G. Bendit, Biopolymers, 5, 525 (1967).

30. K. Krishman and R. A. Plane, Inorg. Chem., 6, 55 (1967).

APPLICATION OF FT-IR TO INORGANIC SPECIES AND PROBLEMS IN INORGANIC CHEMISTRY

A. Müller

Faculty of Chemistry
University of Bielefeld
Bielefeld, West Germany

1. INTRODUCTION

The application of physical methods is extremely important for the inorganic chemist [1]. The fact that computer controlled, automated, four circle diffractometers are commercially available has led to a renaissance in inorganic chemistry in the last two decades. (The structural variety of inorganic compounds is much larger than that of organic ones). Though the knowledge of the geometric structure of a compound is necessary to understand its properties, the important information about the electron distribution in molecules can only be obtained from spectroscopic methods (like NMR, ESR, UP, XP, UV/VIS/NIR, IR, Raman, Mössbauer and, NQR spectroscopy) [1-3]. Before the application of a single crystal structure determination was routine, infrared spectra were mostly used for obtaining information about the structure of compounds. Due to the development of FT-IR techniques, several current problems for inorganic chemists can now be studied easily as the FT-IR techniques have the following advantages over dispersive methods [4-6]:

(1) the resolution is high

(2) the sensitivity permits rapid scanning

(3) it permits one to follow the development of concentrations of different species in a solvent

(4) it makes emission methods feasible

(5) it provides great potential for surface studies

J. R. Durig (ed.), Analytical Applications of FT-IR to Molecular and Biological Systems, 257–280.

(6) the quality of the spectra in the far infrared region is better.

2. STUDIES IN THE MIR

IR spectra of inorganic molecules are mostly measured in the mid IR region to obtain information about functional groups (finger print method) [7,8] and about the overall structure in the case of small species [1,7,8]. Often it is intended to get information about the bond strength from frequency data or from force constant calculations [8]. Rovibrational studies could only be done with commercial dispersive instruments for an exremely small number of small molecules having rather high rotational constants.

A. Studies of Different Problems

According to the advantages of commercially available FT instruments, mentioned in the introduction, it is now possible to study different types of problems being of interest to inorganic chemists (see Table 1), like the characterization of transient species or inorganic species in the atmosphere (see Table 2), the investigation of species on surfaces, and solid state reactions.

B. High Resolution Studies for the Determination of Molecular Constants

For a limited number of rather small and highly symmetrical molecules (spherical and symmetric tops) the band envelope can provide information about first order Coriolis coupling constants though certain serious limitations and sources of error exist. (The band contour is a function of the selection rules, the transition probabilities, the moment of inertia, the temperature, first and second order Coriolis coupling constants, l-type and Fermi resonance). The determination of ζ constants can be improved by measuring isotopically pure compounds in the gas phase at lower temperatures [23-26].

Due to the improved resolution of commercial FT-IR spectrometers it is now possible to measure the rotational fine structure for molecules with rotational constants of about 0.1 cm^{-1} or even less. From this type of studies, molecular constants and ζ values can be obtained which are much more accurate than those from band contours. These values are important for force constant determinations, as it is not possible to determine physically reasonable force constants from frequency data alone. It should be mentioned that the knowledge of force constants plays an important role in the interpretation of a chemical bond.

Figure 1 reproduces the ν_3 fundamental of SiH_3I near 363 cm^{-1}.

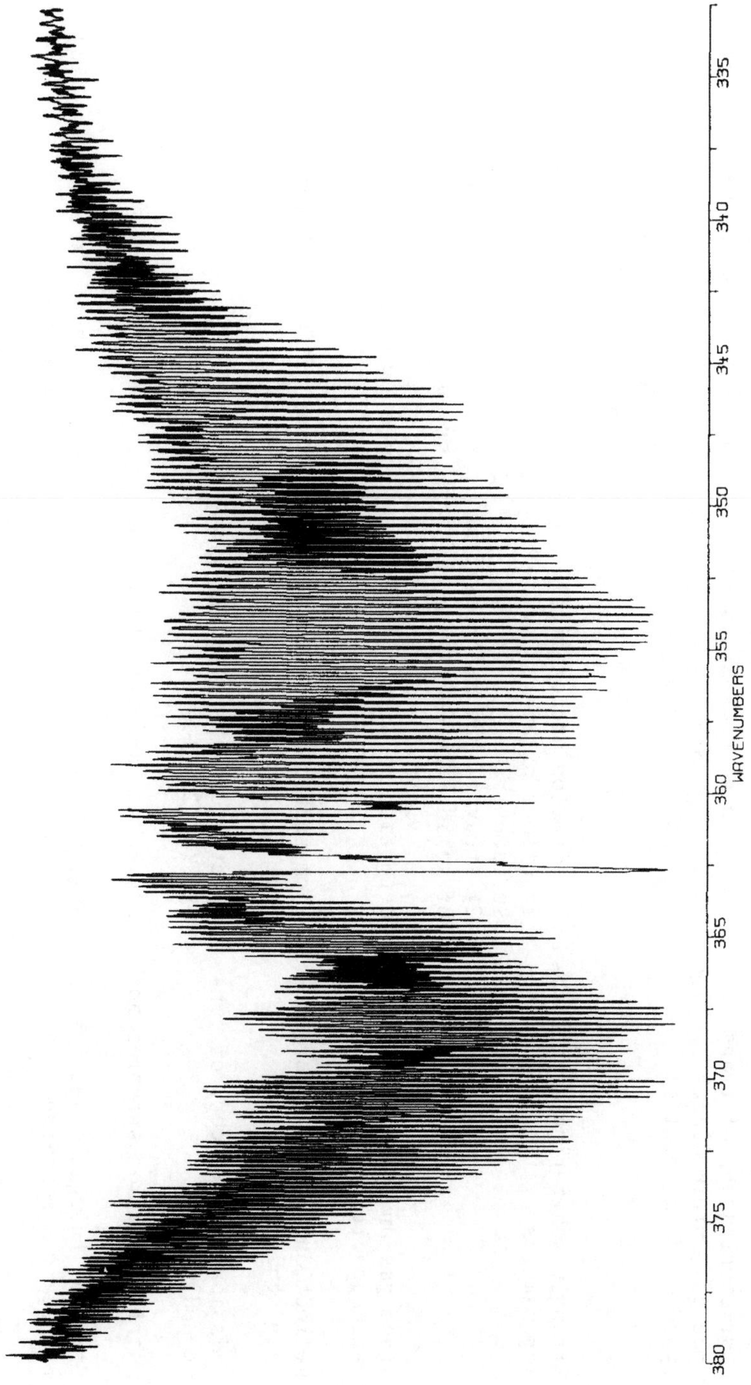

Fig. 1 $\nu_3(A_1)$ of SiH_3I near 363 cm^{-1}.

Table 1 Studies of Different Problems with FT-IR Spectroscopy

Problems	Examples	Results	Ref.
Characterization of transient species (intermediates or highly reactive molecules)	decay in gaseous mixtures of HONO, NO_2, N_2O, N_2 and HONO formation	understanding of the reaction pathway and reactions which play an important role in atmospheric chemistry	[9]
Strongly scattering or absorbing samples	1) sorption and chemical reactions of SO_2 on oxide surfaces	deducing the mechanism	[10-12]
	2) study of the decomposition of solid like $CaSO_3$	kind of decomposition products	
Trace gas analysis of inorganic species	detection of air pollution	see Table 2	
Reflection technique for surface studies	chemisorbed CO on platinum foil	kind of CO-metal interaction	[13]
Emission spectroscopy of condensed phases	spectra of minerals like granite (at 20°Cl.)		[14]
Matrix isolation spectroscopy	see Barnes and Orville-Thomas		
Study of inorganic species in planetary atmosphere	PH_3, GeH_4 (Jupiter)	detection of extremely small amounts	[15]
IR chemoluminescence	reaction $F + C_2H_3X \rightarrow HF + C_2H_2X$	study of the mechanism and partitioning of reaction energy over the product vibrational modes	[16]

Table 2 Detectability Limits of some Pollutant Inorganic Species[a]

Pollutant	Measurement frequency (cm^{-1})	Detectability limit in billionths of an atmosphere (ppb)	Ref.
Carbon monoxide	2170	0.6	[17]
CCl_2F_2	921	0.15	[18]
Hydrogen sulfide	1300	40	[19]
Nitric acid	896	0.3	[20]
Nitric oxide	1900	2.0	[21]
Nitrogen dioxide	1615	0.2	[17]
Ozone	1053	0.6	[22]
Sulfur dioxide	1360		[19]

[a] See also Ref. [6].

The fundamental is accompanied by hot bands belonging to $2\nu_3 - \nu_3$, $3\nu_3 - 2\nu_3$ and $\nu_3 + \nu_6 - \nu_6$. Figure 2 shows, on an expanded scale, the detailed structure and gives an assignment of the J bands, the K structure of which cannot be resolved.

Among others, the following molecular parameters (cm^{-1}) of $H_3{}^{28}SiI$ have been deduced from the spectrum [27].

ν_3	362.795
X_{33}	-1.104
B''	0.10724
α_3^B	$0.4437 \cdot 10^{-3}$
$\bar{D}_J$	$0.036 \cdot 10^{-6}$

The microwave value of B'' is 0.107258 cm^{-1}. If the value of the instrumental resolution is smaller than the quantity $2[A(1 - \zeta) - B]$, Coriolis ζ constants may be derived from the Q_K heads appearing in the perpendicular bands of symmetric tops. Figure 3 reproduces ν_5 and $\nu_2 + \nu_5$ of natural CF_3Br. The strong-weak-weak intensity alternation is well pronounced, the doubling of lines in the $\nu_2 + \nu_5$ spectrum being due to the $^{79/81}Br$ isotopic shift arising from ν_2 [28]. The value $\zeta_5 = -0.789$ is consistently obtained from both spectra.

The spectra have been obtained with a Nicolet Series 8000 Vacuum interferometer equipped with a 6 micron Mylar beam splitter and a TGS detector (for details see Ref. [27]).

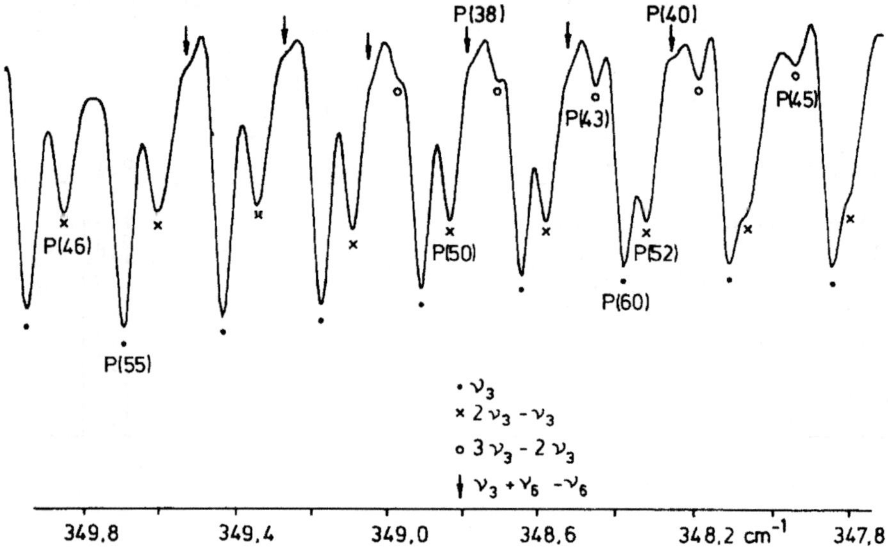

Fig. 2 $\nu_3(A_1)$ on an expanded scale.

3. STUDIES IN THE FIR

A. Studies of Different Problems

The application is mainly concentrated on the following areas

(1) pure rotational spectra

(2) the study of intermolecular vibrations of molecular crystals (including problems related to X-ray structure determination and theory of lattice dynamics),

(3) the observation of weaker bond motions of hydrogen bonded systems A-H..B,

(4) pseudo-lattice vibrations of liquids,

(5) collision induced spectra,

(6) emission studies (submillimeter) of molecules in the stratosphere,

(7) torsional vibrations as well as ring puckering modes,

(8) skeletal-bending modes,

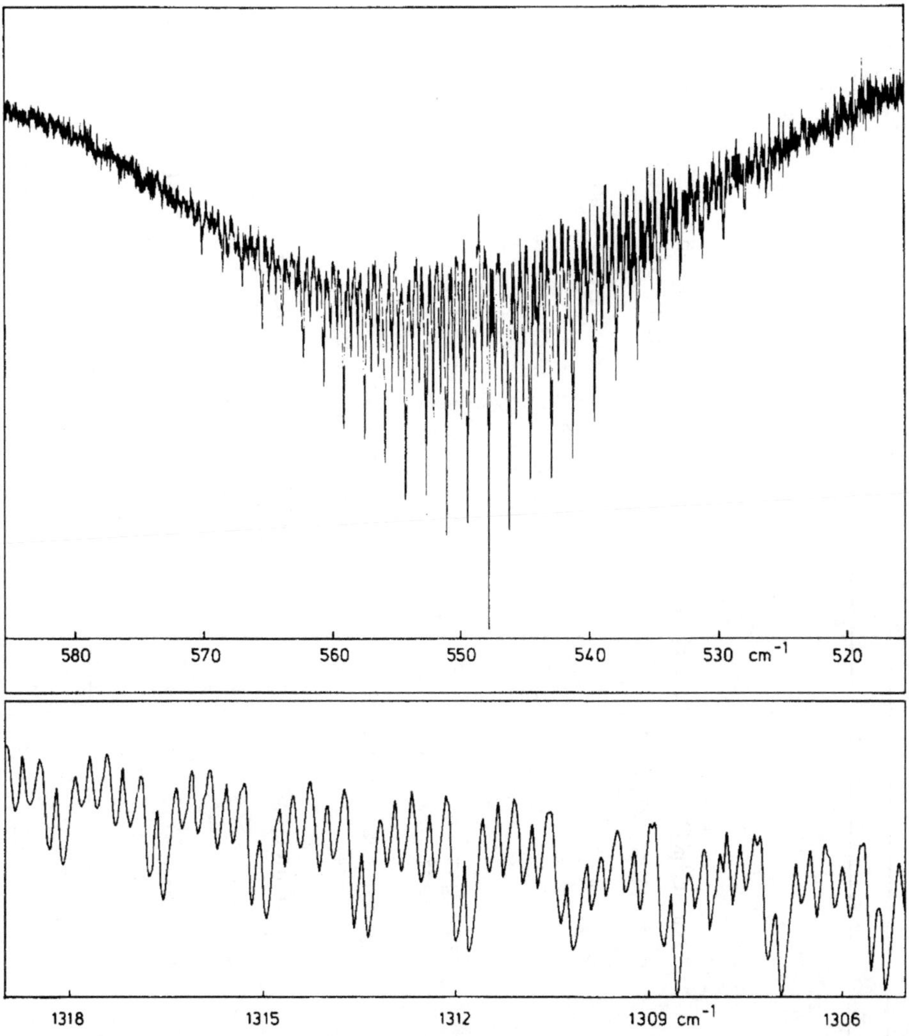

Fig. 3 $\nu_5(E)$ and $\nu_2 + \nu_5$ of CF_3Br.

(9) metal-ligand vibrations, and

(10) studies of polynuclear aggregrates as well as metal-metal-bonded systems (including clusters and cage systems).

As some of the most current topics for inorganic chemists are the two mentioned last [1,7,29-31] we have only given brief examples for the other topics summarized in Table 3. It should

Table 3 Studies of Different Problems with FT Far Infrared Spectroscopy

Problems	Examples	Results	Ref.
Rotational spectra	NO	ΔE between $^2\pi_{1/2}$ and $^2\pi_{3/2}$ (122 cm^{-1}); B_o (±005%); D_o (±2%); H_o (±25%)	[32]
Intermolecular vibrations of solids	X_2 and HX(X = F, Cl, Br, I) H_2S, NH_3, CO_2, COS SO_2 and N_2O	1) distinction between translational and rotational modes 2) verification of the known crystal structure 3) NCA for intermolecular vibrations	[33-38]
Weaker bond motions of H-bonded systems (A-H...B)	straight-chain aliphatic alcohols		[39]
Weak intermolecular interactions in the gas phase	H_2O in the gas phase	detection of a dimer in the earth atmosphere	[40]
Pseudo-lattice vibrations of liquids	CH_3CN and others	broad bands being due to hindered rotations in case of polar molecules	[41-44]
Submillimeter stratospheric emission spectra	O_3 and HNO_3	detection and concentrations of these species in the stratosphere	[45]

Table 3 (Continued)

Problems	Examples	Results	Ref.
Torsional vibrations	species with threefold symmetric tops ($-SiH_3$), twofold tops ($-NO_2$) and asymmetric tops ($-SH$)	determination of the shape of the potential and the barrier height	[46]
Ring-bending motions	four-, five-, six- and seven-membered rings	information about the potential function and on the conformation	[47]
Skeletal-bending modes	ν_7 bending mode about the central C atom of the quasilinear C_3O_2 and C_3S_2		[48,49]
Metal-ligand vibrations	ν(M-N)-vibration of complexes of the type $[M(NH_3)_6]^{n+}$	information about the metal-ligand interaction	[30]

be mentioned that the understanding of intermolecular vibrations and forces becomes more and more interesting for inorganic chemists due to its relation to X-ray investigations and to the possibility of the deliberate precipitation and stabilization of rather unstable ions by counterions.

B. Structural Studies

(1) Mononuclear species. The interpretation of vibrational spectra are very useful for obtaining information concerning the structure and bonding of simple types of compounds. It is very easy to distinguish for example between a planar or pyramidal XY_3 or $X(YZ)_3$ type species by comparing the number of experimentally observed bands with the predicted ones [7,8]. This can be done for the skeletal vibrations as well as for the ligand internal vibrations. Examples of the latter case are carbonyls [1], for which the structures can easily be deduced from vibrational and especially infrared spectroscopy. (Information can be obtained about the existence of bridging CO groups, bond angles (from relative intensities), conformers and force contants (CK-model), also. The skeletal vibrations of the type $\nu(XY)$ (metal-ligand vibrations) have been interpreted with respect to the type of metal-ligand interaction, the understanding of which is very important for a coordination chemist. As most of the mentioned vibrations can be studied within the NIR or MIR region and no high resolution is necessary, dispersive instruments can be used for this type of studies with no disadvantage.

(2) Polynuclear aggregates. These are either polyhedral cages, ring systems, clusters or aggregates of coordination compounds (in which the ligands share corners, edges, or faces of different polyhedrons). The chemistry of cages and clusters (with metal-metal bonds) has practically been discovered only in the last two decades and these types of compounds are found in nearly all parts of the Periodic Table. Whereas in the case of a complex, the ligands are bound to a central atom, there is no central atom in cage or cluster systems, that means each atom is directly bonded to its neighbor in the polyhedron.

The most interesting species are transition metal clusters, because of the so-called metal-cluster/metal-surface analogy [50], as a model for the understanding of catalysis. There are two reasons for the difficulty in using infrared spectroscopy for structural diagnosis:

(1) It is not possible to get information about the structure from the ligand-internal vibrations (at least in the case of liquids and solids). $Os_3(CO)_{12}$ shows for instance the same four band spectrum as a cis-ML_2CO_4 type molecule (as Os has the same local symmetry C_{2v} [1]). That means we do not observe all

possible bands due to the selection rules as several are accidentally degenerate. The reason is that interaction elements of the G and F matrix like $G(CO_M/CO_{M'})$ or $F(CO_M/CO_{M'})$ are nearly zero.

(2) The internal modes of the cluster or cage systems (which are characteristic for its type) have absorption bands (at least for a heavy atom system) in the FIR. These types of vibrations have not been studied systematically. Therefore we are practically confined to discussing important aspects for future studies.

Two broad classes of transition metal cluster compounds can be distinguished [51]: "electron-rich" clusters formed by metals to the right side of the d-block in the Periodic Table (with metal atoms in low oxidation states and π-acceptor ligands) and "electron-poor" clusters formed by the early transition elements, in which the metal atoms are in higher oxidation states with simple donor type ligands. This type of classification is also important for the characteristic feature of the vibrational spectra as the "electron-rich cluster" need not contain bridging ligands, whereas the "electron-poor clusters" always contain bridging ligands. In both cases the most important structures are the dumb-bell, the triangle, the tetrahedron, and the octahedron.

(a) Electron-rich Clusters

Examples for species with no bridging ligands are $M_2(CO)_{10}$ (M = Mn, Tc, Re) with D_{4d}, $M_3(CO)_{12}$ (M = Ru, Os) having D_{3h} and $Ir_4(CO)_{12}$ with T_d symmetry [1]. In Table 4 the idealized point group of cluster systems, the species for the metal-metal vibrations, the corresponding frequencies, the purity of the ν(M-M) vibrations, and bond parameters are given [52]. Though the metal-metal vibrations are not completely pure, it is possible to deduce the type of cluster compound from a vibrational analysis (mainly if metal-isotopic data are available) using FT-IR techniques.

(b) Electron-poor Clusters of the Early Transition Elements

Frequently a d^n configuration leads to the formation of a dumb-bell (n = 1) like $[Mo_2(S_2)_6]^{2-}$ [53], triangle (n = 2) like $[MO_3S_4(CN)_9]^{5-}$ [54], tetrahedron (n = 3) like $[Re_4S_4(CN)_{12}]^{4-}$ [55], or octahedron (n = 4) like $Mo_6Cl_8{}^{4+}$ [1] or $Nb_6Cl_{12}{}^{2+}$ [1]. In these cases it is very difficult to get information about the structure of the metal-cluster from a simple interpretation of the bands in the FIR as the spectrum strongly depends on the ligand. (The corresponding metal-metal vibrations are strongly coupled with metal-ligand vibrations). The difficult situation for octahedral species like $Mo_6Cl_8{}^{4+}$ and $Nb_6Cl_{12}{}^{2+}$ has been discussed and reviewed by Spiro [52] (see also [7]). There is a vibrational assignment for the tetrahedral $[Re_4S_4(CN)_{12}]^{4-}$ [56]

Table 4 Vibrational Properties of Electron Rich Clusters with no Bridging Ligand

	Idealized point group	Species	ν(M-M) [cm^{-1}]	V_{M-M}[a] [perc.]	f_{M-M} [mdyn/Å]
$Mn_2(CO)_{10}$	D_{4d}	A_1	160	71	0.59
$Tc_2(CO)_{10}$	D_{4d}	A_1	148	73	0.72
$Re_2(CO)_{10}$	D_{4d}	A_1	122	77	0.82
$Ru_3(CO)_{12}$	D_{3h}	A_1'	185	81	0.82
		E'	149	63	
$Os_3(CO)_{12}$	D_{3h}	A'	158	80	0.91
		E'	117	52	
$Ir_4(CO)_{12}$	T_d	A_1	207	91	1.69
		T_2	161	88	
		E	131	84	

[a]Contribution to the PED.

and the dumb-bell type compound $Nb_2(S_2)_2Cl_2$ [57]. Typical structural features of bridged metal clusters are given in Fig. 4. These have an important influence on the vibrational spectrum.

As the group of trinuclear clusters of the early transition elements has an exceptionally high formation tendency and stability and is of special significance for the chemistry of these elements in their middle oxidation states [51] (which is not emphasized in the chemical literature), we are currently interested in the vibrational spectra of these types of species. We also intend to understand the vibrations of cluster compounds with bridging ligands, a problem which is still unsolved [7].

There are three types of structures for trinuclear clusters: [51] without any μ_3 ligand (A), with one μ_3 ligand (B) and with two μ_3 ligands (C), with type B taking two forms (B_1 and B_2) which will be the only ones considered here. The coordination shell of each metal atom in the structure $B_1([M_3(\mu_3 - X)(\mu - Y)_3L_9])$ is a more or less distorted octahedron (MXY_2L_3 unit). A simple cluster of this kind, $[Mo_3S_4(CN)_9]^{5-}$ [54], is seen in Fig. 5. There is only one isolated cluster of type B_2 that does not correspond to the general formula $[M_3(\mu_3 - X)(\mu - Y)_3L_9]$, namely $[Mo_3S_{13}]^{2-}$ [51,53], which has the central unit $[Mo_3(\mu_3 - S)(\mu - S_2)_3]^{4+}$, its structure being presented in Fig. 5 also. (The central unit just mentioned occurs also in the solid state structure of $Mo_3S_4Cl_7$ [51,53]. The shape of the coordination shell can be derived from the D_{2d} structure of $[Cr(O_2)_4]^{3-}$ or $[Mo(O_2)_4]^{2-}$, which is formed by two mutually perpendicular $M(O_2)_2$ units. In $[Mo_3S_{13}]^{2-}$ we find a

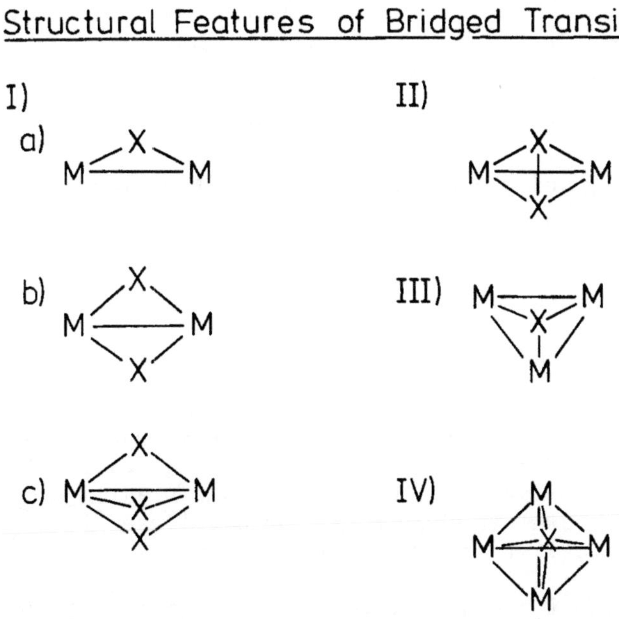

Fig. 4 Structural features of bridged transition metal clusters.

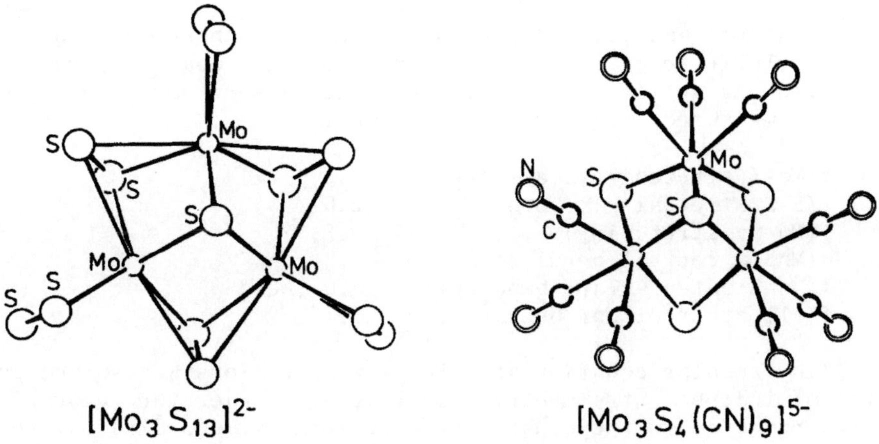

Fig. 5 Structure of $[Mo_3S_{13}]^{2-}$ and $[Mo_3S_4(CN)_9]^{5-}$.

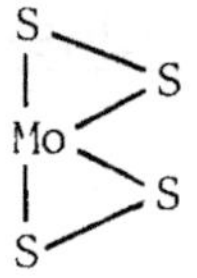

unit analogous to

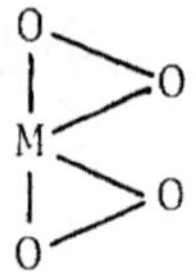

with bridging S_2 groups, and perpendicular to this the planar unit

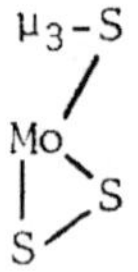

$[Mo_3S_{13}]^{2-}$ [51,53], first prepared by us, is the only known discrete transition metal sulfur cluster. $[Mo_3S_4(CN)_9]^{5-}$ [54], could be obtained from it and CN^- by an activation and sulfur-atom transfer reaction of the cluster-bonded $S_2{}^{2-}$ bridge ligands. This reaction has a biological relevance [54].

We have performed a vibrational analysis of $[Mo_3S_{13}]^{2-}$ [58], $Mo_3S_7Cl_4$ [59], and $[Mo_3S_4(CN)_9]^{5-}$ [60], using ^{92}Mo and ^{100}Mo isotopes and a qualitative NCA [61]. Some of the results are displayed in Fig. 6-8 and Table 5. A simple model calculation for $[Mo_3S_{13}]^{2-}$ was based on an idealized model of C_{3v} symmetry [61]. The distribution of normal modes into different species is:

$$\Gamma = 10A_1 + 4A_2 + 14E.$$

An approximate and very simple force field was chosen as a diagonal force field in terms of valence coordinates including redundancies. The following set of values (all in mdyne/Å) were found to reproduce the measured frequencies rather well:

f(Mo-S stretching) (all types)	1.5,
f(S-S stretching) (both types)	2.0,
f(Mo-Mo stretching)	0.8,
f(MoSMo central bending)	0.0,
f(Mo-S out-of-plane bendings)	0.05, and
f(all other planar bendings)	0.15.

(The secular equation of molecular vibrations was solved in terms of independent symmetry coordinates.) Under the removal of redundancies in the construction of symmetry coordinates, several important types of valence coordinates were omitted. The same technique has been applied to a number of caged systems previously studied (see e.g. [62-66]). It is a disadvantage that the potential energy distribution (PED) terms under these circumstances tend to be useless for the description of normal modes.

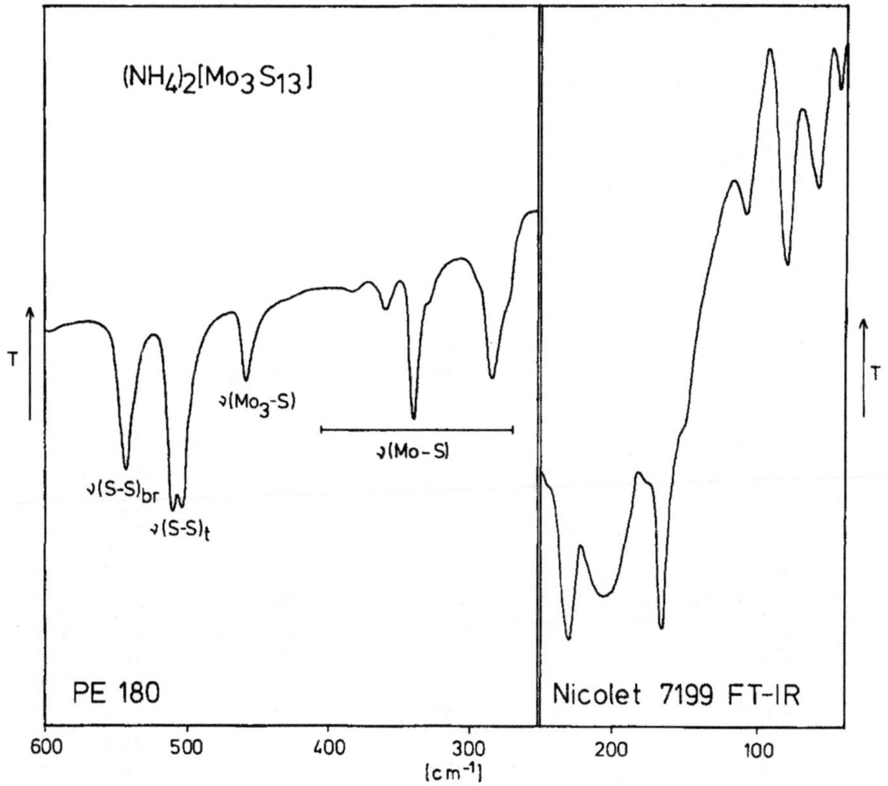

Fig. 6 IR spectrum of $[Mo_3S_{13}]^{2-}$.

The approximations underlying the PED analysis break down, therefore a new approach to the description of normal modes has been tried which may be referred to as being based on numerical Jacobians.

The influence on calculated frequencies from increments on the individual force constants in terms of valence coordinates was studied. Table 5 shows the calculated frequencies for four isotopic ions and two sets of isotopic shifts and some measured frequency data. The following facts can be obtained from a comparison of the spectra of $[Mo_3S_4(CN)_9]^{5-}$ and $Mo_3S_4Cl_7$ as well as from an analysis of calculated and measured isotope shifts. (1) There is a characteristic vibration at ca. 460 cm^{-1} indicating the presence of an apical S above the Mo_3 plane corresponding to type III in all three compounds (see Fig. 4). (2) The ν(S-S) vibrations are rather pure, while ν_{S-S} (inner) > ν_{S-S} (outer).

Table 5 $[Mo_3S_{13}]^{2-}$, Calculated and Measured Frequencies and Isotope Shifts in cm^{-1}

Sym	$^{92}Mo_3{}^{32}S_{13}^{=}$	Assignment	$^{100}Mo_3{}^{32}S_{13}^{=}$	$^{92}Mo_3{}^{34}S_{13}^{=}$	$^{100}Mo_3{}^{34}S_{13}^{=}$	$\Delta\nu^a$	$\Delta\nu^b$
A_1:[c]	543	ν(S-S inner)	542	527	526	1	16
	503	ν(S-S outer)	502	489	488	1	14
	461	ν(Mo-S inner and middle)	455	452	446	6	9
	448	ν(Mo-S)	440	440	432	8	8
	392	ν(Mo-S)	388	382	379	4	10
	350	ν(Mo-S)	349	340	339	1	10
	326	ν(Mo-S)	326	316	316	0	10
	198	ν(Mo-Mo+Mo-S)[e]	196	193	191	2	5
	132		132	128	128	0	4
	63		63	61	61	0	2
A_2:	410	ν(Mo-S middle)	399	407	395	11	3
	299	ν(Mo-S middle)	299	290	290	0	9
	214		213	209	208	1	5
	127		127	123	123	0	4
E:[d]	551	ν(S-S inner)	550	536	535	1	15
	502	ν(S-S outer)	502	488	487	0	14
	457	ν(Mo-S middle)	450	449	442	7	8
	428	ν(Mo-S)	423	418	414	5	10
	415	ν(Mo-S)	407	409	401	8	6
	368	ν(Mo-S)	364	359	356	4	9
	339	ν(Mo-S)	337	330	329	1	9
	307	ν(Mo-S)	306	298	298	1	9
	273	ν(Mo-S)	272	266	265	1	7
	219	ν(Mo-S+Mo-Mo)	217	214	212	2	5
	154		154	150	149	0	4
	134		133	130	129	1	4
	76		75	73	73	1	3
	65		64	64	63	1	1

[a]Isotope shift $^{92}Mo/^{100}Mo$ with ^{32}S
[b]Isotope shift $^{32}S/^{34}S$ with ^{92}Mo
[c]Measured (tentative assignment from the Raman spectrum of $(NH_4)_2[Mo_3S_{13}]$): 550, 510, 460, 385, 362, 290, 230, 199, 168, 86 cm^{-1}.
[d]Measured (tentative assignment from the IR spectrum of $(NH_4)_2[Mo_3S_{13}]$): 544, 510/504, 458, 386, 361, 341, 330, 286, 231, 205, 167, 110, 80, 60 cm^{-1}.
[e]Characteristic breathing motion for the $\{Mo_3S(S_2)_3\}$ central unit; the corresponding intense Raman line at 199 cm^{-1} shows an $^{92/100}Mo$ isotope shift of 2.5 cm^{-1}.

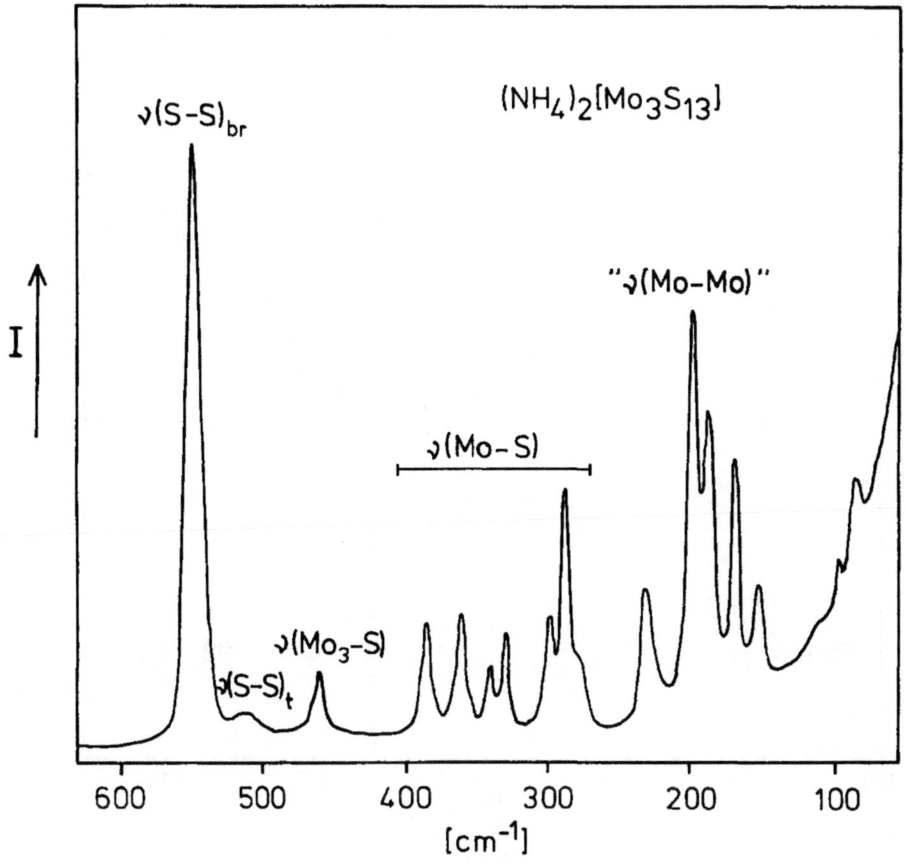

Fig. 7 Raman spectrum of $[Mo_3S_{13}]^{2-}$.

Correspondingly, the lower frequency S-S stretching vibration is missing in $Mo_3S_4Cl_7$ and the higher ν(S-S) frequency nearly identical in both compounds, $Mo_3S_4Cl_7$ and $[Mo_3S_{13}]^{2-}$, having the same central unit $[Mo_3S(S_2)_3]$. (3) The metal-metal vibrations are strongly coupled with metal ligand vibrations but there is a vibration being characteristic for the central cluster unit $[Mo_3S(S_2)_3]$ (see Table 5).

These results demonstrate that some structural features can be obtained from the spectra, but that it is necessary to investigate systematically cluster compounds with different structural features in the FIR region in the future in order to be able to interpret the spectrum characteristic for the skeletal modes in that region.

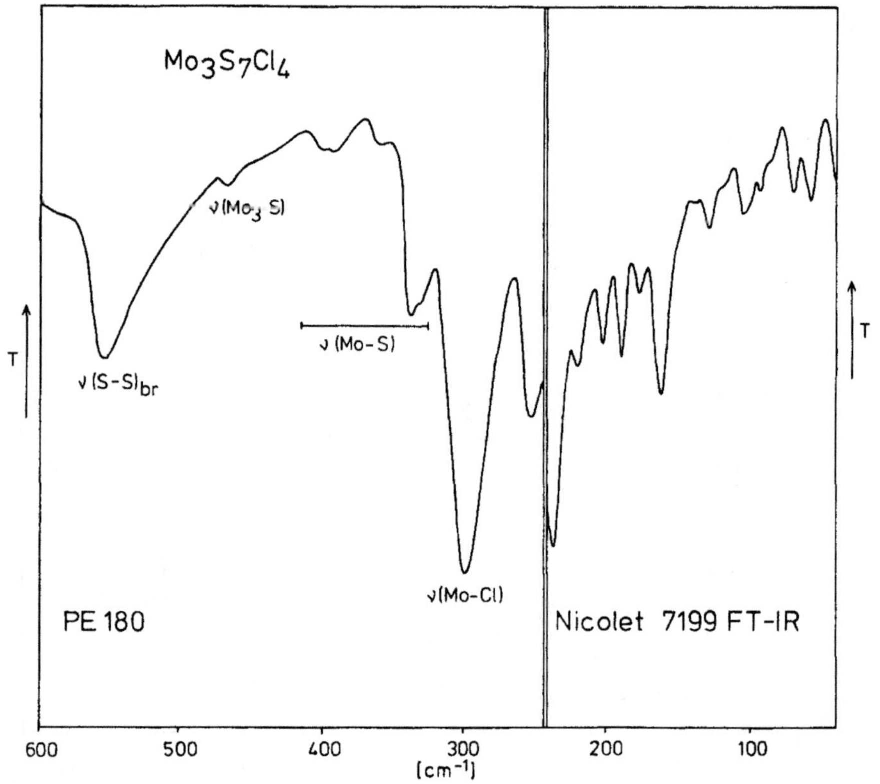

Fig. 8 IR spectrum of $Mo_3S_7Cl_4$.

(c) Polynuclear Coordination Compounds

Important structural classes of this type are the isopolyanions and heteropolyanions of the early transition elements, where the vibrational analysis becomes very difficult, as most of the vibrations are extremely coupled. Therefore it was impossible until now to determine the structure from the vibrational spectra alone though it is easy to determine the existence of some bond types like the terminal M=O groups (with short distances). But nevertheless the first tentative force field study for the fifty-three atomic $[PMo_{12}O_{40}]^{3-}$ [62] with T_d-symmetry and the heptamolybdate $[Mo_7O_{24}]^{6-}$ has been done [66].

(d) Cage-type Compounds Including Those with Different Metals

Cage-compounds of Fe, Cu and Mo with sulfur containing ligands

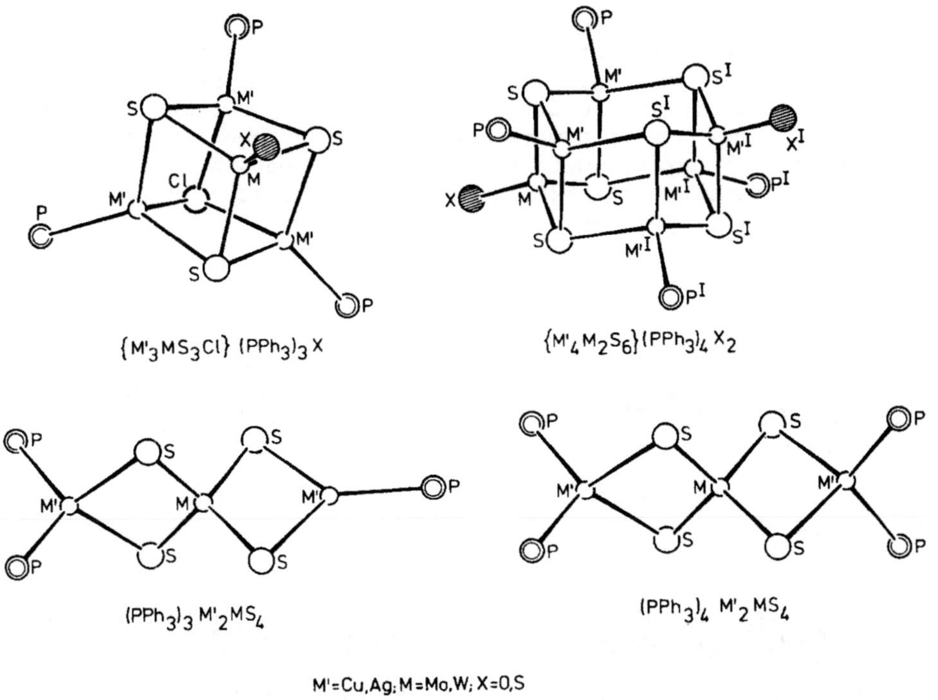

Fig. 9 Different structural types of polynuclear transition metal compounds generated from thiometallates and soft acids (Cu^+, Ag^+).

are of special interest because of their biological relevance (e.g. $[Fe_4S_4(SR)_4]^{n-}$ as a model for rubredoxin [1]). Compounds containing two biometals are also important though these are in general more difficult to prepare. Those containing Fe, S, Mo are interesting with respect to the active sites of nitrogenase and those containing Cu and Mo with respect to the interrelation between these metals in numerous biological processes (the so called Mo-Cu antagonism). Thiometallates like $MoS_4{}^{2-}$ have been used by us as "synthons" for the generation of these types of compounds [67,68], which are also interesting from the structural point of view (see Fig. 9), and others which show reversible electrochemical reduction because of the existence of delocalized molecular orbitals [69]. It would be very interesting to study the skeletal vibrations of the compounds.

4. SUMMARY AND OUTLOOK

The application of FT-IR spectroscopy to inorganic species

and problems in inorganic chemistry has been briefly reviewed. Due to the development of that technique, several interesting problems can now be studied easily. The current problems for inorganic chemists, the understanding of the vibrational spectra of polynuclear transition metal compounds, has been discussed in more detail. Progress in that field is only possible if the skeletal-vibrations of these compounds will be studied systematically with FT-IR spectroscopy in the future.

ACKNOWLEDGMENT

We thank the Deutsche Forschungsgemeinschaft, the Fonds der Chemischen Industrie, and the Minister für Wissenschaft und Forschung (NRW) for financial support and Prof. H. Bürger for supplying us with results prior to publication.

REFERENCES

1. F. A. Cotton and G. Wilkinson, Advanced Inorganic Chemistry, 3rd ed., Wiley-Interscience, New York, 1972.

2. H. A. O. Hill and P. Day (Ed.), Physical Methods in Advanced Inorganic Chemistry, Interscience Publishers, London, 1968.

3. P. Day (Ed.), Electronic States of Inorganic Compounds: New Experimental Techniques, D. Reidel, Dardrecht, Holland, 1975.

4. F. J. Comes, A. Müller, and W. J. Orville-Thomas (Ed.), Spectroscopy in Chemistry and Physics - Modern Trends, Elsevier, Amsterdam, 1980.

5. P. R. Griffith, Chemical Infrared Fourier Transform Spectroscopy, Wiley-Interscience, New York, 1975.

6. J. R. Ferraro and L. Basile (Ed.), Fourier Transform Spectroscopy, Application to Chemical Systems, Vol. 1 and Vol. 2, Academic Press, New York, 1978.

7. K. Nakamoto, Infrared Spectra of Inorganic and Coordination Compounds, 3rd Ed., Wiley-Interscience, New York, 1977.

8. H. Siebert, Anwendungen der Schwingungsspektroskopie in der Anorganischen Chemie, Springer-Verlag, Berlin, 1966.

9. W. H. Chan, R. J. Nordstrom, J. G. Calvert and J. H. Shaw, Env. Sci. Technol., 10, 674 (1976).

10. M. J. D. Low, A. J. Goodsel, and N. Takezawa, Env. Sci. Technol., 5, 1191 (1971).

11. A. J. Goodsel, M. J. D. Low, and N. Takezawa, Env. Sci. Technol., 6, 268 (1972).

12. M. J. D. Low, A. J. Goodsel, and H. Mark, in Molecular Spectroscopy, (P. Hepple, ed.), Inst. of Petroleum, London, 1972.

13. M. J. D. Low and J. C. McManus, Chem. Comm., 1166 (1967).

14. M. J. D. Low and I. Colema, Appl. Opt., 5, 1453 (1966).

15. S. S. Barshay and J. S. Lewis, Icarus, 33, 593 (1978).

16. J. G. Moehlmann, J. T. Gleaves, J. W. Hudgens, and J. D. McDonald, J. Chem. Phys., 60, 4790 (1974).

17. P. L. Hanst, W. E. Wilson, R. K. Patterson, B. W. Gay Jr., L. W. Chaney, and C. S. Burton, Rep. No. EPA-650/4-75-006. Environ. Prot. Agency, Research Triangle Park, North Carolina (1975).

18. D. S. Erley and B. H. Blake, Infrared Spectra of Gases and Vapors, Vol. 2., Dow Chem. Co., Midland, Michigan, 1965.

19. R. H. Pierson, A. N. Fletcher, and E. St. Clair Gantz, Anal. Chem., 28, 1218 (1956).

20. P. L. Hanst, A. S. Lefohn, and B. W. Gay Jr., Appl. Spectrosc., 27, 188 (1973).

21. P. L. Hanst, Adv. Env. Sci. Technol., 2, 91 (1971).

22. J. N. Pitts, J. M. McAfee, W. D. Long, and A. M. Winter, Env. Sci. Technol., 10, 787 (1976).

23. A. Müller, "Isotopic Substitution" in Vibrational Spectroscopy - Modern Trends, (A. J. Barnes and W. J. Orville-Thomas, ed.), Elsevier, Amsterdam, 1977.

24. F. Königer and A Müller, J. Mol. Spectrosc., 65, 339 (1977).

25. A. Müller and F. Königer, J. Mol. Struct., 30, 195 (1976).

26. F. Königer, C. E. Blom, and A. Müller, J. Mol. Spectrosc., 77, 76 (1979).

27. H. Bürger, S. Dobos, P. Schulz, and A. Ruoff, J. Mol. Spectrosc., in press.

28. K. Burczyk, H. Bürger, P. Schulz, and A. Ruoff, to be published.

29. D. M. Adams, Metal-Ligand and Related Vibrations, Arnold, London, 1967.

30. K. H. Schmidt and A. Müller, Coord. Chem. Rev., 19, 41 (1976).

31. N. Mohan, A. Müller and K. Nakamoto, "Metal Isotope Effect on Molecular Vibrations" in Advances in Infrared and Raman Spectroscopy, Vol. 1, R. J. H. Clark and R. E. Hester, (ed.), p. 173, Heyden, London, 1975.

32. R. T. Hall, and J. M. Dowling, J. Chem. Phys., 45, 1899 (1966).

33. A. Anderson, S. H. Walmsley, and H. A. Gebbie, Phil. Mag., 7, 1243 (1962).

34. A. Anderson, H. A. Gebbie, and S. H. Walmsley, Mol. Phys., 7, 401 (1964).

35. S. H. Walmsley and A. Anderson, Mol. Phys., 7, 411 (1964).

36. A. Anderson and S. H. Walmsley, Mol. Phys., 9, 1 (1965).

37. A. Anderson and S. H. Walmsley, Mol. Phys., 7, 583 (1964).

38. A. Anderson and S. H. Walmsley, Mol. Phys., 10, 391 (1966).

39. R. F. Lake and H. W. Thompson, Proc. Roy. Soc. (London), A291, 469 (1966).

40. H. A. Gebbie, W. J. Burroughs, J. Chamberlain, J. E. Harries, and R. G. Jones, Nature, 221, 143 (1969).

41. S. G. Kroon and J. van der Elsken, Chem. Phys. Lett., 1, 285 (1967).

42. H. A. Gebbie, N. W. B. Stone, F. D. Findlay, and E. C. Pyatt, Nature (London), 205, 377 (1965).

43. N. E. Hill, Proc. Phys. Soc., 78, 311 (1961).

44. R. J. Jakobsen and J. W. Brasch, J. Am. Chem. Soc., 86, 3571 (1964).

45. J. E. Harries, N. R. W. Swann, J. E. Beckman, and P. A. R. Ade, Nature (London), 236, 159 (1972).

46. J. R. Durig, S. M. Craven, and W. C. Harris, "Determination of Torsional Barriers from Far-Infrared Spectra" in Vibrational Spectra and Structure, Vol. 1, (J. R. Durig, ed.), Marcel Dekker, New York, 1972.

47. Ibid: C. S. Blackwell and R. C. Lord, "Far-Infrared Spectra of Four-Membered-Ring Compounds"; and J. Laane, "Pseudorotation of Five-Membered Rings".

48. W. H. Smith and G. E. Leroi, J. Chem. Phys., 45, 1778 (1966).

49. L. A. Carreira, R. O. Carter, J. R. Durig, R. C. Lord, and C. C. Milionis, J. Chem. Phys., 59, 1028 (1973).

50. G. A. Ozin, Catal. Rev. Sci. Eng., 16, 191 (1977).

51. A. Müller, R. Jostes and F. A. Cotton, Angew. Chem., (in press).

52. T. G. Spiro, Progress in Inorg. Chem., 11, 1 (1970).

53. A. Müller and W. Jaegermann, Inorg. Chem., 18, 2631 (1979).

54. A. Müller and U. Reinsch, Angew. Chem., 92, 69 (1980); Angew. Chem. Int. Ed. Engl., 19, 72 (1980).

55. M. Laing, P. M. Kieran, and W. P. Griffith, Chem. Comm., 221 (1977).

56. W. P. Griffith, P. M, Kieran, and B. P. O'Hare, J. Mol. Struct., 46, 307 (1978).

57. C. Perrin, A. Perrin, and P. Caillet, J. Chim. Phys., 70, 105 (1972-1973).

58. A. Müller *et al.*, unpublished results.

59. A. Müller *et al.*, unpublished results.

60. A. Müller *et al.*, unpublished results.

61. S. J. Cyvin, A. Müller *et al.*, unpublished results.

62. L. E. Lyhamn, S. J. Cyvin, B. N. Cyvin, and J. Brunvoll, Z. Naturforsch., 31a, 1589 (1976).

63. S. J. Cyvin, B. N. Cyvin, W. Brockner, and A. F. Demiray, Z. Naturforsch., 33a, 714 (1978).

64. A. Müller, W. O. Nolte, S. J. Cyvin, B. N. Cyvin, and A. J. P. Alix, Spectrochim. Acta, 34A, 383 (1978).

65. J. C. Whitmer and S. J. Cyvin, J. Mol. Struct., 50, 21 (1978).

66. L. E. Lyhamn and S. J. Cyvin, Z. Naturforsch., 34a, 867 (1979).

67. A. Müller, H. Bögge, U. Schimanski, and T. K. Hwang, Chem. Comm., 91 (1980).

68. A. Müller, H. Bögge, E. Königer-Ahlborn, and W. Hellmann, Inorg. Chem., 18, 2301 (1979) and references cited therein.

69. A. Müller, R. Jostes, V. Flemming, and R. Potthast, Inorg. Chim. Acta, 44, L33 (1980).

STUDIES OF SPECIES ADSORBED ON CATALYST SURFACES BY FOURIER TRANSFORM INFRARED SPECTROSCOPY

N. Sheppard

School of Chemical Sciences
University of East Anglia
Norwich, England, U. K.

1. INTRODUCTION

One of the important, and still increasingly important, areas of application of infrared spectroscopy during the past twenty-five years has been the study of the structures of species chemisorbed on surfaces. Of particular interest to chemists are those surfaces which are of importance for catalytic purposes, either for the enhancement of the rates of slow but thermodynamically feasible reactions, or for the improvement of selectivity when various parallel or sequential reaction pathways are possible for a given set of reagents. The object of the infrared study in such cases is to attempt to identify the chemical structures of the new species formed by chemisorption, with a view to obtaining insight into the catalytic reaction mechanisms. Other applications relate to the characterization of the catalyst surfaces themselves, or the study of surface reactions that are of importance in corrosion processes.

Three books cover the field until 1971 [1,2,3], and other more recent review articles are concerned with different aspects of the field [4-7]. However, to date no review has discussed specifically the applications of FT infrared methods in this area, although such applications are rapidly increasing.

As will be described below, various experimental difficulties arise in attempting to obtain infrared bands from the minor fraction of a sample that normally constitutes the chemisorbed layer. Except when polymerization occurs on a surface, this consists of at most a molecular monolayer in the presence of a much greater amount of the solid adsorbent. The Fourier transform technique

J. R. Durig (ed.), Analytical Applications of FT-IR to Molecular and Biological Systems, 281–291.

will be shown to be capable of making a variety of different contributions to these problems.

2. APPLICATIONS TO CHEMISORPTION ON FINELY-DIVIDED ADSORBENTS

As an introduction we discuss first the experimental problems associated with the study of chemisorption on a very finely divided adsorbent. Such samples could be studied in the form of a powder on a horizontal infrared-transmitting plate, as was originally done in the pioneering infrared chemisorption studies by Eischens and his colleagues [8,9]. However, it is often possible to make a coherent pressed disc from the catalyst (in a manner analogous to making a pressed KBr disc for solid-state infrared studies) which retains a porosity so that the gas-phase can interact with most of the surface. The latter is nowadays the preferred method as it leads to the use of vertical samples in the standard infrared optics. It also obviously greatly simplifies the construction of the evacuable infrared cell [1,3].

Experimental problems associated with studies of finely-divided adsorbents may be listed as below:

(1) Loss of radiation can occur strongly by sample scattering, particularly by low area materials with high refractive indices such as the heavy-metal oxides. The low area usually implies the presence of large particles, or wide internal pores. When the particle or pore size is greater than the wavelength of the radiation involved, this gives greatly enhanced scattering which is made worse by a large difference of refractive index. The lighter element oxides, such as silica, alumina, or the zeolites, have been particularly widely studied because of their moderate refractive indices and their ready availability in high area, small particle forms.

(2) Strong absorption of energy by the adsorbent often occurs over considerable regions of the infrared spectrum. Figure 1 illustrates the transmission spectrum of a pressed disc of silica, a material often used as a support for metal catalysts or as an adsorbent itself. The absorption bands between 3800 and 300 cm^{-1} originate in individual 'free' or hydrogen-bonded pairs of chemically bound surface OH groups. Those giving a very variable background between 2000 and 1300 cm^{-1} are overtones and combinations of fundamental frequencies of the silica skeleton. Below 1300 cm^{-1} down to 400 cm^{-1} only very weak transmission windows occur in the otherwise 'blackout' region associated with the wavenumber range of the skeletal fundamentals themselves. Zeolites give similar spectra at the pathlengths needed to give spectra from adsorbed species, and aluminas 'blackout' below ca. 1100 cm^{-1}.

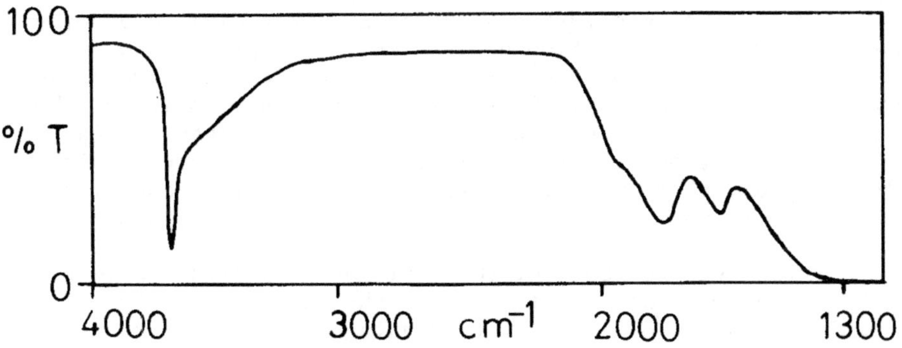

Fig. 1 The infrared transmission spectrum of a porous pressed disc of silica.

(3) Many surfaces of interest, e.g. heavy-metal oxides or oxide-supported metal catalysts, have only limited specific areas of the order of 10 m^2 g^{-1}. High sensitivity is therefore needed to record spectra of adsorbed species. Silicas and aluminas are available with specific areas of ca. 200 m^2 g^{-1} or more.

(4) Many adsorbed species of interest, e.g. adsorbed hydrocarbons, have relatively weak absorption bands requiring high sensitivity. Carbon monoxide is, at the other end of the scale, one of the strongest-adsorbing surface species and is often used for surface characterization [10].

(5) It is often not possible to predict beforehand what surface species are going to be present. Quite unexpected species can frequently be found. This latter factor makes the work difficult and requires the measurement of spectra over as wide a wavenumber range as possible. On the other hand, if the spectral information can be convincingly interpreted, new and valuable chemical information can be obtained. One of the merits of the infrared and other vibrational spectroscopic methods is the comparative ease with which even unexpected spectral features can be interpreted in terms of chemical structures.

(6) More often than not a given finely-divided adsorbent gives rise to several chemisorbed species, presumably on different surface sites, from a single adsorbate. This complicates spectral interpretation. Additional techniques such as variable-temperature studies [11], or the displacement of weaker chemisorbed species by others with known spectra [12] are necessary to separate the overlapping spectra. This is particularly true for chemisorbed hydrocarbon species for which band overlap can readily occur in the νCH bond-stretching region near 3000 cm^{-1}.

A typical chemisorption study on a finely-divided adsorbent then involves the detection of very weak absorption bands over a wide spectral range on a sample of very poor and widely-variable background transmission. These are exactly the conditions where the high signal-to-noise capability of the FT method can make a major contribution. This capability in large measure results from the multiplex or Fellgett Advantage which is proportional to the square root of the number of spectra elements studied.

If the dynamic range of the analogue to digital converter (ADC) is not exceeded - this will be the case for samples of poor overall transmission - then the advantages of Fourier transform spectroscopic methods can be fully exploited.

A minicomputer of considerable power has to be incorporated in a mid infrared FT spectrometer in order to transform the interferogram back into the usual spectral presentation of intensity versus wavenumber. This computational facility can also be used with great advantage for spectral ratioing in transmission units, or its equivalent of spectral subtraction in absorbance units. Thus the spectrum of the catalyst sample itself, i.e. before adsorption of a gas, can be stored in the memory of the minicomputer and then ratioed, point by point, with a new spectrum from (catalyst + adsorbed species). By this means the detailed shapes of adsorption bands from the new chemisorbed species can be more readily evaluated against an effectively flat background. In addition, and more importantly, weak absorption bands can be readily seen that would often not be noticed in the unratioed spectra because of their superimposition on a background from the spectrum of the catalyst which is very variable with wavenumber. In some cases the adsorption process leads to a change in the shape or position of a band of the adsorbent and spectral ratioing then leads to negative contributions to the spectra where a catalyst absorption band has been displaced, and a positive contribution at the position of the displaced band. These several features are illustrated in Figs. 2 and 3.

The ratioing technique with the help of computation facilities is increasingly becoming available with dispersion infrared spectrometers also, but the signal/noise gain is unique to the Fourier transform instruments when overall transmission is low. Although with a dispersion spectrometer, an improvement in signal/noise can often alternatively be made by widening the spectrometer slits, in many cases the consequent loss of resolution affects substantially the quality of the spectra obtained. Nonetheless, if only a narrow spectral range has to be studied with few spectral elements, e.g. when measuring the strong νCO absorption bands resulting from the chemisorption of carbon monoxide on a metal surface which normally occur between 2200 and 1800 cm^{-1}, then comparable results can often be obtained with a dispersion

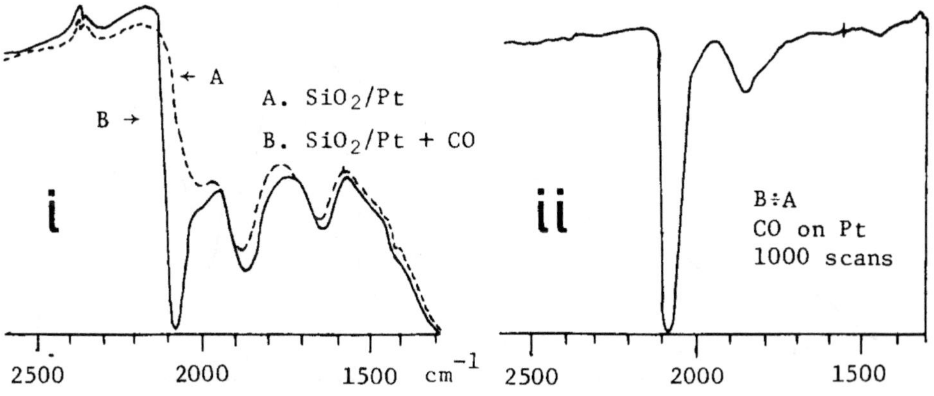

Fig. 2 (i) The infrared transmission spectrum of (A) a Pt/SiO_2 catalyst sample and (B) of the same Pt/SiO_2 sample with chemisorbed CO. (ii) The transmission spectrum of the chemisorbed CO obtained by ratioing spectrum (B) with spectrum (A).

spectrometer.

3. REFLECTION/ABSORPTION INFRARED SPECTROSCOPY AS APPLIED TO MOLECULES CHEMISORBED ON BULK METAL SURFACES INCLUDING SINGLE CRYSTALS

It is often found that several different species result from the adsorption of a single type of molecule on a finely-divided metal catalyst surface. This results in substantial complications in spectral interpretation in structural terms. For example, CO chemisorbed on metal gives both 'linear' species (bonded to a single metal atom) and 'bridged' species (bonded to two or more metal atoms) on the same finely-divided metal catalyst [9,10]. Equally, ethylene on metals appears to give σ-diadsorbed $\underline{M}\ CH_2CH_2\underline{M}$ and π-adsorbed $(C_2H_4...\underline{M})$ species ($\underline{M}$ = metal) plus some additional species in which CH bonds have been broken to give $(C_2H_x)M_y$ surface species where $x < 4$ [11,13].

This type of behavior occurs because a crystalline or quasi-crystalline metal particle has available a number of different sites corresponding to different local arrangements of metal atoms. Figure 4 shows, for example, the different arrangements of metal atoms that are expected to occur on the principal low-index surface planes of a face-central cubic metal. Many of the catalytically active metals, e.g. Ni, Pd and Pt, crystallize in this form. As always in science, when a situation proves to be complex, then progress is made by judicious simplification of the experiment. The obvious approach to obtaining an understanding of the complex

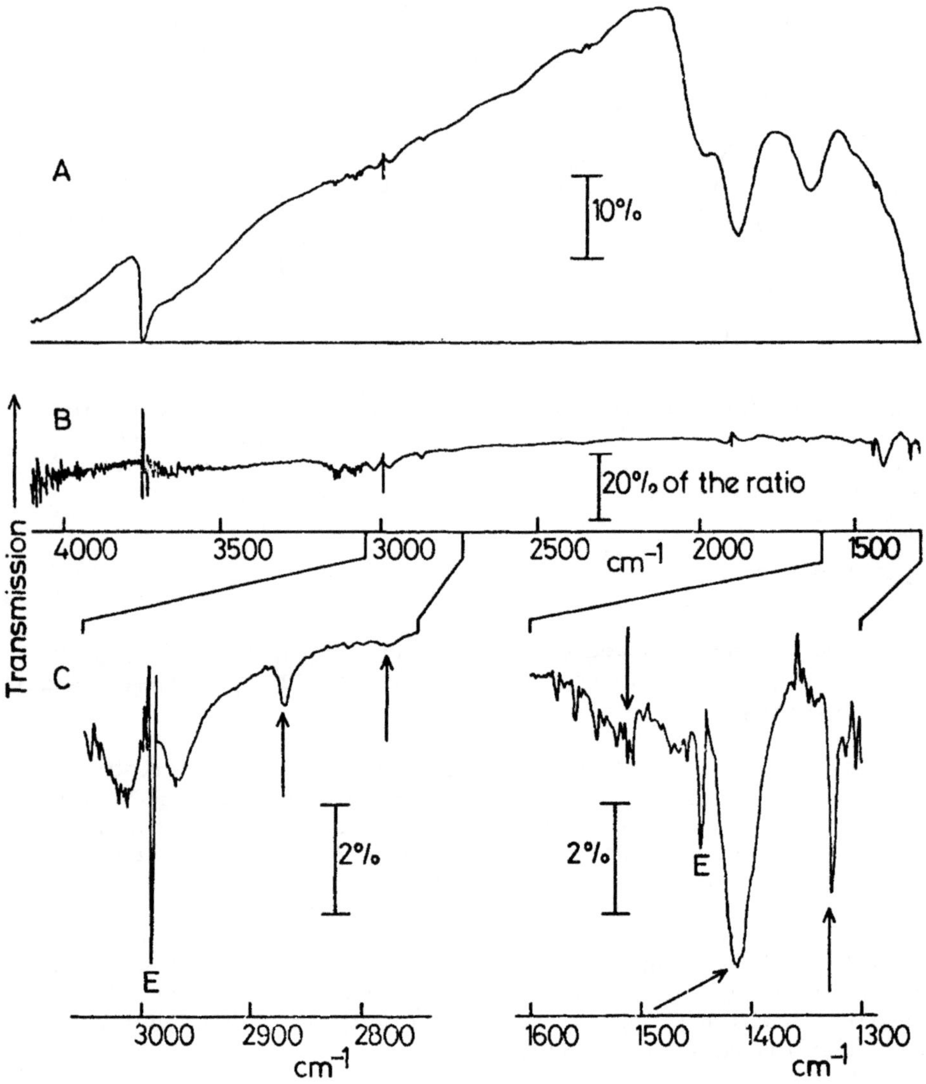

Fig. 3 Typical infrared transmission spectra from a Pd/SiO_2 catalyst in the presence of gas-phase ethylene showing bands from adsorbed species. (A) Overall spectrum dominated by the absorption bands of the catalyst. (B) Spectrum ratioed against that of the base catalyst showing only the bands from gas-phase and adsorbed ethylene. (C) Expanded section of spectrum B. Arrowed bands are from adsorbed species; E marks the Q branches of gas-phase absorptions.

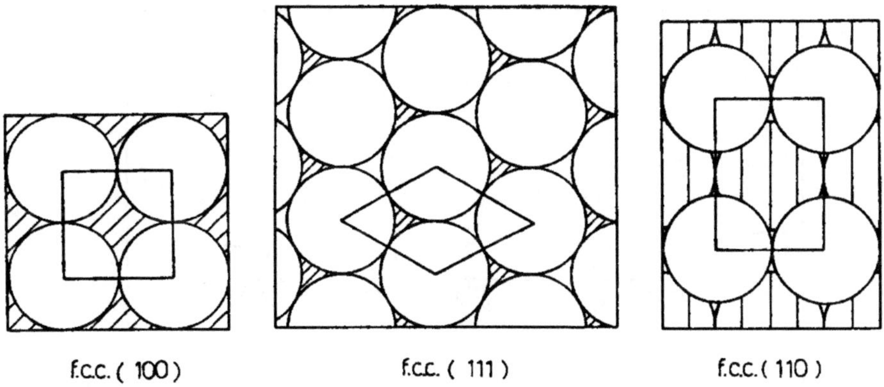

Fig. 4 Atomic arrangements on specified planes on a face centered cubic metal.

spectra from finely-divided metals is to obtain spectra from species adsorbed on specified single crystal surfaces. Here, however, the experimental arrangements are much more demanding because a flat low-area plane of metal is being studied and much sensitivity is lost.

An infrared spectrum of the adsorbed layer is obtained in such cases by reflecting the radiation beam off the metal surface, thereby effectively traversing a single monolayer twice, Fig. 5. This technique is described as reflection/absorption spectroscopy [14]. Greenler [15,16] has shown that for very thin layers on metals the highest sensitivity is obtained when the reflection is made at near-grazing incidence. Even then only those vibrations of the adsorbed species which give dipole changes perpendicular to the surface are observable with appreciable strength. This restriction is conventionally termed the metal surface selection rule and is thought to apply to some degree to spectra of chemisorbed species obtained on the larger metal particles [12]. Pritchard [17,18] has pioneered the use of reflection/absorption infrared spectroscopy to the study of single crystal surfaces, particularly the adsorption of CO on copper.

The strengths of the absorption bands obtained by the reflection/absorption method are in practice one or two orders of magnitude weaker than those obtained from finely-divided catalysts by the transmission method. Hence only intrinsically very strong absorption bands, such as those from chemisorbed CO [17] or carboxylate ions [14] can be easily measured by this technique. Nevertheless some very interesting results have been achieved, and the relatively high sensitivity of the FT method has enabled some spectra to be obtained with good resolution [19]. An example of such spectra from CO chemisorbed on a polycrystalline ribbon

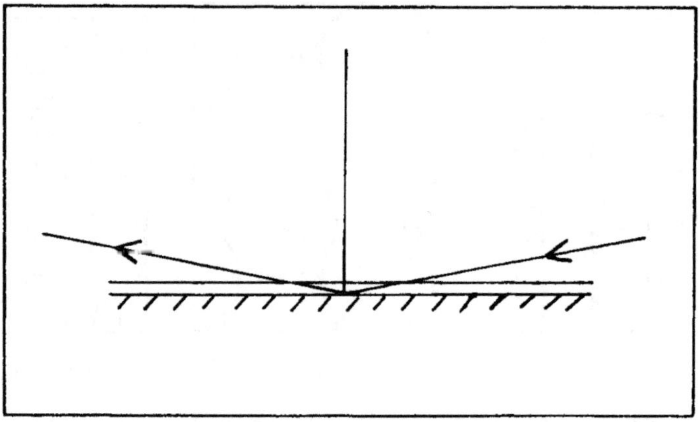

Fig. 5 The experimental arrangement for measuring reflection/absorption infrared spectra for adsorbed monolayers on flat surfaces of metals, including single crystal planes. High intensity comes from reflection at near grazing incidence.

of Pt with predominant (111) surfaces is given in Fig. 6.

Further advances in infrared techniques, such as the advent of high stability tuneable infrared lasers, will be needed to allow such infrared spectra to be obtained on a routine basis. In the meantime a complementary technique involving the inelastic reflection of electrons by the metal surface - electron energy loss spectroscopy [20,21] - is beginning to provide similar results with the required one to two orders of magnitude greater sensitivity. On the other hand, the resolution of this technique (only ca. 50 cm^{-1}) is an order of magnitude worse than that obtained by reflection/absorption infrared spectroscopy and this will lead to limitations when other than adsorbed species of simple structure are studied. Also, in general, the retention of cleanliness of these low-area metal surfaces requires the use of ultra-high vacuum techniques. Electron spectra are difficult to obtain with appreciable pressures of the gas-phase in contact with the metal surface and this makes difficult the study of flowing reacting systems. This limitation is, however, much less important for infrared spectroscopy and hopes are high for the future.

4. STUDIES OF ADSORBED SPECIES BY INFRARED EMISSION SPECTROSCOPY

Applications of Fourier transform methods to study emission spectra in the infrared region are described in my other chapter of this book. Applications to adsorbed species have been made

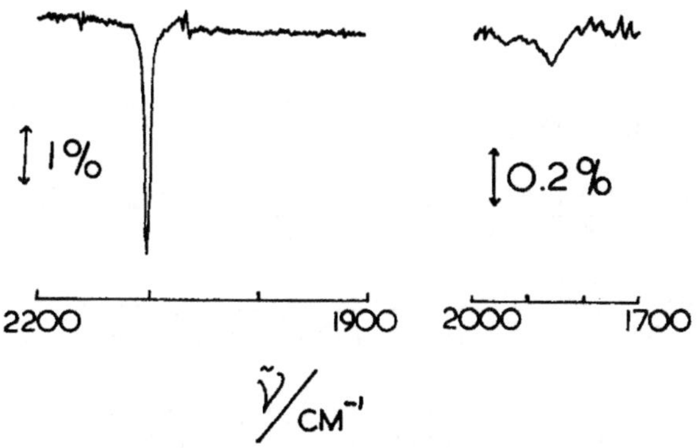

Fig. 6 A reflection/absorption infrared spectrum from CO chemisorbed on a Pt ribbon with predominantly (111) polycrystalline facets. The band between 2060 and 2110 cm^{-1} arises from CO molecules bonded to single metal atoms; the weaker band at ca. 1820 cm^{-1} arises from CO bridge-bonded to two metal atoms.

both to thin overlayers on highly reflecting (and hence weakly emitting) surfaces such as metals [22,23] and to adsorption on finely-divided oxide adsorbents [24,25].

The FT method contributes particularly well to obtaining good results from the weak emission spectra of samples at moderate temperatures. Also the computing facility of the FT spectrometer enables the ready ratioing of the emission spectrum of the sample against that of a blackbody at the same temperature. This provides an emissivity spectrum which is closely related to a percentage absorption spectrum.

5. CONCLUSION

The study of the infrared spectra of chemisorbed species has thrown much light on the structures of surface species of importance in catalysis. Whether the experimental method used involves transmission, reflection/absorption, or emission measurements, Fourier transform techniques contribute strongly to the qualities of the spectra obtained.

REFERENCES

1. L. H. Little, Infrared Spectra of Adsorbed Species, Academic Press, London, 1966.

2. M. L. Hair, Infrared Spectroscopy in Surface Chemistry, Dekker, New York, 1967.

3. A. V. Kiselev and V. I. Lygin, Infrared Spectra of Surface Compounds, Wiley, New York, 1975 (Russian Edition 1972).

4. L. H. Little, in Chemisorption and Reactions on Metallic Films, Vol. I (J. R. Anderson, ed.) Academic Press, London, 1971, p. 489.

5. M. R. Basila, Appl. Spectrosc. Rev., 1, 289 (1968).

6. J. Pritchard, 'Surface and Defect Properties of Solids', (Specialist Periodical Reports, Chemical Society, London), (M. W. Roberts and J. M. Thomas, eds.), Vol. 1, 222 (1972); (same series but name changed to 'Chemical Physics of Solids and Surfaces', Vol. 5, 157 (1979)).

7. C. H. Rochester and M. S. Scurrell, 'Surface and Defect Properties of Solids', (Specialist Periodical Reports, Chemical Society, London), (M. W. Roberts and J. M. Thomas, eds.), Vol. 2, 114 (1973).

8. R. P. Eischens, S. A. Francis, and W. A. Pliskin, J. Chem. Phys., 22, 1786 (1954).

9. R. P. Eischens and W. A. Pliskin, Adv. Catalysis, 10, 1 (1958).

10. N. Sheppard and T. T. Nguyen in Advances in Infrared and Raman Spectroscopy, (R. J. H. Clark and R. E. Hester, eds.), Heyden, London, 1978, Vol. 5, p. 67.

11. B. A. Morrow and N. Sheppard, Proc. Roy. Soc., A311, 391 (1969).

12. H. A. Pearce and N. Sheppard, Surface Science, 59, 205 (1976).

13. J. D. Prentice, A. Lesiunas, and N. Sheppard, Chem. Commun., 76 (1976).

14. S. A. Francis and A. H. Ellison, J. Opt. Soc. Amer., 49, 131 (1959).

15. R. G. Greenler, J. Chem. Phys., 44, 310 (1966).

16. R. G. Greenler, J. Chem. Phys., 50, 1963 (1969).

17. M. A. Chesters, J. Pritchard, and M. L. Sims, Chem. Commun., 1454 (1970).

18. J. Pritchard, T. Catterick, and R. K. Gupta, Surface Sci., 53, 1 (1975).

19. M. Baker, M. A. Chesters, and V. Higgs, unpublished work.

20. H. Ibach, H. Hopster, and B. Sexton, Applic. Surface Sci., 1, 1 (1977).

21. H. Froitzheim, H. Hopster, H. Ibach, and S. Lehwald, Appl. Phys., 13, 47 (1977).

22. D. Kember and N. Sheppard, Appl. Spectrosc., 29, 496 (1975).

23. D. Kember, D. H. Chenery, N. Sheppard, and J. Fell, Spectrochim. Acta., 35A, 455 (1979).

24. M. Primet, P. Fouilloux, and B. Imelik, Surface Sci., 85, 457 (1979).

25. O. Koga, T. Onishi, and K. Tamaru, Chem. Commun., 464 (1974).

VIBRATIONAL SPECTRA AT VARIABLE TEMPERATURE AND THE DETERMINATION OF ENERGIES BETWEEN CONFORMERS

William O. George and Jeanette E. Goodfield

Department of Science
Polytechnic of Wales
Pontypridd, Mid Glamorgan, U. K.

1. INTRODUCTION

This chapter describes experimental methods used to determine the energy differences between stable conformers and discusses the results. A number of physical methods permit these determinations but only results based on vibrational spectroscopy will be reviewed here to provide comparison of a wide range of compounds by those experimental methods which are based on molecular vibrations.

A proper understanding of conformational equilibria requires consideration of many physical and chemical techniques but such studies are best made on selected classes of compounds.

The energy difference is one of a number of thermodynamic parameters which may be determined but a wider coverage of thermodynamic results would be too voluminous in other than a selected group of compounds.

The techniques of vibrational spectroscopy constitute the trinity, Infrared, Raman, and Fourier Transform Infrared. They will be compared to make clear their complementary role particularly in relation to variable temperature studies.

2. CONFORMATIONAL ISOMERISM

If one part of a molecule is rotated with respect to another part of the same molecule about a single chemical bond one or more stable conformational isomers (or rotational isomers) may be

J. R. Durig (ed.), Analytical Applications of FT-IR to Molecular and Biological Systems, 293–332.

formed corresponding to minimum potential energy states. These conformers (or rotamers) are rapidly interconvertible in the fluid phases and cannot be separated into chemically distinct forms (although it will be seen that they have some physically distinct properties). Their existence was established in the mid 1930's for a particularly simple molecule, ethane, by comparison [1] between the entropy, enthalpy and heat capacity values determined firstly by calorimetric methods and secondly by the methods of statistical thermodynamics.

Figure 1 shows the variation of potential energy of ethane with rotation of the CC bond through 360° (i.e. with change in the azimuthal angle). The minima correspond to staggered conformations in which the H atoms have greatest separations and least repulsive interaction energies. The maxima correspond to eclipsed conformations in which the H atoms have least separations and greatest repulsive interaction.

The potential energy function loses its three-fold periodic symmetry if the three-fold symmetry of the methyl is lost. Thus substitution of hydrogen by R (e.g. in n-butane) leads to two types of eclipsed conformers (R, H and R,R) and two kinds of staggered conformers (<u>trans</u> and <u>gauche</u>). The potential energy expression for n-butane is shown in Fig. 2. Consideration of the two types of staggered conformations reveals that the repulsive energies are greater in the <u>gauche</u> form than in the <u>trans</u> form. The difference between these two energy levels shown as minima in Fig. 2 is known as the enthalpy difference (ΔH) between the conformational isomers. The measurement of this difference in various systems is the subject of this chapter. The difference between neighboring maxima and minima is termed the potential barrier hindering internal rotation. This barrier height may be determined in various ways but will not be considered here.

The penetrating insight provided by vibrational spectroscopy into conformational isomerism began to emerge very shortly after the discovery in 1928 of the Raman effect. Kohlrausch [2,3] observed more lines in the Raman spectrum of n-propyl and n-butyl halides than would be expected on the basis of a single conformer and tentatively assigned these to isomerism. With rapid development of vibrational assignments Cabannes [4] was able to identify and assign Raman bands to isomers of 1,2-dihaloethanes in terms of the symmetry of the s-<u>trans</u> form and less symmetrical conformational isomers.

Thus direct measurement was possible of components of a mixture of conformational isomers based on the vibrational molecular property. This provided confirmation for that evidence which was less direct and rested on measurement of bulk thermodynamic properties of the system.

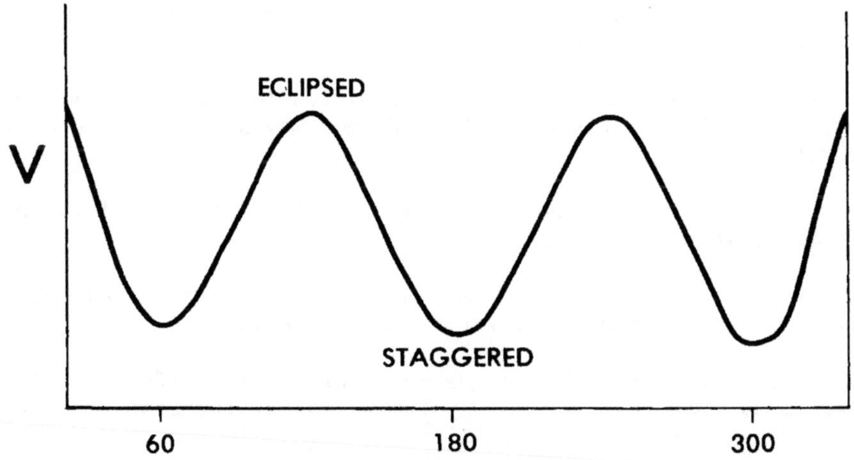

Fig. 1 Variation of potential energy of ethane with rotation of C-C bond between different conformers (3-fold barrier).

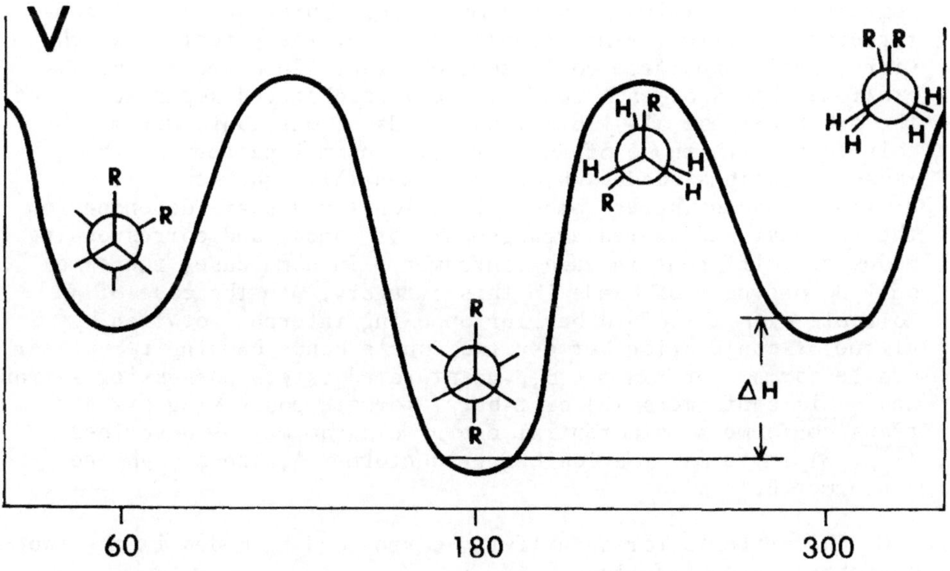

Fig. 2 Variation of potential energy of 1,2-disubstituted ethanes with angle of torsion.

3. CONFORMATIONAL EQUILIBRIA

The study of conformational equilibria requires a method with the following attributes.

(i) Ability to study a system in which the separate existence and structural features of each conformer may be identified qualitatively.

(ii) Ability to measure the concentration of each of two or more conformers quantitatively.

(iii) Ability to study compounds in solid, liquid and gas phases and in solution in a variety of solvents.

(iv) Ability to make measurements within the life-time of short-lived components of the system which may interconvert on a very rapid time-scale.

(v) Ability to make measurements on a system when the conditions affecting the position of equilibrium (for example, temperature) can be varied and measured.

In principle the three principal methods of vibrational spectroscopy possess these various attributes. The vibrational spectra of an equilibrium (or non-equilibrium) mixture of conformers will contain bands arising from the molecular vibrations of all conformers. Practical limitations are set by the extent to which vibrational bands from conformers overlap. In extreme, the vibrational bands of each conformer may be observed separately. At the other extreme all vibrational bands of one conformer may be coincident with those of another. The normal pattern is that those vibrational modes which are essentially the same are not distinguishable whereas other vibrational modes are dependant on the skeletal and stereo arrangements of bonds, and corresponding modes are different in each conformer. In some cases there may only be one pair of bands in this category. In the case of a molecule with a twofold barrier opposing internal rotation by virtue of conjugation between two double bonds causing two planar stable forms, for example α,β-unsaturated esters possessing s-<u>trans</u> and s-<u>cis</u> conformers [5] or t-<u>butyl</u> formate possessing <u>cis</u> and <u>trans</u> conformers, a potential energy diagram may be described (Fig. 3) in terms of a low energy conformer A, and a high energy conformer B.

The criteria for identifying corresponding modes in the two conformers are as follows:

(i) As the temperature of a sample in the fluid form increases the relative intensity of bands due to high energy conformers

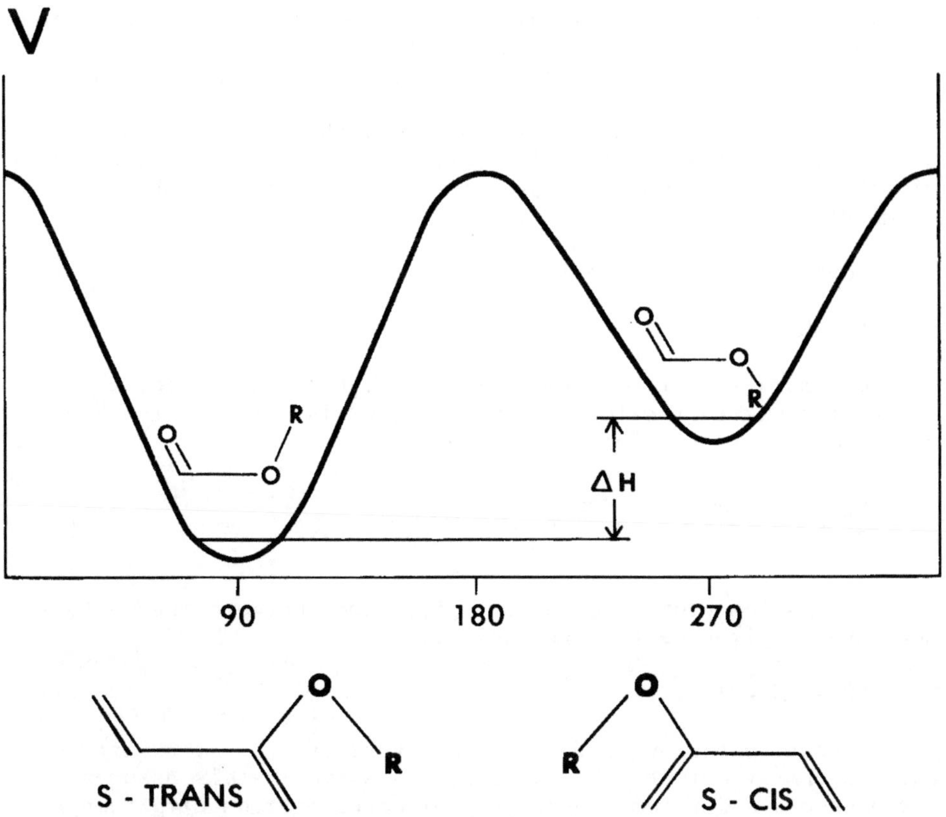

Fig. 3 Variation of potential energy for a system with twofold barriers.

increases (and that of bands due to low energy conformers decreases).

(ii) As the temperature of a sample in the fluid form decreases the relative intensity of bands due to low energy conformers increases (and that of high energy conformers decreases).

(iii) When the sample is cooled in such a way as to form a stable crystalline solid form, bands due to high energy conformers disappear.

(iv) The experimental information obtained for infrared, Raman, or interferometry is consistent taking into account differences in selection rules, intensities, and technical factors which influence limits of detection and spectroscopic range and states of matter.

(v) Conformer bands should be observed in pairs corresponding to modes of similar form but slightly different frequencies in the high energy conformer (B) and low energy conformer (A).

(vi) The concentration of A and B at the conditions of measurement must be sufficient for detection and measurement of the vibrational band under the conditions used (typically about 5% or more).

Consider the equilibrium

$$A = B.$$

The concentration of conformers in the high energy state, C_B relative to that in the low energy state, C_A is given by the van't Hoff isotherm expression.

$$\frac{C_B}{C_A} = e^{-\Delta G/RT}. \tag{1}$$

Where ΔG is the Gibbs free energy difference between the conformers and is given by the expression:

$$\Delta G = \Delta H - T\Delta S. \tag{2}$$

The standard enthalpy difference ΔH may be taken as the difference in energy between the two minima. Strictly this assumes the pressure and volume dependance of concentration with temperature is negligible which is approximately true.

The vibrational band intensities of selected conformer bands may be related to the concentration of each conformer. In the case of absorbance (A) measurement by dispersive or Fourier transform methods, the Beer Lambert Law leads to the relationships

$$A_A = \varepsilon_A C_A \ell \tag{3}$$

$$A_B = \varepsilon_B C_B \ell \tag{4}$$

where ε and ℓ are extinction coefficients of particular conformer bands and the path-length, respectively.

Substitution of (2), (3), and (4) into (1) leads to

$$\frac{A_B \varepsilon_A}{A_A \varepsilon_B} = e^{-\Delta H/RT} e^{\Delta S/R}. \tag{5}$$

This may be more conveniently expressed

$$\ln \frac{A_B}{A_A} = -\frac{\Delta H}{RT} + \frac{\Delta S}{R} - \ln \frac{\varepsilon_A}{\varepsilon_B} . \qquad (6)$$

If it is assumed that ΔS and the ratio $\varepsilon_A/\varepsilon_B$ are constant with temperature it is possible to determine ΔH by making a series of measurements of A_B/A_A at various accurately determined temperature values T.

From Eq. (6) further assumptions have led to the calculation of ΔS^0 either using measurements of a third band common to both conformers [6] or determining ε_A and ε_B by a simple graphical procedure [7].

For Raman spectra, the intensity of a conformer band is linearly dependent on concentration and the corresponding equation to (6) differs by replacing the extinction coefficient by the corresponding Raman emission intensity coefficient.

The contribution of vibrational spectroscopy to the study of conformational isomerism has been reviewed by Mizushima [8], Sheppard [9] and Park, Pethrick, and Thomas [10].

4. VIBRATIONAL SPECTROSCOPIC METHODS AND CONFORMATIONAL ISOMERISM

It has been noted that Raman spectroscopy was the initial method by which conformational isomerism was detected. This accords with the relative ease with which vibrational spectra could be obtained in the 1930's, providing favorable samples were used which would scatter sufficient of the incident radiation from the mercury arc sources available at that time.

Raman spectra are measured in the visible region of the spectrum. The control of and measurement of temperature of the sample over a wide range can be readily achieved using the appropriate optical materials subject to the collection of sufficient incident radiation by the sample.

The vibrations which are weak or inactive in the Raman are usually strong or active in the infrared. The IR region of the electromagnetic spectrum in which absorption of vibrational frequencies takes place posed technical difficulties of source, optics, and detection until advances of electronics in the 1940's made the acquisition of infrared spectra easier than Raman for most samples, particularly if colored or present in small quantities.

The early infrared spectrometers evolved from single beam to double beam utilizing an attenuator in the reference beam which was driven in and out by a servo mechanism to achieve a null balance in comparison with the sample beam. The absorbance measured as a function of wavenumber was proportional to this mechanical movement. Non-linear effects in this movement limited the accuracy of absorbance measurements further. Recent advances in electronics and instrumentation have led to several manufacturers producing instruments which measure the direct ratio of sample to reference beam with considerably more accuracy than is possible by a self-balancing servo mechanism and which is capable of scale expansion and computer enhancement with opportunities to measure bands at low absorbance or against strongly absorbing backgrounds. A further feature of several modern instruments is the ability to chop the radiation before and after the sample. This enables a true measure of absorption of radiation from the source by the sample to be made by isolating effects due to exchange of energy between sample and detector. These effects may be expected particularly when the sample is at a higher or lower temperature than the source. However, when a comparison was made [11] of the calculation of ΔH for methyl crotonate using two infrared spectrometers with the different chopping modes, no significant differences were obtained.

The advent of the Laser in the 1960's enabled Raman spectroscopic measurements to be made with similar ease to infrared. The two methods are complementary both in a fundamental way and in a technical way. Fundamental principles govern which method enables particular vibrational modes to be measured, but consequent technical considerations governing the condition or nature of the sample leads more favorably to one or another method.

The potential application to vibrational spectroscopy of the interferometer designed by Michelson in 1891 lay dormant until the 1960's. Early developments were directed to the measurement of far infrared spectra. The disadvantage of the computer stage to convert the interferogram into an absorption spectrum by Fourier transformation led to parallel developments in dispersive instruments for the far infrared region. This region is important in determining torsional frequencies and hence barrier heights opposing internal rotation.

Improvement in instrumentation led to interferometers for mid infrared spectroscopy. The precision needed for the collection of data and the computing of the Fourier transform stage resulted in the total process being more costly than dispersive methods for infrared. The spectra of conventional samples are essentially the same by interferometry and dispersion. The ability to accumulate interferometric data by rapid scanning of a sample many times leads to high sensitivity. Samples of very high absorbance and samples of very low absorbance may therefore be measured.

The method therefore provides the potential for making variable temperature measurements on samples inaccessible to conventional dispersive spectroscopy.

The ability to measure spectra by Raman, direct infrared absorption, or interferometry (followed by Fourier transformation leading to an infrared spectrum) has led to a healthy rivalry. Raman and infrared are not directly comparable. Raman spectroscopy is a light scattering phenomena in which shifts in visible radiation are a consequence of polarizability changes accompanying molecular vibrations. Infrared spectroscopy is an absorption of vibrational frequencies as a consequence of dipole moment changes accompanying molecular vibrations.

IR and FT-IR are directly comparable leading essentially to the same result. There is therefore scope for enthusiasts to develop each method to demonstrate achievements not possible by the other. Indications are so far that IR has the edge in cost and convenience. FT-IR has the edge in sensitivity and speed.

5. SAMPLE HANDLING

The essential requirement in the determination of enthalpy difference between conformers by vibrational spectroscopy is to make accurate intensity measurements in the fluid phases at measured temperatures over temperature ranges as wide as possible.

Intensity measurements by infrared or Raman of samples at ambient temperatures are well established. The peak height of the band or the area measured by a planimeter can be related to concentration. Comparison of results using both methods tend to show similar final results. When bands overlap computer methods aid the determination of the separate intensities.

There is much literature on measurement of absolute intensities in vibrational spectroscopy. For the determination of ΔH values the relative intensities of two similar adjacent bands is required. Many of the errors of instrumentation and cell imperfections are fortunately cancelled out in the determination of the A_B/A_A in Eq. 6. The nature of these errors and their elimination need therefore not be pursued for the present purposes.

A cell for recording samples at variable temperatures is shown in Fig. 4. Liquids for Raman study may sometimes, with advantage, be distilled into a capillary tube and sealed under vacuum. Small samples contained in this way can be mounted in a system described

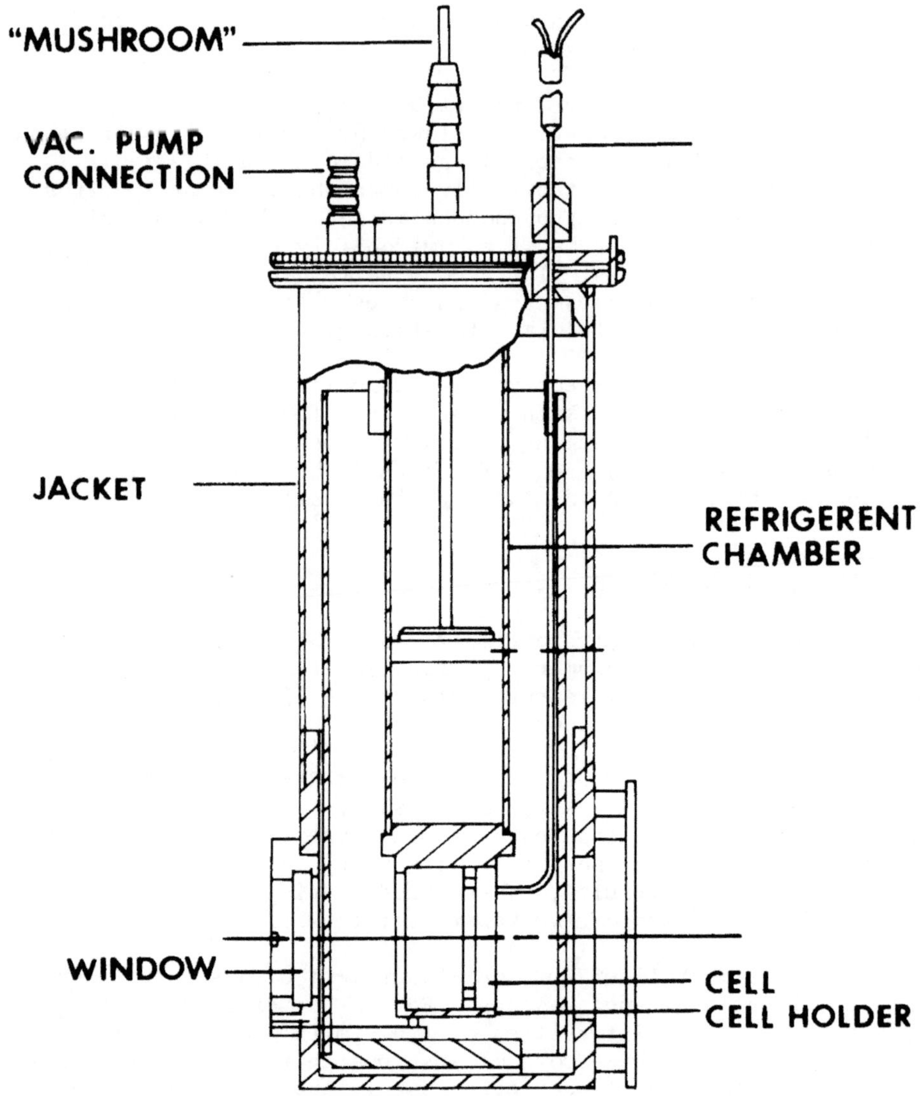

Fig. 4 A variable temperature cell for Raman, infrared, and interferometry.

by Harney and Miller [12] in which a controlled flow of cold nitrogen or heated nitrogen takes the sample to each required temperature as measured by a thermocouple near the sample. Spectra recorded in this manner can give misleading band depolarization data because of multiple reflections from the walls which can change the polarization of a band. It is therefore desirable that any necessary depolarization measurements are taken in conventional cells.

The measurements of spectra in the infrared at variable temperature by either dispersive or interferometric methods require a vacuum tight cell mounted in a block which can be cooled by refrigeration or heated electrically in a controlled manner with temperature measurement from a thermocouple inserted in the cell near to the sample.

The equipment shown in Fig. 4 can be readily mounted in the cell compartment of either a dispersive instrument or an interferometer. A double jacket enables the surroundings to the cell to be evacuated to minimize loss or gain of heat by the sample other than from the controlled source.

Most ΔH measurement reported relate to samples in the liquid phase when Raman spectroscopy may be used or solution phase (usually CS_2) when infrared spectroscopy may be used. Annealing by controlled cooling can lead to stable solid forms. Gas cells which are heatable are readily available or easily constructed for infrared and can be constructed with more difficulty for Raman. If the cell in either technique is multipath via multiple internal reflection by mirrors, there are problems of changing alignments during heating.

The greater sensitivity of interferometry opens up the opportunity to make ΔH measurements under conditions otherwise inaccessible and/or on pairs of bands normally too weak to be used. The remainder of this chapter will indicate the state of the art achieved largely by conventional Raman and IR and serves to point up problems partly or wholly unsolved which require the extra dimension promised by FT-IR for much better understanding.

6. FACTORS GOVERNING RELATIVE STABILITIES BETWEEN PHASES

The ability to determine ΔH values in various phases have provided interesting differences. Thus in the 1,2-dihaloethanes a detailed study suggests that the s-*trans* form is the more stable in the gas phase, whereas the *gauche* form is the more stable in the liquid phase [13]. These and other workers consider that the stability in the gas phase is determined by intramolecular steric interactions but that the stability in the liquid phase is

Table 1 RCH_2-CH_2R Enthalpy Differences

Compounds	Substituents (R)	ΔH (kJ/mol)	Phase	Method	Refs.
1,2-Difluoroethane	F	0	Gas	IR	[14]
		3.8	Liq.	IR	[14]
1,2-Dichloroethane	Cl	4.88±0.4	Gas	IR	[16]
		0.00	Liq.	IR,R	
		1.67	Soln.[a]	IR	[17]
1,2-Dibromoethane	Br	7.4±0.63	Gas	IR	[16]
		3.1±0.5	Liq.	R	[18]
		3.95	Soln.[a]	IR	[17]
		1.67	Soln.[b]	IR	[17]
1,2-Diiodoethane	I	5.77±0.42	Soln.	IR,R	[20]
Succinonitrile	CN	-1.5±0.20	Liq.	IR	[21]
n-Butane	Me	3.18±0.4	Liq.	R	[22]
n-Hexane	Et	2.1±0.4	Liq.	R	[22]
1,6-Dibromobipropargyl	C≡CBr		Liq.	IR	[23]
1,4-Dibromobutane	CH_2Br	-2.1±0.4	Liq.	IR	[24]

[a]Solution in Hexane [b]Solution in Nitromethane

determined by dipolar and quadrupolar interactions. Although some progress was made towards putting these ideas into a quantitative form it is apparent that the forces governing conformational isomerism are only imperfectly understood.

7. SYMMETRICALLY 1,2-DISUBSTITUTED ETHANES

These compounds may be rationalized in terms of an equilibrium between s-*trans* and *gauche* conformers as shown in Fig. 2. A negative value for ΔH implies *gauche* has lower energy than the *trans*.

Table 1 provides data on those compounds which have halogens, cyanide, methyl, ethyl or larger substituents. It is noted [14, 16] that the value of ΔH increases with size of halogen in the gas phase. Vibrational assignments reveal that for the gas phase, bands associated with the low energy form correspond to the *trans* conformer, whereas the bands associated with the high energy form correspond to the *gauche* conformer.

Furthermore, in Table 1 it is noted that the ΔH increases from 0.84, 4.58 to 7.4 for F, Cl and Br. This correlates with the sum of the Van der Waals radii of the two halogen atoms. So far it has not been possible to obtain ΔH values for larger substituents in the gas phase.

Klaboe and Nielsen have demonstrated [14] that the trans and gauche forms of 1,2-difluoroethane are comparable in enthalpy in the gas phase, but the gauche form is considerably more stable in the liquid phase. This change is not completely understood but is in accord with dielectric theory. The polar gauche isomer should have a lower potential energy in the liquid state than in the gas state while the non-polar trans isomer should have similar potential energies in the two states. The authors were able to correlate relative stabilities with dielectric constants of a range of solvents of increasing polarity (cyclohexane, carbon tetrachloride, chloroform, acetonitrile, and nitromethane).

Succinonitrile [21] shows unusual behavior in the solid state in that bands are observed corresponding to two isomers until the temperature is reduced below -44°C, when bands are only observed corresponding to the low energy conformer.

A vibrational study of 1,2-diiodoethane by Tanabe [20] in solution and the solid enabled the ΔH value between the trans and gauche forms to be estimated as 5.77 ± 0.42 by variable temperature. This result is considered more accurate than several previous ones by virtue of the double chopping mode discussed earlier. A method based on measurement of absolute intensities is described which leads to a ΔH value of 5.43 ± 0.42 which is in excellent agreement. Additionally, by computing normal frequencies as a function of the dihedral angle of the gauche conformer and comparing with observed values the author was able to determine a value of $85° \pm 5°$ for the dihedral angle. These parameters have been compared with values determined by calculation based on atomic Van der Waals forces and residual electrical charges in the iodine atoms and from ^{13}C NMR data.

A method based on absorbance, measured as a function of differing solvent polarites by Powling and Bernstein [25], enabled comparison of ΔH values of 1,2-dihalogenated ethanes with temperature measurements in the liquid and gas phase. It was argued that the solvent effect was eliminated in the method and shown that these values correlated better with the gas than the liquid values. This approach to ΔH measurements appears to be little developed in other systems.

8. ASYMMETRICALLY 1,2-DISUBSTITUTED ETHANES

These are also understood in terms of trans and gauche conformers as in Fig. 2, although the substituents differ.

Table 2 consists of mixed 1,2-disubstituents of ethane. The halogen compounds reveal ΔH values generally consistent with the symmetrical disubstituted compounds (Table 1). Thus the value of 1-bromo-2-chloroethane [25], $\Delta H = 6.0 \pm 0.4$, is intermediate be-

Table 2 R^1CH_2-CH_2R^2 Enthalpy Difference

Compounds	Substituents R^1	R^2	ΔH (kJ/mol)	Phase	Method	Refs.
1-Bromo-2-chloroethane	Br	Cl	6.0±0.4	Gas	IR	[25]
			2.05±0.4	Liq.	R	[25,26]
1-Bromo-2-fluoroethane	Br	F	1.25±0.32	Gas	IR	[27]
			3.4±0.8	Liq.	IR	[27]
1-Iodo-2-fluoroethane	I	F	6.30	Gas	IR	[27]
			2.5	Liq.	IR	[27]
β-Bromo-propionitrile	Br	CN	-2.26±0.22	Liq.	IR	[28]
β-Bromo-propionitrile	Br	CN	-2.38	Liq.	IR	[29]
β-Chloro-propionitrile	Cl	CN	-1.85±0.18	Liq.	IR	[28]
β-Chloro-propionitrile	Cl	CN	-1.55	Liq.	IR	[30]
β-Fluoro-propionitrile	F	CN	0.0	Liq.	IR	[28]
β-Iodo-propionitrile	I	CN	-1.59±0.15	Liq.	IR	[31,32]
2-Chloro-ethanol	Cl	OH	4.0±0.38	Gas	IR	[33]
			5.0±0.80	Soln.[a]	IR	[34]
2-Fluoro-ethanol	F	OH	8.7±2.2	Soln.[a]	IR	[34]

[a]Solution in CCl_4

tween that of 1,2-dichloroethane, (4.88 ± 0.4) and that of 1,2-dibromoethane (7.4 ± 0.83) for measurements in the gas phase. These may be correlated with steric interactions as measured by Van der Waals radii.

In the case of 1-fluoro-2-halogenoethanes [27], reversal in stability occurs on passing through the liquid-gas phase boundary. The trans form is again more stable in the gas state, whereas the gauche form is more stable in the liquid state.

The infrared and Raman spectra of the halogenated compounds X CH_2CH_2CN have been studied in the liquid state [28-32] and the gauche form is found to be the low energy form, the difference from the trans increasing with X = F, Cl and Br. The effect noted in FCH_2CH_2F and $CNCH_2CH_2CN$, namely persistence of a conformer mixture until the temperature is lowered to -44°C, is also observed in FCH_2CH_2CN.

The infrared and Raman spectra of 2-chloroethanol [33], 2-fluoroethanol [34], 2-bromoethanol [34], and 2-iodoethanol [34] show the gauche form to be relatively more stable because a strong intramolecular hydrogen bond is favored. In the case of the most polar fluorine compound the ΔH value is largest (8.7 ± 2.2) which correlates with the absence of the unbonded OH stretching mode in solution in carbon tetrachloride and the high stability of the gauche conformer with strong intramolecular hydrogen bonding between OH and F. Conformational equilibria have also been established in compounds with aromatic substituents [35].

9. 1,1,2-TRISUBSTITUTED ETHANES

This constitutes a class of compounds with structures likely to be as indicated in Fig. 5. One nomenclature commonly used is to specify the single substituent as trans and gauche to the two 1-substituents (tg) as in Fig. 5 or as gauche to both substituents (gg).

The simplest molecule within this category may be taken as 1,1,2-trichloroethane. This comparison of gaseous liquid and solid spectra at variable temperatures [36] suggests the less polar form (tg) has less energy than the more polar form (gg). The ΔH values reported for measurements on several pairs of bands averages to 1.1 ± 0.12 for the liquid and 12.0 ± 0.5 for the gaseous state. The large difference between the vapor and liquid state can be related to the very high polarity of this molecule. Following NMR studies, Sheppard and Turner [17] utilize these values to determine various coupling constants in the different conformers.

Studies of the isobutyl halides [37] and cyanide [38] reveal the influence of methyl substituents on the configuration in question. Isobutyl halides [37] have been shown to exist in low energy (tg) forms for which two equivalent possibilities occur and in a high energy (gg) form. The ΔH values for isobutyl chloride and bromide (Table 3) are given for vapor and liquid states. The latter are not in very good agreement with those derived by ultrasonics [37]. The discrepancies were attributed to an accumulation of errors from assumptions made in the ultrasonic method. The ΔH value for isobutyl cyanide was determined for the liquid phase by Raman [38] but the assignments suggest that the gg form is more stable than the tg form.

In view of the similarity in polarity and other properties between the halogen and cyanide substituents, the conflict in relative stabilities between the gg and tg conformers is likely to be more apparent than real and a detailed reexamination of assignments would seem desirable.

An example of a similar configuration is 2,2-dichloroethanol [40]. Here the stable predominant form is the gg by virtue of intramolecular hydrogen bonding, but IR and Raman spectra suggest the tg form also exists in the solid, liquid, and gaseous state.

Secondary butyl halides differ from isobutyl halides by per-

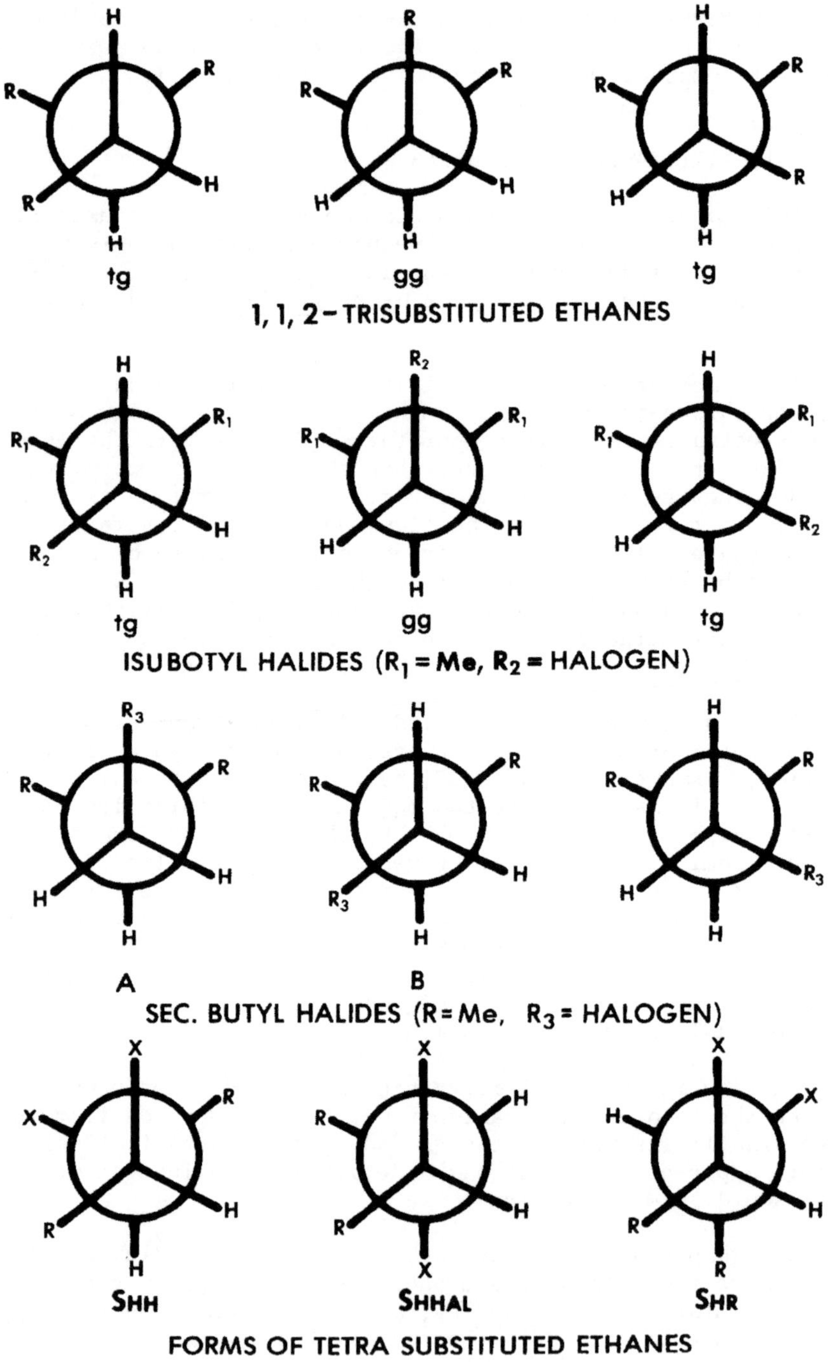

Fig. 5 Substituted ethanes as described in Tables.

Table 3 $R^1R^2CH-CH_2R^3$ Enthalpy Difference

Compounds	Substituents R^1	R^2	R^3	ΔH (kJ/mol)	Phase	Method	Refs.
1,1,2-Trichloroethane	Cl	Cl	Cl	12.0±0.5	Gas	IR	[36]
				1.1±0.12	Liq.	IR	[36]
				1.6	Soln.[a]	IR	[36]
				0.0	Soln.[b]	IR	[36]
Isobutylchloride	Me	Me	Cl	0.96±0.08	Gas	IR	[37]
				1.55±0.6	Liq.	IR	[37]
Isobutylbromide	Me	Me	Br	1.25±0.12	Gas	IR	[37]
				1.10±0.50	Liq.	IR	[37]
Isobutylcyanide	Me	Me	CN	3.0±0.2	Liq.	R	[38]
2,2-Dichloroethanol	Cl	Cl	OH		Liq.	IR	[40]
2-Chlorobutane	Cl	Me	Me	2.5±1.2	Liq.	IR[c]	[39]
				1.6±1.2	Liq.	IR[d]	[39]
				1.0±0.4	Liq.	IR[e]	[39]
2-Bromobutane	Br	Me	Me	2.5±1.2	Liq.	IR[c]	[39]
				2.1±1.0	Liq.	IR[d]	[39]
				0.4±0.4	Liq.	IR[e]	[39]

[a](1:1) Solution in CCl_4 [b](1:1) Solution in CH_3CN
[c]Equilibrium A↔B [d]Equilibrium C↔B [e]Equilibrium A↔C

mitting three rather than two conformers with different energy. Pentin and co-workers [39] have shown that the relative energies can be determined as B < C < A (Fig. 5) and then the ΔH values can be determined between each two states, (Table 3) for secondary butyl chloride and bromide.

Some rationalization to the values of ΔH was provided [37] based on Sheppard's rules [9]. The ΔH is considered in terms of contributions of interactions between non-bonded groups of the same kind and non-bonded groups of a different kind. Comparison of results calculated from these results with those shown in Table 3 are in reasonable agreement and suggest that non-bonded interactions between similar groups do not change greatly from molecule to molecule.

10. 1,1,2,2-TETRASUBSTITUTED ETHANES

The 1,1,2,2-tetrahalogenoethanes (Table 4) have possibilities of existing in trans and gauche forms (Fig. 2). In tetrafluoroethane, IR and Raman studies [15] suggested the trans was more stable by 4.85 kJ mol^{-1} compared to the gauche in the vapor state. Intramolecular HF bonding is considered to stabilize this form.

Tetrachloroethane results have been attributed to Bernstein and Powling [25] using solvent rather than temperature variation. Results for this compound in liquid and solution given in Table 4 are quoted by Sheppard and Turner [17] but taken from earlier work. The tetrabromoethane has been reported [41] as showing total reversal in relative energy between gas and liquid states for the trans and gauche forms, i.e. the gauche is the low energy form in the liquid.

In the case of the mixed tetrasubstituted compound, 1,1-dichloro-2,2-difluoroethane, the ΔH value in the liquid state also follows the trend of revealing a higher energy for the trans conformer than for the gauche.

11. UNSYMMETRICAL - TETRASUBSTITUTED ETHANES

Table 5 contains compounds in which two substituents R^2 and one substituent R^1 are attached to the same carbon. In the case of 1,2-dibromo-1,1-difluoroethane, the ΔH value between the less stable and more stable conformer has the same sign in the gas and liquid phase. The spectroscopic data did not permit the identification of the more stable conformer as trans or gauche.

In the case of the 2-halogeno-2-methylbutanes [44], the ΔH values are determined in the liquid and gas states (except for the iodo compound). Simple models based on steric repulsive forces and Van der Waals radii suggest the ΔH values should decrease from the chloride to the iodide. The probable trend is in the other direction which suggests other interactions additional to steric repulsive forces must occur.

12. DIASTERO FORMS OF TETRASUBSTITUTED ETHANES

The 2,3 dihalogenobutanes listed in Table 6 possess asymmetry at the C-2 and C-3 carbons which enables the molecules to exist in the meso form and the (±) form. Conformational possibilities are different in the two diastereoisomers. In the meso form the low energy form is found to be trans in the vapor state in all cases listed. In the case of 2,3-dibromo-butane, for the liquid phase the low energy state is gauche. The difference between liquid and gas phase enthalpies is explained by dielectric theory. In the (±) form there are two different energy levels corresponding to two identical gauche conformers in the meso form. The three conformers are shown in Fig. 5 using a nomenclature put forward in earlier references. Careful assignments permits the identity of the two levels corresponding to each ΔH value (Table 6). In the gas state the ΔH values reveal a definite pattern. For the (±) form the relative energies are found to be

Table 4 $R^1R^1CH\text{-}CHR^2R^2$ Enthalpy Difference

Compounds	Substituents R^1	R^1	R^2	R^2	ΔH (kJ/mol)	Phase	Method	Refs.
1,1,2,2-Tetrafluoroethane	F	F	F	F	4.85±0.4	Gas	IR	[15]
						Liquid	IR	[15]
1,1,2,2-Tetrachloroethane	Cl	Cl	Cl	Cl	1.2±0.8	Soln.	IR	[25]
					-4.6±0.4	Liquid	IR	[17]
					-3.76	Soln.[a]	IR	[17]
					-5.86	Soln.[b]	IR	[17]
1,1,2,2-Tetrabromoethane	Br	Br	Br	Br	2.85	Gas	IR	[41]
					-3.2	Liquid	IR	
1,1-Dichloro-2,2-difluoroethane	Cl	Cl	F	F	-1.47	Liquid	IR	[42]

[a]Solution in n-Heptane, [b]Solution in CH_3NO_2

Table 5 $R^1R^2R^2C\text{-}CH_2R^3$ Enthalpy Difference

Compounds	Substituents R^1	R^2	R^2	R^3	ΔH (kJ/mol)	Phase	Method	Refs.
1,2-Dichloro-1,1-Difluoroethane	Cl	F	F	Cl	1.8±0.4	Gas	IR	[42]
						Liquid	Raman	[42]
1,2-Dibromo-1,1-Difluoroethane	Br	F	F	Br	3.0±0.2	Gas	IR	[43]
					4.3±0.2	Liquid	Raman	[43]
2-Chloro-2-methylbutane	Cl	Me	Me	Me	4.6	Gas	IR	[44]
					1.5	Liquid	IR	
2-Bromo-2-methylbutane	Br	Me	Me	Me	5.9	Gas	IR	[44]
					1.6	Liquid	IR	
2-Iodo-2-methylbutane	Me	Me	I	Me	2.6	Liquid	IR	[44]

Table 6 $R^1R^2CH\text{-}CHR^1R^2$ (meso and ±) Enthalpy Difference

Compounds	Substituents R^1	R^2	R^3	R^4	ΔH (kJ/mol)	Phase	Method	Refs.
Meso-2,3-Dichlorobutane	Me	Cl	Me	Cl	5.4±1.2	Gas	IR	[45]
					2.7±0.4	Liquid	IR	[45]
Meso-2,3-Dibromobutane	Me	Br	Me	Br	6.0±0.88	Gas	IR	[45]
					-1.82±0.28	Liquid	IR	[45]
(±) 2,3-Dichlorobutane	Me	Cl	Me	Cl				
S_{HCl} S_{HH}					0.18	Gas	IR	[45]
					-0.29±0.9	Liquid	IR	[45]
$S_{HH}\text{-}S_{HCH}$					6.7±0.8	Gas	IR	[45]
					-1.63±0.36	Liquid	IR	[45]
$S_{HCl}\text{-}S_{HCH}$					6.9±0.6	Gas	IR	[45]
					1.75±0.24	Liquid	IR	[45]
(±) 2,3-Dibromobutane	Me	Br	Me	Br				
$S_{HH}\text{-}S_{HCH}$					6.6±0.5	Gas	IR	[45]
					-1.8±0.5	Liquid	IR	[45]
$S_{HBr}\text{-}S_{HCH}$					7.5±1.2	Gas	IR	[45]
					1.9±0.24	Liquid	IR	[45]
$S_{HBr}\text{-}S_{HH}$					0.75±1.5	Gas	IR	[45]
					3.1±0.50	Liquid	IR	[45]
1,2,3,4-Tetrachlorobutane	CH_2Cl	Cl	CH_2Cl	Cl		Liquid	IR,R	[46]

$S_{HH} \simeq S_{HHal} > S_{HCH_3}$. This is consistent with the large steric repulsion arising from the four large substituents in *gauche* positions in the S_{HH} and S_{HHal} but much less steric crowding in the S_{HCH_3} form.

1,2,3,4-tetrachlorobutane has, in principle, fifteen *meso* conformers and eighteen (±) conformers. To a reasonable approximation many of these are similar and only three spectroscopically distinguishable conformers need to be considered for each form [46]. Although it is not feasible to report ΔH values, the stable forms have been identified by vibrational spectroscopy and NMR studies.

13. HEXASUBSTITUTED ETHANES

Table 7 consists of ethane-type compounds where all the hydrogens have been replaced by one or another of the halogens. Considering the 1,2-dihalotetrafluoroethanes, we see that for the gas phase ΔH increases with the size of the halogen in the 1,2 position. i.e. in the order Cl, Br, I. As would be expected, the value for 1-bromo-2-chlorotetrafluoroethane lies between the values for the dichloro compounds and the dibromo compounds and the value for trifluoro compounds is slightly smaller than that of the dichloro compounds. All the gas phase values in this table therefore follow the expected trends of the earlier tables.

The *trans* isomer is the more stable form in both liquid and gaseous phases for each of the compounds included in the table except 1-fluoro-1,1,2,2-tetrachloroethane in which the equilibrium is reversed to favor the *gauche* isomer in the liquid phase.

14. ETHERS

Rotation of one or both of the R-O bonds in ethers may be expected to follow similar patterns as in corresponding C-C bonds with additional features arising from the electronic properties of the oxygen. Vibrational consequences of conformational changes are more readily detected when the R group has marked polarity or lack of symmetry. In other cases conformational equilibria may take place but be difficult to detect. (See table 8).

A number of studies have been made in which a vinyl group is directly attached to an ether oxygen. In the simplest case of methyl vinyl ether, two conformers were shown to exist in liquid and vapor [56]. Rotational structure of bands due to the low energy form indicated this was in planar *cis* conformation (Fig. 6 (II)), whereas the high energy form was *gauche* rather than the

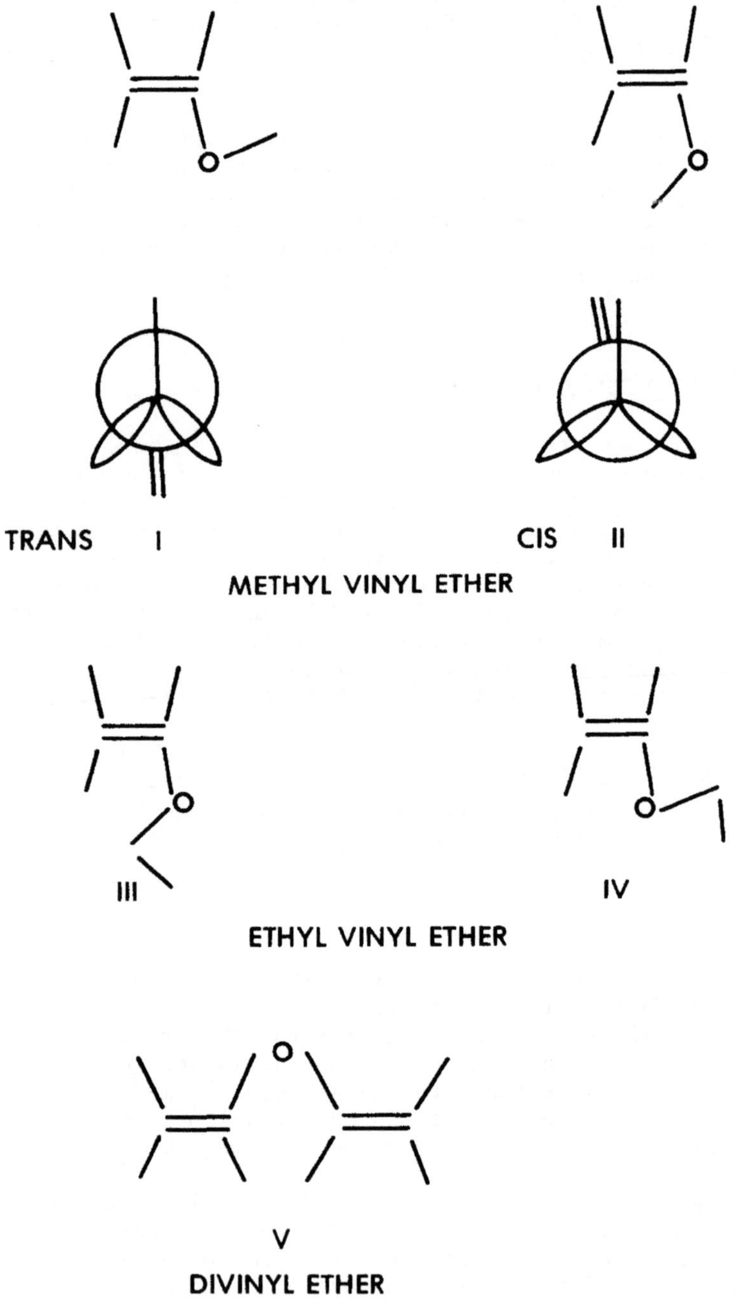

Fig. 6 Ethers as described in Table 9.

Table 7 $R^1R^2R^3C\text{-}CR^4R^5R^6$ Enthalpy Differences

Compounds	Substituents R^1	R^2	R^3	R^4	R^5	R^6	ΔH (kJ/mol)	Phase	Method	Refs.
1,2-Dibromotetrafluoroethane	Br	F	F	Br	F	F	3.92±0.2	Gas	IR	
									Raman	[47]
							3.85±0.2	Liquid	IR	
									Raman	[47]
							3.90±0.3	Liquid	IR	
									Raman	[48]
1-Bromo-2-Chlorotetrafluoroethane	Br	F	F	Cl	F	F	2.5±0.8	Gas	IR	
	Br	Cl	F	F	F	F	2.5±0.8	Gas	IR	[49]
							2.18±0.8	Liquid	IR	[49]
1,2-Dichlorotetrafluoroethane	Cl	Cl	F	F	F	F	2.1±0.8	Gas	IR	[49]
1,1,2-Trichlorotrifluoroethane	Cl	Cl	Cl	F	F	F	1.47±0.63	Gas	IR	
									Raman	[50]
1,2-Diiodotetrafluoroethane	I	I	F	F	F	F	7.70±0.4	Gas	IR	[51]
									Raman	[52]
1-Fluoro-1,1,2,2-Tetrachloroethane	F	Cl	Cl	Cl	Cl	H	3.35±0.4	Gas	IR	[49]
							-1.60	Liquid	IR	[50]

TABLE 8 R^1-O-R^2 Enthalpy Differences

Compounds	Substituents R^1	R^2	ΔH (kJ/mol)	Phase	Method	Refs.
Methylvinylether	CH_2=CH	Me	4.81±1.0	Gas	IR	[56,57]
			2.76±0.84	Soln.[a]	IR	[56,58]
Ethylvinylether	CH_2=CH	Et	3.3±0.8	Soln.[a]	IR	[54]
Alkylvinlether	CH_2=CH	Me	6.02±0.67	Liq.	R	[53]
	CH_2=CH	Et	6.53±0.41	Liq.	R	[53]
2,2,2-Trifluoroethylvinylether	CH_2=CH	CH_2CF_3		Gas, Liq.	IR	[59]
Divinylether	CH_2=CH	CH_2=CH	2.51	Liq.	IR,R	[55]
Propargylmethylether	HC≡CCH_2	Me		Liq.	IR	[60]
Methoxyacetonitrile	N≡CCH_2	Me		Liq.	IR	[60]
Bischloromethylether	$ClCH_2$	$ClCH_2$		Liq.	IR	[61]
2-Haloethylalkylether	CH_2CH_2	R		Liq.	R	[62]

[a]Solution 1,1,2-Trichloroethane.

TABLE 9 $R^1C(O)OR^2$ Enthalpy Differences

Compounds	Substituents R^1	R^2	ΔH (kJ/mol)	Phase	Method	Refs.
Methylacetate	CH_3	CH_3				[64,70]
Propargylformate	H	CH_2C≡CH	~0	Liq. & Gas	IR	[66]
t-Butylformate	H	$C(CH_3)_3$	0.63±0.08	Soln.[a]	IR	[68]
Ethylacetate	CH_3	C_2H_5	1.32±0.06	Soln.[a]	IR	[63]
Isobutylformate	H	$CH(CH_3)_2$	3.12±0.17	Soln.[a]	IR	[68]
Methylpropionate	C_2H_5	CH_3	4.6±1.25		IR	[69]
Methylchloroacetate	$ClCH_2$	CH_3			IR	[67]
Dimethylcarbonate	OCH_3	CH_3	10.9±2.1	Liq.	IR,R	[65]

[a]Solution in CS_2

trans form (Fig. 6 (I)) unless the trans (Fig. 6 (II)) is stabilized by a particularly bulky substituent.

The larger value of ΔH in the gas compared to the solution state is consistent with results described previously for halogenated ethanes. The stability of the cis form is attributed to electrostatic attraction between the polar O-CH_3 group and the polarizable C = C group and between the α vinyl C-H bond and the oppositely orientated dipole of the oxygen lone pair orbitals. Both of these interactions would be a maximum in the cis structure (Fig. 6 (II)). Higher alkyl vinyl ethers exist in similar equilibria, the low energy conformer of ethylvinylether is considered 'sickle' shaped [54] as in Fig. 6 (III)). High energy conformers are judged to be non-planar from the poorly defined contours of bands. In both compounds the proportion of the high energy conformer increases in solvents of increasing polarity which suggest it is the C-O bond which is rotated out of the plane of the vinyl group. Further vibrational analysis [53] of alkylvinylether supports the conclusions that the stable form is cis and that in the case of ethylvinylether, two skew high energy forms exist in the liquid and gaseous states.

In this more recent work the ΔH values for alkylvinylethers in the liquid state, determined by Raman, appear significantly greater than the values determined in the vapor or solution state by infrared methods, although the authors claim they are consistent within experimental error. Conformations of n-butylvinylether were suggested [53] but no quantitative measurement were possible.

The general conclusions are further supported by studies of 2,2,2-trifluoroethylvinylether [59], in which the rotation was found to take place at the bond between the oxygen and the vinyl group rather than between the oxygen and the ethyl group.

In the case of divinylether [55], two isomers are indicated by Raman and infrared. Observed spectra indicate from symmetry consideration that the more stable form is planar trans-trans (Fig. 6 (V)), whereas the less stable form is non-planar. The results were supported qualitatively by Molecular Orbital calculations, although a somewhat higher value for the ΔH value between the conformers was indicated.

Ethers with other groups bonded to oxygen have been shown to exhibit conformational isomerism. In the case of the groups propargyl [60], acetonitrile [60] and halosubstituted dialkylethers, the stable conformer may be determined in each case, but the nature of the spectra render the enthalpy determination difficult.

15. ESTERS

The OR^2 group of esters is likely to generate features in common with ethers and additional features stemming from the R^1CO group. The former features are isolated in formates (R^1 = H). Methyl and ethyl formates are isoelectronic with methyl [56] and ethyl [54] vinylether. Spectroscopic evidence suggest a cis skeletal form (Fig. 7 (I)) unless the trans (Fig. 7 (II)) is stabilized by a particularly bulky substituent (e.g. R^2 = t-butyl [68] or R^2 = propargyl [66]) which leads to steric repulsive forces between lone pair electrons on the carbonyl and R^2. These forces result in an equilibria between I and II. Evidence based on vibrational spectroscopy [68] suggests conformers arise from rotation about the HCO-O bond in t-butylformate for which ΔH has been estimated as 0.63 ± 0.08 and from rotation about the HCOO-R group in iso-butylformate for which ΔH was estimated as 3.1 ± 0.2. This latter rotation is also considered present in a series of ethylesters and $^2H_5$ ethylesters of saturated acids [63]. The ΔH values provided in Table 9 are attributed to the equilibrium III ⇆ IV in Fig. 7.

Studies of alkyl-n-butyrate [63] suggested equilibria of the type V ⇄ VI. For methyl propionate some comparative evidence [69] supports the more stable conformer as being cis V, in which the methyl group eclipses the carbonyl, and the less stable conformer as gauche where equivalent forms provide a favorable entropy factor. A more sensitive compound for detecting this equilibrium is methylchloracetate [67] because the polar chlorine group can exert interactions through space which split the carbonyl according to whether the chlorine group is cis or gauche. Raman and infrared studies [65] also enable conformations of dimethylcarbonate to be deduced.

16. KETONES AND ALDEHYDES

The conformers which exist in esters from rotation of the group bonded to the carbonyl may be expected also in ketones. Conformers are detected in butanone [71] and in diethylketone [72], but the lack of difference in polarity between conformers results in lack of sufficient resolution between bands belonging to different conformers for quantitative measurements. In chloroacetone [73] the ΔH value has been obtained both by the temperature variation method and by the absolute intensity method which enables the dihedral angle of the gauche form to be determined as 150° ± 10°. The two conformers have well resolved carbonyl bands as in esters with similar substituents [67], the stable conformer being trans in which the chlorine eclipses the carbonyl in each case. A detailed study of propanal revealed confirmation of the greater stability of the cis form compared to the skew [74]. (See Table

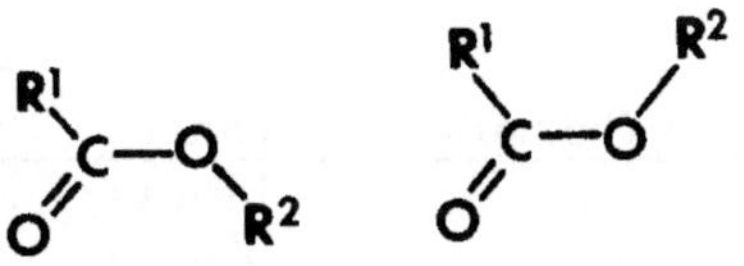

I II

CIS TRANS

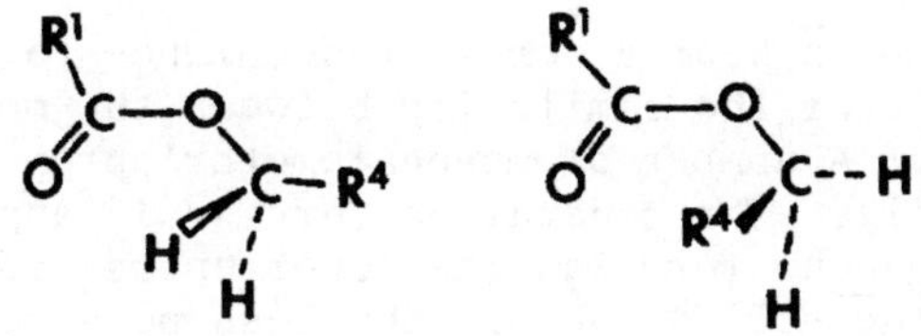

III IV

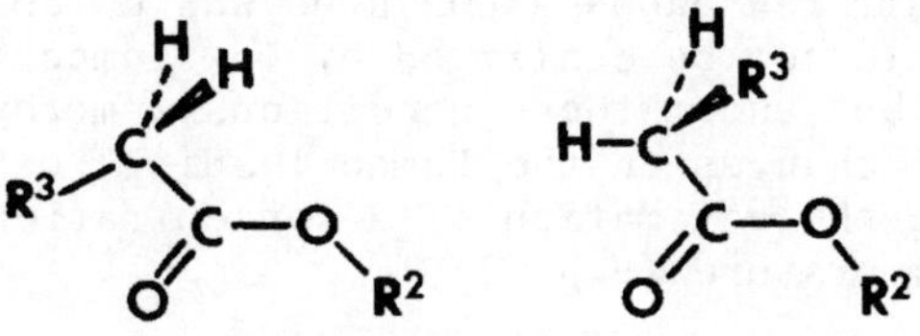

V VI

Fig. 7 Esters as described in Table 10.

Table 10 R^1-C(O)-R^2 Enthalpy Differences

Compounds	Substituents R^1	R^2	ΔH (kJ/mol)	Phase	Method	Refs.
Butanone	Et	Me		Liquid	Raman	[71]
Diethylketone	Et	Et		Liquid	IR	[72]
				Soln.[a]	Raman	
				Soln.[b]		
Chloroacetone	CH_2Cl	Me	1.84±0.42	Soln.[b]	IR	[73]
Propanal	Et	H	4.9±1.0	Gas	IR	[74]
			4.9±0.5	Liquid	Raman	[71,74]

[a]Solution in CCl_4; [b]Solution in CS_2

10).

17. SUBSTITUTED PROPENES

These compounds bear a formal resemblance to ketones and aldehydes by virtue of the similarity between the methylene and carbonyl groups. A number of monohalomethyl propenes have been studies (Table 11). The general pattern would appear to show that the more polar gauche form was the lower energy form which may be isolated in the solid, whereas the less polar cis form was the higher energy form which was present to an increasing extent at higher temperatures in the liquid and gas state. The ΔH values have not been reported. Studies of allylchloride [78] and cyanide favor a gauche stable conformer.

18. SUBSTITUTED 1,3-BUTADIENES

1,3-butadiene has been shown to be predominantly in the s-trans form. High temperature vapor studies [79] led to changes attributed to s-trans and s-cis equilibria and from intensity measurement of infrared bands a ΔH value was calculated (Table 12). These changes could not be confirmed by low temperature studies in solution [80] but under these conditions 2-methyl-1,3-butadiene revealed spectral changes in the Raman leading to the determination of ΔH and to the estimation of a concentration of s-cis of 11% at normal temperature (Fig. 8).

The 1,1-dibromo-3-fluoro-1,3-butadiene and the corresponding 1,1-dichloro compounds have been shown [7] to possess a higher proportion of the s-cis conformer leading to ΔH values which were also obtained by NMR in $CFCl_3$ as solvent to give similar values to those determined by infrared in CS_2.

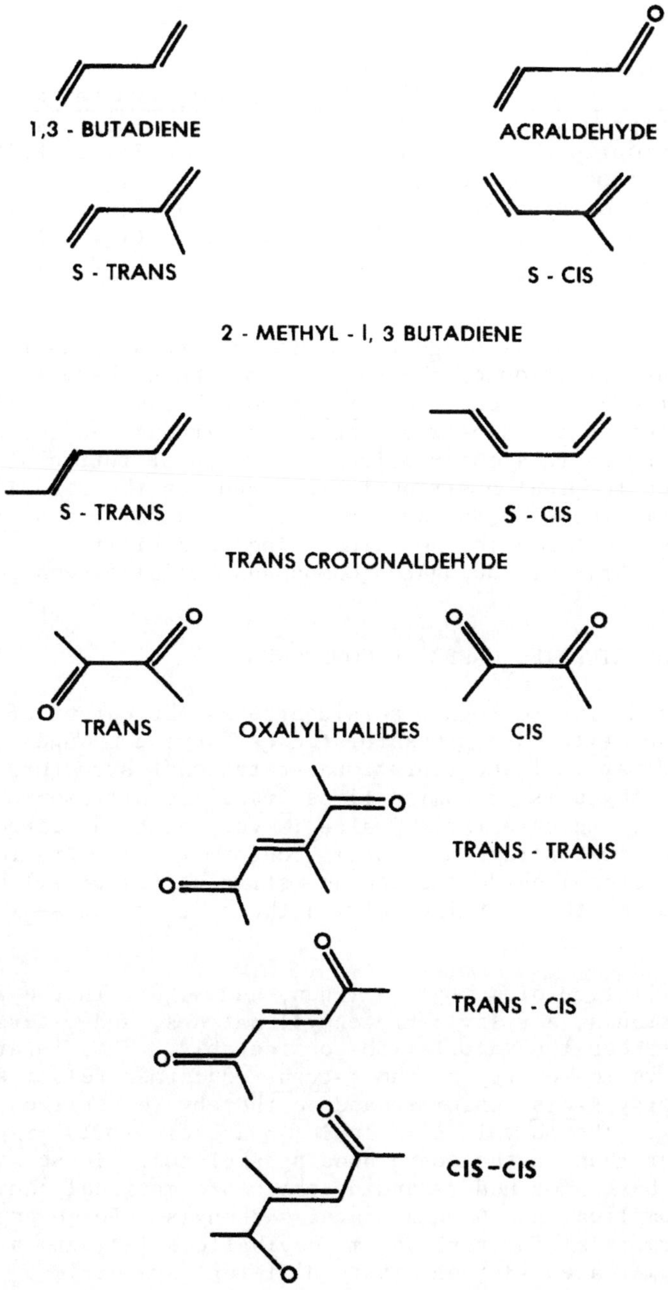

Fig. 8 Conjugated systems as described in Tables 12 and 13.

Table 11 R^1-C(CH_2)-R^2 Enthalpy Differences

Compounds	Substituents R^1	R^2	ΔH (kJ/mol)	Phase	Method	Refs.
2,3-Dibromopropene	CH_2Br	Br		Soln.	IR	[75]
2,3-Dichloropropene	CH_2Cl	Cl		Soln.	R,IR	[75,76]
2-Chloro-3-fluoropropene	CH_2F	Cl		Liq.	R,IR	[77]
2-Bromo-3-fluoropropene	CH_2F	Br		Liq.	R,IR	[77]
Allylchloride	CH_2Cl	H	2.1±0.2	Liq.	IR	[78]
Allylcyanide	CH_2CN	H	1.9	Liq.	IR	[78]

Molecular models suggest 2,3 steric interactions in butadiene inhibit the formation of the s-cis form. When there is a substituent at the 2-position, the s-trans form becomes less stable because of interactions between the 2 substituent and the nearest vinylichydrogen (2,4 interactions). The substitution of a halogen atom in the terminal position further reduces the stability leading to an increase in the energy towards that of the s-cis conformer and a consequent smaller ΔH value. The possibilities of gauche non-planar forms of conjugated compounds has also been considered.

19. α,β UNSATURATED CARBONYL COMPOUNDS

It is useful to regard acraldehyde as the parent of a wide range of substituted α,β unsaturated carbonyl compounds. Evidence based on rotational and vibrational-rotational structure has shown that acraldehyde is predominantly s-trans and ultrasonic relaxation studies indicate the ΔH value is very high and corresponds to very low concentration of s-cis conformer. A vibrational study of trans-crotonaldehyde enabled an estimate [81] of 7.1 kcal mol^{-1} to be made for the ΔH value between the s-trans and s-cis forms (Fig. 8).

Substitution of methyl, halogen, and alkoxy in the 2-position to form ketones, acryloyl (crotonoyl) halides, and esters, respectively, reduces the value of ΔH considerably. This is attributed to a rise in the energy of the s-trans conformer relative to the higher energy s-cis conformer and to thereby lessen their energy difference. The ΔH values in Table 13 are of similar magnitude but smaller than in the corresponding skeleton. These studies provide a basis for understanding the conformational equilibria in more complicated α,β unsaturated carbonyls. For example, various conformers of fumaroyl and maleoylhalides [84] and alkylfumarates and maleates [85] have very different symmetries. In the case of the fumaroyl halides, vibrational spectra support at least two forms of approximately equal energies having C_s (cis-trans) and C_{2h} (trans-trans) symmetries in solution in CS_2. In the case

Table 12 Substituted 1,3-Butadienes

Compounds	Substituent	ΔH (kJ/mol)	Phase	Method	Refs.
1,3-Butadiene	H	7.1±0.8	Gas	IR	[79]
2-Methyl-1,3-butadiene	Me	4.59±0.6	Liq.	R	[80]
1,1-Dibromo-3-fluoro-1,3-butadiene	F	3.53	Soln.[a]	IR	[78]
1,1-Dichloro-3-fluoro-1,3-butadiene	F	2.95	Soln.[a]	IR	[78]

[a]Solution in Cs_2.

of fumaroyl fluoride, ^{19}F, NMR provides evidence for the three isomers shown in Fig. 8, but it was not possible to determine the relative energies of ΔH values. Similar results were obtained in the maleoylhalides.

Dialkylfumarates were considered [85] to exist in equilibrium between trans-trans and trans-cis. The latter being the high energy form. The ΔH value for the dibutylfumarate was found to be 1.55 ± 0.08 in CS_2 solution by infrared measurements and 1.66 ± 0.17 by Raman measurements. The corresponding values for the diethyl and dimethylfumarates by Raman were somewhat lower and higher, respectively. The stabilizing features which cause this ester to exist as a solid at room temperature may also perturb the relative conformational energies. The Raman and infrared studies of alkylmaleates suggested [85] only one conformer, but a later study [86] supported at least two conformers.

20. OXALYL HALIDES

These represent the extension of the trend of compounds in Fig. 8 and provide opportunities to obtain wide general comparisons in conjugated systems.

The compounds studied are oxalylchloride and bromide [87] shown to be more stable in the trans than cis by 9.2 and 12.1 ± 0.4, respectively, based on low and high temperature Raman studies of liquids.

21. RING COMPOUNDS

Table 14 is made up of ring compounds, mainly substituted cyclohexanes and cyclohexenes. The energy values for cis and trans-2-chloromethylcyclohexanes are just about equal and are opposite in sign. The cis favors the equatorial methyl conformer

Table 13 2-Substituted Acraldehydes and 2-Substituted trans-Crotonalydehyde

Compound	Substituent	ΔH (kJ/mol)	Phase	Method	Refs.
trans-Crotonaldehyde	H	7.1	Liq.	R	[81]
Methylvinylketone	CH_3	1.93±0.10	Soln.[a]	IR	[82]
		2.36±0.21	Gas	IR	
Ethylideneacetone	CH_3	2.14±0.21	Soln.[a]	IR	[82]
		2.45±0.19	Gas	IR	
Acryloylfluoride	F	0.63	Soln.[a]	IR	
Acryloylchloride	Cl	2.50	Gas	IR	[83]
Methylacrylate	OMe	1.36±0.08	Soln.[a]	IR	[11]
Methyl-trans-crotonate	OMe	1.27±0.12	Soln.[a]	IR	[11]
		1.58±0.23	Liq.	R	[5]

[a]Solution in CS_2.

Table 14 Ring Compounds

Compound	ΔH (kJ/mol)	Phase	Method	Refs.
cis-2-chloromethylcyclohexane	3.34		IR	[88]
trans-2-chloromethylcyclohexane	-3.55		IR	[88]
cis-4-chloromethylcyclohexane			IR	[89]
trans-4-chloromethylcyclohexane			IR	[89]
Iodocyclohexane	1.67	Liq.	IR	[90]
1-iodo-1-methylcyclohexane	2.80	Liq.	IR	[90]
4-methylcyclohexene	3.59	Liq.	IR,R	[91]
4-bromocyclohexene	1.6	Liq.	IR	[92]

Table 15 Miscellaneous

Compound	ΔH (kJ/mol)	Phase	Method	Refs.
Dialkyldisulphide	3.76	Liq.	IR,R	[93]
	2.50			
Isobutylmethylsulphide		Liq.	IR,R	[94]
Bis(alkylthio)methane		Liq.	IR	[95]
Ring sulphites	2.42		IR	[96]
Alkylsulphides		Liq.	IR	[97]
Chlorophosphines		Liq.	IR	[98]
Dichlorophenoxysilane	3.34	Liq.	IR	[99]
Ethylchlorosilanes	1.67	Liq.	IR	[100]
	-1.17		R	
Methylvinylsilane	2.76±0.21	Liq.	IR,R	[101]
Allylsilane		Liq.	IR,R	[101]

and the trans favors the diequatorial conformation. The most stable conformer of cis or trans-1-chloro-4-methylcyclohexane is the conformer in which the Me group is in equatorial conformations.

The energy differences for iodo conformations of cyclohexane and 1-iodo-1-methyl-cyclohexane are 1.67 and 2.80 kJ mol^{-1}, respectively, in the former case favoring the equatorial conformer and in the latter favoring the axial conformer. Dilution of either compound with polar or non-polar solvents does not change the energy differences appreciably.

4-methylcyclohexene exists mainly as the equatorial conformer in the liquid state with energy difference of ~1.59 kJ mol^{-1}, and 4-bromocyclohexene also exists mainly as the equatorial conformer in the liquid state, the energy difference in this case is 3.59 kJ mol^{-1}; in both cases the equatorial form only was present in the solid phase.

22. MISCELLANEOUS

In Table 15 there is no structural similarity between the compounds considered. Considering conformations of dialkyldisulphides, for ethylmethyldisulphide, the gauche-gauche isomer was found to be more stable than the trans-gauche isomer by ~3.76 kJ mol^{-1} and for the diethyldisulphide the ggg conformer was found to be more stable than the tgg conformer by about 2.5 kJ mol^{-1} in the liquid state.

Of the three types of ring sulphite molecules considered, form I exists mainly as the chair conformation, form II as the twist conformations and form III as the chair twist conformation.

In the case of dichlorophenoxysilane the gauche conformer is predominant in the liquid phase while the anti-conformer is favored in the crystalline phase, the energy difference of the two conformers is ~3.34 kJ mol^{-1}.

23. CONCLUSION

The measurement of the energy difference between conformers, ΔH, and other thermodynamic parameters is a fundamental probe into chemical and physical behavior. Rotation of a chemical bond leads to chemical and physical change in the same way as the breaking of a chemical bond. Vibrational spectroscopy serves as an analytical method to determine the way proportions of the components change with temperature or other constraints. The factors which determine change and the position at which equilibrium takes place are twofold: tendency to achieve greater stability or lower energy and a tendency to increase the entropy of the system. This statement is represented by Eq. (2).

This section surveys results obtained by vibrational spectroscopy. Within the various classes of compounds studied it is possible to examine the structural factors which influence equilibrium and change.

Many of the results were measured many years ago when instrumentation was not adequate for accuracy. Improvements have been made in recent times in sensitivity, accuracy, resolution, and speed in dispersive infrared, Raman, and Fourier Transformation Infrared following interferometry measurements. Additionally, advances have been made in the determination of intensities of bands which are partially overlapped and in computer handling of data. Using faster automated methods, it is likely that the ΔH values in families of compounds can be measured rapidly and accurately leading to improved correlations with structural factors and a better understanding of the process of chemical and physical change.

REFERENCES

1. J. D. Kemp and K. S. Pitzer, J. Chem. Phys., 4, 749 (1936).

2. K. W. Kohlrausch, Z. Physik. Chem., 18B, 61 (1932).

3. K. W. Kohlrausch, and G. P. Ypsilanti, Z. Physik. Chem., 29B, 79 (1935).

4. J. Cabannes, J. Chim. Phys., 35, 1 (1938).

5. W. O. George, D. V. Hassid, W. C. Harris and W. F. Maddams, J. Chem. Soc., Perkin II, 392 (1975).

6. S. Mizushima, T. Shimanouchi, K. Kuranti and T. Miyazawa, J. Amer. Chem. Soc., 74, 1378 (1952).

7. K. O. Hartman, G. L. Carlson, R. E. Witkowski and W. G. Fateley, Spectrochim. Acta, 24A, 157 (1968).

8. S. Mizushima, Structure of Molecules and Internal Rotation, Academic Press, New York 1954.

9. N. Sheppard, Adv. Spectroscopy, 1, 288 (1959).

10. P. J. D. Park, R. A. Pethrick and B. M. Thomas, "Infrared and Raman Band Intensities and Conformational Change" in

Internal Rotation of Molecules, (W. J. Orville-Thomas, Ed.) W. J. Wiley, Chichester, England, 1974.

11. W. O. George, D. V. Hassid and W. F. Maddams, J. Chem. Soc., Perkin Trans., II, 400 (1972).

12. F. A. Miller and B. M. Harney, Appl. Spect., 24, 291 (1970).

13. R. J. Abraham, L. Cavalle and K. G. Pachler, Mol. Phys., 11, 471 (1966).

14. P. Klaboe and J. R. Nielson, J. Chem. Phys., 33, 1765 (1960).

15. P. Klaboe and J. R. Nielson, J. Chem. Phys., 32, 899 (1960).

16. Y. A. Pentin and V. M. Tatetevskii, Dokl. Akad USSR, 108, 290 (1956).

17. N. Sheppard and J. J. Turner, Proc. Roy Soc. (London), A252, 506 (1959).

18. R. E. Kagarise and D. H. Rank, Trans. Faraday Soc., 48, 394 (1952).

19. S. Mizushima, Y. Morino, I. Wanutabe, T. Shimanouchi and S. Yamaguchi, J. Chem. Phys., 17, 591 (1949).

20. K. Tanabe, Spectrochim. Acta, 30A (10), 1901, (1974).

21. W. E. Fitzgerald and J. G. Janz, J. Mol. Spectry., 1, 49 (1957).

22. N. Sheppard and G. J. Szusz, J. Chem. Phys., 17, 86 (1949).

23. D. L. Powell, P. Klaboe, B. N. Cyvin, H. Hopf, J. Mol. Struct., 41 (2), 215, (1977).

24. T. Koide, T. Oda, K. Ezumi, K. Iwatoni and T. Kubota, Bull. Chem. Soc. Japan, 41, 307 (1968).

25. J. Powling and H. J. Bernstein, J. Am. Chem. Soc., 73, 1815 (1951).

26. J. K. Wilmshurst and H. J. Bernstein, Can. J. Chem., 35 734 (1957).

27. M. F. El Bermani and N. Jonathan, J. Chem. Phys., 49, 340 (1968).

28. M. F. El Bermani and N. Jonathan, J. Chem. Soc. (A), 1711 (1968).

29. K. Tanabe, J. Mol. Struct., 25 (2), 259 (1975).

30. K. Tanabe, J. Mol. Struct., 28 (2), 329 (1975).

31. E. Wyn-Jones and W. J. Orville-Thomas, J. Chem. Soc., (A), 5855, (1964).

32. E. Wyn-Jones and W. J. Orville-Thomas, J. Chem. Soc. (A), 101 (1966).

33. S. Mizushima, T. Shimanouchi, T. Miyazawa, K. Abe and M. Yasume, J. Chem. Phys., 19, 1477 (1951).

34. P. J. Krueger and H. D. Methec, Can J. Chem., 42, 326 (1964).

35. J-S. Chen, W-C Lin, Hua Hsueh, 1, 1 (1976).

36. K. Kuratani and S. Mizushima, J. Chem. Phys., 22, 1403 (1954).

37. E. Wyn-Jones and W. J. Orville-Thomas, Trans. Faraday Soc., 64, 2907 (1968).

38. J. Som and D. K. Mukherjee, Indian J. Phys., 48 (10), 887, (1974).

39. Yu. A. Pentin, L. P. Melikhova and O. D. Ul'Yanov. Zh. Strukt. Khim., 4, 535 (1963).

40. T. Hirokawa, Y. Sumida, M. Hayashiand M. Murata, J. Sci., H. Hiroshima Univ., Ser. A: Phys. Chem., 38 (2-3), 281 (1974).

41. G. L. Carlson, W. G. Fateley and J. Hiraishi, J. Mol. Struct., 6, 101 (1970).

42. H. P. Bucker and J. R. Nielson, J. Mol. Spectry., 11, 47 (1963).

43. R. E. Kagarise, J. Chem. Phys., 24, 1264 (1956).

44. P. J. D. Park and E. Wyn-Jones, J. Chem. Soc. (A), 2944 (1968).

45. P. J. D. Park, and E. Wyn-Jones, J. Chem. Soc. (A), 422 (1969).

46. A. B. Dempster, K. Price and N. Sheppard, Spectrochim. Acta, 31A, (4), 331 (1975).

47. R. E. Kagarise and L. W. Daasch, J. Chem. Phys., 23, 113 (1955).

48. P. J. D. Park, K. R. Crook and E. Wyn-Jones, J. Chem. Soc. (A), 2910 (1969).

49. R. E. Kagarise, J. Chem. Phys., 26, 380 (1957).

50. P. Klaboe and J. R. Nielson, J. Mol. Spectry., 6, 520 (1961).

51. C. Serbali and B. Minasso, Spectrochim. Acta, 24A, 1813 (1968).

52. D. L. Powell, J. E. Gustavsen, P. Klaboe and C. J. Nielson, J. Raman. Spectrosc., 7 (2), 111, (1978).

53. M. Sakakilara, F. Inagaki, I. Harasla and T. Shimanouchi, Bull. Chem. Soc. Japan., 49, 46 (1976).

54. N. L. Owen and N. Sheppard., Spectrochim. Acta, 22, 1101 (1966).

55. B. Cadioli, B. Fortunato, E. Gallinella, P. Mirone and U. Pinelli,Gazz. Chim. Ital., 104 (3-4), 369 (1974).

56. N. L. Owen and N. Sheppard, Trans. Faraday Soc., 60, 634 (1964).

57. N. L. Owen and N. Sheppard, Proc. Chem. Soc., 264 (1963).

58. G. E. McGraw, D. L. Burnitt and I. C. Hisatsune, J. Chem. Phys., 45, 1392 (1966).

59. S. W. Charles, F. C. Cullen, and N. L. Owen, J. Chem. Soc., Faraday Trans. 2., 70 (3), 483 (1974).

60. S. W. Charles, F. C. Cullen, G. I. L. Jones and N. L. Owen, J. Chem. Soc., Faraday Trans., 2, 70 (4), 758 (1974).

61. S. W. Charles, F. C. Cullen and N. L. Owen, Spectrochim. Acta, 32A, 5, 1171 (1976).

62. A. Maradi-Araghi and K. E. Hansen, Spectrosc. Lett, 11 (8), 645 (1978).

63. W. O. George, D. V. Hassid and W. F. Maddams, J. Chem. Soc., Perkin Trans., II, 12, 1978 (1972).

64. W. O. George, T. E. Houston and W. C. Harris, Spectrochim. Acta, 30A (4), 1035 (1974).

65. J. E. Katon and M. D. Cohen, Can. J. Chem., 53 (9), 1378 (1975).

66. G. I. L. Jones, D. G. Lister and N. L. Owen, J. Chem. Soc., Faraday Trans., 2, 71 (7), 1330 (1975).

67. J. E. Katon and V. Sinha, Spectrochim. Acta, 33A (1), 45 (1977).

68. W. O. George, D. V. Hassid and W. F. Maddams, J. Chem. Soc., Perkin II, 952 (1973).

69. R. M. Moravie and J. Corset, J. Chem. Phys. Lett., 26 (2), 210 (1974).

70. J. K. Wilmshurst, J. Mol. Spectrosc., 1, 201 (1957).

71. A. Chapput, B. Roussel and G. Fleury, J. Raman Spectrosc., 2 (2), 117 (1974).

72. Z. Burio and P. J. Krueger, Spectrochim. Acta, 30A (11), 2069 (1974).

73. K. Tanabe and S. Saeki, J. Mol. Struct., 25 (2), 243 (1975).

74. S. G. Frankiss and W. Kynaston, Spectrochim. Acta, (A), 28 (11), 2149 (1972).

75. T. Torgrimsen and P. Klaboe, J. Mol. Struct., 20 (2), 229 (1974).

76. G. A. Crowder, J. Mol. Spectrosc., 20 (4), 430 (1966).

77. P. Klaboe, T. Torgrimsen and D. H. Christensen, J. Mol. Struct., 23 (1), 15 (1974).

78. C. Sourisseay and B. Pasquier, J. Mol. Struct. 12, 1 (1972).

79. E. B. Reznikova, V. I. Tulin and V. M. Tatevskii, Opt. I. Spektro., 13, 364 (1962).

80. D. A. C. Compton, W. O. George and W. F. Maddams, J. Chem. Soc., Perkin II, 1666 (1976).

81. J. R. Durig, S. C. Brown, V. F. Kalasinsky and W. O. George, Spectrochim. Acta, 32A, 807 (1976).

82. A. J. Bowles, W. O. George and W. F. Maddams, J. Chem. Soc., (B), 810 (1969).

83. J. E. Katon and W. R. Feairheller, J. Chem. Phys., 47, 1248 (1967).

84. D. F. Koster, T. P. Vasileff and G. L. Carlson, Spectrochim. Acta, 27A, 1633 (1971).

85. D. A. C. Compton, W. O. George and A. J. Porter, J. Chem. Soc., Perkin Trans., VII, 400 (1975).

86. H. G. Buge, P. Reich and E. Steger, J. Mol. Struct., 35 (2), 175 (1976).

87. J. R. Durig and S. E. Hannum, J. Chem. Phys., 52, 12 (1969), ibid. 54, 6 (1971).

88. U. Kh. Agaev, A. T. Alyev, S. Z. Rizaeva and Yu. A. Pentin, Zh. Prikl. Spectrosk., 19 (4), 739 (1973).

89. U. Kh. Agaev, A. T. Alyev, S. Z. Rizaeva and Yu. A. Pentin, Zh. Khim., 48 (4), 945 (1974).

90. U. Kh. Agaev, S. Z. Rizaeva, A. T. Alyev and Yu. A. Pentin, Zh. Fiz. Khim., 47 (4), 1027 (1973).

91. G. F. Fernandez, V. I. Lysenkov, O. D. Ul'Yanova, Yu. A. Pentin and I. I. Bardyshev, Zh. Fiz. Khim, 51 (10), 2710 (1977).

92. G. F. Fernandez, O. D. Ul'Yanova and Yu. A. Pentin, Zh. Fiz. Khim., 51 (4), 1021 (1977).

93. H. Sugeta, A. Go and T. Miyazawa, Bull. Chem. Soc. Japan, 46 (11), 3407 (1973).

94. N. Nogami, H. Sugeta and T. Miyazawa, Bull. Chem. Soc. Japan, 48 (9), 2417 (1975).

95. M. Ohsaku, Spectrochim. Acta, 31A (9-10), 1271 (1975).

96. H. Faucher, Tetrahedron Lett., 20, 1743 (1977).

97. M. Ohsaku, H. Murata and Y. Shiro, Spectrochim. Acta, 33A (3-4), 467 (1977).

98. A. I. Fishman, A. B. Remisov, I. Ya. Kuramshin and I. S. Pominov, Spectrochim. Acta, 32A (3), 651 (1976).

99. V. I. Fedoseeva and T. L. Krasnova, Zh. Obshch. Khim., 45 (9), 2108 (1975).

100. L. V. Khristenko, S. E. Rudakova and Yu. A. Pentin, Zh. Fiz. Khim., 40 (9), 2313 (1966).

101. K. Ohno, K. Taga and H. Murata, Bull. Chem. Soc. Japan, 50 (11), 2870 (1977).

CONFORMATIONAL ANALYSIS BY FAR INFRARED FT SPECTROSCOPY

J. R. Durig and D. J. Gerson

Department of Chemistry
University of South Carolina
Columbia, South Carolina, U. S. A.

1. INTRODUCTION

The high-resolution and low-energy detection capabilities of the Fourier Transform interferometer allow molecular spectroscopists greater access to the far infrared region of the spectrum. In particular, the low-frequency vibrations corresponding to the normal motion interconverting two conformers, i.e. an asymmetric torsion, can be observed with greater ease than previously with dispersive grating spectrometers [1]. In previous chapters the specific instrumental capabilities of the Fourier Transform spectrometer, as well as the determination of conformers and molecular structures, have been covered. In this chapter the methods of computational analysis of the vibrational data will be examined in some detail with emphasis on the practical usage, rather than the theoretical development.

One of the major goals of conformational analysis is the calculation of the energy difference between the two conformers, ΔE, as well as the energy necessary for interconversion. Once these data are available, calculation of the thermodynamic functions, G, H, S and C_p, is possible. The calculation of these energy quantities is facilitated by using a potential function which describes the vibrational motion, or internal rotation (torsion), as a function of the dihedral angle, α. The potential function is called asymmetric because both the frame and the top portions of the molecule have no symmetry element higher than a plane. This can be illustrated by 1-butene where the frame portion of the molecule is the ethyl moiety, C_2H_5, and the lighter top portion is the vinyl group.

J. R. Durig (ed.), Analytical Applications of FT-IR to Molecular and Biological Systems, 333–360.

In order to characterize an asymmetric potential function, four types of information are required: (1) the approximate dihedral (torsional) angle of each conformer, because the number of torsional energy levels is directly related to the number of potential minima; (2) the approximate relative enthalpy difference between the high and low energy conformers, since this is one of the contraints defining the potential function; (3) the change in molecular kinetic energy as a function of torsional angle; and (4) the accurate observation of torsional transition frequencies. The first two factors have been examined in other chapters and will only be mentioned in this chapter. The last two factors will be examined in detail later in this chapter.

2. EXPERIMENTAL CONSIDERATION

A. Collecting Torsional Data

FT-IR techniques allow a measurement precision of better than 0.1 cm^{-1}, provided that sufficient data points are sampled on the interferogram [2]. Torsional transitions are usually very sharp when compared to other fundamental vibrations, thus high resolution spectroscopy (0.25 to 0.10 cm^{-1} spectral band width) may be required in order to observe each band belonging to a torsional series. However, increased resolution will decrease the signal-to-noise ratio, S/N, so many more scans (usually ~2000) must be co-added in order to give a S/N ratio equivalent to that in the mid infrared spectral region.

It is also important to purify the sample and to remove all interfering substances such as low molecular weight gases, i.e. H_2O, H_2S, PH_3, HCl, etc. This is especially true for H_2O since its rotational spectrum is very strong in the far infrared and extends to well above 300 cm^{-1}. Excess water can frequently be removed from the sample by passing the compound through 3Å molecular sieves or by storing the compound over anhydrous P_2O_5, NaH, BaO, or $LiAlH_4$, provided these materials do not decompose the sample.

B. Symmetry Considerations

Several important consequences of molecular symmetry properties should be mentioned since they have a bearing on the observed data necessary to calculate the potential function. The most common type of molecule being studied has at least one planar conformation, which has at least C_s symmetry. For the conformer with C_s symmetry the normal vibrations are either in-plane or out-of-plane. The out-of-plane vibrations, of which the asymmetric torsion is one, have the dipole change along the principal axis out-of-plane. When this corresponds to a B type contour, problems arise, since usually only the torsional fundamental is observed. However, no such problems arise for A or C type band contours. In the case of non-planar (gauche) conformers, C_1 symmetry is usually present, giving rise to A/B/C hybrid band contours for the asymmetric torsion. In summary, bands which give rise to A or C type contours are most desirable as they exhibit strong Q branches, whereas B type bands may have no Q branches and thus only allow the location of the fundamental torsional frequency at the band center. In fact it is sometimes difficult to find the band center because of the overlapping contours of the hot bands.

3. THEORY

A molecule undergoing hindered internal rotation about a single bond changes its overall energy as the torsional angle, α, changes. This energy change is manifested in the torsional vibrational energy as well as the overall molecular energy. The change in the molecular rotational energy arises from the alteration in geometry of the molecule as α changes, which in turn alters the principal moments of inertia which are dependent upon α. In the analysis of the torsional motion, calculation of the asymmetric potential function is a necessary requirement for the understanding of the energy restrictions and molecular dynamics involved in hindered internal rotation. The theoretical treatment used most often is a rigorous numerical method involving the calculation of torsional eigenfunctions and corresponding eignevalues which are then iterated, in a least-squares manner, to give the best possible fit to the experimentally observed torsional transition frequencies [3]. This method requires the researcher to have access to the appropriate computer programs and a high-speed-large access memory computer. Another method for calculating the asymmetric potential function has been discussed in detail by Fateley et al. [4]. This method was developed especially to handle molecular systems existing as a mixture of planar conformers, where as few as one torsional transition is observed for each conformer. In the treatment of hindered internal rotation we will develop the rigorous approach and show what effect each of the potential coefficients has on the shape of the asymmetric potential function.

In the case of isolated internal rotation about a single bond, the Hamiltonian can be written as a sum of molecular kinetic and potential energy terms:

$$H_\tau = T(\alpha) + V(\alpha) \tag{1}$$

where α is the one-dimensional torsional motion variable. The term $V(\alpha)$ is the asymmetric potential function for the internal rotation and can be expressed in a number of different ways. The model used to determine $V(\alpha)$ depends on the geometry of the molecule of interest. If the molecule is totally asymmetric, with no plane or rotational axis present in any conformer, then the potential function must be expressed as a Fourier series in both sine and cosine terms:

$$V(\alpha) = V_o + \sum_i a_i \sin i\alpha + \sum_i b_i \cos i\alpha \tag{2}$$

where V_o is an arbitrary zero of potential energy defined at the torsional angle of zero. V_o has no effect on the shape of the potential function. A potential of the type in Eq. (2) results in all the minima and all the barriers being nonequivalent. This potential requires considerable torsional data in order that the function be accurately described. In general, this type of potential function has seen more utility in the analysis of low frequency, large amplitude vibrations in small ring molecules [3] than in molecules which exist as a mixture of conformers. This is probably due to the lack of sufficient experimental data required for an accurate analysis rather than a lack of suitable molecular systems.

Simpler potential functions result when both the asymmetric top and the molecular frame have planes of symmetry. In these types of systems at least one value of α results in the entire molecule having a plane of symmetry. A much more insignificant consequence of symmetry in these systems is that a rotation of $+\alpha$ or $-\alpha$ away from the planar configuration results in energetically indistinguishable conformations, since the potential function itself now has a plane of symmetry. A Fourier series expansion in one term is now adequate to describe the potential function. The most common type of potential function, where at least one of the minima coincides with a plane of symmetry, is of the type

$$V(\alpha) = (1/2)\sum_i V_i(1-\cos i\alpha). \tag{3}$$

In general it has been found that a six term expansion in V_i is sufficient to handle almost any molecular system [3]. This is because V_1, V_2, and V_3 are in general larger than V_4, V_5, and V_6; and thus terms greater than V_6 should be negligible.

The kinetic energy term, $T(\alpha)$, is extremely complicated. Several different approaches have been made to solve this expression using either ab-initio or semi-empirical quantum mechanical methods [5]. In general, if the only data being used to calculate the potential function are torsional transitions, and if one continues within the boundary conditions of a one-dimensional problem, then the following treatment is adequate. Calculation of the kinetic energy term requires evaluation of the internal rotation constant, F, commonly called the "F number," which is related to the reduced moment of inertia of the top, I_r, by

$$F = h/8\pi^2 c I_r. \quad (4)$$

The calculation of I_r is complicated, and it must take into account the relationship between the moments of inertia of the asymmetric top and the overall principal moments of inertia of the molecule. As α changes, this relationship changes, and thus I_r and F are functions of α and should more appropriately be referred to as $I_r(\alpha)$ and $F(\alpha)$. In order to determine the angular dependence of $F(\alpha)$, F is calculated at various values of α, the results of which are then curve fitted to a Fourier expansion in α via a least-squared method. As we have already stated, if a cosine expansion in $V(\alpha)$ is adequate, then a cosine expansion in $F(\alpha)$ should also be adequate:

$$F(\alpha) = F_o + \sum_i F_i \cos i\alpha. \quad (5)$$

The kinetic term, $T(\alpha)$, can now be written as

$$T(\alpha) = P_\alpha F(\alpha) P_\alpha \quad (6)$$

where P_α is the conjugate momentum operator given by

$$P_\alpha = -\frac{1}{i}\left(\frac{\partial}{\partial\alpha}\right). \quad (7)$$

The two Fourier expansions, one in $F(\alpha)$ and one in $V(\alpha)$, are now substituted into the Schrodinger equation for the general torsional Hamiltonian,

$$H_\tau \Psi = E_\tau \Psi \quad (8)$$

where

$$H_\tau = P_\alpha F(\alpha) P_\alpha + V(\alpha). \quad (9)$$

The solution to the Schrodinger equation (Eq. 8) is obtained by solving the secular equation for the Hamiltonian matrix, to give both the torsional eigenvalues and eigenfunctions. In order to symmetrize the Hamiltonian matrix, Eq. (9) is rewritten using the definition of P_α (Eq. 7):

$$H_\tau = (1/2)[P_\alpha^2 F(\alpha) + F(\alpha) P_\alpha^2] + f(\alpha) + V(\alpha). \quad (10)$$

The new variable $f(\alpha)$, called the "pseudopotential" [3] can be expressed as

$$f(\alpha) = (\partial^2/\partial\alpha^2) \cdot F(\alpha). \quad (11)$$

This term is always assumed to be negligible and is dropped from the calculations, since the angular dependence of F is very small [3].

Once these substitutions have been completed, every element in the Hamiltonian matrix is then set up using a free-rotor basis set,

$$\Psi_m = (2\pi)^{-1/2} e^{im\phi}, \text{ where } m = 0, \pm 1, \pm 2 ... \tag{12}$$

In the case of the cosine-based potential function, Eq. (3), each eigenfunction is either symmetric or asymmetric with respect to rotation about $\alpha = 0$. These eigenfunctions are more commonly denoted as even (+) or odd (-) for symmetric and asymmetric reflection about $\alpha = 0$, respectively. To preserve these symmetry properties, a Wang transformation [6] is applied to the free-rotor basis functions (Eq. (12)) to express them as cosine or sine functions of even or odd symmetry, respectively:

$$\Psi_n^{even} = \pi^{-1/2} \cos n\phi \qquad n = 1,2... \tag{13}$$

$$\Psi_o^{even} = (2\pi)^{-1/2} \tag{14}$$

$$\Psi_n^{odd} = \pi^{-1/2} \sin n\phi \qquad n = 1,2... \tag{15}$$

In general it has been found that a maximum of fifty basis functions is sufficient for a normal potential function [3].

The Hamiltonian matrix elements derived using the symmetry basis set (Eqs. (13-15)), are divided into two blocks, even and odd. These matrix elements can be used to calculate the infrared intensity of a band due to a transition from one eigenvalue, v'', to a higher level, v'. The relative intensity of this band is

$$I_{v'',v'} = (E' - E'') < v' |M| v'' > [g'' e^{E''/kT}], \tag{16}$$

where E' and E'' are the energies of the v' and v'' levels, respectively, $< v'|M|v'' >$ is the dipole transition matrix element, and g' and g'' are the degeneracies of the v' and v'' levels, respectively. It should be pointed out that these intensities are relative, and hold only for transitions of the same conformer. The selection rules for infrared and Raman activity are:

Infrared $+ \leftrightarrow -$
Raman $+ \leftrightarrow +,\ - \leftrightarrow -$.

The approximate potential function is used when there is not sufficient data to support a rigorous treatment. In this type of approximation the torsional motion is assumed to be harmonic, i.e.

a very high barrier is present, and hence for small values of α, cos α can be replaced by a power series in α, terminated at α^4,

$$V_\alpha = [V_1 + 4V_2 + 9V_3]\frac{\alpha^2}{4} - [V_1 + 16V_2 + 81V_3]\frac{\alpha^4}{48}. \quad (17)$$

Because α is small, α^4 is insignificant for initial calculations, leaving

$$V_\alpha = [V_1 + 4V_2 + 9V_3]\frac{\alpha^2}{4} = V^*\alpha^2/4. \quad (18)$$

Equation (16) now has the form of a harmonic oscillator, where V^* approximates the height of the barrier for the potential well. This barrier height can be related to the torsional frequency by the approximation

$$V = \frac{\nu^2}{F} = V_1 + 4V_2 + 9V_3. \quad (19)$$

If one now observes a second torsional frequency within the well, more information can be input via the anharmonicity of the torsional vibration, X_{11}. The difference between the $1 \leftarrow 0$ and $2 \leftarrow 1$ transitions is denoted as $2X_{11}$. Fateley et al. [4] have derived the following relationship for X_{11},

$$-\frac{8\nu^2 X_{11}}{F^2} = V_1 + 16V_2 + 81V_3 \quad (20)$$

or the V^* term in Eq. (18).

As stated earlier, if one confines the molecular system to planar s-cis and s-trans configurations, $\alpha = 0°$ and $\alpha = \pm 180°$, respectively, then the following relationship holds provided that terms $V_n > 3$ are ignored

$$\Delta E = V_1 + V_3. \quad (21)$$

In practice ΔE is not known; however, it is assumed to be equal to ΔH, which is the difference between the lowest torsional levels in the two wells.

The solutions for V_1, V_2, and V_3 follow by combining equations (19) through (21), and they are

$$V_1 = \Delta H - V_3$$

$$24\, V_2 = (\nu^2/F)(10 + \frac{8X_{11}}{F}) - 9\Delta H$$

$$48\, V_3 = 3\Delta H - (4\nu^2/F)(2X_{11} + F)/F, \quad (22)$$

where ΔH has been substituted for ΔE. The experimental determination of ΔH has been given in another chapter. It cannot be em-

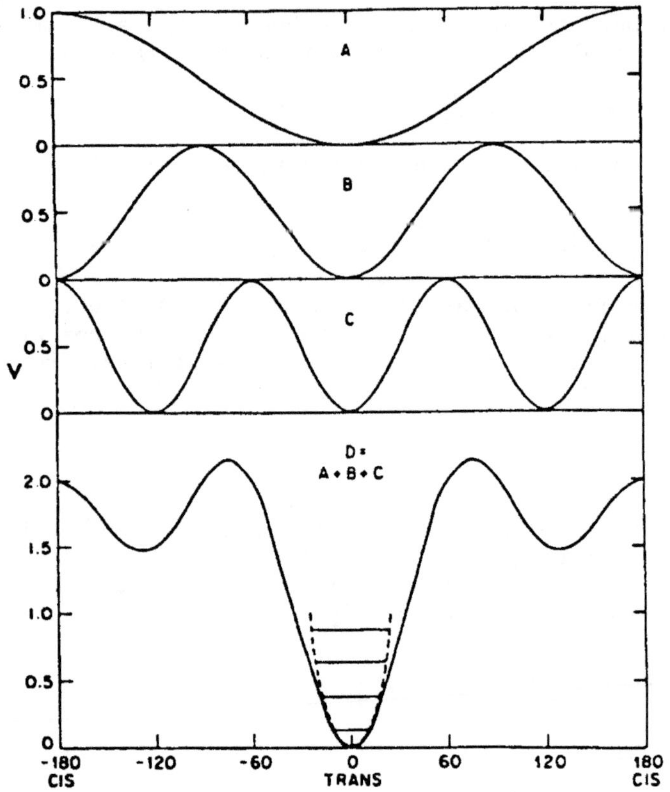

Fig. 1 Components of the potential function $V(\alpha) = \Sigma\ (1 - \cos n\alpha)/2$. A: $(1 - \cos\alpha)/2$; B: $(1 - \cos 2\alpha)/2$; C: $(1 - \cos 3\alpha)/2$; D: A + B + C. In this figure, $V_1 = V_2 = V_3 = 1$, and α is taken as zero for the trans form, (used by permission, Ref. [4]).

phasized enough that ΔH is probably one of the most important pieces of datum necessary for the calculation of the potential function with the others being F and ν_τ observed.

In evaluating the effect of each potential coefficient, V_1 through V_6 of Eq. (3), and ΔH on the final form of the potential function, a qualitative assessment is the easiest to provide. In general, the subscript of the potential function indicates the periodicity of the potential provided it was only made up of that term as shown in Fig. 1. For those systems where there is an equilibrium between a planar form and two nonplanar forms, V_3 is usually the predominant term. This is because V_3 presents a three maxima/three minima potential per $2\pi(360°)$ internal rotation. In a similar fashion, V_2 is found to be the predominant term in the

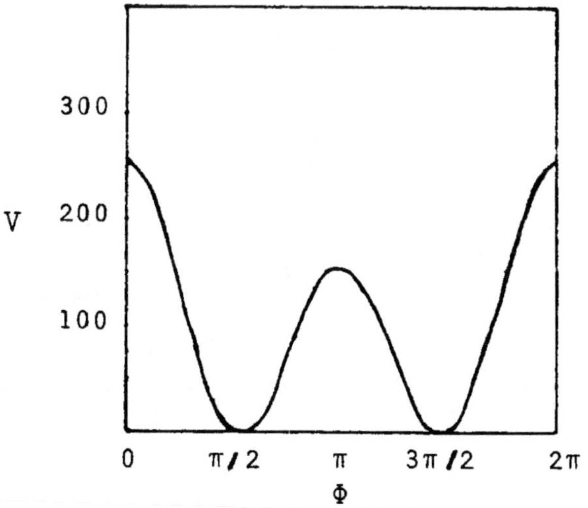

Fig. 2 A potential energy function with V_1 = -100 and V_2 = +200.

potential function for a system consisting of two equivalent nonplanar conformers, such as diemthyldisulfide (Fig. 2). For the qualitative explanation of V_1, a two conformer potential function must be invoked. For this we have chosen a system consisting of a planar, s-trans form and two equivalent nonplanar gauche conformers. In such a system V_3 gives the potential its overall three-foldness, and V_2 determines the placement of the two gauche wells as well as the height of the gauche → s-trans barrier, depending upon its sign and magnitude. As shown in Fig. 3, a positive V_2 will raise the energy of the s-trans wells relative to a pure V_3 and lower the gauche ↔ gauche barrier relative to a pure V_2 term. Constrastingly, a negative V_2 will raise the energy of the gauche wells relative to the s-trans wells. In a system where V_3 is positive and V_2 is smaller in magnitude and negative, V_1 will raise or lower the relative height of the gauche ↔ gauche barrier and the relative energy of the gauche wells relative to the s-trans wells depending upon its sign and magnitude. As shown in Fig. 4, a positive V_1 will lower the gauche ↔ gauche barrier as well as have a small but noticeable effect on lowering the energy of the gauche wells. A negative V_1 will have an opposite effect.

At this point a note on the use of ΔH is in order. As stated earlier, ΔH is the energy difference between the lowest energy level of the high-energy conformer and the lowest energy level of the low-energy conformer. This is shown for a hypothetical s-trans/gauche system in Fig. 5.

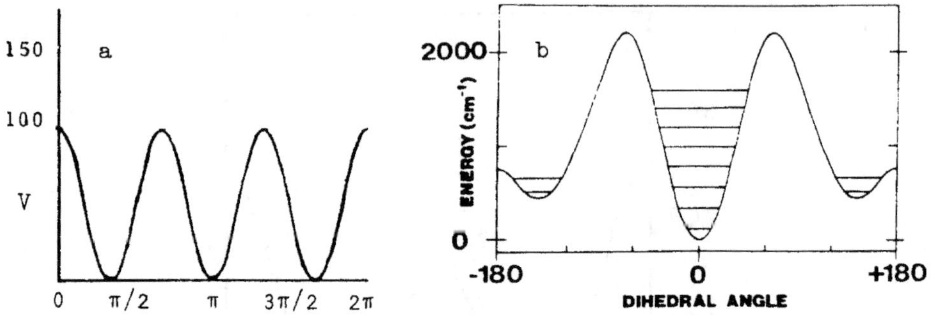

Fig. 3 (a) A potential function with V_3 = +100 and V_2 = 0. (b) Torsional potential function, including observed energy levels for an asymmetric torsion with $V_3 \simeq 1000$ and $V_2 \simeq 1000$.

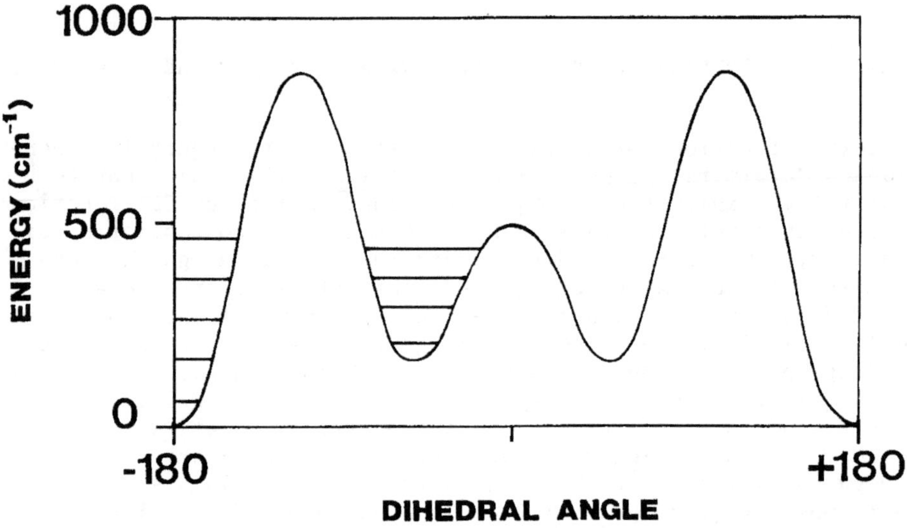

Fig. 4 Potential function for an s-trans/gauche system with V_2 = -200, V_3 = +800, V_6 = -60 and ΔH = +200.

In the calculation of the asymmetric potential function, ΔH is input to the computer program [3] as a symmetry-forbidden s-trans ↔ gauche transition between the lowest s-trans and the lowest gauche energy level. It is ΔH that will have the largest effect upon the energy of the gauche wells relative to the s-trans well.

In summary, the effects of V_1, V_2, V_3, and ΔH are:

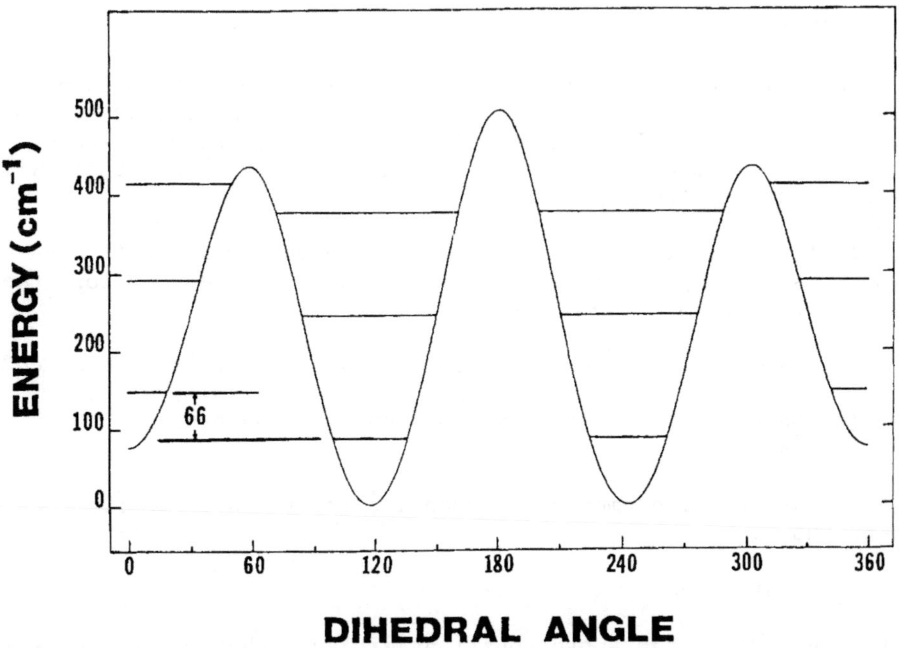

Fig. 5 Potential function for a trans/gauche molecule with ΔH indicated.

(1) V_3: determines the overall height of the planar ↔ non-planar barrier.

(2) V_2: determines the position (in radians) of the non-planar (gauche) wells and is the major component of the gauche ↔ gauche barrier.

(3) V_1: acts as an adjustment coefficient in determining the position of the gauche wells and the height of the gauche ↔ gauche barrier.

(4) ΔH: acts to fix the energy of the gauche wells relative to the planar conformer. ΔH does not have any effect on the conformer ↔ conformer barriers.

Several attempts have been made to explain the physical significance of the potential coefficients V_4, V_5, V_6 [3], however, our experience has shown these coefficients to be fitting parameters rather than terms having any real physical significance [6].

4. EXAMPLES

In this section a selected set of examples will be presented. The order has been chosen to first present those molecular systems where the vibrational assignment is straightforward, followed by successively more difficult systems. In this way each example will build on the previous.

In the analysis of a molecule existing as a mixture of conformers, several preliminary questions must be answered before the analysis can begin. These are:

(1) Is the structure (symmetry) of each conformer determined?

(2) Have the low frequency fundamentals (those below 600 cm^{-1}) been assigned?

(3) Are the fundamental frequencies for the high-energy conformer known?

(4) What is ΔH?

If the answers to the above questions are yes, then a final question is asked:

(5) Are these data correct?

Providing that the answer to this last question is yes, analysis of the torsional data can proceed. However, if the answer is no, then any observed discrepancy must be corrected before continuing with the analysis.

For clarity we will refer only to the torsional study under review, with the inferred assumption that all the necessary data (answers to 1-5) is contained within.

A. Propenoyl Fluoride

The infrared spectrum of propenoyl fluoride [8] is shown in Fig. 6. The high energy conformer was identified (with the use of microwave spectroscopy) to be s-cis and the low energy conformer to be s-trans. ΔH was calculated to be 90 $\pm$ 100 cal/mole. Based upon ΔH, both conformers should exhibit torsional transitions, especially since the torsional mode in both conformers possesses a C type infrared band contour. Careful analysis of the infrared spectrum indicates two sets of transitions with a Boltzman-like distribution. The first series, at higher frequency, is due to the low energy conformer and the second series is due to the high energy conformer. At this point, assignment of the Q branches in Fig. 6 is straightforward. The highest frequency Q branch for

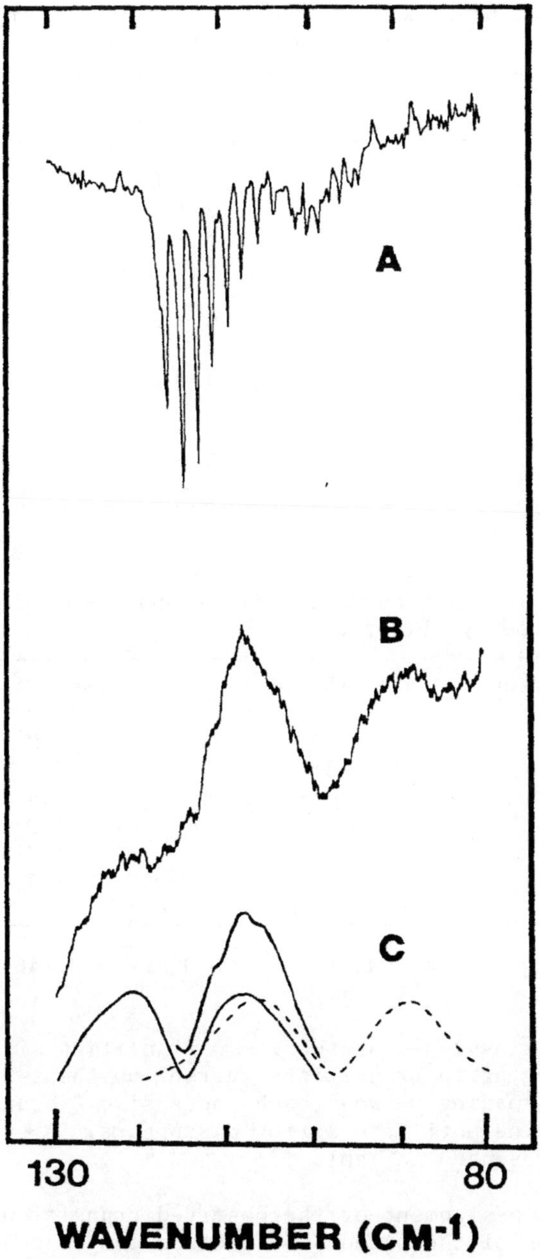

Fig. 6 (a) Far infrared spectrum, (b) Raman spectrum, and (c) band resolution of the Raman spectrum of gaseous propenoyl fluoride between 130 and 80 cm^{-1} (used by permission, Ref. [8]).

Table 1 Observed Bands and Proposed Vibrational Assignment of the Spectrum of Gaseous Propenoyl Flouride

Infrared Frequency	Conformer	$v'' \leftarrow v'$	$\Delta\nu$ Obs.-Calc.
116.7	t	$1 \leftarrow 0$	0.2
114.9	t	$2 \leftarrow 1$	0.0
113.1	t	$3 \leftarrow 2$	-0.1
111.4	t	$4 \leftarrow 3$	-0.1
109.4	t	$5 \leftarrow 4$	-0.3
107.7	t	$6 \leftarrow 5$	-0.1
105.8	t	$7 \leftarrow 6$	0.1
104.0	t	$8 \leftarrow 7$	0.4
101.5	c	$1 \leftarrow 0$	0.0
100.2	c	$2 \leftarrow 1$	0.0
99.8	c	$3 \leftarrow 2$	-0.1
96.4	c	$4 \leftarrow 3$	0.1

t = s-trans, c = s-cis

Table 2 Potential Function Coefficients and Dispersions in cm^{-1} for Propenoyl Fluoride[a]

Coefficient	Value	Dispersion
V_1	-108	1.9
V_2	1859	7.7
V_3	168	3.4
V_4	-80	4.2
V_5	-16	1.7
V_6	3	1.4
ΔH	36	

[a]Calculated using F_0 = 1.804, F_1 = -0.044, F_2 = 0.146, F_4 = -0.009.

each conformer is assigned as the $1 \leftarrow 0$ transition with successive Q branches belonging to transitions further up the s-trans well. Analysis of the spacing between each successive Q branch shows a decrease in spacing until the series disappears. A summary of the assignment is presented in Table I.

Now that an assignment of the observed transitions has been made, calculation of the potential function can be conducted. Since this molecule exists as an s-cis $\leftrightarrow$ s-trans mixture, V_2 should be the predominant term, with V_1, V_3 and ΔH (input as 0(trans) $\rightarrow$ 0(cis)) having a less significant role in shaping $V(\alpha)$. Once a satisfactory fit was obtained, in other words, once the frequency difference, $\Delta\nu$, and the dispersions on V_i were minimized,

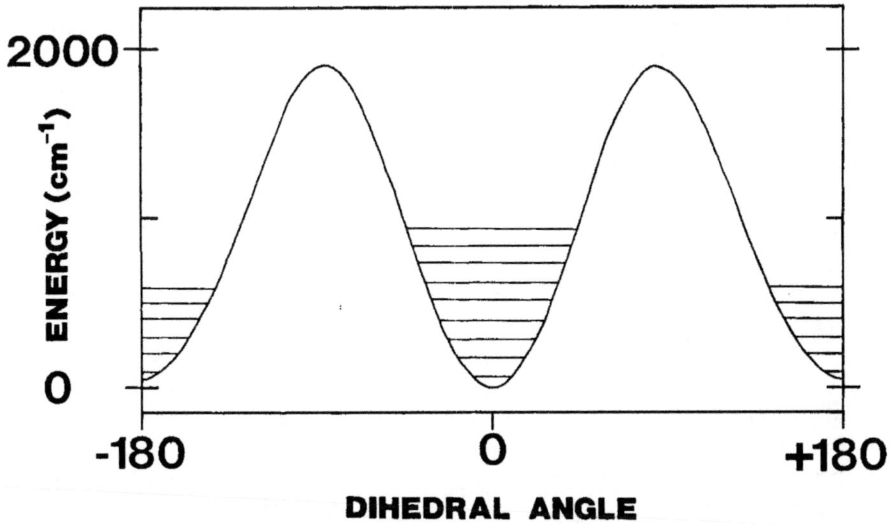

Fig. 7 Torsional potential function, including the observed transitions, for the C-C torsion of propenoyl fluoride. The dihedral angle of zero corresponds to the s-trans conformer (used by permission, Ref. [8]).

ΔH was allowed to "float" until the best fit was obtained. The final values of $V_1(\alpha)$ are listed in Table 2 and the final form of $V(\alpha)$ is shown in Fig. 7.

B. Ethyl Selenol-d_1

In ethyl selenol-d_1 [9] a mixture of s-trans and gauche conformers is present. The Newman projections of these conformers are shown in Fig. 8. Analysis of the microwave spectroscopic data indicated that $\Delta E = 146$ cm^{-1}, with the gauche conformer more stable at room temperature. The far infrared spectrum of ethyl selenol-d_1 is shown in Fig. 9. Several transitions are observed around 120 cm^{-1}. Since the gauche conformer is more stable, its transitions are at a higher frequency than those of the s-trans conformer. The assignment of these transitions is given in Table 3.

The calculation of the potential function was carried out using a dominant V_3 term and a smaller V_2 term. As mentioned earlier, V_2 and V_3 are necessary to shape the potential of an s-trans ↔ gauche system. Again, as in propenoyl fluoride, ΔH was allowed to float in the final calculations. The results of the calculation are given in Table 4.

Table 3 The Observed Torsional Frequencies and Assignments of Ethyl Selenol-d_1

Frequency	Conformer	Assignment	$\Delta\nu$
123.3	g	$1^{\pm} \leftarrow 0^{\pm}$	0.01
111.4	g	$2- \leftarrow 1+$	-0.05
110.8	g	$2+ \leftarrow 1-$	0.04
(110.3)[a]	t	$1 \leftarrow 0$	--
96.3	t	$2 \leftarrow 1$	-0.01

[a]Calculated infrared frequency.

Table 4 Potential Coefficients and Dispersions in cm^{-1} for Ethyl Selenol-d_1

Coefficient	Value	Dispersion
V_2	-91.7	0.4
V_3	408.6	0.2
V_6	-7.8	0.2
ΔH	66	4

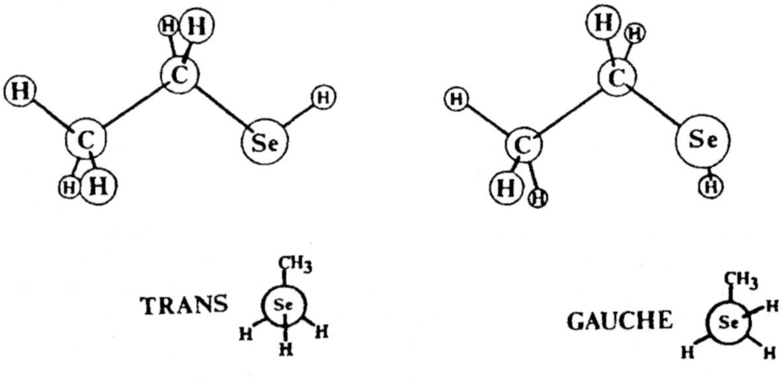

Fig. 8 Drawing of gauche and trans EtSeH (used by permission, Ref. [9]).

C. Methylphosphinosilane ($CH_3SiH_2PH_2$)

In methylphosphinosilane [10] the phosphino torsion results in two conformers. The s-trans conformer has C_s symmetry with the

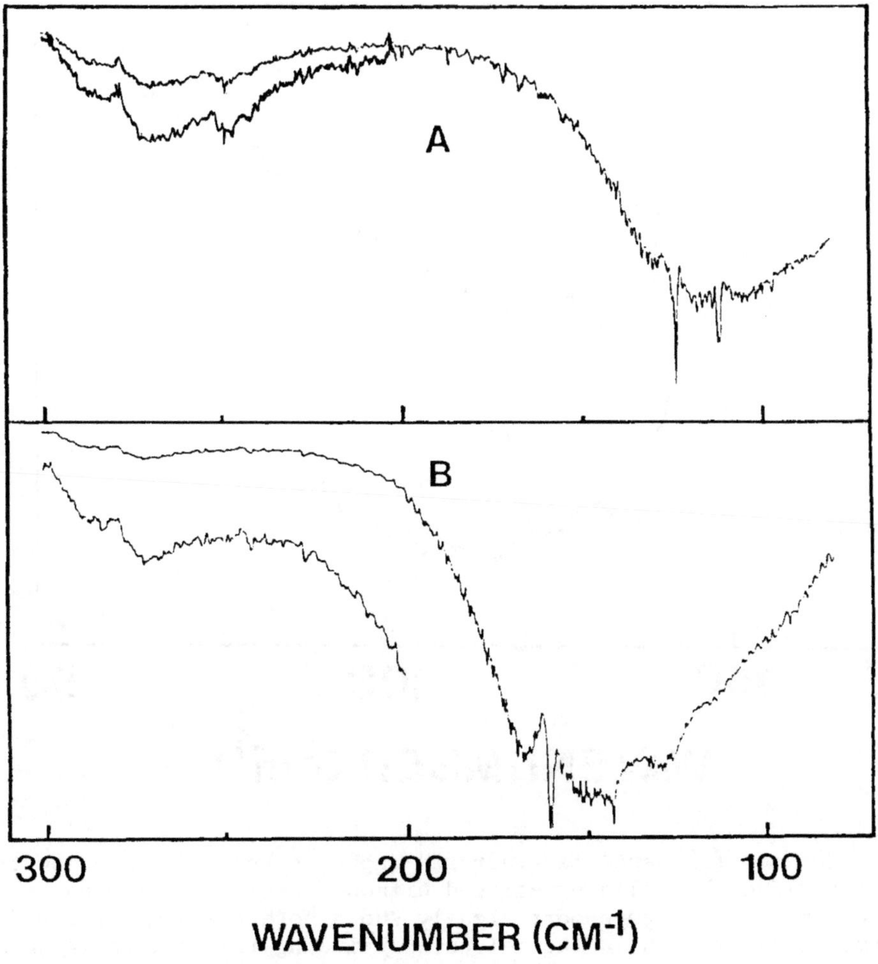

Fig. 9 Far infrared spectra of gaseous selenols. (a) EtSeD, 135 mm sample pressure, 0.5 cm^{-1} resolution, and triangular apodization; (b) EtSeH, 170 mm sample pressure, 0.5 cm^{-1} resolution and triangular apodization (used by permission, Ref. [9]).

torsion being of A" symmetry with strong infrared Q branches. The gauche conformer has C_1 symmetry with the torsion having an A/B/C hybrid infrared band contour and weaker Q branches. The low frequency spectra of methylphosphinosilane are shown in Fig. 10. The s-trans conformer which is the more stable rotamer has its 1 ← 0 transition at 138.1 cm^{-1}, while the gauche transition which is stronger in the Raman spectrum, is at 127.2 cm^{-1}. The assignment is given in Table 5. Calculation of the asymmetric potential func-

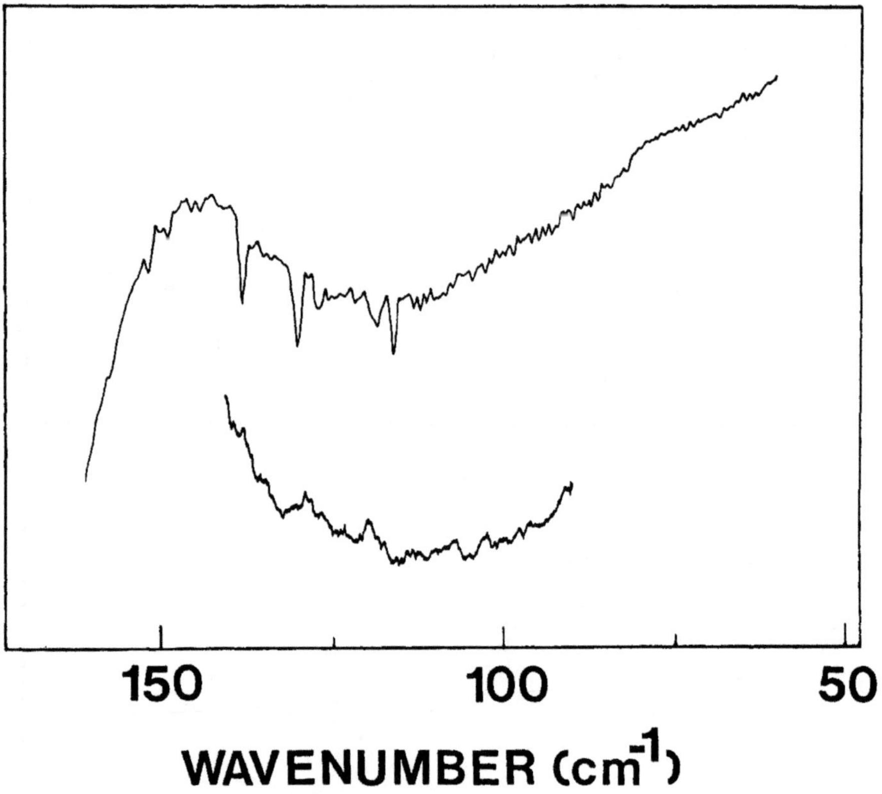

Fig. 10 Low frequency far-infrared (upper scan) and Raman spectra of methylphosphinosilane retraced without the intervening phosphine spectrum. This figure more clearly shows both the symmetric and asymmetric torsions in the molecule including transitions in the trans and gauche wells. The Raman spectra was recorded with a resolution of 2 cm^{-1} and the infrared with a resolution of 1.0 cm^{-1} (used by permission, Ref. [10]).

tion can now be conducted. The value of V_3 can be approximated using $V_3 = (\nu_{1 \leftarrow 0})^2/9F$ [4], leading to a final potential of the form $V(\alpha) = \Sigma\ (1 - \cos i\alpha)$, where $V_2 = 81\ cm^{-1}$, $V_3 = 523\ cm^{-1}$, $V_6 = -23\ cm^{-1}$ and $\Delta H \sim 58\ cm^{-1}$.

D. Methylvinlyether

The analysis of the far infrared spectra of methylvinylether-d_0, d_3 [11],shown in Fig. 11, is the most complex of all the examples given. This is because the asymmetric torsion for the planar s-cis conformer occurs near several hot bands of a skeletal

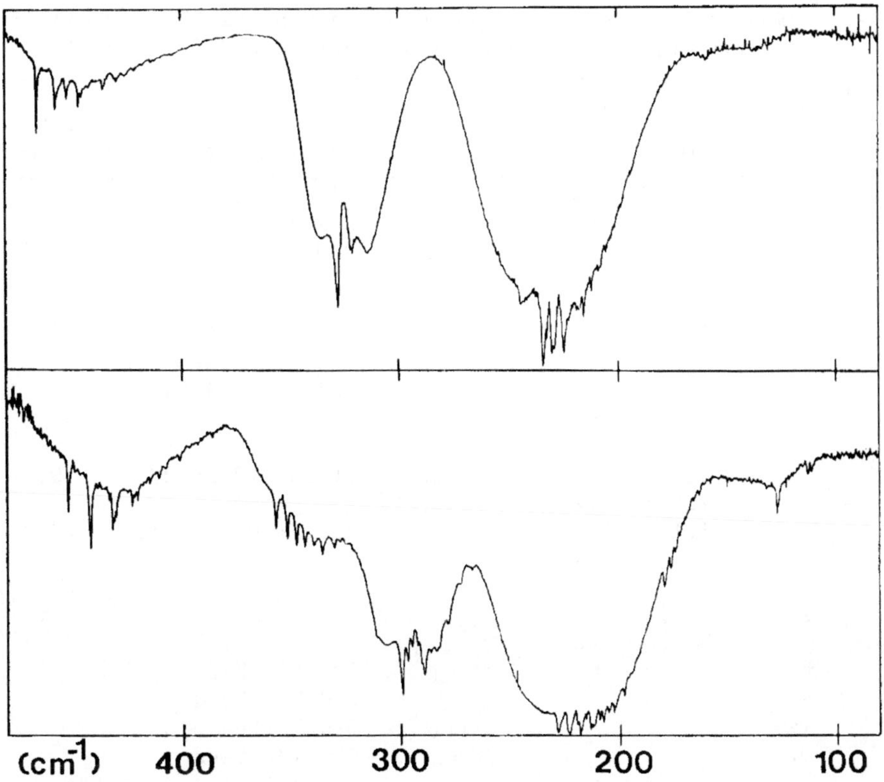

Fig. 11 Far infrared spectra of gaseous methylvinylether-d_0 (top) and -d_3 (lower) between 480 and 80 cm^{-1} (used by permission, Ref. [11]).

Table 5 Vibrational Assignment of the Torsional Transitions of Methylphosphinosilane

Conformer	Frequency	Transition
s-trans	138.1	$1 \leftarrow 0$
	130.1	$2 \leftarrow 1$
	116.4	$3 \leftarrow 2$
gauche	127.2	$1^{\mp} \leftarrow 0^{\pm}$
	118.7	$2^{\pm} \leftarrow 1^{\mp}$

bending motion near 300 cm^{-1} in the light molecule. Further complication is observed in the spectrum of the heavy molecule, where the asymmetric torsional transitions were observed for both the ground and first excited state of the methyl torsion.

Table 6 Observed Bands and Proposed Assignment of the Vibrational Spectrum of Gaseous Methylvinylether Below 500 cm^{-1}

Raman		Infrared				
cm^{-1}	Rel. Int.	cm^{-1}	Rel. Int.		Assignment[a] (cm^{-1})	$\Delta\nu$ (cm^{-1})
478	vw	477.8	vw		244.3+232.7=477.0	
467	vw	465.5	mw	2 ← 0	234.5+230.4=464.9	0.3
457	w	456.9	w	3 ← 1	230.4+224.8=455.2	2.1
453	vw	451.7	w		229.0+222.6=451.6	
		446.4	w			
		445.1	w		232.7+212.2=444.9	
					229.0+215.8=444.8	
		441.4	vw		224.8+215.8=440.6	
		435.3	w			
		434.4	vw			
		429.0	w	2 ← 0*	215.8+212.2=428	0.1
		425.3	vw			
		421.0	vw	3 ← 1*	212.2+209.0=421.2	-0.8
		416.8	vw			
		413.3	vw			
327	s	327.9	s	δCOCC		
		322.3	vw			
321	m	320.9	w	δCOCC	1st hot band	
312	w			δCOCC	2nd hot band	
303	vw			δCOCC	3rd hot band	
294	vw			δCOCC	4th hot band	
		244.3	w			
238	w,br	234.5	s	1 ← 0		-0.2
		232.7	w,sh			
		230.4	m	2 ← 0		-0.1
		229.1	w,sh			
		224.8	m	3 ← 2		0.5
221	w,br	222.6	w,sh			
		217.8	vw			
		215.8	w	1 ← 0*		0.6
		212.2	w	2 ← 1*		-1.5
		209.0	vw	3 ← 1*		0.8
		208.0	vw			
		205.6	vw			
		159.4	vw	g 1 ← 0*		0.1
144	w			g 1± ← 0±		±0.2
		137.9	vw	g 2 ← 1*(E ← E)		-1.4
		135.9	vw	g 2 ← 1*(A ← A)		1.1

[a]Torsions labeled with * refer to methyl torsion, other transitions are due to asymmetric torsion. g refers to gauche conformer, all others are s-cis. Abbreviations used: s, strong; m, medium; w, weak; v, very; sh, shoulder.

Table 7 Observed Bands and Proposed Assignment of the Vibrational Spectrum of Gaseous Methylvinylether-d_3 Below 500 cm^{-1}

Raman		Infrared				
cm^{-1}	Rel. Int.	cm^{-1}	Rel. Int.		Assignment[a] (cm^{-1})	Δν (cm^{-1})
452	w	452.0	w	cg 2 ← 0	228.4+223.0=451.4	0.9
				ce 2 ← 0	227.8+223.5=451.3	1.1
		442.2	w,sh	ce 3 ← 1	223.5+219.5=443.0	-2.3
441	w	441.7	mw	cg 3 ← 1	223.0+218.2=441.2	-1.9
		443.4	vw	ce 4 ← 2	219.5+215.5=435.0	-2.0
431	w	432.0	w	cg 4 ← 2	218.2+213.5=431.7	-1.7
		431.1	w,sh		218.2+212.7=430.9	
		430.3	w,sh		218.2+212.1=430.3	
		425.0	vw	ce 5 ← 3	215.5+209.4=424.9	1.0
422	w	423.0	w	cg 5 ← 3	213.5+207.5=421.1	1.5
		420.4	w		212.7+207.6=420.3	
415	vw	414.9	vw	ce 6 ← 4	209.4+205.4=418.8	b
410	vw	410.8	vw	cg 6 ← 4	207.6+202.8=410.4	b
		407.6	vw	ce 7 ← 5	205.4+201.8=407.2	b
401	vw	401.5	vw	cg 7 ← 5	202.8+198.1=400.9	b
393	vw	392.9	vw	cg 8 ← 6	198.1+194.2=392.3	b
		386.3	vw	cg 9 ← 7		
357	w	357.0	w	c 2 ← 0*	179.5+176.7=356.2	3.6
351	w	351.7	w	c 3 ← 1*	176.7+174.3=352.0	-3.0
347	w	347.5	w	c 4 ← 2*	174.3+172.1=346.4	-5.3
343	w	343.5	w	c 5 ← 3*	172.1+170.3=342.4	-4.6
338	vw	339.4	vw	c 6 ← 4*	170.3+169.2=339.5	-0.9
335	w	335.6	w	c 7 ← 5*	169.2+165.7=334.9	6.1
329	w	330.0	vw	c 8 ← 6*	165.7+163.0=328.7	b
325	vw	325.0	vw			
322	vw	321.5	vw			
309	w			g δCOCC		
298	s	299.2	m	cg δCOCC		
		296.5	w	cg δCOCC	hot band A	
		294.4	w	cg δCOCC	hot band B	
		292.2	vw	cg δCOCC	hot band C	
		290.4	vw	cg δCOCC	hot band D	
288	m	288.9	m	cg δCOCC	1st hot band	
		286.0	w	cg δCOCC	hot band E	
283	w	283.2	w	cg δCOCC	2nd hot band	
		282.0	vw	cg δCOCC	hot band F	
277	w	278.0	w	cg δCOCC	3rd hot band	
272	w	272.2	vw	cg δCOCC	4th hot band	
268	vw	267.0	w	cg δCOCC	5th hot band	
254	vw	254.0	vw	cg δCOCC	6th hot band	
227[c]	vw	228.4	s	cg 1 ← 0		1.3
		227.8	w,sh	ce 1 ← 0		1.1
		223.5	w,sh	ce 2 ← 1		-0.7
		223.0	s	cg 2 ← 1		-1.0

Table 7 continued

Raman		Infrared			
cm^{-1}	Rel. Int.	cm^{-1}	Rel. Int.	Assignment[a] (cm^{-1})	Δν (cm^{-1})
		219.5	m,sh	ce 3 ← 2	-0.8
		218.2	s	cg 3 ← 2	-1.5
		215.5	m	ce 4 ← 3	0.4
		213.5	m	cg 4 ← 3	-0.6
		212.7	m,sh		
		212.1	w,sh		
		209.4	m	ce 5 ← 4	0.5
		207.6	m	cg 5 ← 4	0.2
		205.4	w	ce 6 ← 5	4.3
		202.8	m	cg 6 ← 5	3.4
		201.8	w,sh	ce 7 ← 6	b
		198.1	m	cg 7 ← 6	b
		194.2	vw	cg 8 ← 7	b
		179.5	w	c 1 ← 0*	3.6
		178.8	w,sh		
		176.7	w	c 2 ← 1*	-0.9
		174.3	w	c 3 ← 2*	-3.0
		172.1	w	c 4 ← 3*	-3.5
		170.3	w	c 5 ← 4*	-2.2
		169.2	w	c 6 ← 5*	1.4
		165.7	w	c 7 ← 6*	4.0
		162.9	w	c 8 ← 7*	b
		160.6	vw	c 9 ← 8*(A ← A)	b
		160.0	vw	c 9 ← 8*(E ← E)	b
138	w			g 1± ← 0±	±1.8
		127.4	w	g 1 ← 0	0.0
		113.3	vw	g 2 ← 1*(E ← E)	0.8
		111.4	vw	g 2 ← 1*(A ← A)	-0.7

[a]Torsions labeled with * refer to methyl torsions, other transitions are due to asymmetric torsions. g refers to gauche conformer, c to s-cis, and cg and ce to the ground and excited asymmetric torsional states of the s-cis conformer, respectively. For abbreviations used, refer to Table 6.

[b]Transition not used in fitting the potential function.

Table 8 Potential Function Coefficients and Dispersions, in cm^{-1}, Calculated for Methylvinylether and Methylvinylether-d_3

	$CH_3OCH=CH_2$ [a]		$CD_3OCH=CH_2$ [b]			
	Ground state		Ground state		Excited state	
Coefficient	Value	Dispersion	Value	Dispersion	Value	Dispersion
V_1	-222	13.1	-413	6.3	-424	7.2
V_2	1073	39.2	1434	21.3	1484	37.6
V_3	1058	8.9	1177	9.7	1174	13.0
V_6	-90.8	5.3	-96.0	2.8	-104	6.1
ΔH	402		402		402	

[a]Calculated using F_0=2.881, F_1=0.662, F_2=0.294, F_3=0.083, F_4=0.026, and F_5=0.008

[b]Calculated using F_0=2.496, F_1=0.449, F_2=0.219, F_3=0.054, F_4=0.017, and F_5=0.005

In the light molecule, the hot bands were assigned to the evenly spaced bands at 321, 312, 303 and 294 cm^{-1}, which were also observed in the Raman effect. The torsional transitions were then assigned to the three relatively strong bands at 234.5, 230.4, and 224.8 cm^{-1}, using the guidelines previously mentioned in the earlier examples. Analysis of the spectrum of the d_3 molecule was, as stated above, complicated. As in the light molecule, the hot bands were assigned to Raman/far infrared pairs. The ground state transitions were assigned to the relatively strong bands at 228.4 ($1 \leftarrow 0$), 223.0, 218.2, 213.5, 207.6, 202.8, 198.1, and 194.2 cm^{-1}. The excited state transitions were assigned to shoulders present on the ground state transitions, with an average ground state-excited state splitting of ~0.6 cm^{-1}. As in the previous examples, the gauche torsional transitions were observed in the Raman effect, near 140 cm^{-1}. The full vibrational assignments for both molecules are given in Tables 6 and 7.

Using the frequencies assigned to the s-cis and gauche conformers, torsional potential functions were calculated for the light and heavy molecules in both the ground and first excited states of the methyl torsion. The potential functions thus obtained have several common features: large values of V_2 and V_3, large s-cis $\leftrightarrow$ gauche barriers and small gauche $\leftrightarrow$ gauche barriers. A complete comparison is given in Table 8.

From the potential function data for methylvinylether-d_0, -d_3, and -d_3-excited states, several conclusions were reached, and they are: (1) the low gauche $\leftrightarrow$ gauche barrier indicates that the forces favoring the gauche conformers are relatively weak, leading to the wide gauche angle of 144°; (2) the high s-cis $\leftrightarrow$ gauche barrier indicates strong non-bonded in-plane π-type interactions through the oxygen lone-pairs which must be broken by the asymmetric torsion; and (3) the higher d_3-molecule barriers relative to those of the d_0-molecule are probably due to steric interactions between the methyl and vinyl hydrogens.

5. SUMMARY AND CONCLUSIONS

It is clear from the few examples given above that Fourier transform far infrared spectroscopy can provide a wealth of information on the torsional transitions of a large number of asymmetric rotors. Usually the spectra are very "rich" with many "hot bands" and frequently overtones with accompanying "hot bands" can also be observed. The transitions may yield very sharp Q branches when the asymmetric torsions give rise to C type band contours. When two or more conformers are present in appreciable quantities it may be difficult to distinguish which Q branch series should be assigned to which conformer. In many cases the Raman spectrum of the vapor can provide the polarization data necessary for making

Table 9 Calculated Values for the Potential Coefficients for Internal Rotation of an Asymmetric Rotor, in cm^{-1}

Compound	ΔH	V_1	V_2	V_3	V_4	V_5	V_6	Low[a] Energy Form	High[a] Energy Form	Ref.
1-Fluoropropane	1344	1126	-1067	2266	-437	129	-266	g	t	12
1,2-Difluoroethane	692	-1018	143	787				g	t	13
1,1,2,2-Tetrabromoethane	238	765	-985	2025				g	t	14
Butane	310	418		639			136	t	g	15
3-Fluoropropene	58	-247	185	857	188	7	-93	c	g	16
Propanoyl fluoride	451	325	252	402	45			c	g	17
1-Butene	26	-243	185	921	165	-36	-76	c	g	18
Propanal	241		559	651	-105		-46	c	g	19
2-Methyl-1-butene	61		79	1484	108	-95	-29	t	g	20
3-Methyl-1-butene	145	-129	361	627			30	t	g	21
Glyoxal	1182	1182	1114		-56			t	c	22
Acrolein	584	306	1919	338	-96	-57		t	c	23
1,3-Butadiene	873	600	2068	273	-49			t	c	24
Propenoyl Chloride	215	96	1734	134	-123			t	c	8
Isoprene	384	112	2742	294	-150			t	c	25
Ethylmethylether	390	562	-26	897			-5	t	g	26
Methylvinylether	402	-222	1073	1058			-91	c	g	11
Ethylmethylsulfide	16	-45	24	1230			-125	c	g	27
Isopropylamine	156	65	30	979	228	-25	-143	t	g	28
Ethylamine	217	317	-11	714	-25	25	-4	t	g	29,30

[a] g = gauche; t = s-trans; c = s-cis.

the correct assignment. For example, if one of the conformers has a symmetry plane and the asymmetric torsional mode is an A'' vibration giving rise to a depolarized Raman line, and if the other conformer has no symmetry so that the asymmetric torsion gives rise to a polarized Raman line, then one can readily distinguish between the two torsions in the Raman effect. Also for some molecules the asymmetric torsional overtones are observed in the Raman effect; these data can provide a check of the assignments made on the basis of the infrared data. Nevertheless we have found that the far infrared spectra usually contain many more transitions than the Raman spectra and that they are much more valuable for studying asymmetric torsions.

From these torsional transitions it is not only possible to obtain the potential functions governing the internal rotation but it is also possible to obtain the energy differences between conformers and the various barriers to interconversion. Additionally, under favorable circumstances, it may be possible to obtain the dihedral angle and identify the symmetry of the most stable conformer. Only recently has far infrared FT spectroscopy been used to study asymmetric torsions.

Examination of the above examples shows that the asymmetric potential function may reflect molecular interactions. In the cases of methylphosphinosilane and ethyl selenol-d_1, the potential functions reflect the involvement of the "gauche effect." Thus the low-energy conformer in methylphosphinosilane is planar while that in ethyl selenol-d_1 is non-planar. In the cases of propenoyl fluoride and methylvinylether the s-cis - planar form is more stable presumably due to pseudoconjugation between the vinyl and oxygen components in the planar configuration.

Thus by examination of the available data from the literature (see Table 9), the asymmetric potential function can be used to determine: (a) conformer structure and stability, (b) ΔH and (c) barriers to internal rotation of molecules existing as multiple rotamers. The resulting potential functions will reflect intramolecular interactions such as steric repulsion, lone-pair interactions (the "gauche effect") or inductive electronic effects that one would expect to see based on the approximate molecular structure.

REFERENCES

1. J. R. Durig and W. Cox, Jr. in Fourier Transform Infrared Spectroscopy, Volume I, (J. R. Ferraro, L. J. Basile, eds.) Academic Press, New York, 1978.

2. P. R. Griffiths, Chemical Infrared Fourier Transform Spectroscopy, Wiley-Interscience, New York, 1975.

3. J. D. Lewis, T. B. Malloy, Jr., T. H. Chao and J. Laane, J. Mol. Struct., 12, 427 (1972).

4. W. G. Fateley, R. K. Harris, F. A. Miller, and R. E. Witkowski, Spectrochim. Acta, 21, 231 (1965).

5. N. D. Epiotis, W. R. Cherry, S. Shark, R. Yates, and F. Bernardi, Structural Theory of Organic Chemistry, Springer-Verlag, Berlin, 1977.

6. H. C. Allen and P. C. Cross, Molecular Vib-Rotors, Wiley, New York, 1963.

7. D. A. C. Compton in Vibrational Spectra and Structure, Vol. X, (J. R. Durig, ed.), Elsevier, Amsterdam, in press.

8. J. R. Durig, J. S. Church, and D. A. C. Compton, J. Chem. Phys., 71, 1175 (1979).

9. J. R. Durig and W. E. Bucy, J. Mol. Spectrosc., 64, 474 (1977).

10. R. O. Carter, W. J. Natter, and J. R. Durig, J. Phys. Chem., 82, 2728 (1978).

11. J. R. Durig, and D. A. C. Compton, J. Chem. Phys., 69, 2028 (1978).

12. E. Hirota, J. Chem. Phys., 37, 283 (1962).

13. W. C. Harris, J. R. Holtzclaw, and V. F. Kalasinsky, J. Chem. Phys., 67, 3330 (1977).

14. G. L. Carlson, W. G. Fateley, and J. Hiraishi, J. Mol. Struct., 6, 101 (1970).

15. J. R. Durig, and D. A. C. Compton, J. Chem. Phys., 83, 265 (1979).

16. P. Meakin, D. O. Harris, and E. Hirota, J. Chem. Phys., 51, 3775 (1969).

17. O. L. Stiefvater, and E. B. Wilson, Jr., J. Chem. Phys., 50, 5385 (1969).

18. J. R. Durig, and D. A. C. Compton, J. Phys. Chem., in press.

19. J. R. Durig, A. Q. McArver, and D. A. C. Compton, J. Chem. Phys., in press.

20. J. R. Durig, D. J. Gerson, and D. A. C. Compton, J. Phys. Chem., in press.

21. J. R. Durig, and D. J. Gerson, J. Phys. Chem., in press.

22. J. R. Durig, W. E. Bucy, and A. R. H. Cole, Can. J. Phys., 53, 1832 (1975).

23. L. A. Carriera, J. Phys. Chem., 80, 1149 (1976).

24. L. A. Carriera, J. Chem. Phys., 62, 3850 (1975).

25. L. A. Carriera, Private Communication.

26. J. R. Durig, and D. A. C. Compton, J. Chem. Phys., 56, 4713 (1978).

27. J. R. Durig, D. A. C. Compton, and M. R. Jalilian, J. Phys. Chem., 83, 511 (1979).

28. J. R. Durig, G. A. Guirgis, and D. A. C. Compton, J. Phys. Chem., 83, 1313 (1979).

29. M. Tsuboi, K. Tamagake, A. Y. Hirakawa, J. Yamaguchi, H. Nakagawa, A. D. Manocha, E. C. Tuazon, and W. G. Fateley, J. Chem. Phys., 63, 5177 (1975).

30. J. R. Durig, and Y. S. Li, J. Chem. Phys., 4110 (1975).

CURRENT RESEARCH IN FAR INFRARED INTERFEROMETRY

J. R. Durig and J. F. Sullivan

Department of Chemistry
University of South Carolina
Columbia, South Carolina, U. S. A.

1. INTRODUCTION

The historical development of interferometry and Fourier transform infrared spectroscopy has been presented earlier in this volume; however, it should be noted that the first chemical applications of FT-IR spectroscopy were made in the far infrared region of the spectrum (~400 to 10 cm^{-1}) because the instrumentation required to study this region was much simpler than that needed for the mid or near infrared regions. The instrumental advantages in the far infrared region included a lower tolerance for the mirror drive of a Michelson interferometer, a smaller dynamic range of the interferogram, and a longer sampling time which resulted in a reduced number of data points. However, far infrared spectroscopy also had its inherent difficulties such as low-energy sources, poor detectors, a lack of suitable optical materials, and spectral interference of water. Researchers in this area have overcome most of these difficulties and therefore investigations in the far infrared spectral region have become widespread. In the following sections, some of the applications of Fourier transform far infrared spectroscopy will be examined and, when applicable, they will be compared with the more conventional grating spectrometers. It seems appropriate to begin the discussion with a brief outline of some of the initial applications of far infrared interferometry.

2. INITIAL APPLICATIONS OF FOURIER TRANSFORM FAR INFRARED SPECTROSCOPY

The three areas in which research was initially carried out

J. R. Durig (ed.), Analytical Applications of FT-IR to Molecular and Biological Systems, 361–389.

in Fourier transform far infrared spectroscopy were investigations of pure rotational spectra, studies of molecules in the solid state, and studies of heavy atom vibrations and skeletal bending modes. Diatomic or symmetric top molecules yield the simplest pure rotational spectra. Since the spacing between rotational energy levels is quite small for most molecules, the rotational transitions usually occur in the microwave region. However, if the molecule is very "light" the rotational transitions can occur in the far infrared region and the resulting spectrum consists of an evenly spaced series of absorptions which correspond to the different J-level transitions. The pure rotational spectrum of NO and CO have been studied [1,2] using a lamellar grating interferometer and the NO spectrum was especially interesting since the NO molecule contains an odd number of electrons. The resulting spectrum is a series of doublets from which the rotational constant B_0 was obtained to an accuracy of ±0.005% for each state [2]. The rotational spectra of several symmetric top molecules such as CH_3Cl [3], NH_3 [4], PH_3 [5], CH_3CCH [6], CH_3F [6], $(CH_3)_3N$ [7] have been studied and the interested reader can obtain additional details in Refs. [7], [8] and [9].

In the case of an asymmetric rotor, the resulting spectrum is rather complex since the rotational lines are not evenly spaced but are in fact rather randomly distributed. Water is one such molecule which strongly absorbs in the far infrared and since it is a common atmospheric component, one must take great care in eliminating all traces of water from the spectrometer housing and the gaseous sample to be investigated in order to obtain good far infrared data. The interferometrically measured H_2O spectra by Hall et al. [10] show superior resolution over a wide spectral range compared to the spectra reported earlier [11,12] which were recorded with dispersive instruments. From such work it has been possible to obtain very accurate values of the rotational and centrifugal distortion constants but it is not possible to take the molecular constants and calculate the observed frequencies exactly [9]. However, the spectrum is well known and it is often used as a calibration and resolution standard. The O_3 and SO_2 molecules are examples of other asymmetric tops whose far infrared spectra have been obtained interferometrically [13].

The above mentioned studies did not show a significant departure from the mainstream of far infrared research since a majority of it could be carried out utilizing dispersive instrumentation, but in the field of the solid state, the interferometrically obtained spectra provided data which were not easily obtained by grating instruments. For example, Sievers and Tinkham [14] observed the ferromagnetic absorption line of ytterbium-iron garnet at zero-field conditions with a lamellar grating interferometer at a resolution of better than 0.1 cm^{-1}. Such absorptions are generally quite weak and occur below ten wavenumbers at very low temperatures (≃1°K). Thus, physicists have put interferometers to use in the study of the solid state to determine optical constants such as the index of refraction, complex indices, phase angle transmission coefficients, and the electronic processes in insulating crystals as well as the intermolecular vibrations of molecular crystals. Such studies have been important in the development of interpretive theories of solid state behavior, and this area of far infrared studies will be covered in another chapter.

Among the first molecules to be studied in the far infrared spectral region for purposes of obtaining frequencies for intramolecular normal modes were the inorganic salts and complexes which contain organic ligands. In the mid infrared spectra of such compounds, one usually obtains only the vibrations of the ligands or the stretching vibrations of the very strongly bound atom-to-metal modes. The metal-ligand stretching and bending modes as well as most of the metal-halogen or metal heavy atom stretching and bending modes are found in the far infrared spectral region, and therefore many chemists have studied such molecules to obtain information on bonding, structures, force constants, substituent effects, or simply for completing vibrational assignments.

3. SOME CURRENT CHEMICAL APPLICATIONS OF FAR INFRARED INTERFEROMETRY

For the sake of brevity, we shall mainly discuss the current uses of far infrared interferometry for the study of molecules in the gas phase which contain very anharmonic vibrations, such as torsional vibrations and ring-puckering motions in some small ring compounds. In order to interpret the far infrared data it will be necessary to review the theory in some detail. However examples will be provided and some trends will be illustrated.

A. Torsional Vibrations

It has long been established that while a double bond does not allow the rotation of a group of atoms at one end with respect to the group at the other, a single bond permits such rotation, at least to the extent that it is not generally possible for isomers to be isolated. Such rotation about the single bond was initially thought to be essentially free but, by the mid-1930's, it had been verified that barriers had to be surmounted in going from one conformation to another by rotation around a single bond. Although these potential barriers are only usually a few thousand calories, there are a number of thermodynamic properties which are markedly affected by them. Thus, the heat capacity, entropy, and equilibrium constants contain an appreciable contribution from hindered internal rotation.

Initially, and for several years, torsional barriers were determined from what can be described as indirect measurements, of which thermodynamics and microwave studies were the most prolific. The barriers were calculated from thermodynamic data by relating the difference in the observed and statistical entropies by tables involving the barrier height and the reciprocal of the partition function for free rotation. However discrepancies often arose because many of the barriers were calculated from erroneously assigned or assumed normal vibrational frequencies. In the microwave spectrum, the observed perturbations on the pure rotational transitions are correlated to the torsional barrier height by either the splitting or intensity methods. The splitting method is by far the most exact and usually gives barriers to 1% but it is only applicable to rather low barriers and "light" rotors. The intensity method may give barrier heights to 10 to 15% accuracy but it is difficult to obtain quantitative intensities from the microwave spectra.

Far infrared spectroscopy is probably the most convenient method for studying barrier heights since one is actually obtaining the energy level separations by vibrational spectroscopy. Experimentally one wishes to obtain the fundamental frequency for the torsional mode for the gaseous molecule so that the barrier in the isolated molecule can be ascertained. However, frequently the dipole moment change associated with the torsional mode may be quite small and consequently the resulting infrared band may often be quite weak. Therefore, assignments of the torsional modes from the infrared spectra of molecules in the gas phase were frequently in error in the initial investigations and it is often necessary to use isotopic substitution to verify the torsional assignment for many molecules which have other low frequency vibrations. Also, it should be pointed out that torsional modes which give rise to B type band contours may have bands which are very broad and ill-defined because of unresolved excited state transitions. However, in favorable cases where the torsional mode gives rise to A or C type band contours with relatively strong Q branches, several excited state transitions may be resolved and not only the barrier height obtained but also the detailed shape of the potential well may be ascertained.

Raman spectroscopy can also be complementary to the far infrared studies on torsional modes since the overtones of these motions are symmetric and give rise to Q branches which are frequently several times more intense than the fundamentals. In the Raman spectrum, both mechanical and electrical anharmonicity allow the torsional overtones to be active and since they are totally symmetric, only isotropic polarizability terms can exist which lead to stronger Q branches than for anisotropic fundamentals. Finally, since $\Delta v = 2$ transitions are usually studied, the frequency spacing between peaks in a Raman torsional overtone series is approximately twice that in the infrared where $\Delta v = 1$ selection rules are obeyed, and therefore resolution rarely presents a problem in the Raman spectrum. Therefore when the torsional modes give rise to band contours in the infrared which contain no Q branches it may be possible to obtain reliable frequencies from the Raman data.

Finally, it should be pointed out that when a molecule has a too complicated far infrared spectrum for quantitative interpretation because of the difficulty in establishing the origin of the Q branches for the torsional mode, reliable barrier heights may still be obtained by studying the infrared and Raman spectra of the sample in the crystalline state. At liquid nitrogen temperature, the upper vibrational states are effectively depopulated and only the $1 \leftarrow 0$ transition is observed. Although it is possible to obtain the barrier height from this single piece of experimental datum, a detailed analysis of the shape of the potential well is not possible. Additionally, it is sometimes possible to observe the torsional mode in the solid state when it is forbidden for the gaseous molecule [15]. However, it has been found that barriers obtained in the solid state are usually 10 to 15% higher than the ones for the gaseous molecules [16].

Types of internal rotors fall into two categories - symmetric tops and asymmetric tops. For symmetric tops, a rotation about the top-frame bond of $2\pi/n$ (where n is an integer) will bring the top to a position symmetrically equivalent to, or indistinguishable from, the original configuration. Therefore it is usual to speak of the foldness of the rotor in terms of n. For example, a silyl group (SiH_3) is a threefold symmetric top (local C_{3v} symmetry) since a rotation about the silicon-to-frame bond of 120° will result in an orientation superimposable on the initial orientation. Twofold rotors include $-NO_2$ $-BF_2$, and phenyl groups (local C_{2v} symmetry). When a rotation of 360° (i.e. when n = 1) is the only operation that results in a symmetrically equivalent position for the top, the rotor is known as an asymmetric top. Examples of asymmetric tops include alcohol (-OH), thiol (-SH), amino ($-NH_2$), and phosphino ($-PH_2$) groups when they are bonded to an asymmetric frame, and these asymmetric tops will be covered in another chapter. For the case of a symmetric frame, the top with the highest degree of symmetry prevails, and when two tops of different foldness are bonded directly to one another, the resultant foldness is the product of the two individual tops' foldness. For instance, CH_3BH_2 would be classified as a sixfold internal rotor whereas ethane, CH_3CH_3, would be threefold.

The energy minima and maxima for a symmetrical threefold group ($-CH_3$) are 60° apart (Fig. 1). The simplest mathematical

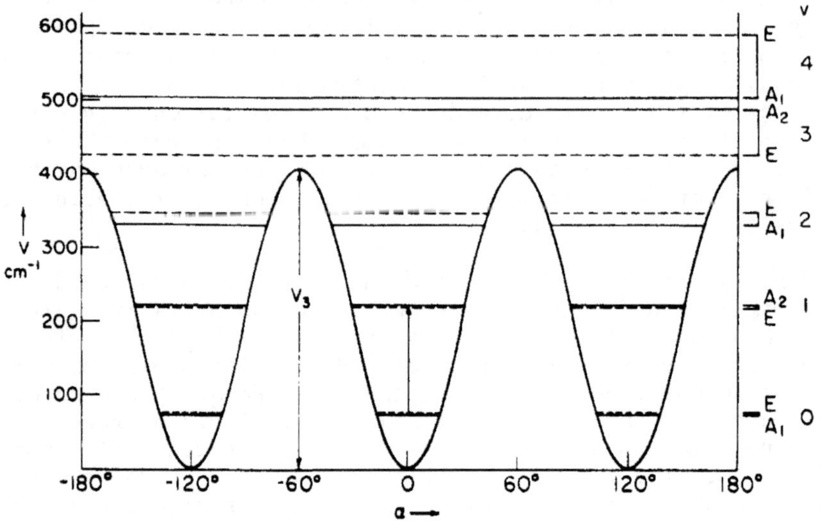

Fig. 1 Three rotor potential function (Used by permission, Ref. [19]).

function which will reproduce such a potential variance upon rotation is a cosine function. By assuming the problem to be one-dimensional, the quantum mechanical energy solutions are readily obtainable, and they have been discussed in detail by Kilpatrick and Pitzer [17], Herschbach [18], and Fateley and Miller [19-21]. The eigenvalue problem will be outlined and the interested reader is referred to these cited references for additional details. The model employed is a rigid symmetric top (CH_3 group) attached to a rigid frame which may be completely asymmetric. There are four rotational degrees of freedom, three for overall rotation and one for the hindered rotation of the two groups. The axis of internal rotation coincides with the unique axis of the symmetric top. Since the top has a threefold symmetry axis, the potential energy hindering rotation may be expressed by a Fourier expansion:

$$V(\phi) = (V_3/2)(1 - \cos 3\phi) + (V_6/2)(1 - \cos 6\phi) + \ldots \qquad (1)$$

where V_3 is the height of the threefold barrier; V_6, of the sixfold; and so forth. The V_6 term does not affect the barrier height because the V_6 term is zero at both the potential minima and maxima. The barrier height is therefore given by V_3; V_6 merely changes the shape of the potential well. A positive V_6 makes the minima narrower and the maxima broader which results in the energy levels near the bottom of the well becoming somewhat more widely separated. A negative V_6 term has the opposite effect. Thus the well is broader at the bottom and the levels are more closely spaced than when $V_6 = 0$ (see Fig. 2). Experimentally it has been found that $0 \leq V_6/V_3 < 0.05$ and $V_6 >> V_9 >>$ higher order terms [19]. Thus to a good approximation

$$V(\phi) = (V_3/2)(1 - \cos 3\phi). \tag{2}$$

This potential function can be expressed in a more general manner so as to be applicable to an n-fold barrier, thus

$$V(\phi) = (V_n/2)(1 - \cos n\phi) \tag{3}$$

where V_n is the height of the n-fold barrier.

The Hamiltonian for this system may be expressed as

$$H = H_r + F(p - P)^2 + (V_n/2)(1 - \cos n\phi) \tag{4}$$

where H_r = the rigid rotor Hamiltonian; $F = h^2/8\pi^2 I_r$; I_a = moment of inertia of the internal top; $I_r = I_a[1 - \Sigma_g \lambda_g{}^2(I_a/I_g)]$ = the reduced moment of inertia for internal rotation; I_g = gth principal moment of inertia of the entire molecule; λ_g = cosine of the angle between the axis of the internal top and the gth principal axis of inertia of the entire molecule; p = the angular momentum of the internal top; and (p - P) = the relative angular momentum of the top and the frame.

The eigenvalue problem associated with the torsional motion may be transformed into the well-established Mathieu equation, if the cross term $-2FP_p$ is neglected:

$$(d^2\phi_{\nu\sigma}/dx^2) + (b_{\nu\sigma} - s \cos^2 x)\phi_{\nu\sigma} = 0 \tag{5}$$

where $b_{\nu\sigma} = (4/n^2)E_{\nu\sigma}/F$ = an eigenvalue of the Mathieu equation; $s = (4/n^2)V_n/F$; $2x = n\phi + \pi$; n = foldness of barrier; ν = torsional quantum number; and σ = sublevel index. The energy levels for the hindered-rotational mode are

$$E_{\nu\sigma} = \tfrac{1}{4}n^2 F b_{\nu\sigma}. \tag{6}$$

These levels are n-fold degenerate and yield two sublevels; one is labeled +, and the other -, the signs corresponding to the symmetry of the wavefunction.

The observed vibrational frequency, $\bar{\nu}$, is the difference between two of the energy levels:

$$\bar{\nu} = E_{\nu'\sigma'} - E_{\nu\sigma} = \tfrac{1}{4}n^2 F(b_{\nu'\sigma'} - b_{\nu\sigma}) \ (\text{cm}^{-1}). \tag{7}$$

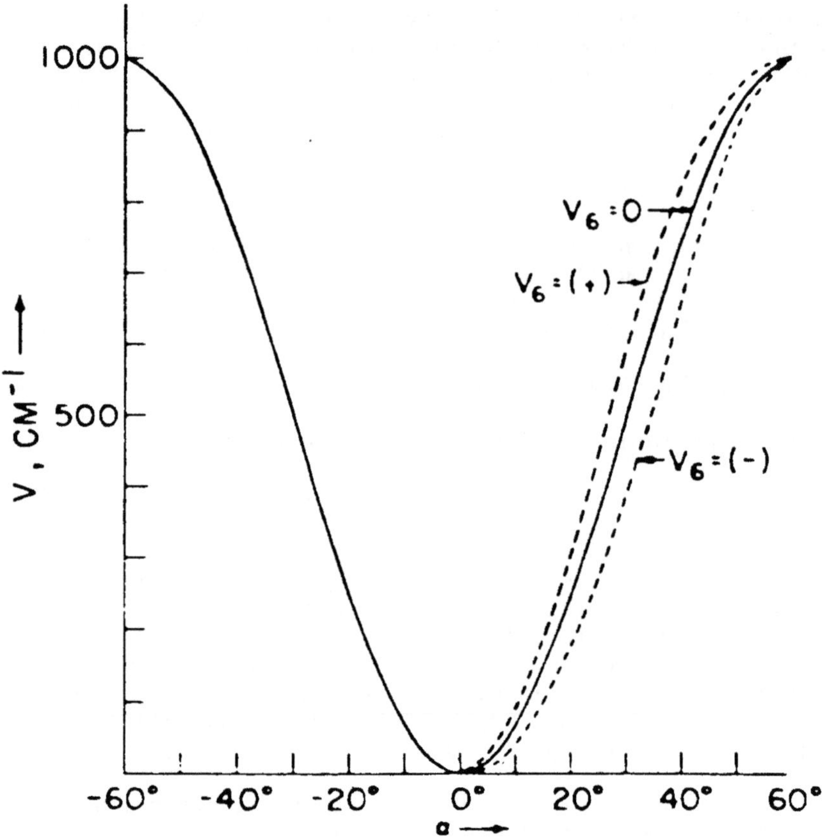

Fig. 2 Effect of V_6 on the shape of the potential curve, $V(\alpha) = (V_3/2)(1 - \cos 3\alpha) + (V_6/2)(1 - \cos 6\alpha)$; α is the angle of rotation; $V_3 = 1000$, $V_6 = +100$, 0, and -100 cm^{-1} (Used by permission, Ref. [21]).

Clearly if the $1 \leftarrow 0$ frequency is known and if there is enough structural information to determine F, $\Delta b_{v\sigma}$ can be calculated. From $\Delta b_{v\sigma}$ a dimensionless parameter s can be obtained from tables of solutions for the Mathieu equation. By employing the definition of s, the barrier height may be calculated:

$$V_n = \tfrac{1}{4} n^2 F s \ (cm^{-1}). \tag{8}$$

If the vibrational frequencies are known and have been assigned correctly, the principal source of error in the calculated barrier height will be in the value of F. This is quite obvious since V_n is a linear function of F and an error in the latter is reflected directly in V_n.

The barrier height may also be obtained by assuming that the torsional oscillation is harmonic. This assumption is valid only if the first excited state is well below the top of the barrier; consequently this method is applicable to problems where the barrier is ≥2.5 kcal/mole. The mathematical treatment of this case is quite straightforward once the cosine term is expressed in series form

$$1 - \cos x = \frac{1}{2!} x^2 - \frac{1}{4!} x^4 + \ldots \qquad (9)$$

where $x = n\phi$. The x^4 and higher order terms are zero for small amplitudes, and the potential energy expression may now be expressed as

$$V(\phi) = \tfrac{1}{4}n^2\phi^2 V_n. \qquad (10)$$

The frequency of the one-dimensional oscillator is

$$\nu = (1/2\pi c)(k/I_r)^{\frac{1}{2}} \qquad (11)$$

and upon rearrangement one observes

$$k = 4\pi^2 c^2 \nu^2 I_r. \qquad (12)$$

The force constant, k, is by definition equal to the second derivative of the potential energy

$$k = d^2V(\phi)/d\phi^2 = \tfrac{1}{2}n^2 V_n. \qquad (13)$$

If (12) is substituted into (13), the resulting expression becomes

$$V_n = \frac{8\pi^2 I_r c^2 \nu^2}{n^2} \text{ (ergs)}. \qquad (14)$$

Division of (14) by hc will convert the units to wavenumbers

$$V_n = \frac{8\pi^2 c I_r \bar{\nu}^2}{n^2 h}. \qquad (15)$$

This expression is quite simple and an estimate of the barrier to internal rotation may be easily computed if the torsional fundamental is observed and the molecular geometry is reasonably well characterized so that F may be ascertained. One should be aware that two assumptions are inherent in these calculations: (1) the torsional mode is not mixed with other vibrations, and (2) the barrier is symmetrical. The experimental evidence accumulated thus far indicates that methyl torsions interact very little with the other fundamentals. Since the torsional fundamental is generally at a very low frequency, any perturbation would probably result in the absorption appearing at a higher wavenumber. This, of course, would result in an apparent higher barrier. It should be reiterated that the torsional modes are apparently not mixed appreciably, particularly since they often shift by a factor of approximately 1.4 upon deuteration. If the barrier is not symmetric, in other words, if all the barriers of the particular torsion are not the same, the potential barrier can no longer be expressed by one parameter. The problem then becomes one of evaluating the contributions of the V_1, V_2, and V_3 terms. Obviously this requires three pieces of experimental data and this topic is extensively covered in another chapter and will not be treated further in this chapter.

We shall now demonstrate how far infrared interferometry is utilized to obtain barriers to internal rotation. The spectrum of ethylchloride is shown in Fig. 3 and the $1 \leftarrow 0$, $2 \leftarrow 1$, $3 \leftarrow 2$ transitions are clearly discernable. Additionally, the $4 \leftarrow 3$ transition shows the A-E splitting expected as the energy levels approach the top of the barrier. From these transitional frequencies, one calculates the V_3 to be 1291 cm^{-1} and the V_6 to be -12 cm^{-1}. Please note that the V_6 term is quite small and that it may either be positive or negative. The observed and calculated transitions are shown in Table 1. This example demonstrates the utility of far infrared data for the determination of barriers to internal rotation. Several compounds are listed in Table 2 for which the torsional barriers have been obtained and it should be noted that trends appear as additional halogens are added to one end of ethane. The addition of one chlorine atom raises the barrier 0.79 kcal/mole whereas the second raises the barrier 0.41 kcal/mole and the third one by 1.27 kcal/mole. Similar effects are observed in the bromine series, where the addition of one, two, and three bromine atoms to ethane raises the barrier 0.78, 0.62 and 1.45 kcal/mole, respectively. Thus the addition of the third chlorine or bromine raises the barrier considerably more than either of the first two halogens, but each halogen does raise the barrier. This observation disproved the belief which apparently arose from the trends in the fluorine series that, after the addition of one halogen, the barrier would not be drastically affected by the substitution of the second or third halogen. Barriers have also been obtained [29] for some of the mixed trihaloethanes, i.e. CH_3CCl_2F and CH_3CBr_2Cl, and these fall at predictable values. Therefore with the increased amount of reliable torsional data available, trends appear to be emerging which are consistent with theoretical predictions.

The internal rotational problem where a molecule contains two C_{3v} rotors is much more complicated and it will be treated in a separate chapter. However it should be mentioned that the

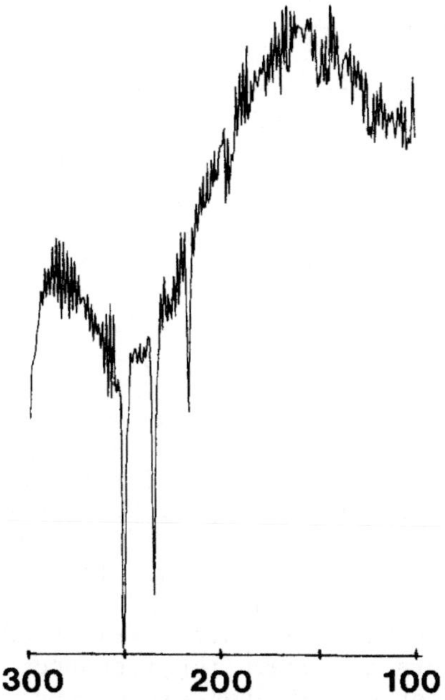

Fig. 3 Far infrared spectrum of gaseous ethylchloride recorded on a Digilab FTS-15B.

Table 1 Torsional Transitions for Ethylchloride

Observed Frequency (cm^{-1})	Assignment $v'' \leftarrow v'$	Calculated Frequency (cm^{-1})	Obsd-Calc
251.0	$1 \leftarrow 0$	251.5	0.5
235.0	$2 \leftarrow 1$	235.5	0.5
217.3	$3 \leftarrow 2$	217.0	0.3
198.0	$4E \leftarrow 3$	197.0	1.0
196.0	$4A \leftarrow 3$	194.0	2.0

far infrared spectra are usually much more complex than those for a single rotor and the correct assignment of the observed transitions is frequently not straightforward.

B. Ring-Bending Motions in Small Molecules

Chemists have been interested in the equilibrium conformation

Table 2 Torsional Barriers (kcal/mole) for Some Haloethanes

Molecule	X = F	X = Cl	X = Br	X = CH_3
CH_3CH_3	2.93 [22][a]	2.93	2.93	2.93
CH_3CH_2X	3.33 [23]	3.72 [24]	3.71 [24]	3.28 [25]
CH_3CHX_2	3.18 [26]	4.13 [27]	4.33 [27]	3.94 [28]
CH_3CX_3	3.19 [29]	5.40 [27]	5.78 [27]	4.30 [30]

[a]Numbers in brackets indicate references.

of four-, five-, six-, and seven-membered rings for some time. Many of the ring-bending vibrations in such molecules have normal modes in the far infrared region of the spectrum and, thus, there are several vibrational energy levels populated at room temperature. From the analysis of the frequencies for these ring-bending motions it is possible to obtain the potential surfaces for the interconversion of the different conformers. Fourier transform spectroscopy has been extensively used along with conventional spectrometers to obtain the frequencies for these ring-bending modes. The methods of analysis of such data along with some examples will now be presented in some detail.

a. Four- and unsaturated five-membered rings

The equilibrium conformation of four- and unsaturated five-membered rings is determined by two major opposing forces. The first is the ring strain which tends to keep the ring skeleton planar. Puckering the ring decreases the already strained angles, thereby increasing the angle strain. The second is the torsional forces. The torsional repulsions of adjacent groups are at a maximum for a planar ring since the groups are eclipsed; therefore, bending the ring out of the plane will reduce these repulsive forces. It is the delicate balance between these two large forces which determine the ground state structure of the molecule.

Bell [31] predicted that the potential function of the out-of-plane bending vibration in a four-membered ring should be quartic in nature, by assuming that the potential energy of this mode is due only to a quadratic potential for bending of the ring angles. This treatment implies that the vibration is one-dimensional and that there is free rotation about a single bond. The first implication is quite reasonable since all other vibrational modes are very much higher in energy and their normal coordinates do not mix appreciably with the ring-puckering motion; however, the assumption of free rotation about a single bond is incorrect. Thus, the addition of a quadratic term to the potential expression to represent the torsional interactions should provide the most accurate representation of the potential energy for this vibration. The resultant expression has the form $V = ax^4 + bx^2$ and is called a mixed quartic-harmonic potential function. A five-membered ring which contains a single double bond is often termed a "pseudo" four-membered ring because the double bond portion of the ring moves as a unit during the ring-puckering vibration; consequently, the mixed quartic-harmonic potential function should also provide a quantitative representation of the potential energy for the ring-bending mode in this system. It will be shown that the ring-puckering potential function and hence the skeletal structure of a small ring can be ascertained by an analysis of the absorption spectrum resulting from the ring-puckering vibration.

The one-dimensional Schrödinger equation for the ring-puckering vibration is given by

$$-(\hbar^2/2\mu)(d^2\psi/dx^2) + (V - E)\psi = 0$$

where

$V = ax^4 + bx^2$ = the potential function

x = puckering coordinate

μ = reduced mass for the vibration in atomic mass units.

Since the amplitude of the puckering depends on the reduced mass, the constants a and b are not transferable from one system to another. The above expression is transformed to a reduced form which is independent of the molecule by the following equations

$$z = (2\mu/\hbar^2)^{1/6}a^{1/6}x,\quad B = (2\mu/\hbar^2)^{1/3}a^{-2/3}b,$$

$$\lambda = (2\mu/\hbar^2)^{2/3}a^{-1/3}E.$$

The resulting equation in the reduced form is

$$(d^2\psi/dz^2) + (\lambda - z^2 - Bz^2)\psi = 0. \qquad (17)$$

The potential function in the dimensionless coordinate z is

$$V = A(z^4 + Bz^2) \qquad (18)$$

where

$$A = (\hbar^2/2\mu)^{2/3}a^{1/3}\ (\mathrm{cm}^{-1}).$$

The potential function which best fits the spectral data is

determined by finding a value of B for which the differences between the eigenvalues are most nearly in the same proportion as the observed frequencies. Once B has been evaluated, the scaling coefficient A can be determined.

Planar rings have ring-puckering potential functions which are either quadratic, quartic, or mixed quartic-quadratic in form and thus consist of only a single minimum. The eigenvalue structure of a harmonic oscillator is well established and corresponds to energy levels which are equally spaced. Consequently, the ground and upper state transitions will occur at the same energy. The far infrared spectrum resulting from a harmonic ring-puckering vibration will therefore consist of only a single, well-defined band. A quartic or mixed quartic-harmonic potential function results when b in the expression above is zero or positive, respectively. The eigenvalues for both expressions diverge to higher energies; however, the larger the a/b ratio, the greater the divergence in the case of a mixed quartic-harmonic oscillator. For these potential functions, the absorption spectrum is characterized by a series of transitions which decrease in relative intensity upon going to higher energies.

A nonplanar ring has a barrier to the planar conformation and this results in the potential function having a double minimum. An inversion barrier is mathematically introduced when b in the potential expression has a negative value, because the quadratic term will dominate the potential function for small x. The barrier height is defined as the difference between the maximum point in the hump and the minimum point in the well. The minimum and maximum points may be determined by differentiating the expression (for the potential function given above) with respect to x and setting the result equal to zero:

$$x(4ax^2 - 2b) = 0. \tag{19}$$

Thus, $x = 0$ at the maximum point and $\pm(b/2a)^{\frac{1}{2}}$ at the minimum. The barrier height then becomes

$$V = b^2/4a \ (\mathrm{cm}^{-1}). \tag{20}$$

Although the barrier pushes all the eigenvalues to higher energies, the symmetric levels are perturbed more than the antisymmetric levels. This perturbation difference results because the symmetric levels have a maximum probability at the center of the well, whereas the antisymmetric levels have nodes at the well center. The levels then appear to approach each other in pairs and in the limit of an infinite barrier the levels are observed to be degenerate. The far infrared spectrum resulting from the ring-bending mode of a nonplanar ring is often extensive and the energy level pattern is quite complex. Thus, the appearance of an irregular progression of Q branches is immediately suggestive of a puckered ring. Additionally, relatively high resolution is needed and FT spectroscopy is a convenient way of obtaining the desired data.

The far infrared spectrum of 1,3-disilacyclobutane [32] recorded with an FT vacuum spectrometer is shown in Fig. 4. The assignment for the observed bands are listed in Table 3. Additionally, Irwin et al. [32] obtained the far infrared spectrum of 1,3-disilacyclobutane-1,1,3,3-d_4 as well as the gas phase Raman spectrum of both compounds. The observed transitions were reproduced with the reduced potential $V = (15.2)[Z^4 - 4.79Z^2]$ cm^{-1} which gave a barrier of 87 cm^{-1} and from the dimensional potential function they [32] obtained a dihedral angle of 24 ± 2°. This potential function is shown in Fig. 5.

The inversion barriers and dihedral angles for several four-membered rings and cyclopentene are listed in Table 4. The parameters a and b in the potential functions for these molecules have been shown to arise primarily from angle strain and torsional forces, respectively, and for an in-depth discussion the interested reader should consult Refs. [40] and [41].

A single substituent on the cyclobutane ring leads to conformational isomers and introduces asymmetry into the potential function for ring inversion. As the ring inverts, the substituent goes from the axial conformation to the equatorial, or vice versa. These two conformations will of course have different energies. The observed spectra for this type of conformational inversion can be interpreted using a potential function of the form: $V(x) = ax^4 - bx^2 - cx^3$, where x is the coordinate of the ring inversion. For the four-membered rings, 2x corresponds to the distance between the two ring bisectors. The cubic term is added to the usual quartic-harmonic potential function because of the asymmetry. For a review of the interpretation of the spectra of such molecules, the reader should see Refs. [8] and [42].

b. Five-membered rings

In five-membered ring molecules, there are two low-frequency out-of-plane ring motions. These are usually qualitatively described as the ring-twisting (radial mode) and the ring-puckering (pseudorotational mode) motions. Initially, in order to handle the interpretation of the low-frequency far infrared data, assumptions were usually made as to the form of these normal vibrations. If the five-membered ring contains an endocyclic double bond, it has usually been assumed that there is no interaction between the "high" frequency ring-twisting mode which falls around 400 cm^{-1} and the "low" frequency ring-puckering mode which is near 100 cm^{-1}. Therefore, the two out-of-plane ring modes are handled as two

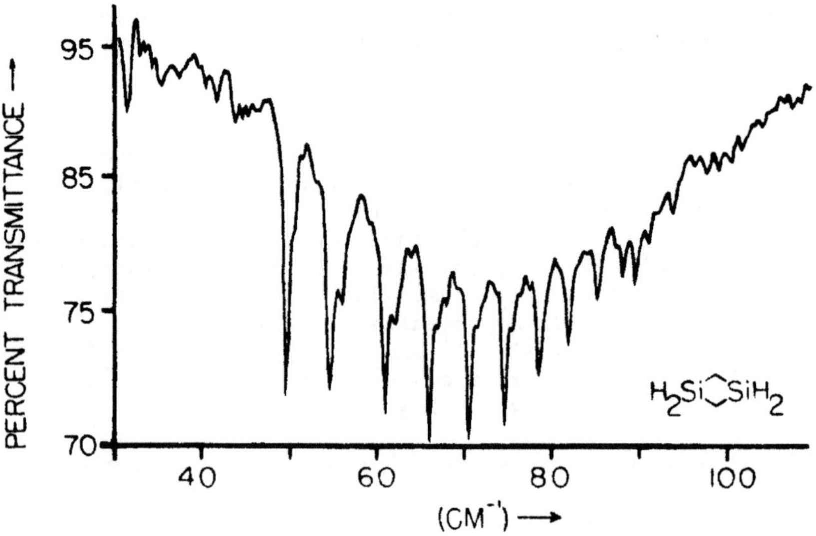

Fig. 4 Far infrared spectrum of 1,3-disilacyclobutane: 140 Torr vapor pressure, 15 cm pathlength, 0.5 cm^{-1} resolution (Used by permission, Ref. [32]).

one-dimensional problems with the anharmonic, low-frequency ring-puckering transitions interpreted in terms of a one-dimensional potential function of the form $V(X) = AX^4 \pm BX^2$. A review of the spectral studies of these molecules has recently been given by Wurrey _et al_. [43].

If the five-membered ring contains no endocyclic double bonds, then the two out-of-plane ring motions can usually couple and the Hamiltonian for this motion is generally written as a two-dimensional problem in Cartesian coordinates. The two-dimensional Cartesian Hamiltonian is then transformed to polar coordinates, q and ϕ. If one assumes that the barrier to planarity is very high, this two-dimensional Hamiltonian can be simplified by separating it into two one-dimensional Hamiltonians as suggested by Kilpatrick, Pitzer and Spitzer [44]. One of these coordinates is essentially a ϕ-dependent pseudorotational motion and the other coordinate is a q-dependent radial mode which is a pseudorotational amplitude coordinate and initially assumed to be harmonic. The anharmonic low-frequency transitions observed in the far infrared spectral region for this type of ring have been explained in terms of a ϕ-dependent Hamiltonian of the form: $H = \beta P_\phi^2 + \Sigma_n(V_n/2)(1 - \cos n\phi)$ and these transitions are commonly referred to as pseudorotational transitions.

Table 3 Ring-Puckering Spectra of 1,3-Disilacyclobutane[a]

	Frequency (cm^{-1})		Relative Intensity	
Transition	Observed	Calculated	Observed	Calculated
1 ← 0	--	2.99	--	0.01
2 ← 1	56.0	56.1	0.56	0.50
3 ← 2	-31.6	31.3	~0.4	0.38
4 ← 3	49.5	50.0	1.33	0.86
5 ← 4	54.5	55.0	1.10	0.95
6 ← 5	61.0	61.1	(1.00)	1.00
7 ← 6	66.0	66.1	0.98	0.95
8 ← 7	70.6	70.6	0.87	0.84
9 ← 8	74.7	74.7	0.77	0.71
10 ← 9	78.5	78.3	0.56	0.58
11 ← 10	82.0	81.7	0.46	0.45
12 ← 11	85.3	84.9	0.28	0.34
13 ← 12	88.2	87.8	0.21	0.25
14 ← 13	91.2	90.6	0.11	0.18
15 ← 14	93.9	93.2	0.11	0.13
3 ← 0	89.6	90.4	0.27	0.40

[a]Data taken from Ref. [32].

Carreira _et al._ [45] have shown that the assumption of separability of the twisting and puckering modes is not entirely valid even when the ring has an endocyclic double bond. They have shown that the fourth power cross term connecting the two vibrations is significantly large and should probably be used for all cyclopentene-like rings. Ikeda _et al._ [46] have investigated the effects of a finite barrier to planarity on unhindered pseudorotation in saturated five-membered rings. They concluded that a finite barrier would show negative curvature from a straight line in a plot of the pseudorotational energy level spacings versus the pseudorotational quantum number. Also the q-dependent radial mode should show negative anharmonicity.

The setup of the Hamiltonian for the two-dimensional problem is the same as that presented by Carreira _et al._ [45] and it will only be briefly outlined here. The Hamiltonian for the two out-of-plane ring modes is written as:

$$H = (1/2m_1)P_{R_1}^{\,2} + (1/2m_2)P_{R_2}^{\,2} + V(R_1,R_2). \qquad (21)$$

Here R_1 is the ring-puckering coordinate and R_2 is the ring-twisting coordinate, P_{R1} and P_{R2} are the corresponding moments, and m_1 and m_2 are the effective masses. Allowing the possibility of a double minimum potential function for each of the two vibrational modes, a potential function for at least C_{2v} symmetry can

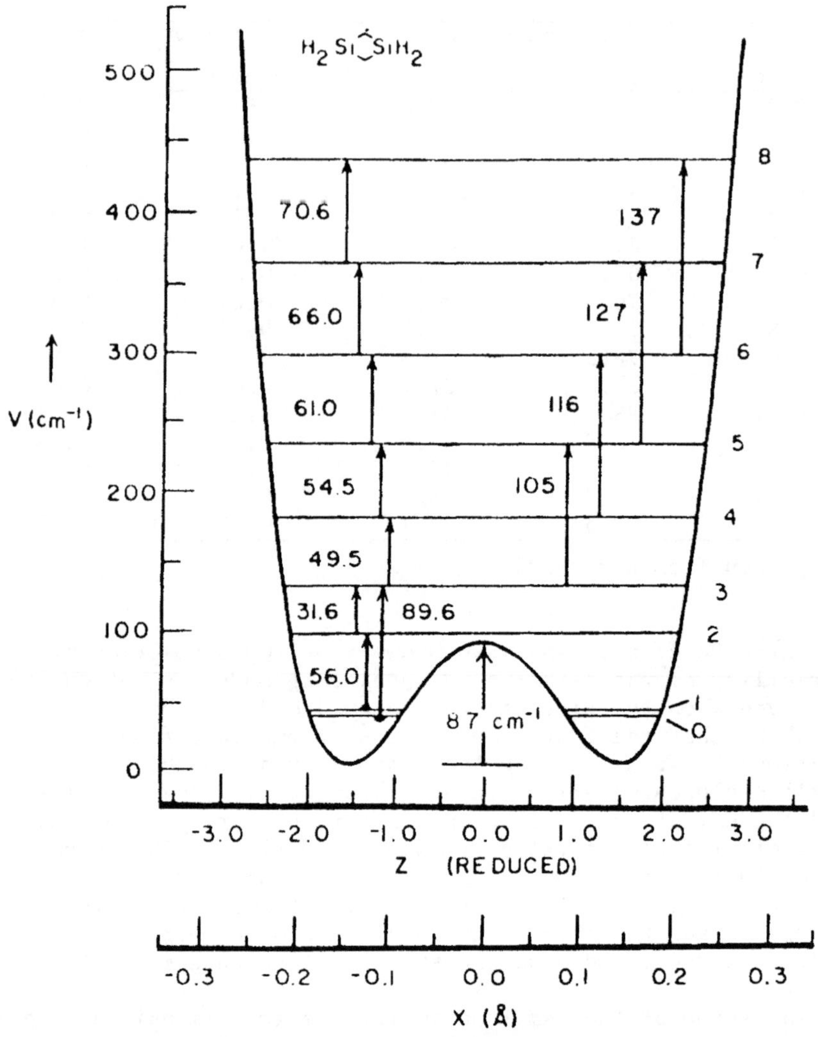

Fig. 5 Potential energy function showing the infrared ($\Delta v = 1$ or 3) and Raman ($\Delta v = 2$) transitions of 1,3-disilacyclobutene (Used by permission, Ref. [32]).

be written as

$$V(R_1,R_2) = A_1R_1^4 + A_2R_2^4 + B_1R_1^2 + B_2R_2^2 + C_{12}R_1^2R_2^2 \tag{22}$$

Table 4. Ring-Puckering Potential Energy Functions: $V = aX^4 - bX^2$

Molecule	$a(cm^{-1}/A^4)$	$b(cm^{-1}/A^2)$	Barrier	Dihedral angle	Ref.
Cyclobutane	7.6×10^5	39.6×10^3	515	33°	[33]
Silacyclobutane	3.5×10^5	24.9×10^3	440	37°	[34]
1,3-Disilacyclobutane	2.3×10^5	9.0×10^3	87	24°	[32]
Cyclobutanone	4.3×10^5	2.9×10^3	5	~0°	[35]
Methylene cyclobutane	4.1×10^5	15.2×10^3	141		[36]
Oxetane	7.2×10^5	6.6×10^3	15	~0°	[37]
Thietane	5.1×10^5	23.5×10^3	274	33°	[35]
Selenetane	4.3×10^5	25.4×10^3	378	32.5°	[38]
Cyclopentene	12.34×10^5	33.8×10^3	232	23°	[39]

where A_1, A_2, B_1 and B_2 are the potential constants in dimensional form and C_{12} is the cross-term potential constant.

Transforming first to mass-weighted coordinates, Q_i,

$$R_1 = m_1^{-\frac{1}{2}}Q_1 \tag{23}$$

$$R_2 = m_2^{-\frac{1}{2}}Q_2$$

the Hamiltonian becomes

$$H = (\tfrac{1}{2})P_1^2 + (\tfrac{1}{2})P_2^2 + a_1Q_1^4 + a_2Q_2^4 + b_1Q_1^2 + c_{12}Q_1^2Q_2^2 + b_2Q_2^2. \tag{24}$$

For computational convenience one transforms to dimensionless coordinates, q_i,

$$q_1 = \gamma_1^{\frac{1}{2}}Q_1 \tag{25}$$

$$q_2 = \gamma_2^{\frac{1}{2}}Q_2$$

and the Hamiltonian then becomes

$$H = (\tfrac{1}{2})\hbar^2\gamma_1p_1^2 + (\tfrac{1}{2})\hbar^2\gamma_2p_2^2 + \alpha_1q_1^4 + \alpha_2q_2^4 + \beta_1q_1^2 + \beta_2q_2^2 + \delta_{12}q_1^2q_2^2 \tag{26}$$

or

$$H = (\tfrac{1}{4})\hbar^2(\gamma_1 + \gamma_2)(p_1^2 + p_2^2) + (\tfrac{1}{4})\hbar^2(\gamma_1 - \gamma_2)(p_1^2 - p_2^2) + V(q_1,q_2)$$

where p_i is the momentum conjugate to q_i, and

$$V(q_1,q_2) = A(q_1^2 + q_2^2)^2 + B(q_1^2 + q_2^2) + C(q_1^2 - q_2^2) + D(q_1^2 + q_2^2)(q_1^2 - q_2^2) + E[2(q_1^2 - q_2^2)^2 - (q_1^2 + q_2^2)^2]. \tag{27}$$

When equation (27) is written in terms of polar coordinates, q and ϕ,

$$q_1 = q \cos \phi \tag{28}$$

$$q_2 = q \sin \phi$$

the potential expression takes the form

$$V(q,\phi) = Aq^4 + Bq^2 + Cq^2 \cos 2\phi + Dq^4 \cos 2\phi + Eq^4 \cos 4\phi \qquad (29)$$

where

$$A = (1/8)[3(A_1 m_1^{-2} \gamma_1^{-2} + A_2 m_2^{-2} \gamma_2^{-2} + C_{12}(m_1 m_2 \gamma_1 \gamma_2)^{-1}]$$

$$B = (1/2)[B_1 m_1^{-1} \gamma_1^{-1} + B_2 m_2^{-1} \gamma_2^{-1}]$$

$$C = (1/2)[B_1 m_1^{-1} \gamma_1^{-1} - B_2 m_2^{-1} \gamma_2^{-1}]$$

$$D = (1/2)[A_1 m_1^{-2} \gamma_2^{-2} - A_2 m_2^{-2} \gamma_2^{-2}]$$

$$E = (1/8)[A_1 m_1^{-2} \gamma_1^{-2} + A_2 m_2^{-2} \gamma_2^{-2} - C_{12}(m_1 m_2 \gamma_1 \gamma_2)^{-1}].$$

The potential constants in the two-dimensional potential function (Eq. 9) are numerically evaluated by use of the computer program developed and described by Carreira et al. [45]. The dimensionless potential constants are varied in a least-squares manner in order to simultaneously fit the observed transitions for the ring-puckering and twisting modes in these five-membered rings.

The ring-puckering vibrations are usually observed in the far infrared spectra whereas the ring-twisting modes are observed in the Raman effect. The first saturated five-membered rings to be studied were 1,3-dioxolane [47] and tetrahydrofuran [48]. In these early studies it was assumed that the barrier to pseudo-rotation was zero or near zero and as previously pointed out, the radial and angular equations were separable. Several recent studies have shown that even though the twisting and bending motions are fairly independent, the one-dimensional approximation leads to a much too high energy for interconversion and also exaggerates the actual molecular path for interconversion.

The transitions for the ring-puckering mode of thiacyclopentane are shown in Fig. 6 and the observed frequencies are listed in Table 5. The original data were treated [49] on the basis of the one-dimensional model which gave a barrier of 773 cm^{-1}. At this time excited states of the twisting mode have not been observed so the two-dimensional treatment has not been carried out. However the two-dimensional analysis has been given for cyclopentane [50], silacyclopentane [51], germacyclopentane [52] selenacyclopentane [53], and cyclopentanone [54] and a comparison of the two different treatments is given in Table 6. One can then see that, by a combination of far infrared data ($\Delta v = 1$) for the ring-puckering vibration (pseudorotational mode) and the $\Delta v = 1$ transitions for the ring-twisting mode (radial mode)

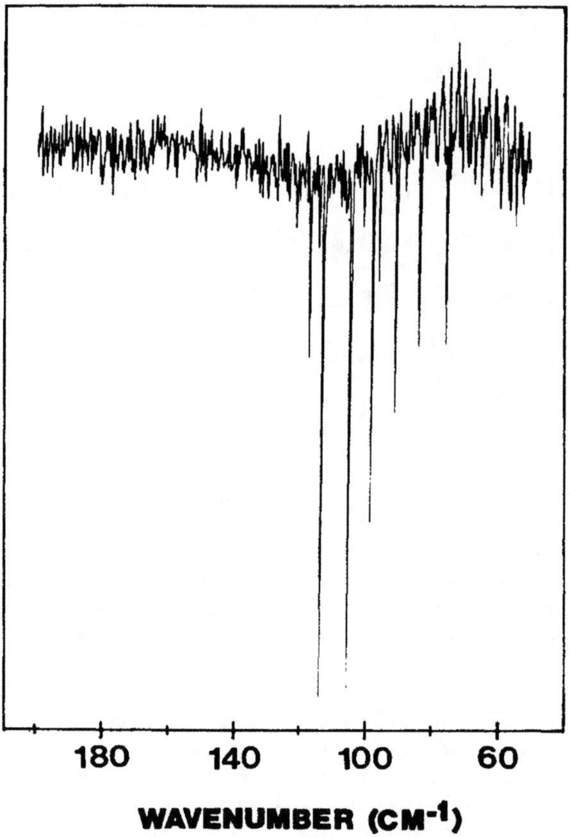

Fig. 6 Far infrared spectrum of gaseous thiacyclopentane recorded on a Digilab FTS-15B.

obtained from the Raman effect for saturated five-membered rings, it is possible to observe the effects of a finite barrier to planarity and to calculate the two-dimensional potential surface for the two low-frequency out-of-plane modes for these molecules. On the basis of a very limited number of cases it would seem that the one-dimensional model breaks down as the path to interconversion approaches one which involves a planar intermediate. Unfortunately, it is usually not possible to determine a priori if a twisted molecule interconverts through a planar or puckered intermediate. For a more detailed discussion on this topic, the reader is referred to Ref. [55].

c. Six- and higher-membered rings

Six-membered rings which contain one double bond have two

Table 5 Observed versus Calculated Frequencies for the Pure Pseudorotational Transitions[a]

Assignment	Observed	Calculated[b]	Δ
±1 ← 0	112.3	110.9	+1.4
±2 ← ±1	104.2	104.8	-0.6
±3 ← ±2	97.5	98.3	-0.8
±4 ← ±3	90.6	91.3	-0.7
±5 ← ±4	83.4	83.6	-0.2
±6 ← ±5	75.4	75.0	+0.4
±7 ← ±6	65.4	65.1	+0.3
-8 ← -7	53.1	53.4	-0.3
+8 ← +7	50.6	51.6	-1.0
-9 ← -8	44.7	42.2	+2.5
±2 ← 0	216.4	216.5	-0.1

[a]Taken from Ref. [49].

[b]$V(\phi) = -\frac{1}{2}[772.7)(1 - \cos 2\phi) + \frac{1}{2}(169.1)(1 - \cos 4\phi)$ and $B = 2.35\ cm^{-1}$.

large-amplitude, low-frequency ring deformations: a ring bend and a ring twist, along with a third out-of-plane motion which is a torsion about the double bond. This torsion has a much higher frequency and therefore, does not couple with the other two out-of-plane ring motions. However the two out-of-plane ring modes are strongly coupled and the two-dimensional treatment must be used. Six-membered rings with two double bonds or one double bond and a fused epoxide ring have one low-frequency puckering mode which does not appreciably couple with the other ring-bending modes so the one-dimensional Hamiltonian is appropriate for the ring inversions. There have been many studies of such molecules utilizing FT spectroscopy and usually the ring-puckering motion is relatively strong in the far infrared. However, the higher frequency ring modes may not be observed in the far infrared but difference bands with the low-frequency ring-puckering motions may give rise to far infrared absorptions. From such studies it has been possible to determine the most stable conformation as well as the barriers to interconversion and the lowest energy path for inversion. For a more complete discussion of this topic see Ref. [8].

For seven- and eight-membered rings, most of the far infrared studies have been done by Strauss and co-workers [56] at Berkeley utilizing an FT spectrometer. For example, cycloheptanone was found to be nonrigid with a pseudorotational energy surface whereas cycloheptane, oxepane, and 1,3-dioxepane showed no noticeable effects of nonrigidity. For the eight-membered rings, the far infrared spectra have helped to confirm the boat-chair conformations determined by other methods but the far infrared

Table 6 Comparison of Barriers Between the One-Dimensional and Two-Dimensional Treatments

Molecule	One-dimensional barrier (cm^{-1}) to interconversion	Two-dimensional barrier (cm^{-1}) to planarity	Two-dimensional Barrier (cm^{-1}) to interconversion	Ref.
Cyclopentane	--	1824	--	[50]
Silacyclopentane	1390	1558.6	1414	[51]
Germacyclopentane[a]	2043	1474.4	1454	[52]
Selenacyclopentane[a]	1882	1693	1693	[53]
Cyclopentanone[a]	1303	750		[54]

[a] The interconversion for these molecules is through the planar conformation whereas the pathway for interconversion for the other molecules takes place through a slightly puckered configuration.

data provide a means for obtaining the parameters for a model of the conformations. For further details on the stable conformation of seven- and eight-membered rings, the interested reader should refer to the recent review article on this subject by Rounds and Strauss [56].

C. OTHER APPLICATIONS

The previous discussions on the applications of FT-IR spectroscopy to the far infrared spectral region have undoubtedly reflected the authors' personal interests. There are several other areas of spectroscopy whre the far infrared region yields unique information on:

1. nonspecific far infrared absorption in liquids
2. weak, intermolecular interactions
3. intermolecular vibrations of molecular crystals
4. ionic crystals
5. low-temperature effects

The latter three topics will be covered in other chapters. In the area of weak intermolecular interactions, most of the studies have been on hydrogen bonded molecules or charge-transfer complexes. Griffiths [7] has recently discussed this topic. There have been several far infrared studies of polar liquids which show strong, broad absorptions below 100 cm^{-1}. This absorption cannot be attributed to intramolecular vibrations or pure rotations of the molecules. However other explanations have been proposed such as: (1) stretching modes of dipole-dipole complexes; (2) vibrations of a "liquid lattice"; (3) hindered rotation of the molecule in a cage of its neighbors. The first explanation seems unlikely but the latter two have considerable merit and studies are continuing in this area.

Clearly the increasing application of Fourier transform spectroscopy has resulted in a marked improvement in the far infrared data obtainable. With such data it has been possible to obtain more accurate barriers to internal rotation, better parameters for ring inversions, and provided the necessary information for the interpretive theories of many solid-state phenomena.

REFERENCES

1. R. T. Hall and J. M. Dowling, J. Mol. Spectrosc., 19, 108 (1966).

2. R. T. Hall and J. M. Dowling, J. Chem. Phys., 45, 1899 (1966).

3. K. D. Möller, V. P. Tomaselli, L. R. Skube, and B. K. McKenna, J. Opt. Soc. Amer., 35, 1233 (1965).

4. R. B. Barnes, Phys. Rev., 47, 658 (1935).

5. N. Wright and H. M. Randall, Phys. Rev., 44, 391 (1933).

6. P. R. Griffiths and H. W. Thompson, Spectrochim. Acta, 24A, 1325 (1968).

7. P. R. Griffiths, Chemical Infrared Fourier Transform Spectroscopy, Wiley (Interscience), New York, (1975) p. 229.

8. J. R. Durig and A. W. Cox, in Fourier Transform Infrared Spectroscopy, Vol. 1, J. R. Ferraro and L. J. Basile, editors, New York, Academic Press (1978).

9. K. D. Möller and W. G. Rothschild, Far-Infrared Spectroscopy, Wiley (Interscience), New York (1971).

10. R. T. Hall, D. Vrabec, and J. M. Dowling, Appl. Opt., 5, 1149 (1966).

11. H. Yoshinaga, S. Fujitz, S. Minami, A. Mitsuishi, R. A. Oetjen, and Y. Yamada, J. Opt. Soc. Amer., 48, 315 (1958).

12. D. W. Robinson, J. Opt. Soc. Amer., 49, 966 (1959).

13. H. A. Gebbie, N. W. B. Stone, G. Topping, E. K. Gora, S. A. Clough, and F. X. Kneizys, J. Mol. Spectrosc., 19, 7 (1966).

14. A. J. Sievers and M. Tinkham, Phys. Rev., 129, 1995 (1963).

15. J. R. Durig, S. M. Craven, and J. Bragin, J. Chem. Phys., 52, 2046 (1970).

16. J. R. Durig, S. M. Craven, and W. C. Harris in Vibrational Spectra and Structure, Vol. 1, J. R. Durig, editor, Marcel Dekker, Inc. (1972), p. 147.

17. J. E. Kilpatrick and K. S. Pitzer, J. Chem. Phys., 17, 1064 (1949).

18. D. R. Herschbach, J. Chem. Phys., 31, 91 (1959).

19. W. G. Fateley and F. A. Miller, Spectrochim. Acta, 17, 857 (1961).

20. W. G. Fateley and F. A. Miller, Spectrochim. Acta, 18, 977 (1962).

21. W. G. Fateley and F. A. Miller, Spectrochim. Acta, 19, 611 (1963).

22. S. Weiss and G. E. Leroi, J. Chem. Phys., 48, 962 (1968).

23. G. Sage and W. Klemperer, J. Chem. Phys., 39, 371 (1963).

24. J. R. Durig, W. E. Bucy, L. A. Carreira, and C. J. Wurrey, J. Chem. Phys., 60, 1754 (1974).

25. J. R. Durig, P. Groner, and M. G. Griffin, J. Chem. Phys., 66, 3061 (1977).

26. D. R. Herschbach, J. Chem. Phys., 25, 359 (1956).

27. J. R. Durig, W. E. Bucy, and C. J. Wurrey, J. Chem. Phys., 60, 3293 (1974).

28. J. R. Durig, S. M. Craven, and J. Bragin, J. Chem. Phys., 53, 38 (1970).

29. J. R. Durig, W. E. Bucy, and C. J. Wurrey, J. Chem. Phys., 63, 5498 (1975).

30. J. R. Durig, S. M. Craven, and J. Bragin, J. Chem. Phys., 52, 2046 (1970).

31. R. P. Bell, Proc. Roy. Soc. London, A183, 328 (1945).

32. R. M. Irwin, J. M. Cooke, and J. Laane, J. Amer. Chem. Soc., 99, 3273 (1977).

33. J. M. R. Stone and I. M. Mills, Mol. Phys., 18, 631 (1970).

34. J. Laane and R. C. Lord, J. Chem. Phys., 48, 1508 (1968).

35. T. R. Borgers and H. L. Strauss, J. Chem. Phys., 45, 947 (1966).

36. J. R. Durig, A. C. Shing, L. A. Carreira, and Y. S. Li, J. Chem. Phys., 57, 4398 (1972).

37. S. I. Chan, T. R. Borgers, J. W. Russell, H. L. Strauss, and W. D. Gwinn, J. Chem. Phys., 44, 1103 (1966).

38. A. B. Harvey, J. R. Durig, and A. C. Morrissey, J. Chem. Phys., 50, 4949 (1969).

39. J. Laane and R. C. Lord, J. Chem. Phys., 47, 4941 (1967).

40. J. Laane, J. Chem. Phys., 50, 1946 (1969).

41. J. Laane, Quant. Rev., 25, 533 (1971).

42. C. S. Blackwell and R. C. Lord, Vibrational Spectra and Structure, Vol. 1, J. R. Durig, editor, Marcel Dekker, Inc., New York (1972) pp. 1-24.

43. C. J. Wurrey, J. R. Durig and L. A. Carreira, Vibrational Spectra and Structure, Vol. 5, J. R. Durig, editor, Elsevier Scientific Publishing Co., Amsterdam (1976) pp. 176-221.

44. J. E. Kilpatrick, K. S. Pitzer and R. Spitzer, J. Amer. Chem. Soc., 69, 2483 (1947).

45. L. A. Carreira, I. M. Mills, and W. B. Person, J. Chem. Phys., 56, 1444 (1972).

46. T. Ikeda, R. C. Lord, T. B. Malloy, Jr., and T. Ueda, J. Chem. Phys., 56, 1434 (1972).

47. G. G. Engerholm, A. C. Luntz, W. D. Gwinn, and D. O. Harris, J. Chem. Phys., 50, 2446 (1969).

48. J. A. Greenhouse and H. L. Strauss, J. Chem. Phys., 50, 124 (1969).

49. D. W. Wertz, J. Chem. Phys., 51, 2133 (1969).

50. L. A. Carreira, G. J. Jiang, W. B. Person, and J. N. Willis, Jr., J. Chem. Phys., 56, 1440 (1972).

51. J. R. Durig, W. J. Natter, and V. F. Kalasinsky, J. Chem. Phys., 67, 4756 (1977).

52. J. R. Durig, Y. S. Li, and L. A. Carreira, J. Chem. Phys., 58, 2393 (1973).

53. J. R. Durig and W. J. Natter, J. Chem. Phys., 69, 3714 (1978).

54. T. Ikeda and R. C. Lord, J. Chem. Phys., 56, 4450 (1972).

55. L. A. Carreira, R. C. Lord and T. M. Malloy, Jr., Topics in Current Chemistry, Vol. 82, Springer-Verlag, Berlin (1979) pp. 1-95.

56. T. C. Rounds and H. L. Strauss, in Vibrational Spectra and Structure, Vol. 7, J. R. Durig, editor, Elsevier Scientific Publishing Co., Amsterdam (1978) pp. 238-268.

REVIEW ON BASIC CONCEPTS OF LOW FREQUENCY MODES IN CRYSTALS

A. Hadni

Laboratoire Infrarouge Lointain
Universite de Nancy I
Nancy, France

1. INTRODUCTION

Before considering the results obtained on different cyrstals, a summary on the interactions of far infrared with lattice vibrations introducing the notions of soft-mode, Debye relaxation and polaritons is needed. We shall also look carefully at the experimental methods which up to now have severely limited the far infrared investigation of crystals. This kind of spectroscopy is sometimes a problem with highly absorbing materials, large damping, and very low frequencies. It is so difficult that recently Raman spectroscopy has been more and more used to study low frequency modes in crystals. Raman spectroscopy has been so dramatically improved by the use of lasers (especially multiwatts argon lasers) that it is much faster than far infrared spectroscopy. It needs samples of much smaller areas and avoids the problem of making very thin plates for the study of transmission spectra when the absorption coefficient is high. However the Raman spectroscopy of low frequencies ($\sigma < 20\ cm^{-1}$) is still uncertain because of the proximity of the Rayleigh diffusion line and, more definitely, the information of infrared and Raman spectroscopy are not always equivalent but are often complementary. We shall see that the far infrared techniques can be improved considerably, and far infrared study of low frequency modes in crystals is still a worthwhile challenge.

2. SUMMARY OF ELECTRO-MAGNETIC WAVES PROPAGATION

The Maxwell field equations can be written in a simplified way when dielectric and magnetic permeabilities ε and μ respec-

J. R. Durig (ed.), Analytical Applications of FT-IR to Molecular and Biological Systems, 391–431.

tively are both constants relative to time or space. We shall also confine our study to a place devoid of free charges ($\rho = 0$, and hence div E = 0):

$$\text{curl } \hat{E} = -\mu \frac{\partial \hat{H}}{\partial t} \tag{1}$$

$$\text{curl } \hat{H} = \sigma \hat{E} + \varepsilon \frac{\partial \hat{E}}{\partial t} \tag{2}$$

$$\text{div. } \hat{E} = 0 \tag{3}$$

$$\text{div. } \hat{H} = 0 \tag{4}$$

Combining Eqs. (1) and (2) and taking account of Eqs. (3) and (4) we get the partial derivative equation:

$$\frac{\partial^2 u}{\partial x^2} + \frac{\partial^2 u}{\partial y^2} + \frac{\partial^2 u}{\partial z^2} = \sigma\mu \frac{\partial u}{\partial t} + \varepsilon\mu \frac{\partial^2 u}{\partial t^2} \tag{5}$$

where u is any component of either $\hat{E}$ or $\hat{H}$. In the case of an electro-magnetic source far away in the x direction it simplifies to:

$$\frac{\partial^2 u}{\partial x^2} = \sigma\mu \cdot \frac{\partial u}{\partial t} + \varepsilon\mu \frac{\partial^2 u}{\partial t^2} \tag{6}$$

and a plane wave solution is written:

$$u = u_o \, e^{j\omega(t - \frac{\bar{n}x}{c})} \tag{7}$$

$c = (\varepsilon_o \mu_o)^{\frac{1}{2}}$ is the phase velocity in vacuum; $\bar{n} = n - jk$ is the complex refractive index, with

$$\bar{n}^2 = \varepsilon_R \mu_R - (j\sigma\mu_R/\varepsilon_o\omega) \tag{8}$$

and recalling that ε_R and σ may be complex numbers, we have:

$$u = u_o \, e^{j\omega(t - \frac{njc}{c})} \, e^{-\frac{\omega k x}{c}}$$

or

$$u = u_o\, e^{-\frac{Kx}{2}} \cdot e^{j\omega(t - \frac{nx}{c})}$$

with

$$K = 2\,\frac{\omega k}{c} \tag{9}$$

being the absorption coefficient which may be expressed in cm^{-1} while k is devoid of any dimension.

Note that in the case of a non conducting material, with $\mu_R = 1$, we get:

$$\bar{n}^2 = \varepsilon_R. \tag{10}$$

From Eqs. (1) and (2) it is also shown* that the plane electromagnetic wave is generally a transverse wave (Fig. 1),

with $E = E_o\, e^{j\omega t}$, $H = H_o\, e^{j\omega t}$ and $\frac{E_o}{H_o} = Z$,

Z being the characteristic impedance of the medium:

$$Z = \frac{\mu}{\varepsilon - \frac{j\sigma}{\omega}} \tag{11}$$

This is an exception for $\varepsilon = 0$, in that case the wave is purely electric ($H = 0$), and E is longitudinal (Fig. 2) with $\hat{P} = -\varepsilon_o \hat{E}$ which is the equivalent (Fig. 3) to the perfect diamagnetism of superconductors. It will be shown later that such a null value for ε (and also $n = 0$, and $k = 0$) is only possible for a few

*See Ref. [6], p. 149 for establishing first

$$\hat{E} = \frac{-\hat{k}\,(\hat{P} \cdot \hat{K}) + \hat{P}\,\frac{\omega^2}{c^2}}{\varepsilon_o(k^2 - \frac{\omega^2}{c^2})},$$

and then considering the case $\hat{P} \cdot \hat{K} = 0$ (transverse wave), $\hat{P}$ being the electric polarization, and $\hat{k}$ the wave vector, with $k = 27k/\lambda$.

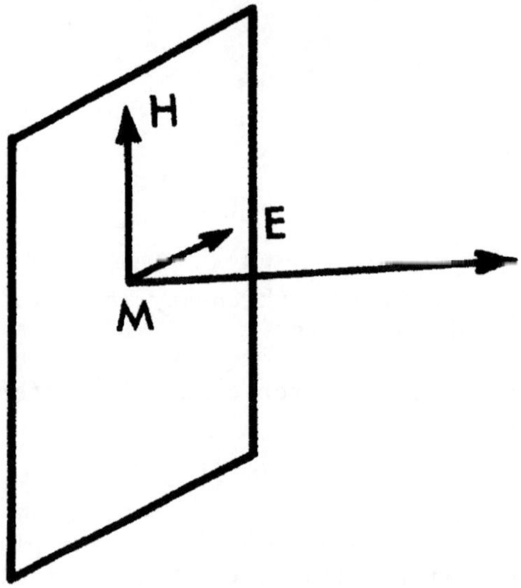

Fig. 1 Structure of a plane transverse electro-magnetic wave.

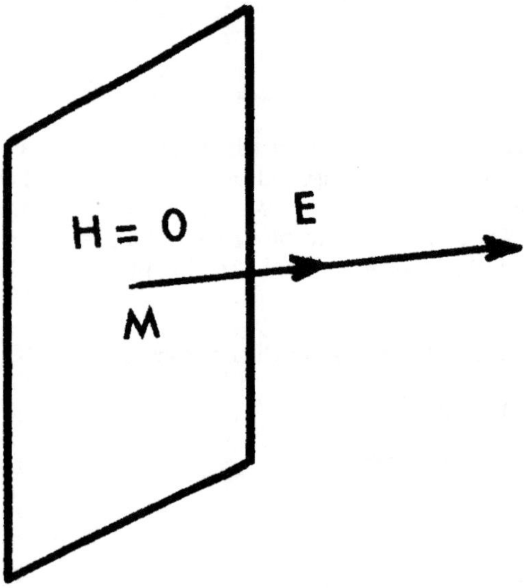

Fig. 2 Structure of a plane longitudinal electro-magnetic wave.

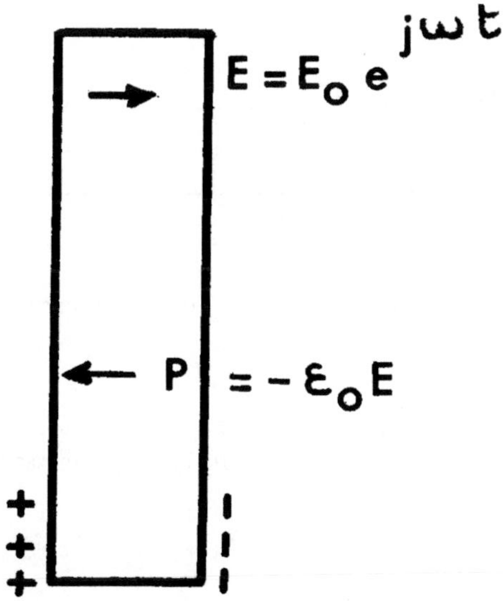

Fig. 3 The case of a longitudinal electro-magnetic wave needs $\varepsilon = 0$, i.e. $n = k = 0$: all atoms are vibrating in phase in the medium.

frequencies ω_L which can be calculated from the microscopic structure of the medium. Such a longitudinal electric wave can be effectively excited when irradiating a crystal plate for non normal incidence with frequency ω_L (see next chapter).

3. LATTICE VIBRATIONS

A. Wavelengths Supporting the Elastic Vibrations

Let us consider a crystal made of N atoms i.e. possessing three N degrees of freedom. Because of thermal agitation each atom will exhibit small displacements around its equilibrium position. It is assumed that this thermal agitation is the addition of elastic standing waves in resonance within the solid boundaries. In the simplest case of a linear chain of molecules (Fig. 4) we write that the length L is equal to an integer number of half wavelengths Λ:

$$L = n \frac{\Lambda}{2}, \text{ with } n = 0; 1; \ldots.$$

For a continuum there is an infinite number of resonant wavelengths. For a chain made of N' molecules (or unit cells) we have only to

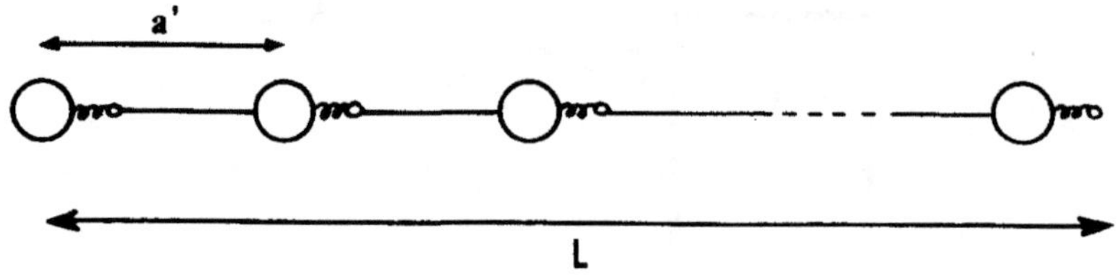

Fig. 4 Linear chain of diatomic molecules separated by distance a'.

consider N' wavelengths i.e. n = 0; 1; ... N' - 1. Effectively the phase difference $\Delta\phi$ between two neighboring molecules is written

$$\Delta\phi = \frac{2\pi a'}{\Lambda}, \quad \Delta\phi = n \cdot \frac{\pi}{N' - 1}. \tag{12}$$

It is clear that after considering the N' first values of n, n will have increased from 0 to N' - 1 and $\Delta\phi$ from 0 to π by increments equal to $\pi/N' - 1$. Considering higher values of n will not give any new standing wave because

$$\cos\left(\omega t + \pi + \frac{\pi}{N' - 1}\right) = -\cos\left(\omega t + \frac{\pi}{N' - 1}\right).$$

The effect of the increasing phase difference is considered on Fig. 5 at time t = 0. The wave vector is defined as $\hat{q} = 2\pi/\Lambda\ \hat{q}_o$, $\hat{q}_o$ is the unit vector directed along the direction of propagation. Thus we have

$$\hat{q} = n \frac{\pi}{L} \hat{q}_o \tag{13}$$

The maximum value of q is $q_{max} = (N' - 1)\ \pi/L$

$$q_{max} = \frac{\pi}{a'} \tag{14}$$

with a' = 2a (Fig. 6). That result is easy to generalize for the case of a tridimentional crystal made of N' primitive unit cells. The number of wavelengths will be limited to N'. Each wavelength will carry cn three n vibrations (n being the number of atoms in the unit cell), hence a total of three N'n standing waves. That is also the number of degrees of freedom for the total number of atoms in the crystal (N = N'n atoms). It is more difficult to get the three N'n different elastic frequencies supporting the thermal agitation in the crystal. To go farther we shall consider

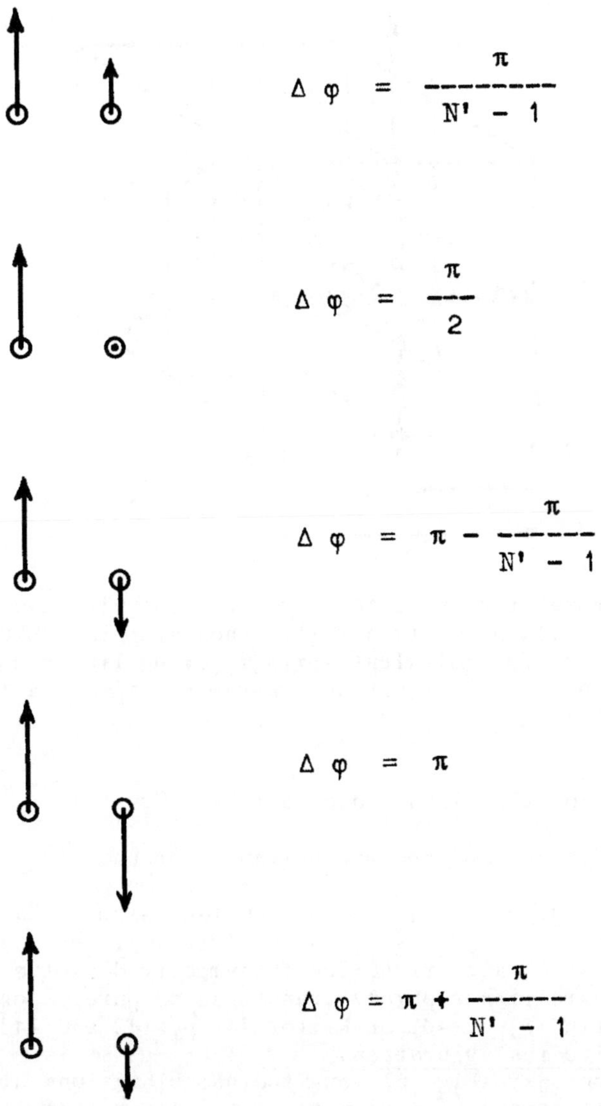

Fig. 5 When the phase difference between two neighboring molecules is increased up to π all situations have been described.

the case of a cubic crystal and we shall assume a very simple model.

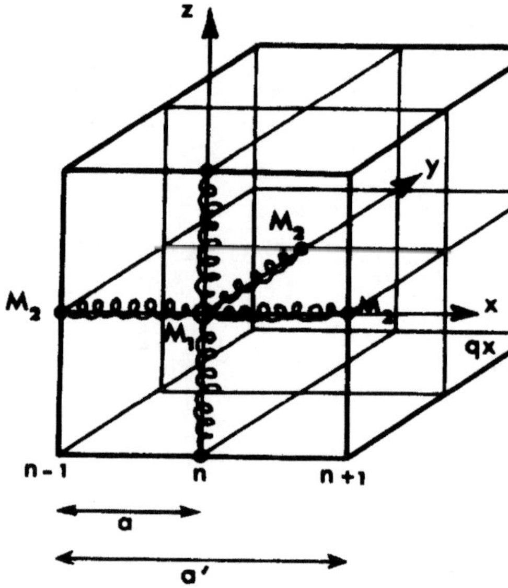

Fig. 6 Structure of a cubic NaCl type crystal with two types of atoms M_1 (mass m_1) and M_2 (mass m_2). Each atom is linked to its six neighbors by six equivalent springs giving linear compression and shearing forces. The unit cell parameter a' = 2 a (adapted from Barker [1]).

B. Lattice Vibration in a Cubic NaCl Type Crystal

1. Uncharged atoms and nearest neighbor forces only

Following Barker [1,2], let us consider a cubic NaCl type crystal with atoms of masses m_1 and m_2 (Fig. 6). We shall assume that plane elastic waves traveling in symmetry directions in a cubic crystal are either purely transverse or purely longitudinal [i.e. see Martin [3] p. 40, or Kittel [4] p. 142 and 148].

a. Longitudinal vibrations

Let us consider (Fig. 6) longitudinal vibrations along x [i.e. (001) direction] of wave vector q_x. The x axis is parallel to the cubic unit cell edge and the atoms along the x axis are numbered n - 1, n, n + 1, etc. ..., and also the atoms located in a plane perpendicular to the x axis and passing by atom n - 1, n, or n + 1 of the x axis. A plane wave of wave-vector q_x means that all atoms of type M_1 vibrate in phase with same displacement $x_1(n)$, and also atoms of type M_2 with displacement $x_2(n)$.

Let us assume nearest neighbor forces only. The displacement along x of the central atom M_1 in Fig. 1, will first develop two forces corresponding to the compression or extension, respectively, of the bonds on each side of M_1 along x: $-A[x_1(n) - x_2(n+1)]$ and $-A[x_1(n) - x_2(n-1)]$. It will also rise four forces generally weak corresponding to the shearing along x of the four other bonds: $-B[x_1(n) - x_2(n)] \times 4$. The equations of motions are written:

$$m_1 x_1''(n) = -A[x_1(n) - x_2(n+1) + x_1(n) - x_2(n-1)]$$
$$-B[\{x_1(n) - x_2(n)\} \times 4]$$

$$m_1 x_1''(n) = -A[2x_1(n) - x_2(n+1) - x_2(n-1)]$$
$$- 4B[x_1(n) - x_2(n)] \qquad (15)$$

$$m_2 x_2''(n) = -A[2x_2(n) - x_1(n+1) - x_1(n-1)]$$
$$- 4B[x_2(n) - x_1(n)] \qquad (16)$$

Let us try a traveling wave solution of the form:

$$x_j(n) = x_j \, e^{i\omega t} \, e^{inqa}$$

where a is half the primitive cell parameter, $\omega = 2\pi/T$ the time frequency, $q = 2\pi/\Lambda$ the spatial frequency. We obtain the secular equation:

$$(-\omega^2 m_1 + 2A + 4B)(-\omega^2 m_2 + 2A + 2B)$$
$$- (4B - 2A \cos qa)(-4B - 2A \cos qa) = 0 \qquad (17)$$

For each allowed value of $q \equiv q_x$ there are two values of ω^2. Fig. 7a gives ω vs q in the x direction for B = 0 and Fig. 7b for B > 0. The upper branch is called the optic branch, and $w_o^2 = (2A + 4B)/\mu$, with $1/\mu = 1/m_1 + 1/m_2$, or $\mu = m_1 m_2/m_1 + m_2$ (reduced mass of the pair of atoms).

The lower one is the acoustic branch, it goes to zero for $q = \pi/a$, when B = 0. The assumption of shearing forces ($B \neq 0$), when two atoms in a direction perpendicular to $\vec{q}$ have a relative displacement parallel to q, is thus a need to prevent this collapse. It is also shown that $\omega'_{LA} \sim 16B/(m_1 + m_2)\omega_o$.

b. Transverse vibrations

With the same wave-vector, the atoms are now assumed to move along y, and we have:

$$m_1 y_1''(n) = -2A[y_1(n) - y_2(n)] - 2B[y_1(n) - y_2(n)] - B[2y_1(n) - y_2(n-1) - y_2(n+1)]$$

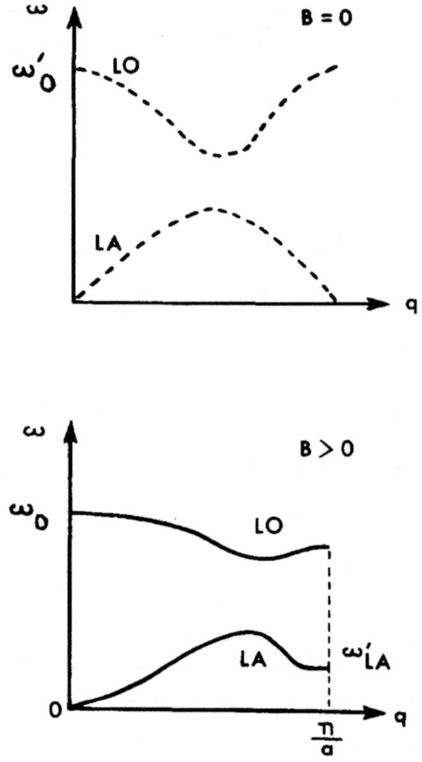

Fig. 7 Longitudinal vibrations in a cubic crystal of uncharged atoms when shearing forces are neglected (B = 0), and when they are introduced (B ≠ 0); adapted from Barker [1].

$$m_2 y_2''(n) = -2A[y_2(n) - y_1(n)] - 2B[y_2(n) - y_1(n)]$$
$$- B[2y_2(n) - y_1(n-1) - y_1(n+1)]$$

We try again a traveling wave solution which is now written:

$$y_j(n) = y_j\, e^{i\omega t} \cdot e^{inqa}$$

and we get:

$$(-\omega^2 m_1 + 2A + 4B)(-\omega^2 m_2 + 2A + 4B) -$$
$$(2A + 2B + 2B \cos q\, a)(2A + 2B + 2B \cos q\, a) = 0.$$

In the case B = 0, the secular equation simplifies to $\omega^4 - 2A(1/m_1 + 1/m_2)\,\omega^2 = 0$. The two solutions are $\omega^2 = 0$,

(TA branch); and $\omega_o'^2 = 2A(1/m_1 + 1/m_2)$; (TO branch). They are q independent and the TA branch has collapsed to zero (Fig. 8). For B > 0 we have Fig. 8 where ω_o is the same as for longitudinal vibration and the value of ω_{TA}' can be calculated for B << A

$$\omega_{TA} \simeq \frac{8B}{(m_1 + m_2)\omega_o} .$$

Fig. 9 shows the displacement patterns for $q \simeq 0$.

2. Charged atoms

We have seen that for q = 0, both the LO and TO branch have the same ω_o frequency. This is not confirmed by experiments on ionic crystals for instance, and the explanation lies on the charges of the ions. Let them be z_1 and $z_2 = -z_1$ for ions of masses m_1 and m_2, respectively. We have to introduce electric forces $z_1 E_{loc}$ and $-z_1 E_{loc}$ which must be added on the right hand side of the equations of motions [i.e. Eq. (15) and (16) in the case of longitudinal waves]. E_{loc} is the local electric field at each ion, i.e. the total field at the center of the ion. The calculation of E_{loc} is a difficult problem. It is the sum of the microscopic Lorentz field ($E_{Lorentz} = P/3\varepsilon_o$), plus the macroscopic electric field which may be associated with the elastic wave (Born and Huang [5], p. 228).

a. Macroscopic electric field issued from the lattice vibrations in a cubic diatomic crystal

(i) Polarization waves

For an elastic wave, we have a displacement

$$\hat{u} = \hat{u}_o\, e^{j\omega t - j\frac{\omega x n}{c}} \quad \text{or} \quad \hat{u} = \hat{u}_o\, e^{j\omega t - j\hat{q}\hat{r}}$$

Now with charged atoms, for TO and LO vibrations, a dipole moment arises in the unit cell, hence a polarization wave:

$$\hat{P} = \hat{P}_o\, e^{j\omega t - j\hat{q}\hat{r}} \tag{18}$$

which may radiate an electric field:

$$\hat{E} = \hat{E}_o\, e^{j\omega t - j\hat{q}\hat{r}} \tag{19}$$

and also a displacement field:

$$\hat{D} = \hat{D}_o\, e^{j\omega t - j\hat{q}\hat{r}}. \tag{20}$$

(ii) Polarization waves are generally not electro-magnetic waves

Could we relate $\hat{P}$ and $\hat{E}$ by the relation $P = \varepsilon_o(\varepsilon_R - 1)E$

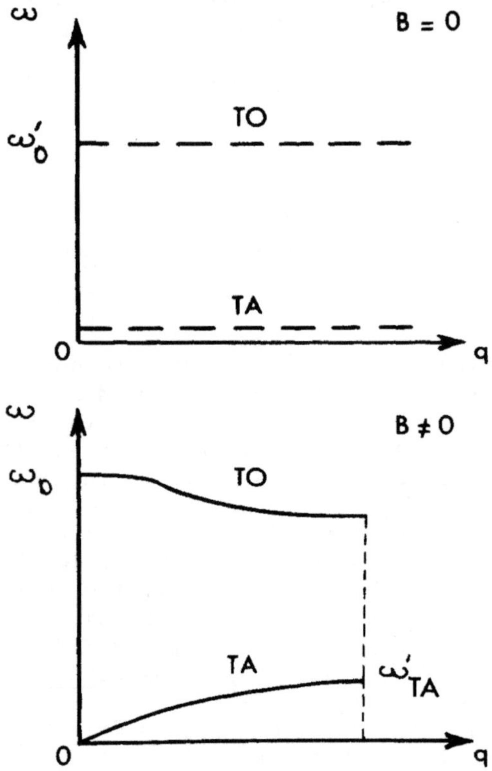

Fig. 8 Transverse vibration in a cubic crystal of uncharged atoms when shearing forces are neglected (B = 0), and when they are considered (B ≠ 0); from Barker [1]. The shearing forces introduce some dispersion both on the TO and TA branch. The TO frequency for q = 0 is the same as the LO frequency because of the assumption of uncharged atoms.

established in electro-static? We should have to consider a dielectric constant depending on frequency [i.e. $\varepsilon_R(\omega)$]. That is done in Maxwell theory where the wave vector k is written $k = \bar{n}\omega/c$, with $\bar{n} = n - jk$, $(\bar{n})^2 = \varepsilon_R(\omega)$, and the phase velocity $v = c/n$.

However, the wave-vector in Eq. (18)(19)(20) is one of the elastic waves whose dispersion $q(\omega)$ is shown in Fig. 10. If we write $q(\omega) = k(\omega)$, we choose a few phonons at the intersection of the dispersion curves of elastic waves and electro-magnetic waves. These ones are polaritons. We shall see later there is in fact a coupling between elastic and electro-magnetic waves,

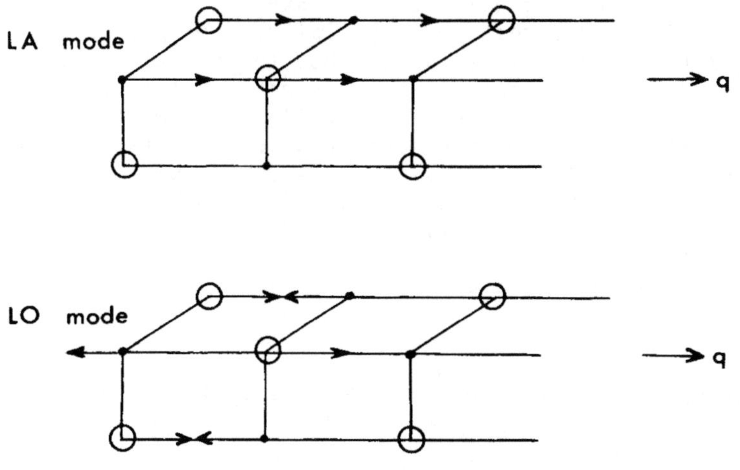

Fig. 9 Displacement pattern in a cubic crystal for q≃ 0.

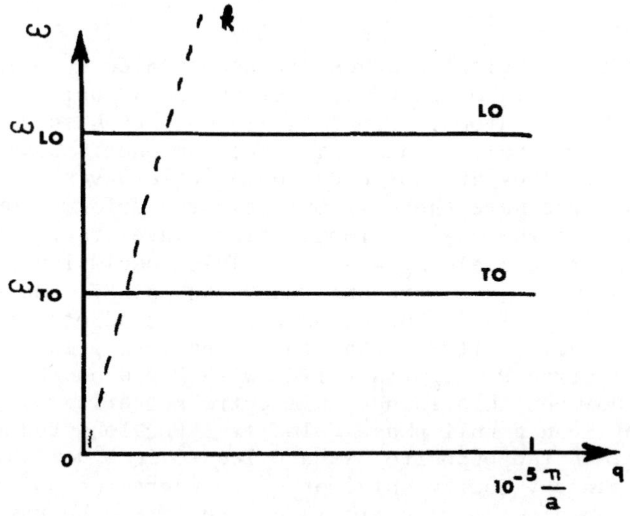

Fig. 10. Optical phonons in the case of one undamped oscillator, dispersion neglected (small values of q).

a polariton dispersion curve has to be considered, but its extension is limited to a very short wave-vector range close to the origin.

(iii) Structure of polarization waves

We shall discard this very special case of polaritons and assume $q \neq n\omega/c$. This is to write that the elastic wave, and the accompanying polarization wave, is not an electro-magnetic wave, and its phase velocity is generally much smaller. The characteristic of an electro-magnetic wave being the association of an electric and magnetic field, we shall assume H = 0. Now $\widehat{\text{Curl}}\ \hat{E} = -\mu\ \partial\hat{H}/\partial T$, hence $\widehat{\text{Curl}}\ \hat{E} = 0$, and $\text{div}\ \hat{D} = \rho = 0$ (no free charges in a perfect dielectric). Now $\text{div}\ \hat{D} = \hat{q} \cdot \hat{D}$, hence $\hat{q} \cdot \hat{D} = 0$.

(a) Longitudinal waves:

$$\hat{q} \cdot \hat{D} = q\ D$$

Let us assume $q \neq 0$, then $D = 0$. Now $\hat{D} = \varepsilon_0\ \hat{E} + \hat{P}$, and $\hat{E}_L = -\hat{P}_L/\varepsilon_0$; $\hat{E}_L = -\hat{P}_0/\varepsilon_0\ e^{j\omega t} - j\hat{q}\hat{r}$.

(b) Transverse waves:

$$\hat{q} \cdot \hat{D} = 0 \text{ and also } \hat{q} \cdot \hat{P} = 0$$

now $\text{div}\ \hat{D} = \varepsilon_0\ \text{div}\ \hat{E} + \text{div}\ \hat{P}$, hence $\text{div}\ \hat{E} = -1/\varepsilon_0\ \text{div}\ \hat{P}$; $\text{div}\ \hat{E} = -\varepsilon_0\ \hat{q}\ \hat{D}$, and thus $\text{div}\ \hat{E} = 0$, but also $\widehat{\text{Curl}}\ \hat{E} = 0$. Hence $E_T = 0$.

In conclusion, elastic waves are accompanied by polarization waves in the case of TO and LO vibrations in the unit cell containing two charged atoms. For these waves the phase velocity is generally much smaller than for electro-magnetic waves at the same frequency. They are not electro-magnetic waves. In the case of transverse wave there is only elastic deformation and polarization. In the case of longitudinal waves there is a depolarizing electric field $E_L = -P/\varepsilon_0$. This conclusion can be seen different ways, see Born and Huang [5], p. 87, and Poulet and Mathieu [6], p. 163. Our demonstration is adapted from Frohlich. The difficulty is that for transverse polarization waves, the relation $P = \varepsilon_0\ (\varepsilon_R - 1)E$, with $P \neq 0$ and $E = 0$, lead to $\varepsilon_R = \infty$. However this is not an electro-magnetic wave, and that does not mean a null phase velocity. For longitudinal polarization waves, on the opposite, we are led to $\varepsilon_R = 0$. This means an infinite phase velocity only for $q = 0$ where the polarization wave is also an electro-magnetic wave (see later the paragraph on polaritons).

(b) Transverse vibrations for $q \approx 0$

In this case, $y_2(n) = y_2(n - 1) = y_2(n + 1)$ and we have:

$$m_1 y_1'' = -2A(y_1 - y_2) - 4B(y_1 - y_2) + z_1\ E_{loc}. \tag{21}$$

$$m_2 y_2'' = -2A(y_2 - y_1) - 4B(y_2 - y_1) - z_1\ E_{loc}. \tag{22}$$

For transverse waves, we simply take:

$$E_{loc} = E + \frac{P}{3\varepsilon_o} \, . \tag{23}$$

The Eq. (22) can be subtracted from Eq. (21):

$$m_1 y_1'' - m_2 y_2'' = -y(4A + 8B) + 2z_1 E_{loc}$$

with $y = y_1 - y_2$. Now let us call G the center of mass of the pair of atoms M_1M_2 (Fig. 11) and put:

$$\overline{GM_1} = r_1 ; \ \overline{GM_2} = r_2 ; \ \overline{M_2M_1} = r ,$$

then

$$m_1 r_1 = -mr_2$$

$$r_1 - r_2 = r ;$$

and

$$r_1 = \frac{m_2}{m_1 + m_2} r$$

$$r_2 = \frac{-m_1}{m_1 + m_2} r$$

Now, assuming G at rest,

$$y_1 = \overline{\Delta r_1} ; \ y_2 = \overline{\Delta r_2} ; \ y = y_1 - y_2 = \Delta r ;$$

then

$$y_1 = \frac{m_2}{m_1 + m_2} y ; \qquad y_2 = \frac{-m_1}{m_1 + m_2} y ;$$

and

$$m_1 y_1'' - m_2 y_2'' = 2\mu y'' , \qquad \text{with } \mu = \frac{m_1 \, m_2}{m_1 + m_2} ;$$

finally:

$$\mu y'' = -y(2A + 4B) + z_1 E_{loc} \, . \tag{24}$$

Now

$$P = \frac{y_1 \, z_1}{v} + \frac{y_2(-z_1)}{v} + \frac{\alpha E_{loc}}{v}$$

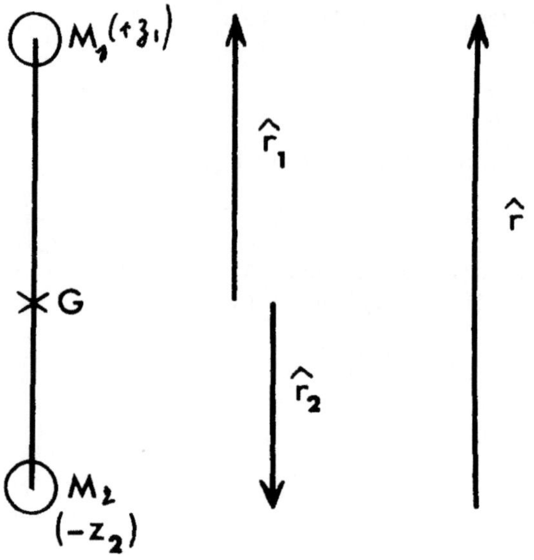

Fig. 11 In an optical vibration the mass center G of the unit cell G does not move.

α being the electronic polarizability and v the volume occupied by two ions. So we have

$$P = \frac{+z_1 \, y}{v} + \frac{\alpha \, E_{loc}}{v}$$

(i.e. for $y = 0$ and $E_{loc} = 0$, $P = 0$; a cubic ionic crystal has no static polarization). Remembering that $E_{loc} = E + P/3\varepsilon_o$, we have

$$E_{loc} = E + \frac{1}{3\varepsilon_o} \left(\frac{z_1 \, y}{v} + \frac{\alpha \, E_{loc}}{v}\right)$$

$$E_{loc} \left(1 - \frac{\alpha}{3v\varepsilon_o}\right) = E + \left(\frac{z_1 \, y}{3\varepsilon_o v}\right)$$

$$E_{loc} = \frac{E}{1 - 3v\varepsilon_o} + \frac{z_1 \, y}{3v\varepsilon_o \, (1 - 3v\varepsilon_o)}$$

hence the equation of motion

$$\mu y'' = -y\left[2A + 4B - \frac{z_1^2/3v\varepsilon_o}{1 - \frac{\alpha}{3v\varepsilon_o}}\right] + \frac{z_1}{1 - \frac{\alpha}{3v\varepsilon_o}} E.$$

The bracketed expression is the effective force constant for the $q = 0$ TO mode. For uncharged atoms ($z_1 = 0$) we have the same force as above. For $z_1 \neq 0$ the effective force constant is reduced.

To get the natural TO mode we set $E = 0$:

$$\omega_{TO}^2 = \frac{1}{\mu}\left(2A = 4B + \frac{z_1^2/3v\varepsilon_o}{1 - \alpha/3v\varepsilon_o}\right). \qquad (25)$$

For $z_1 = 0$, we come back to Eq. (18). For $z_1 \neq 0$, the TO mode frequency for $q = 0$ is lowered. In NaCl or KBr the third term in the bracket is about 30% of the first two. Thus they have moderately high frequency. For some ferroelectrics this third term becomes as large as the first two at the transition temperature and the TO mode at $q = 0$ shows a frequency decreasing down to zero.

(c) Longitudinal vibration for $q \simeq 0$

The LO mode at $q \simeq 0$ is affected in a different manner. Following Barker [1,2] for the calculation of the local field we must now include the depolarization field* $- P/\varepsilon_o$, hence

$$E_{loc} = E + \frac{P}{3\varepsilon_o} - \frac{P}{\varepsilon_o}$$

$$E_{loc} = E - \frac{2P}{3\varepsilon_o} \qquad (26)$$

and we get

$$\omega_{LO}^2 = \frac{1}{\mu}\left(2A + 4B + \frac{2z_1^2/3v\varepsilon_o}{1 + 2\alpha/3v\varepsilon_o}\right). \qquad (27)$$

*In fact, it is the macroscopic electric field of the longitudinal electro-magnetic wave, and indeed it is opposite to the polarization. For $z_1 \neq 0$, the LO frequency is higher than the TO frequency and there is no cancellation of the short range forces A and B by the local field effects. From Eq. (27) we get:

$$\omega_{LO}^2 = \frac{1}{\mu}(2A + 4B) + \frac{1}{\mu}\,\frac{2z_1^2/3v\varepsilon_o}{1 + 2\alpha/3v\varepsilon_o} =$$

$$\omega_{TO}^2 + \frac{1}{\mu}\left(\frac{z_1^2/3v\varepsilon_o}{1 - \alpha/3v\varepsilon_o} + \frac{2z_1^2/3v\varepsilon_o}{1 + 2\alpha/3v\varepsilon_o}\right).$$

Now let us define an effective charge z such as:

$$z^2 = \frac{z_1^2}{1 - \dfrac{\alpha}{3v\varepsilon_o}}$$

$$\omega_{LO}^2 = \omega_{TO}^2 + \frac{z^2}{3v\varepsilon_o\mu}\left[\left(1 - \frac{\alpha}{3v\varepsilon_o}\right) + \frac{\left(1 - \frac{\alpha}{3v\varepsilon_o}\right)}{1 + \frac{2\alpha}{3v\varepsilon_o}}\right]$$

let us put

$$\varepsilon_{RV} = 1 + \frac{\alpha}{v\varepsilon_o}$$

$$\omega_{LO}^2 \simeq \omega_{TO}^2 + \frac{z^2}{3v\varepsilon_o\mu}\,\frac{\left(1 - \frac{\alpha}{3v\varepsilon_o}\right)\left(1 + \frac{2\alpha}{3v\varepsilon_o}\right) + 1 - \frac{2\alpha}{3v\varepsilon_o}}{1 + \frac{2\,\alpha}{3v\varepsilon_o}}$$

$$\omega_{LO}^2 \simeq \omega_{TO}^2 + \frac{2\mu z^2/3v\varepsilon_o}{1 + 4\alpha/3v\varepsilon_o}$$

$$\omega_{LO}^2 \simeq \omega_{TO}^2 + \frac{z^2}{\mu v\varepsilon_o\varepsilon_{RV}} \qquad (28)$$

and

$$\omega_{TO}^2 = \omega_0^2 - \frac{z^2}{3\mu v\varepsilon_o} \qquad \text{(Eq. (25))}.$$

3. Lattice mode instability

a. Lattice distortion

It is the close cancellation of the short range forces by the local field correction which makes

$$k_{eff} = 2A + 4B - \frac{z_1^2/3v\varepsilon_o}{1 - \alpha/3v\varepsilon_o} \rightarrow 0$$

or $\omega_{TO} \rightarrow 0$. This means that the natural motion of the TO mode is the solution of $\mu y'' = 0$ and thus $y = a\ t + b$. The ions move apart linearly in time until they approach other ions which bring strong repulsive forces. It is shown that the displacement is limited to a distortion y_S.

The result of this distortion may be an electric dipole in

the primitive unit cell. In that case we have a ferroelectric transition.

b. Explanation of the temperature sensitive distortion

The cancellation may occur by different processes, for instance by a decrease of temperature which generally decreases v. This can occur at temperatures above the melting point of the crystal so that it is already ferroelectric when it is grown from the melt ($LiNbO_3$ is close to this situation). Other crystals may be approaching a ferroelectric transition on cooling. In a few cases the volume does not decrease enough and a crystal may be heading towards an instability but never make it. $SrTiO_3$ and $KTaO_3$ are good examples of such "nearly ferroelectric crystals".

4. EXCITATION OF THE CRYSTAL BY AN INFRARED WAVE

A. The Dielectric Response Without Damping

Now that we know the relative displacement of the ions in the unit cell we can get the unit cell dipole and hence the polarization. The polarization is directly linked to the dielectric constant which rules everything in optics. Equation (14) can be written:

$$\mu y'' = -k_e y + z E$$

with k_e and z modified by the local field correction:

$$k_e = 2A + 4B - \frac{z_1^2/3v\varepsilon_o}{1 - \frac{\alpha}{3v\varepsilon_o}} ; \tag{29}$$

$$z = \frac{z_1}{1 - \frac{\alpha}{3v\varepsilon_o}} \tag{30}$$

Now

$$P = \frac{z_1 y}{v} + \frac{\alpha E_{loc}}{v}$$

and

$$E_{loc} = E + \frac{P}{3\varepsilon_o}$$

so that

$$P = \frac{z_1 y}{v} + \frac{\alpha E}{v} + \frac{\alpha}{v} \frac{P}{3\varepsilon_o};$$

$$P(1 - \frac{\alpha}{3v\varepsilon_o}) = \frac{z_1 y}{v} + \frac{\alpha E}{v}$$

$$P = \frac{z_1}{1 - \frac{\alpha}{3v\varepsilon_o}} \frac{y}{v} + \frac{\alpha E}{v}$$

$$P = z \frac{y}{v} + \frac{\alpha E}{v}$$

which is the response of the crystal to a transverse electromagnetic wave of frequency ω traveling with a wave vector $\hat{k}$ parallel to x' x. We have to put $E = E_o e^{i\omega t}$ in Eqs. (18) and (20).

Trying solution of the type $y = y_o e^{i\omega t}$ and $P = P_o e^{i\omega t}$ we find

$$y_o = \frac{z/\mu}{k_{eff}/\mu - \omega^2} E_o$$

$$P_o = \frac{z^2/\ \mu v}{k_{eff}/\mu - \omega^2} E_o + \frac{\alpha E_o}{v}$$

Now

$\varepsilon_R = 1 + P/\varepsilon_o E$ (relative dielectric constant for angular frequency ω)

so that

$$\varepsilon_R(\omega) = 1 + \frac{\alpha}{v\varepsilon_o} + \frac{z^2/\mu v\varepsilon_o}{k_{eff}/\mu - \omega^2}$$

Now

$$k_{eff}/\mu = \omega_{TO}^2 \text{ and } 1 + \frac{\alpha}{v\varepsilon_o} = \varepsilon_{RV}$$

so that

$$\varepsilon_R(\omega) = \varepsilon_{RV} + \frac{z^2/\mu v \varepsilon_o}{\omega_{TO}^2 - \omega^2} . \tag{31}$$

In our model (undamped harmonic oscillator ω_{TO}), $\varepsilon_R(\omega)$ is real, either positive, negative or null. In Fig. 13:

$$\varepsilon_{RS} = \varepsilon_{RV} + \frac{z^2/\mu v \varepsilon_o}{\omega_{TO}^2} .$$

The static dielectric constant ε_{RS} is thus increasing to infinity when $\omega_{TO} \to 0$.

$$\varepsilon_R \to \pm \infty \text{ when } \omega \to \omega_{TO}$$

$$\varepsilon_R = 0 \text{ for } \omega_0 = \omega_0'' \text{ such that } \frac{z^2/\mu v \varepsilon_o}{\omega_{TO}^2 - \omega_0''^2} = -\varepsilon_{RV}$$

$$\omega_0''^2 = \omega_{TO}^2 + \frac{z^2/\mu v \varepsilon_o}{\varepsilon_{RV}}$$

and from Eq. (17), $\omega_0''^2 = \omega_{LO}^2$. It is thus demonstrated that $\varepsilon_R = 0$ for the LO branch frequency at $q = 0$. Now

$$\frac{\varepsilon_S}{\varepsilon_V} = 1 + \frac{z^2/\mu v \varepsilon_o}{\varepsilon_V \omega_{TO}^2};$$

$$\frac{\varepsilon_S}{\varepsilon_V} = \frac{\omega_{TO}^2 + z^2/\mu v \varepsilon_V \varepsilon_o}{\omega_{TO}^2}$$

$$\frac{\varepsilon_S}{\varepsilon_V} = \frac{\omega_{LO}^2}{\omega_{TO}^2} \tag{32}$$

ω_{TO} and ω_{LO} being the values of the TO and LO branch frequencies for $q = 0$. This is the Lyddane-Sachs-Teller relation for undamped oscillators. The importance of this relation can be shown in the case where ε_S follows the Curie-Law above the transition temperature:

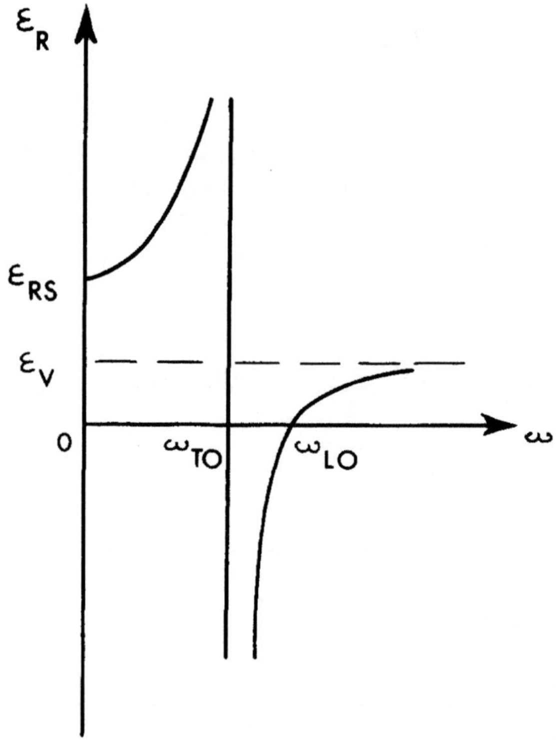

Fig. 12 In the case of one undamped oscillator the dielectric constant is a real number either positive, negative or null. It shows a pole at ω_{TO}, and a zero at ω_{LO}.

$\varepsilon_{RS} \simeq C/T\text{-}T_c$. For $T \to T_c$, $\varepsilon_S \to \infty$. Now ε_r does not change significantly. Then $\omega_{LO}^2/\omega_{TO}^2 \to \infty$. From our discussion of local field effects $\omega_{TO}^2 \to 0$ is much more likely than $\omega_{LO}^2 \to \infty$. Thus $\omega_{TO}^2 = \omega_{LO}^2\ \varepsilon_{RV}\ T\text{-}T_c/C$. Figure 13 is an extension of Fig. 12 in the case of three undamped oscillators.

B. The Dielectric Response With Damping

We have seen in Eqs. (21) and (22) that without damping, y_o and P_o should go to infinity at the resonance. To avoid it, we have to introduce a damping force in the equation of motion.

$$\mu y'' = -k_{eff}\ y - \mu\gamma y' + z\ E$$

and we obtain

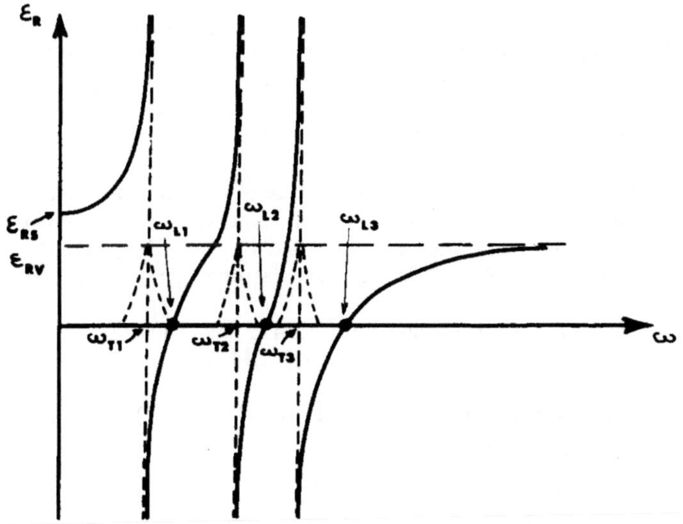

Fig. 13 The case of three umdamped oscillators (adapted from Ref. [12]).

$$\varepsilon_R(\omega) = \varepsilon_{RV} + \frac{z^2/\mu v \varepsilon_o}{\omega_{TO}^2 - \omega^2 + i\omega\gamma}; \tag{33}$$

with

$$\omega_{TO}^2 = \frac{k_{eff}}{\mu} .$$

Now

$$\varepsilon_R = \varepsilon_R' - i\varepsilon_R''$$

$$\varepsilon_R' = \varepsilon_{RV} + \frac{(\omega_{TO}^2 - \omega^2)z^2}{\mu v \varepsilon_o[(\omega_{TO}^2 - \omega^2)^2 + \omega^2\gamma^2]} \tag{34}$$

$$\varepsilon_R'' = \frac{\gamma z^2 \omega^2}{\mu v \varepsilon_o[(\omega_{TO}^2 - \omega^2)^2 + \omega^2\gamma^2]} \tag{35}$$

It is seen that ε_R'' is always positive (except for $\omega = 0$ where it is null), but except for $\omega \approx \omega_{TO}$, it is very small (Fig. 14a). Now ε_R' is the sum of two terms, the first one ε_{RV} is positive and constant, the second one is positive for $\omega^2 < \omega_{TO}^2$, null for $\omega^2 = \omega_{TO}^2$; and negative for $\omega^2 > \omega_{TO}^2$. It is thus possible for ε_R' to be null and even negative (Fig. 14a).

It is left as an exercise to consider $1/\varepsilon_R$ and show that its real part is null for ω_{TO}, and its imaginary part is maximum for ω_{LO} (Fig. 14b). The conclusion is that by looking at the maximum of either Im (ε) or Im ($1/\varepsilon$) we get, respectively, ω_{TO} and ω_{LO}.

In the case of the undamped harmonic oscillator the pole and zero of ε_R were, respectively, ω_{TO} and ω_{LO}.

For the damped harmonic oscillator, Eq. (33) can be written

$$\varepsilon_R(\omega) = \varepsilon_{RV} \frac{\frac{\varepsilon_S}{\varepsilon_V} \omega_{TO}^2 - \omega^2 + i\omega\gamma}{\omega_{TO}^2 - \omega^2 + i\omega\gamma}.$$

The zero of ε_R occurs when the numerator is zero. The real and imaginary parts are zero for different real frequencies. To get one frequency, we must consider ω as a complex number:

$$-\omega^2 + i\omega\gamma + \frac{\varepsilon_S}{\varepsilon_V} \omega_{TO}^2 = 0$$

$$\Delta = -\gamma^2 + 4 \frac{\varepsilon_S}{\varepsilon_V} \omega_{TO}^2.$$

Let us assume $\Delta > 0$ (small damping), then

$$\omega = \frac{i\gamma \pm \sqrt{\Delta}}{2}, \text{ or } \omega_1 = \frac{\sqrt{\Delta}}{2} + \frac{i\gamma}{2}$$

and

$$\omega_2 = \frac{-\sqrt{\Delta}}{2} + \frac{i\gamma}{2}, \text{ hence } \omega_2 = -\omega_1^*.$$

There are two frequencies giving a null value for ε_R. In the complex frequency plane ($\omega = \omega' - i\omega''$), they are represented by points Z_1 and Z_2 which are symmetric relative to the ε'' axis (Fig. 15). They are called by Barker [2], Transverse Phonon Frequencies. The pole of ε_R occurs when the denominator is zero and this also happens for two frequencies which are complex

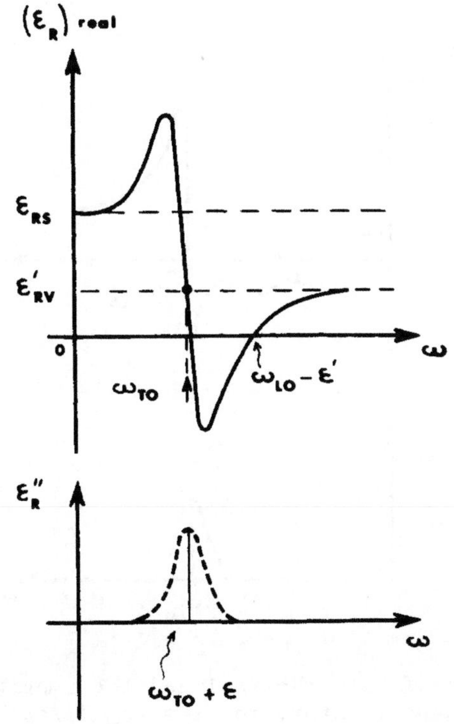

Fig. 14a Variation of both the real and the imaginary parts of ε for one damped oscillator; for $\omega \simeq \omega_{TO}$, ε'' is maximum.

numbers. They are represented by points P_1 and P_2 in the complex frequencies plane and it is easy to show they are also symmetric relative to the ω'' axis.

C. Polaritons

Now that we have $\varepsilon_R(\omega)$ from the study of forced lattice vibrations in a crystal where phonons are known (Fig. 16), we can look at the propagation of an electro-magnetic wave through the crystal. The radiation wave-vector is written:

$$k = \frac{2\pi}{\lambda}; \quad k = \frac{n\omega}{c}; \text{ or } k^2 = \varepsilon_R \frac{\omega^2}{c^2} \tag{36}$$

For very low frequencies $\varepsilon_R = \varepsilon_{RS}$ and $k = \sqrt{\varepsilon_{RS}}\ \omega/c$, for very high frequencies $\varepsilon \simeq \varepsilon_{RV}$ and $k = \sqrt{\varepsilon_{RV}}\ \omega/c$.

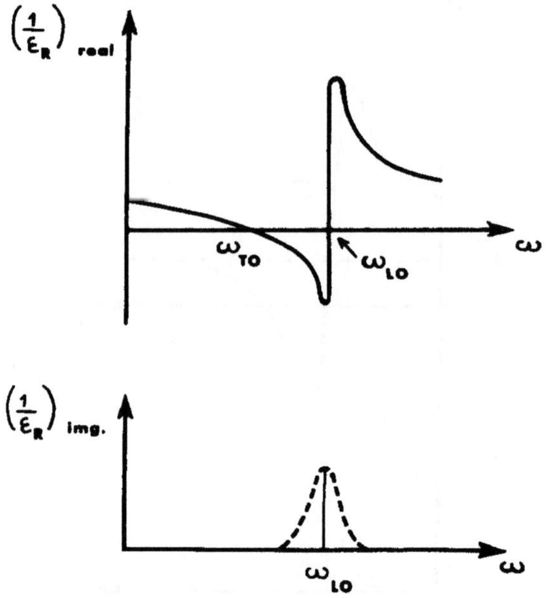

Fig. 14b Variation of both the real and the imaginary parts of $1/\varepsilon_R$ for one damped oscillator; for $\omega \simeq \omega_{LO}$ $(1/\varepsilon_R)_{Imag}$ is maximum (from Ref. [1]).

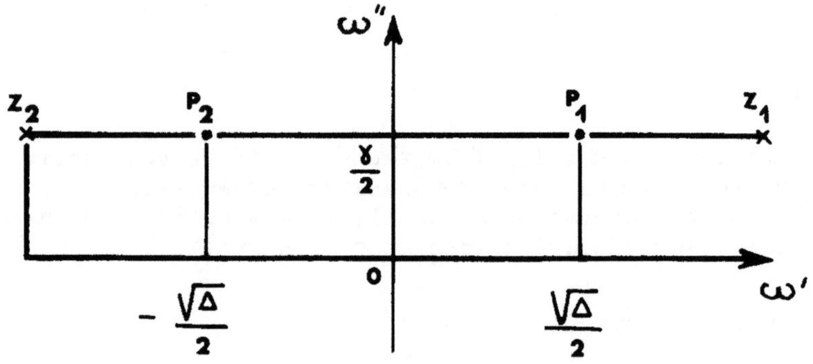

Fig. 15 The complex frequencies plane with the poles and zeroes of ε_R.

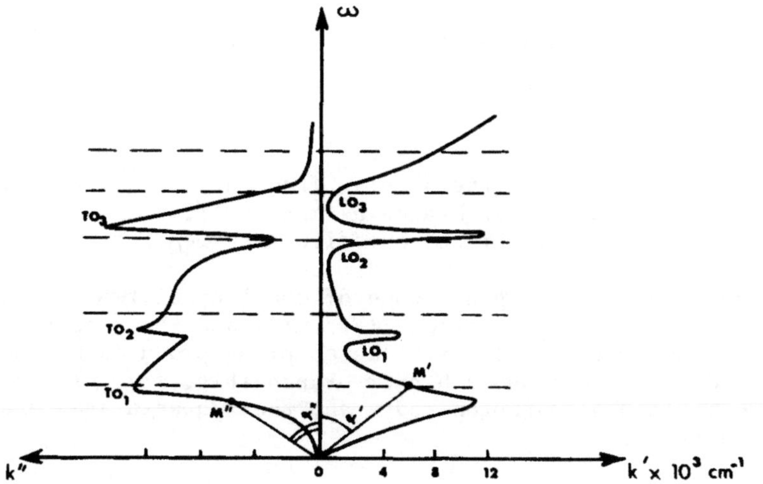

Fig. 16 Case of three damped oscillators (Barker [1]). (1) The values of k' and k" are strictly limited to very small values. (2) tg α" = k/c for the left hand side curve, and tg α' = n/c for the right hand side curve. The maximum on the k"(ω) curve has the abscissa close to the TO frequencies and the minimum on the k'(ω) curve has the abscissa close to the LO frequencies. (3) The transverse vibration modes exist at all frequencies. There are no completely non-propagation gaps though there are still regions of high damping.

1. Damping neglected

We put

$$\varepsilon_R = \varepsilon_{RV} + \frac{\varepsilon_{RS} - \varepsilon_{RV}}{1 - \omega^2/\omega_{TO}^2}$$

in Eq. (36) and get:

$$\omega^2[\varepsilon_{RV} + \frac{\varepsilon_{RS} - \varepsilon_{RV}}{1 - \omega^2/\omega_{TO}^2}] = c^2k^2.$$

For a given k, there are two real frequency roots, giving two different dispersion equations. They are represented on Fig. 17a, where $k = (\omega/c)n$. As an exercise, check that

(1) for $k = 0$, we have either $\omega = 0$ or ω_{LO}

(2) for $k \to \infty$, we have either $\omega \to \omega_{TO}$ or $\omega^2 \varepsilon_{RV} \simeq c^2k^2$;

$$\omega \simeq \frac{ck}{n_V}$$

(3) for $k \simeq 10^4\ \text{cm}^{-1}$ the angular frequency ω is within 1% close to the upper branch asymptote (i.e. $\omega \simeq c\ k/n_V$); or to the lower branch one (i.e. $\omega \simeq \omega_{TO}$).

In fact this is a typical case of coupling of two collective modes: the TO lattice vibrations (Fig. 10) and the electro-magnetic transverse vibrations. The coupling produces new modes represented by a lower and a higher branch (Fig. 18). These modes are mixed phonon-photon modes and their quanta are referred to as polaritons.

In a linear chain the wave-vectors of both phonons and photons increase by increments equal to $\pi/N' - 1$. However for phonons, the wave-vector q can only increase up to π/a while for photons, k can increase to infinity. The dispersion curve for polaritons is given in Fig. 18. There is no modification for LO phonons, and for TO phonons the bending of the branch down to zero is affecting only very small values of q. Figure 17 is an extension in the case of three undamped oscillators. For these small values, Born has shown that the energy of the polaritons is mainly electro-magnetic and the correction for specific heat calculation is negligible. There are gaps between pairs of transverse and longitudinal phonons where ε_R is negative and ω imaginary: the energy at such frequencies cannot exist in the lattice.

In conclusion if we consider a cubic crystal with charged ions, not only the TO and LO frequencies in the (100) direction will be different but the TO branch will be bent down to zero for $q \to 0$ and a new branch will start from LO. There will be more possibilities of excitation of modes but their energy will be more or less electro-magnetic. These modes are called polaritons. If we are mainly interested in elastic waves the scheme of pure phonons (Fig. 10) is mainly correct except for very small values of q. On the opposite if we are illuminating the crystal with electro-magnetic wave of frequency ω in a suitable way, we have to use Figs. 18 or 17. For $\omega < \omega_{TO}$ and $\omega > \omega_{LO}$ there is excitation of polaritons in the crystal whose wave-vectors are given in Figs. 18 or 17. The energy of polaritons is quantized and mainly electro-magnetic except for ω close to ω_{LO} or to ω_{TO} where TO and LO elastic waves are excited, the dielectric constant

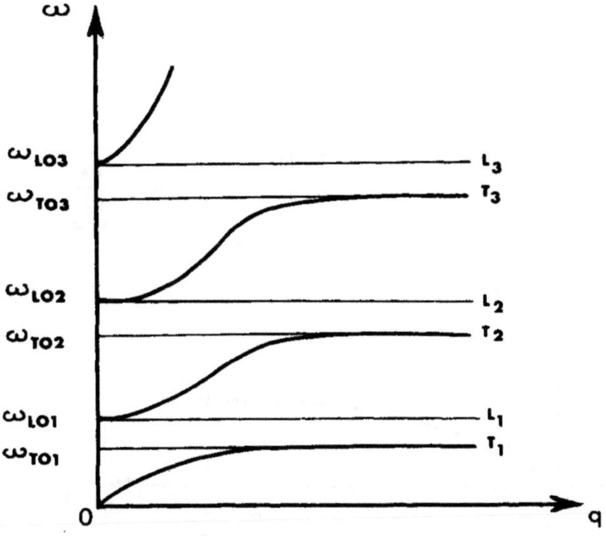

Fig. 17 Dispersion of phonon frequencies in the case of three undamped oscillators.

being, respectively, infinite and null (or very large and null when damping is introduced, see below).

2. Case of damping

We write again Eq. (36), i.e.

$$k^2 = \varepsilon_R \frac{\omega^2}{c^2}$$

Now ε_R is complex*, $\varepsilon_R = (n - j\mathscr{k})^2$; and k is also complex

$$k = k' - i\,k'' = \pm(n - j\mathscr{k})\,\frac{\omega}{c}$$

$$k' = \pm n\,\frac{\omega}{c};\; k'' = \pm\mathscr{k}\frac{\omega}{c}.$$

Let us consider real frequencies ω, Fig. 16 gives |k'| and |k"| vs. ω. It is seen that the wave-vectors of electro-magnetic waves in crystals are limited to around 12,000 cm^{-1} while the wave-vectors

*Note that k is the electro-magnetic wave-vector modulus, while $\mathscr{k}$ is the index of absorption.

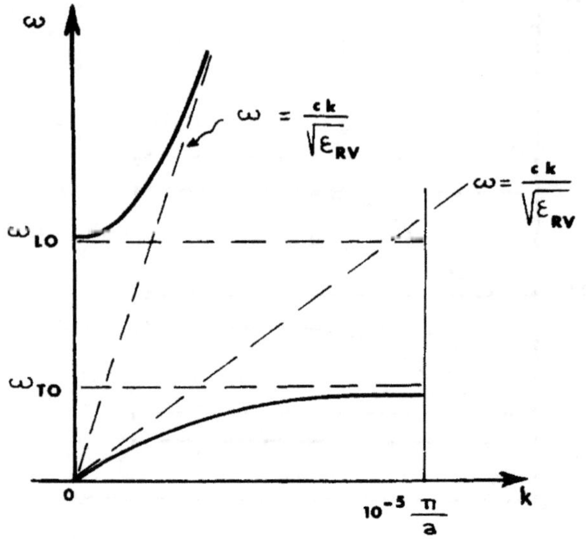

Fig. 18 Dispersion of phonon frequencies in a crystal made of one type undamped oscillators.

of phonons can increase up to $\pi/a \simeq \pi.10^8\ cm^{-1}$, five orders of magnitude higher.

D. Lyddane-Sachs-Teller Relation With Damping

Let us look at the roots of $\varepsilon = 0$ (or the poles of $1/\varepsilon$) in the case of n modes

$$\varepsilon_R(\omega) = \varepsilon_{RV} + \Sigma \frac{S_j\ \omega_j^{\ 2}}{\omega_j^{\ 2} - \omega^2 - i\omega\gamma_j} \tag{37}$$

with

$$S_j = \frac{z_j^{\ 2}}{\mu v \varepsilon_o\ \omega_j^{\ 2}} \tag{38}$$

S_j is a dimensionless mode strength. For instance,

$$\varepsilon_{RO} = \varepsilon_{RV} + \Sigma\ S_j \tag{39}$$

We multiply Eq. (37) by all the resonant denominators and

to Barker [1], consider the roots of the following form:

$$\varepsilon_R \prod_{j=1}^{n} (\omega_j^2 - \omega^2 + i\omega\gamma_j) = \omega_1^2\, \omega_2^2 \ldots \omega_n^2 (S_1 + S_2 + \ldots$$

$$+ S_n + \varepsilon_{RV}) + i\omega(\ldots) - \omega^2(\ldots) - i\omega^3(\ldots) + \omega^4(\ldots) +$$

$$i\omega^5(\ldots) - \ldots(-1)^n\, \omega^{2n}\, \varepsilon_{RV}. \tag{40}$$

There are 2 n roots, they make $\varepsilon_R = 0$, and from Maxwell equations we know that they are the frequencies ω_L which can propagate longitudinal modes in the crystal. Let us note that when ω_L is a root so is its negative complex conjugate $-\omega_L^*$. Finally we get:

$$\frac{|\omega_{L1}|^2 \ldots\ldots |\omega_{Ln}|^2}{\omega_1^2\, \omega_2^2 \ldots \omega_n^2} = \frac{\varepsilon_{RS}}{\varepsilon_{RV}}. \tag{41}$$

This is the Lyddane-Sachs-Teller relation. For our model which includes damping this Lyddane-Sachs-Teller relation connects the absolute value of the longitudinal mode frequencies with the force constant parameters ω_j ... In the special case of no damping the ω_j are the TO frequencies and the ω_{Lj} will be real.

E. Frequency Dependent Damping

The contribution of type j optical phonons to the dielectric constant is given by

$$\frac{S_j\, \omega_j^2}{\omega_j^2 - \omega^2 + i\gamma_j\omega}.$$

In certain crystals the variation of the damping parameter γ with frequency is important. In fact, the damping of the optical mode arises from coupling to other phonons. A high density of one of these phonon combinations (i.e. ω or 2ω) centered at a frequency ω_1 will increase the value of γ_j for this frequency, reflecting the enhanced leaking of the energy from ω_j near this resonance, this will lead to some increase of ε'' at ω_j. A transmission experiment may thus reveal a weak combination band mode at ω_j near ω_1. Reflection experiments are often mainly determined by ε' and cannot detect a wiggle in ε''.

1. Coupling with a nearly harmonic oscillator ω_1

To treat $\gamma(\omega)$ analytically, Barker [2] has chosen a simple form which respects the causality relation:

$$\varepsilon_R(\omega) = \varepsilon_{RV} + \frac{S_1\,\omega_1^2\,(1 - S_2)}{\omega_1^2 - \omega^2 - \dfrac{S_2\,\omega_1^2\,\omega_2^2}{\omega_2^2 - \omega^2 + i\omega\gamma_2}}$$

$$\varepsilon_R(\omega) = \varepsilon_{RV} + \frac{S_1\,\omega_1^2\,(1 - S_2)}{\omega_1^2 - \omega^2 + i\omega\Gamma_1(\omega)} \qquad (42)$$

with

$$\omega\Gamma_1(\omega) = \frac{\omega_1^2\,\omega_2^2}{\omega_2^2 - \omega^2 + i\omega\gamma_2}\,. \qquad (43)$$

The factor $1 - S_2$ is included to simplify the form of ε_{RS}:

$$\varepsilon_{RS} = \varepsilon_{RV} + \frac{S_1\,\omega_1^2\,(1 - S_2)}{\omega_1^2} = \varepsilon_{RV} + S_1 - S_1\,S_2$$

$$\varepsilon_{RS} = \varepsilon_{RV} + S_1$$

and S_2 controls the coupling of the optical mode to other phonons whose density of states has a peak at ω_1.

Now, $\varepsilon_R(\omega)$ has two double poles but its behavior is different than that obtained by simply adding two oscillators with constant damping.

It is shown that the Lyddane-Sachs-Teller relation is now written

$$\frac{\varepsilon_{RO}}{\varepsilon_{RV}} = \frac{|\omega_{L1}|\;|\omega_{L2}|^2}{\omega_1^2\,(1 - S_2)\omega_2^2} \qquad (44)$$

and the effective restoring force in the low frequency limit is not $\omega_1{}^2$ but

$$\omega_1^2\ (0) = \omega_1^2\ (1 - S_2).$$

2. Coupling with a broad continuum

Now the line width of ν_1 is much larger than $\nu_j - \nu_i$. That case has been treated by Breit and Wigner [7] and leads to an asymmetric line shape sometime called "Fano lineshape".

3. Fermi resonance in solids

Fermi resonance is the coupling between two excitations in a molecule, one of which is a one-quantum level, and the other of which is a two-quantum state of another eigen mode. That is occurring for instance in CO_2 where $\nu_1 \approx 2\nu_2$.

The idea that Fermi resonance would exist between phonons in crystals was first put forth by Scott [8] in the case of the soft-mode of quartz.

F. Debye Relaxation

1. The case of triglycine sulphate (TGS)

For some ferroelectrics like TGS, there is no soft mode in the far infrared but there is a broad absorption in the micro-waves. It can be fitted by a simple Debye function with one relaxation time τ_0 which increases drastically when $T \rightarrow T_c$

$$\tau_0 \propto \frac{1}{T - T_c} \qquad (45)$$

2. Debye's equations

Let us recall Debye's equations,

$$\varepsilon' = \varepsilon_{RI} + \frac{\varepsilon_{RS} - \varepsilon_{RI}}{1 + \omega^2\tau^2} \qquad (46)$$

$$\varepsilon'' = \frac{\omega\ \tau}{1 + \omega^2\tau^2}\ (\varepsilon_{RS} - \varepsilon_{RI}) \qquad (47)$$

$\varepsilon'(\omega)$ and $\varepsilon''(\omega)$ are represented in Figs. 19a and 19b, respectively, and $\varepsilon''(\varepsilon)$ is represented in Fig. 20: it is the Cole and Cole

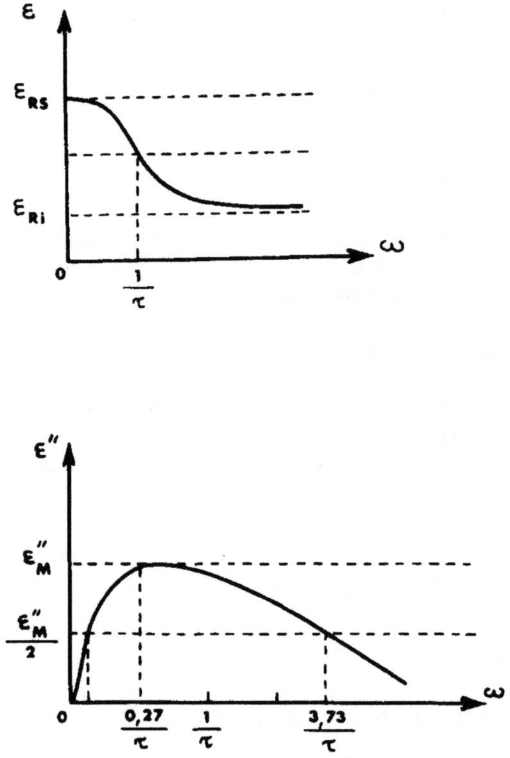

Fig. 19 (a) ε' vs ω in the case of one Debye relaxation time τ.
(b) ε" vs ω in the case of one Debye relaxation time τ.

diagram. The absorption index k can be computed. It is given in Fig. 21. The absorption curve is highly asymmetrical and its half-width is much larger than $1/\tau$.

3. Lyddane-Sachs-Teller

The description of dynamic response in terms of damped oscillators reponse is not valid for most order-disorder systems at low frequencies and a description in terms of Debye relaxation is often needed as in the case of TGS. However the above demonstration of Lyddane-Sachs-Teller relation relies on the assumption of harmonic oscillators and cannot be applied. It has been generalized on the assumption of simple causality to the case of Debye relaxation.

G. Domain Wall Motions

Besides Debye's relaxation there is another mechanism by

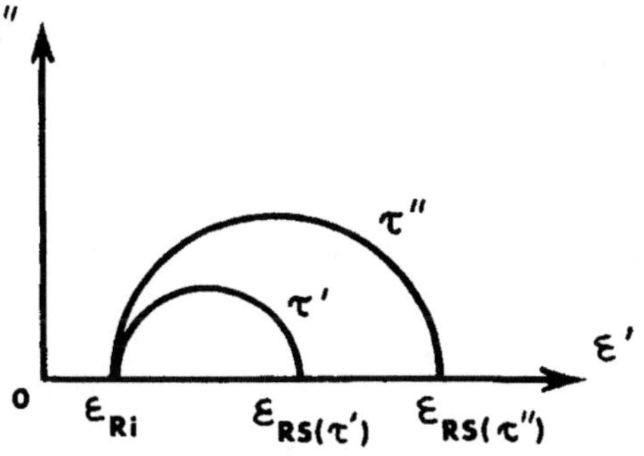

Fig. 20 ε'' vs ε' for two different relaxation times τ' and τ''.

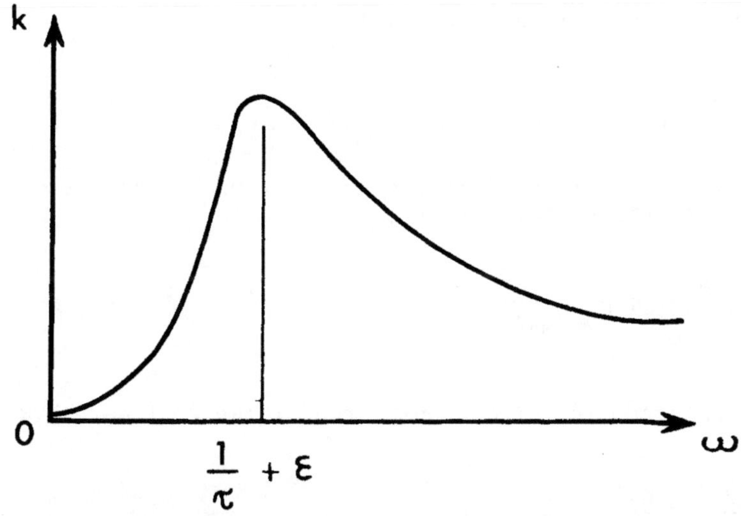

Fig. 21 k vs ω for one Debye relaxation time τ.

which microwave absorption can occur in ferroelectrics. It is the domain boundary motion, because of the inertia of the ions which change position slightly as their dipole moments change direction on passage of the wall.

Kittel [9] has calculated the surface inertia of the boundary in barium titanate on a model having a boundary, N lattice constants thick, separating domains polarized in opposite direction.

It is supposed that the Ti ion changes position within the unit cell by a distance δ between adjacent domain (Fig. 22). In uniform wall motion the average velocity of the Ti ions v_{Ti} is proportional to the wall velocity v_w

$$v_{Ti} = \frac{\delta}{\Delta t} = \frac{\delta}{Na/v_w}; \quad v_{Ti} = \frac{\delta}{Na} \; v_w \tag{48}$$

a being the lattice constant.

The corresponding kinetic energy is written:

$$\frac{1}{2} \, m \, N \left(\frac{\delta}{Na}\right)^2 \; v_w$$

It is related to a unit cell wall area, i.e. a^2. Hence the kinetic energy per unit area of wall:

$$w_o = \frac{1}{2} \, m \, \frac{N}{a^2} \left(\frac{\delta}{Na}\right)^2 \; v_w^{\,2} \tag{49}$$

and the definition of an effective mass per unit area for displacements of wall at velocity v_w:

$$\rho = \frac{w_o}{1/2 \; v_w^{\,2}} \; ; \quad \rho = \frac{m \, \delta^2}{Na^4} \tag{50}$$

The equation of motion of one unit area of the wall may be written:

$$\rho \frac{d^2 x}{dt^2} + r \frac{dx}{dt} + q \, x = 2 \, P_s \, E. \tag{51}$$

P_s is the spontaneous polarization, E the applied field, r represents damping effects due to phonons, impurities etc .... and q the restoring force associated with local trapping. Let us take r = 0 for convenience,

$$E = E_o \, e^{j\omega t},$$

and we look for a solution

$$x = x_o \, e^{j\omega t}, \quad x'' = -\omega^2 \, x$$

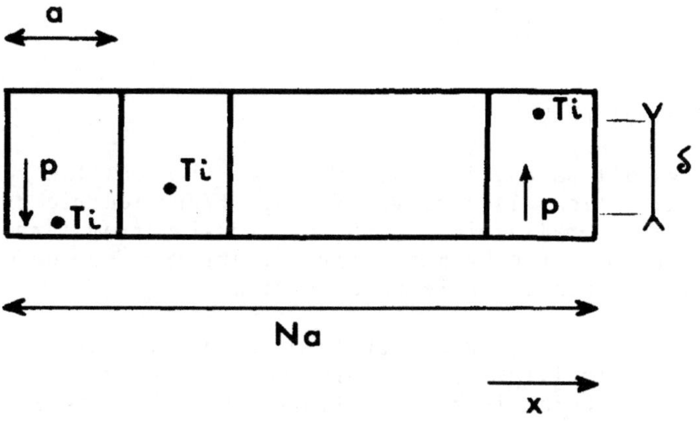

Fig. 22 Domain wall motion in $BaTiO_3$.

and so:

$$x(-\rho\omega^2 + q) = 2P_s E_o e^{j\omega t};$$

$$x = \frac{2 P_s E_o}{q - \rho\omega^2} e^{j\omega t}$$

so there is a resonance at ω_o such that

$$q - \rho\omega_o^2 = 0 \qquad \omega_o = (\frac{q}{p})^{1/2}. \tag{52}$$

Now

$$q = \frac{4 P_s^2}{D \chi d^o} \tag{53}$$

D being the width of a domain and χd^o the part of the static dielectric susceptibility arising from domain boundary displacements. Finally

$$\omega_o = \frac{4 P_s^2}{(\rho D \chi d^o)^{1/2}} . \tag{54}$$

Taking P_s = 50 000 ues; χd^o = 100; D = 10^{-2} cm; δ = 0.2 Å; a = 4 A; N = 1, we get

$$f_o = \frac{\omega_o}{2\pi}; \quad f_o = 2.10^9 \text{ cps.}$$

H. The Central Peak

Rayleigh diffusion, i.e. diffusion of light at the same frequency as excitation, is observed for all ferroelectric crystals. Its origin is generally simple, it is diffusion from ferroelectric domains and it may even be more trivial, arising from deformations, dislocations and other static imperfections.

More interesting is a theory from Cowley [10] or Dorner [11] showing the possibility for the soft mode frequency to shift down to a small value which is however finite when $T \to T_c$ while the intensity of the Rayleigh diffusion is increasing. This additional Rayleigh diffusion due to dynamics effects is called critical central mode or central peak, or critical opalescence. The theory is essentially assuming deviations from the molecular field theory. Quantitative measurements of the width and intensity of the diffusion line are required.

Because of the important static diffusion centers, such definitive observations are rare and the story of the central peak is a frustrating one. Lines and Glass [12] can only cite the example of lead germanate where central peak width and line-shape measurements are reported (Fleury and Lyons [13]).

REFERENCES

1. A. S. Barker, "Infrared Dielectric Behavior of Ferroelectric Crystals", in Ferroelectricity, Elsevier, Amsterdam, 1967, p. 221.

2. A. S. Barker, "Dispersion and Spectra of Ferroelectric Crystals", in Far Infrared Properties of Solids, Plenum Press, New York, 1970, p. 247.

3. D. H. Martin, "The Study of the Vibration of Crystal Lattices by Far Infrared Spectroscopy", in Advances in Physics, 14, 39 (1965).

4. C. Kittel, Introduction to Solid State Physics, John Wiley, New York, 1968.

5. M. Born and K. Huang, Dynamical Theory of Crystal Lattices, Oxford University Press, Oxford, 1954.

6. H. Poulet and J. P. Mathieu, Spectres de Vibration et Symétrie des Cristaux, Gordon and Breach, New York, 1970.

7. G. Breit and E. Wigner, Phys. Rev., 49, 519 (1936).

8. J. F. Scott, Phys. Rev. Lett., 21, 907 (1968).

9. C. Kittel, Phys. Rev. Lett., 4, 458 (1951).

10. R. A. Cowley, G. J. Coombs, R. S. Katiyar, J. F. Ryan, and J. F. Scott, J. Phys., C4, L203 (1971).

11. B. Dorner, J. D. Axe, and G. Shirane, Phys. Rev., B5, 1950 (1972); see also Phys. Rev. Lett., 26, 519 (1973).

12. M. E. Lines and A. M. Glass, Principles and Applications of Ferroelectrics and Related Materials, Oxford University Press, Oxford, 1977.

13. P. A. Fleury and K. B. Lyons, Phys. Rev. Lett., 37, 1088 (1976).

BIBLIOGRAPHY

T. A. Bak, Phonons and Phonon Interactions, Benjamin, New York, 1964.

A. S. Barker and M. Tinkham, J. Chem. Phys., 38, 2257 (1963).

J. C. Burfoot, Ferroelectrics, Van Nostrand, Princeton, 1967.

E. Burstein, "Polaritons, Coupled Photon, Polarization Excitation Modes.I", Comments on Solid State Physics, 1, 202, 1968.

A. J. Dekker, Solid State Physics, Prentice Hall, 1960.

A. Delahaigue, Etude d'un Cristal Moléculaire de thiourée par Spectroscopie Infrarouge et Raman. Thèse de Doctorat d' Etat, Rennes, 1977.

R. Dupeyrat "Raman Spectra of Thin Films," Proceedings of the 6th International Conference on Raman Spectroscopy, Heyden, London, 1978.

J. R. Durig and D. J. Gerson "Historical Survey of the Infrared and Raman Spectroscopic Study of Biological Molecules," in Infrared and Raman Spectroscopy of Biological Molecules, (Theo M. Theophanides, Ed.) Reidel Publishing Co., Dordrecht, 1979.

W. G. Fateley, F. R. Dollish, M. McDevitt, and F. F. Bentley, Infrared and Raman Selection Rules for Molecular and Lattice Vibrations, Wiley, New York, 1972.

B. Fisher, D. Bauerle, and W. J. Buckel, Comm. Solid State Phys., 14, 291 (1974).

X. Gerbaux, A. Hadni, and J. M. Waldschmidt, Opt. Comm., 21, 425 (1977).

M. Gerl, Transitions de Phase Structurales, Ecole d'été d'Aussois, 1978.

A. Hadni, Essentials of Modern Physics Applied to the Study of the Infrared, Pergamon Press, New York, 1967.

A. Hadni, G. Morlot, X. Gerbaux, D. Chanal, F. Brehat and P. Strimer, C. R. A. S., 260, 4973 (1965).

C. Hass, Vibration Spectra of Crystal, Thèse, Amsterdam, 1956.

B. Khelifa, Contribution à l'étude de la Ferroelectricité d'un Cristal Anharmonique par Spectroscopie Infrarouge et Raman, Thèse de Doctorat d'Etat, Reims, 1977.

J. L. Koenig, "Vibrational Spectroscopy of Polypeptide and Proteins, Vibrational Spectroscopy of Carbohydrates", in Infrared and Raman Spectroscopy of Biological Molecules, (Theo. M. Theophanides, Ed.), Reidel Publishing Co., Dordrecht, 1979.

R. A. Levy, Principles of Solid State Physics, Academic Press, New York, 1968.

B. N. Mavrin, T. E. Abramovitch, and K. E. Sterin, Sov. Phys. Solid State, 14, 1562 (1974).

H. D. Megaw, Ferroelectricity in Crystals, Methuen, London, 1957.

M. A. Omar, Elementary Solid State Physics, Addison, New York, 1975.

S. E. Polchlopek," Attenuated Total Reflection", in Applied Infrared Spectroscopy, Reinhold, New York, 1966, p. 462.

J. F. Scott, Review of Modern Physics, 46, 83 (1974).

W. G. Spitzer, R. C. Miller, D. Z. Kleinman, and L. E. Howarth, Phys. Rev., 126, 1710 (1962).

E. F. Steigmeier and G. Harbeke, Solid State Comm., 8, 1275 (1970).

J. A. Stratton, Electromagnetic Theory, McGraw Hill, New York, 1942.

M. Tinkham, Science, 145, 240 (1964).

D. Vergnat-Grandjean, P. Vergnat, X. Gerbaux, and A. Hadni, J. Phys., C12, 3343 (1979).

P. A. Wilks, "A Practical Approach to Internal Reflection Spectroscopy in Laboratory Methods in Infrared Spectroscopy, Heyden and Son, London, 1972, p. 205.

J. M. Ziman, Principles of the Theory of Solids, Cambridge University Press, Cambridge, 1964.

LATTICE MODES OF MOLECULAR CRYSTALS

John E. Bertie

Department of Chemistry
University of Alberta
Edmonton, Alberta, Canada

1. INTRODUCTION

The lattice vibrations of molecular crystals arise from the translational and rotational motions of the molecules, which are unhindered motions in the gas phase. They are intermolecular vibrations and are sometimes called external vibrations, while the vibrations that exist in the gas phase are intramolecular vibrations and are sometimes called internal vibrations. The lattice vibrations can be usefully thought of as translational vibrations and rotational vibrations, although the two types of motion often contribute to the same vibration. Translational and rotational vibrations are sometimes called translatory and rotatory vibrations, and rotational vibrations are frequently called librations.

As is the case for all vibrations, the frequencies of lattice vibrations depend on the inertial mass and the forces which oppose the motion. The inertial mass is the molecular mass for translational vibrations and the moment of inertia for rotational vibrations, and the restoring forces are simply the intermolecular forces in the crystal. Thus, the frequencies of the lattice vibrations give a direct measurement of the intermolecular forces after they have been corrected for the inertial mass. If the intermolecular forces are increased by increasing the pressure or by lowering the temperature, the frequencies of the lattice vibrations increase.

If it can be shown that certain lattice vibrations are purely translational vibrations, or if the molecules are non-polar, the absorption intensities of the translational vibrations are also

J. R. Durig (ed.), Analytical Applications of FT-IR to Molecular and Biological Systems, 433–465.

due to the intermolecular forces. The absorption intensity is proportional to $(\partial\mu/\partial Q)^2$, where $\partial\mu/\partial Q$ is the change in the dipole moment, μ, of a molecule as it translates off of its lattice site, and all of the other molecules translate off their sites, during the vibration, Q. This change in dipole moment is due to the distortion of the electron distribution in the molecule by the intermolecular forces and is, thus, proportional to these forces through the polarizability of the molecule.

The absorption intensity of purely rotational vibrations of non-polar molecules has a similar origin, except that it is the anisotropy of the intermolecular forces at the molecular site that is important in this case. For dipolar molecules, however, the intensities of the rotational vibrations which reorient the dipole moment contain a contribution which is proportional to μ^2/I, where I is the moment of inertia, which is usually much larger than the contribution from the distortion of the electron distribution by the intermolecular forces. Thus, all lattice vibrations of dipolar molecules which involve some rotation of the molecular dipole are usually much stronger than purely translational vibrations, or than lattice vibrations of non-dipolar molecules.

In order to obtain quantitative information about the intermolecular forces from the frequencies and intensities of lattice vibrations, it is necessary to assign the observed absorption lines to particular vibrations and to attempt to reproduce the frequencies and intensities by calculations based on model force fields. Such work constitutes a current research area for physical chemists, physicists and theoreticians and I will not discuss it in detail in this chapter.

The qualitative relationships discussed above can be nicely illustrated by the spectra of the clathrate hydrates of oxetane (trimethylene oxide, $(CH_2)_3O$) and cyclopropane, compared to the spectra of pure oxetane. These two molecules are the only ones that form clathrate hydrates of both the common types, called structure I hydrates and structure II hydrates [1]. The simplest formulae of these hydrates are $6G \cdot 46H_2O$ for structure I and $G \cdot 17\ H_2O$ for structure II, where G is the guest molecule, oxetane or cyclopropane in these cases. In these clathrates, the water molecules are four-coordinated and hydrogen-bonded into three-dimensional networks which form pentagonal dodecahedral cages as well as 14-hedral, or tetrakaidecahedral, cages in the structure I hydrates and 16-hedral, or hexakaidecahedral, cages in the structure II hydrates. In the hydrates of oxetane and cyclopropane the dodecahedral cages are empty, but each 14-hedral cage contains a guest molecule in the structure I hydrate and each 16-hedral cage contains a guest molecule in the structure II hydrate [2,3]. The oxygen atoms of the water molecules which form these occupied cages are shown in Fig. 1. The 14-hedral cage has

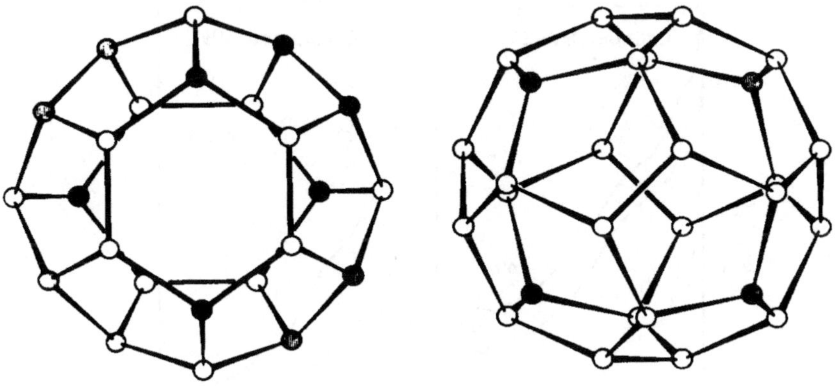

Fig. 1 The oxygen atoms of the water molecules of 14-hedral (left) and 16-hedral (right) cages in the structure I (left) and II (right) clathrate hydrates.

two opposing hexagonal faces joined by 12 pentagonal faces [4,5] and can encage molecules whose van der Waals diameters are 5.5 Å or less [1]. The 16-hedral cage is larger, and has four tetrahedrally disposed hexagonal faces which are joined by twelve pentagonal faces [6]. It can encage molecules whose van der Waals diameters are 6.6 Å or less [1]. The guest molecules are not hydrogen-bonded to the cage, even though oxetane dissolves in water because it can form such bonds.

Thus, the intermolecular forces which act on the oxetane or cyclopropane guest molecules are expected to be much larger in the structure I hydrate than in the structure II hydrate, and this has been confirmed by the dielectric relaxation and N.M.R. studies of Davidson and his coworkers ([1], plus specific citations in Refs. [2] and [3]). The far-infrared spectra of the structure I and II hydrates of cyclopropane and oxetane are shown in Figs. 2 and 3. The comparison of the spectra of the structure I hydrate and deuterate of cyclopropane, (Fig. 2) and the comparison of the spectrum of cyclopropane hydrate with that of Xenon structure I hydrate [7], indicates that no features in the spectrum of the structure I hydrate of cyclopropane can be assigned to the vibrations of cyclopropane. Thus, the intermolecular vibrations of encaged cyclopropane absorb too weakly to be observed above the absorption by the vibrations of the water molecules. The same conclusion was reached [3] for cyclopropane in its structure II hydrate. In contrast, two features in the spectrum of oxetane (TMO) structure I hydrate (Fig. 3) were assigned [2] to the vibrations of oxetane, a peak at 51 cm^{-1} and a shoulder at 68 cm^{-1} at 4.3°K; and two distinct features, and a possible third, were so assigned [3] in the spectrum of the structure II hydrate (Fig. 3),

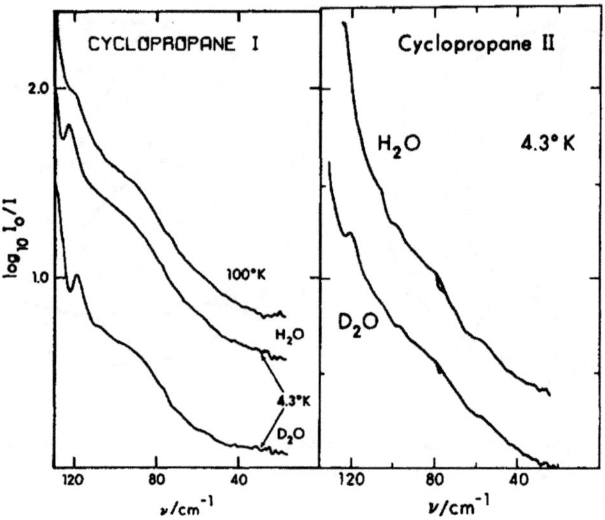

Fig. 2 Far-infrared spectra of the clathrate hydrates and deuterates of cyclopropane; left-structure I, right-structure II; resolution 2 cm^{-1}; samples were powders dispersed in adamantane pellets.

a peak at 34.0 cm^{-1}, a distinct shoulder at 27 cm^{-1}, and a possible faint shoulder at ~20 cm^{-1} at 4.3°K. The two pronounced features in each case arise from the two rotational vibrations that reorient the dipole moment of oxetane, and the higher frequencies in the structure I hydrate reflect the stronger intermolecular forces due to the smaller cage. The restoring forces which act on the oxetane molecule in its structure I hydrate are, however, weaker than the intermolecular restoring forces in crystalline oxetane. This is evident from Fig. 4, which shows a (poor) far-infrared spectrum [7] of crystalline oxetane at 4.3°K, which contains peaks at frequencies up to 165 cm^{-1}. These peaks reduce in frequency without drastic modification of their intensities [7], when the temperature is raised, indicating that they are due to lattice vibrations and not to the other possibility, a ring puckering vibration [8].

2. VIBRATIONAL COUPLING

In the above discussion of the spectra of the clathrate hydrates, it was tacitly assumed that the absorptions by all of the guest molecules in the sample coincide to produce just two spectral features. In other words it was assumed that, for example, the rotational vibration about the largest axis of inertia

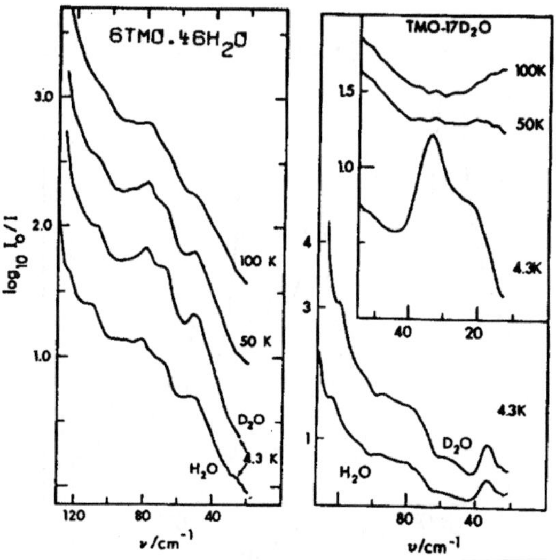

Fig. 3 Far-infrared spectra of the clathrate hydrates and deuterates of trimethylene oxide (TMO); resolution 2 cm^{-1}; left-structure I, right-structure II; samples were powders dispersed in adamantane pellets.

has essentially the same frequency for all oxetane molecules in the crystal. This represents a classic case of quantum mechanical resonance, and the degeneracy is destroyed if one molecule is influenced by the state of vibration of other molecules. The restoring force for the intermolecular vibrations of the guest molecules in the clathrate hydrates is almost entirely due to the water molecules, and the motion of one guest molecule produces a very small force on its neighbors [2,3]. Thus the vibrations of the guest molecules are practically uncoupled which means that they do all have essentially the same frequency. If this were generally the case for molecular crystals, one could assign the spectrum of crystalline oxetane (Fig. 4) by saying that each of the six main peaks, at 165, 122, 99, 85, 77 and 67 cm^{-1} arose from one of the six intermolecular modes of motion of the oxetane molecule. Thus one could assign, for example, the 165 cm^{-1} peak to the rotational vibration about the smallest axis of inertia, the 67 cm^{-1} peak to the translation in the 'plane' of the ring perpendicular to the dipolar axis, etc.

In fact this is totally invalid for molecular crystals, because the restoring force to the intermolecular motion of one molecule arises from its neighbors, so the intermolecular force acting on each molecule depends greatly on the state of vibration of its neighbors. Thus the intermolecular vibrations of molecules

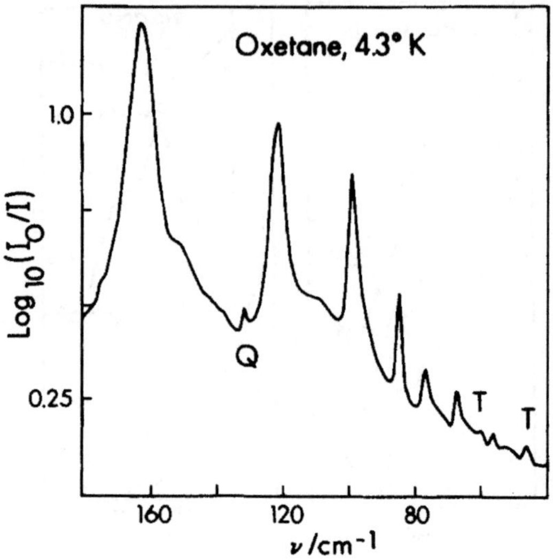

Fig. 4 Far infrared spectrum of crystalline trimethylene oxide (TMO) at 2 cm^{-1} resolution. Sample was a thin film crystallized between TPX windows. Q and T indicate peaks due to crystal quartz and teflon.

in molecular crystals are strongly coupled, which means that in a mole of crystal there are not just three frequencies due to translational vibrations and three frequencies due to rotational vibrations, with the vibrations of the 6 x 10^{23} different molecules being independent but at the same frequency. The vibrational coupling ensures that there are 3 times 6 x 10^{23} frequencies of translational vibrations and 3 times 6 x 10^{23} frequencies due to rotational vibrations, each of which is different except for degeneracies imposed by symmetry. Further, during each normal vibration of the crystal each molecule may be undergoing translational motion of all three types and rotational motion of all three types simultaneously.

The way in which vibrational coupling affects the far-infrared spectrum depends on the nature of the crystal, in particular on the symmetry of the crystal. A molecular crystal can be ordered, which means that each atom in the molecule is repeated at exactly equal distances along any line drawn through the crystal. This means that molecules with a given orientation are repeated at exactly equal distances and the crystal has translational symmetry. The spectrum due to intermolecular vibrations in such a crystal is like that of oxetane (Fig. 4) and consists of a small number of sharp lines. This arises because the symmetry of the crystal provides selection rules that allow only a very small number of vibrations to be infrared-active. Further, this symmetry may mean that the infrared-active vibrations are either translational vibrations or rotational vibrations, rather than vibrations of mixed character.

A molecular crystal can also be orientationally-disordered. This means that each atom of the molecule is not repeated at exactly equal distances along any line drawn through the crystal, but the center of mass of the molecule is repeated at very nearly equal distances along any line, so that the positions of the molecules show translational symmetry but their orientations do not. Ice is the classic example of such a crystal [9], and the clathrate hydrates are also orientationally-disordered crystals [1]. The orientational disorder ensures that the crystal has no symmetry on the molecular level, so there can be no selection rules based on symmetry. All of the 36×10^{23} lattice vibrations per mole of crystal are infrared-active, and very broad absorption is seen. An example of such a spectrum is in Fig. 2, which shows the absorption by the translational vibrations of the water molecules in the cyclopropane hydrates, and in Fig. 5 which extends the absorption to high frequencies for the structure II deuterate [3]. Fairly sharp features can be seen in the spectra of such solids, but they are superimposed on continuous absorption which can show the sort of structure that is evident in Figs. 2 and 5.

A molecular crystal can also be disordered with respect to the positions and orientations of the molecules, but then it is a glass, or amorphous solid, not a crystal. The far-infrared spectra of molecular glasses have not been extensively studied by intention, although the spectra of several glassy crystal samples have been reported as I will mention later. The absorption is expected to be broad because of the lack of symmetry, and to show little structure because of the wide variations in intermolecular distances and orientations that are believed to occur in glasses. To illustrate this, and to terminate the discussion of glasses, Fig. 6 shows the far-infrared spectrum of a sample of vitreous ice [10](some prefer the name 'amorphous solid water' [11]) compared with that of the orientationally-disordered ice Ih or normal ice [10,12].

3. SELECTION RULES FOR ORDERED CRYSTALS

To understand the infrared spectra of lattice vibrations it is necessary to have a language which enables the 36×10^{23} lattice vibrations per mole of crystal to be described concisely. This is only possible for ordered crystals and, to some extent and approximation, for orientationally disordered crystals. This language is based on the translational periodicity of ordered

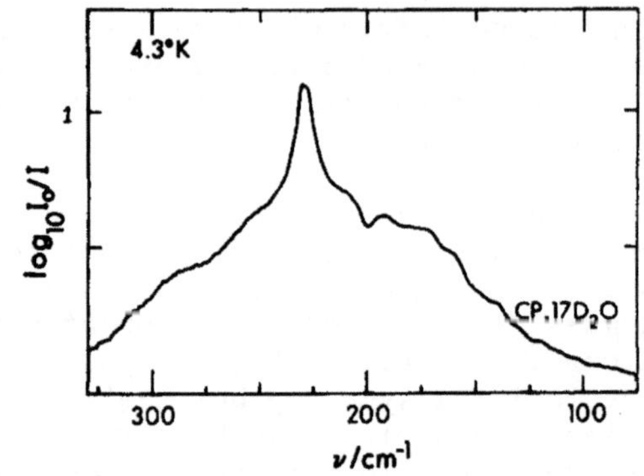

Fig. 5 Far infrared spectrum of the structure II deuterate of cyclopropane at 2 cm^{-1} resolution. The sample was powder dispersed in 3-methyl pentane glass.

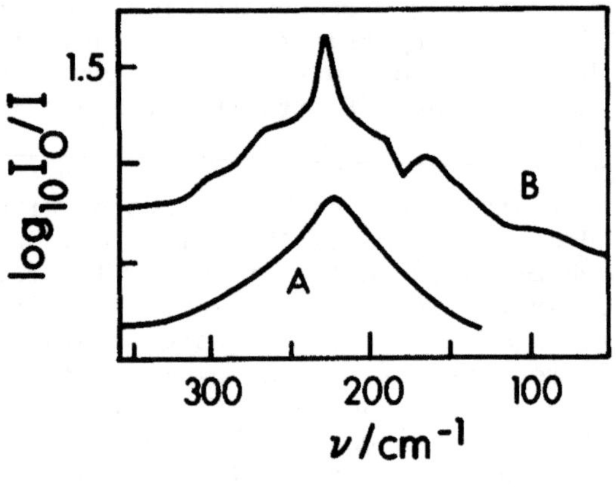

Fig. 6 Far infrared spectrum of vitreous ice (curve A) on a polyethylene window, and of ice 1h (curve B) mulled in propylene, at 100°K and 2.5 cm^{-1} resolution.

crystals, which means that a crystal is composed of a large number of identical units, or unit cells. The smallest unit cell from which the entire crystal can be built is called the primitive unit cell. The relevant consequence of translational periodicity for this discussion is that the vibrations can be described as displacement waves which propagate through the crystal with a wavelength $\lambda = |k|^{-1}$ and frequency f in the direction specified by the wave-vector $\underline{k}$, whose magnitude $|k|$ is λ^{-1}. Only certain wave-vectors are possible, but there is a sufficiently large number of them to ensure that the number of intermolecular displacements that are possible in one unit cell (the number of intermolecular degrees of freedom per unit cell) times the number of wave-vectors per mole of crystal equals the 36×10^{23} intermolecular vibrations per mole of crystal.

Such displacement waves are illustrated for a one-dimensional lattice in Fig. 7. The circles represent two particles, of mass M_1 and M_2, in each primitive unit cell. In the top half of the picture are drawn the two translational vibrations with longitudinal displacements (along the line of the lattice) that are possible if the displacements in one unit cell are duplicated in all other unit cells. The wavelength of the displacement wave is infinite so that the wave-vector $\underline{k}$ is zero for these vibrations. It is clear that the lower of the two vibrations is simply a translation of the entire lattice and that its frequency is zero. This vibration is said to be an acoustic vibration, and in a three-dimensional crystal there are always three acoustic vibrations, i.e. vibrations with zero wave-vector and zero frequency, corresponding to the translation of the entire crystal in the three cartesian directions. The upper of the two $\underline{k} = 0$ vibrations clearly does not have zero frequency because the intermolecular distances are changed by it. It is called an optic vibration. If M_1 and M_2 are two atoms in one molecule, then there is one molecule per unit cell, and the $\underline{k} = 0$ optic vibration is an intramolecular vibration while the only $\underline{k} = 0$ intermolecular vibration is the acoustic mode with zero frequency. If M_1 and M_2 represent two molecules in the unit cell, then there is one optic and one acoustic intermolecular translational vibration with zero wave-vector, the latter having zero frequency. In a three dimensional crystal with m molecules per primitive cell there are 3m - 3 optic and three acoustic translational vibrations with $\underline{k} = 0$. There are also 3m optic rotational vibrations (2m for linear molecules) with $\underline{k} = 0$.

The lower half of Fig. 7 shows the two translational vibrations that are possible if the displacements in one unit cell are the opposite of those in the adjoining cells. The curve shows the form of the displacement wave, with the ordinate showing the actual displacement of the M_1 - M_2 pair for the lower vibration and the relative displacement of the two particles in one unit cell for the upper vibration. The wavelength is 2a where a is the length of the primitive cell or the magnitude of the primitive translation vector, so that the wave-vector's length is $(2a)^{-1}$. The relation between these vibrations and the upper two is clear and the optic or acoustic label for these vibrations follows from it. Neither of the vibrations with $|k| = (2a)^{-1}$ has zero frequency. Other vibrations exist for the 1-dimensional lattice of Fig. 7 with values of $|k|$ between 0 and

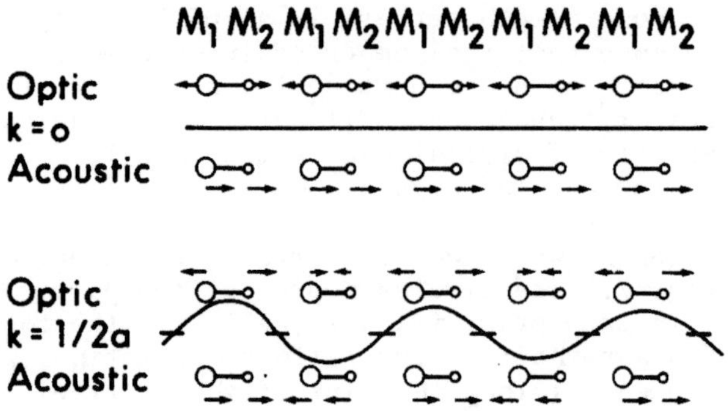

Fig. 7 Longitudinal vibrations with $\underline{k}$ = 0 and $(2a)^{-1}$ of a linear lattice of unit cell length a. The circles are the masses, the arrows are displacements and the solid line or curve is the displacement wave.

$(2a)^{-1}$ and the displacements in one unit cell equal to cos $(2\pi ka)$ times those in the neighboring cell to the left. A graph of the frequency of the vibrations against their wave-vector is shown in Fig. 8, and is said to show the dispersion curves. The upper branch of the dispersion curves is the optic branch and, with a nearest neighbor force constant f, contains frequencies $2\pi\nu$ ranging from $[2f(M_1^{-1} + M_2^{-1})]^{\frac{1}{2}}$ down to $[2fM_2^{-1}]^{\frac{1}{2}}$. The lower branch is the acoustic branch and contains frequencies from $[2fM_1^{-1}]^{\frac{1}{2}}$ to zero. The vibrations of a real, three-dimensional crystal are described in the same way, but are, of course, complicated by the possibility of molecular motion and wave propagation in three-dimensions. These ideas are presented in more detail in several texts [13-17].

In an ordered crystal, the change in dipole moment during a vibration can be described by a wave with the same wave-vector as the displacement wave. Absorption of electromagnetic radiation by the vibration can only occur if the energy, $\underline{E}.\underline{\mu}$, of interaction between the electric vector of the radiation, $\underline{E}$, and the dipole moment change, $\underline{\mu}$, is not zero, and this is the case only if the wave-vector, $\underline{K}$, of the radiation in the crystal equals the wave-vector, $\underline{k}$, of the vibration. Thus, the selection rule based on translational periodicity is $\underline{k} = \underline{K}$, which prevents all but a handful of the 36×10^{23} lattice vibrations in a mole of crystal from absorbing radiation.

The simplifying approximation that is always used for this selection rule is that $\underline{k} = 0$. Its utility is based on the fact that for infrared radiation $\underline{k}$ is about 10^3 to 10^4 times smaller than $(2a)^{-1}$ and the frequency of the vibration with $\underline{k} = \underline{K}$ is

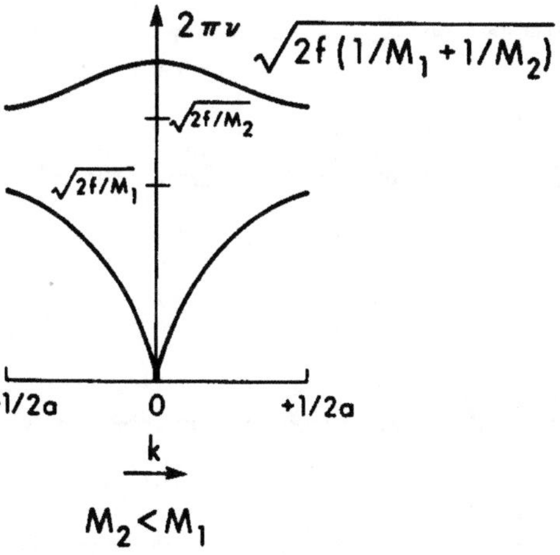

Fig. 8 Dispersion curve for the longitudinal vibrations of the lattice of Fig. 7, assuming harmonic forces between nearest neighbors only.

extremely close to that of the $\underline{k} = 0$ vibration. Further, if $\underline{k} = 0$ is used, only one unit cell of the crystal need be considered for prediction of the intermolecular (or other) vibrations that will be active as fundamental transitions. This, of course, is the tremendous simplification that has led to the use of the $\underline{k} = 0$ approximation.

It should be noted that the $\underline{k} = \underline{K}$ selection rule does very little to restrict the vibrations that can absorb radiation through overtone or combination (2-phonon, 3-phonon, etc.) transitions, because the rule in these cases is $\underline{K} = \Sigma \underline{k}_i$ and for each vibration with $\underline{k} = \underline{k}_i$ there is one with $\underline{k} = -\underline{k}_i$ and the same frequency. Consequently, overtone or combination bands are generally very broad. They will not be discussed further in this chapter.

In another lecture Dr. Califano has discussed crystal symmetry, so the remainder of this discussion of selection rules will be brief and illustrated by the case of crystalline chlorine. Chlorine crystallizes [18] in the orthorhombic system, and has the Hermann-Mauguin symbol Cmca, or the Schonflies symbol D_{2h}^{18}, which indicates that the unit cell group or factor group is isomorphous with the point group D_{2h} [19-21]. Figure 9 shows the diagram given in the International Tables for X-ray Crystallography [21] of the unit cell of the space group Cmca, with the solid circles representing the chlorine atoms and the numbers above these circles indicating the distance, in Å, of the atoms above the bottom of the unit cell. The crystallographic a-axis is vertical, the b-axis is horizontal and the c-axis is perpendicular

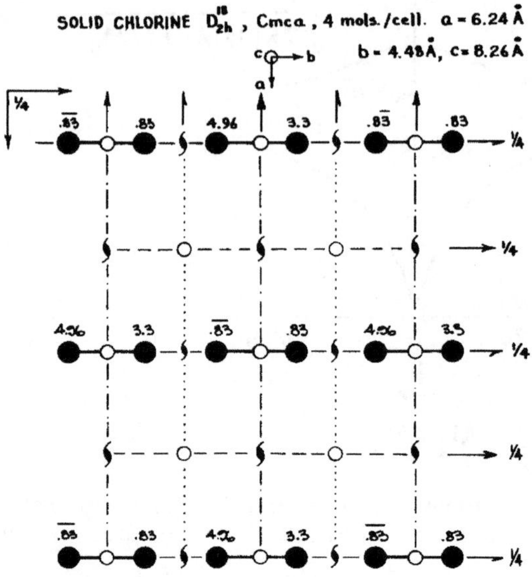

Fig. 9 The crystallographic C-centered unit cell of crystal chlorine. The solid circles are Cl atoms and the numbers above them are the distances in Å from the bottom of the unit cell. Other symbols are described in the text. Space group Cmca, D_{2h}^{18}, Z = 4 molecules; axis lengths a = 6.24 Å, b = 4.48 Å, c = 8.26 Å.

to the plane of the paper. The half-arrows represent 2-fold screw-axes either in the basal plane and in the plane half-way along the c-axis of the unit cell or, if the arrow is marked with 1/4, in the planes parallel to a and b one quarter and three quarters of the way along the c-axis of the cell. The whole arrows represent two-fold rotation axes with the presence or absence of '1/4' meaning the same as for screw-axes. The solid lines parallel to b at the top, center and bottom of Fig. 9 represent mirror planes parallel to c. The dashed, dotted, and dash-dotted lines indicate glide planes parallel to the bc, ac and ac axes, respectively, with translations parallel to b (b-glide), c (c-glide) and a + c (diagonal glide), respectively. The double-arrow symbol at the top left indicates a + b glide planes at one quarter and three quarters of the way along the c-axis of the cell. Finally, ʃ indicates a 2-fold screw axis parallel to the c-axis and the open circle indicates the center of symmetry at the center of each chlorine molecule.

The unit cell in Fig. 9 contains four molecules but it is not the unit cell required for spectroscopic purposes, which is the primitive unit cell. The symbol C in the space group designation indicates that it is a C-centered cell, which is used by the crystallographers because it has orthogonal axes. It is not the primitive cell because the pair of molecules at the top left and top middle can be generated from the pair at the center and center right of the diagram simply by a diagonal translation, and no molecules in a primitive cell are related by a simple translation. Each of these pairs of molecules can be regarded as constituting a lattice point centered on the first molecule, so the C-centered unit cell in Fig. 9 has a lattice point at each corner and one in the center of its base and top planes, the C-faces of the cell. It therefore contains two lattice points and a primitive cell only contains one [17,19,20].

The primitive cell can be drawn in an infinite number of ways, but a standard form of primitive cell is the Wigner-Seitz cell [15, 17, 22]. This cell is the volume enclosed by the planes which bisect, and are perpendicular to, the lines joining one lattice point to its first- and second-nearest neighbors, and to third- etc. neighbors if such planes are needed to form a closed volume. The Wigner-Seitz cell of chlorine is shown in Fig. 10. It was constructed from Fig. 9 as described above, then slid sideways so that the two molecules are totally within the cell. The $\underline{k} = 0$ lattice vibrations can be drawn using this cell.

To determine the representation formed by the lattice vibrations under the unit cell group, (i.e. the symmetry of the lattice vibrations) the site group is useful. It is the group formed by the symmetry elements that leave invariant the site that is occupied by the molecules [14, 15-16]. For chlorine it is (Fig. 9) C_{2h}, the 2-fold axis parallel to $\underline{a}$ ($C_2(a)$), the mirror plane parallel to b and c ($\sigma(bc)$), and the center of symmetry, all leaving the site invariant. The symmetry under the site group of the translations parallel to a, b and c and of the rotations about the a and c axes can be determined from the character tables for the C_{2h} point group and the representations that they generate under D_{2h} can be determined from correlation tables [15,16,23, 24], to give the correlation diagram shown in Table 1. This method, the correlation diagram method, of determining the symmetries of the vibrations of molecular crystals has been simply and well described by W. G. Fateley <u>et al.</u> [24,25] and by others [14-16,26], and is based on the early work of Bhagavantam and Venkatarayudu [27], Hornig [28] and Halford [29].

Figure 10 shows that one molecule in the cell is generated from the other by the symmetry operation $C_2(c)$, the 2-fold screw axis parallel to c, which is (Table 1) $C_2(z)$ of D_{2h}. The character table for D_{2h} reveals that the species A_g, B_{1g}, A_u and B_{1u} are symmetric with respect to $C_2(z)$ while B_{2g}, B_{3g}, B_{2u} and B_{3u} are antisymmetric with respect to $C_2(z)$. With this information the symmetry coordinates for the $\underline{k} = 0$ vibrations based on translational displacements parallel to a and rotational displacements

Table 1 Correlation Diagram for the Intermolecular Vibrations of Crystalline Chlorine

Mode of Motion	Site Group[a] C_{2h}	Unit Cell Group[b] D_{2h}
T_a	A_u	A_u + B_{3u}
T_b	B_u	B_{1u} + B_{2u}
T_c	B_u	B_{1u} + B_{2u}
R_a	A_g	A_g + B_{3g}
R_c	B_g	B_{1g} + B_{2g}

a) Symmetry elements; $C_2(a)$, $\sigma(bc)$, i, and E.
b) The C_2 axis of C_{2h} is $C_2(x)$ of D_{2h}, the b-crystal axis corresponds to y and the c-axis to z of D_{2h}[15].
c) T_a, R_c etc. mean translation parallel to a, rotation about c, etc.

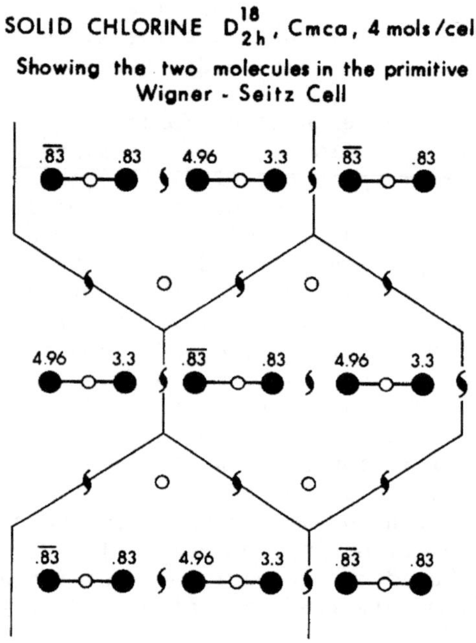

Fig. 10 The Wigner-Seitz, primitive unit cell of crystal chlorine, containing two molecules.

about a can be drawn [23,30], and are shown in Figs. 11 and 12. The B_{3u} mode of motion based on T_a is clearly the translation of

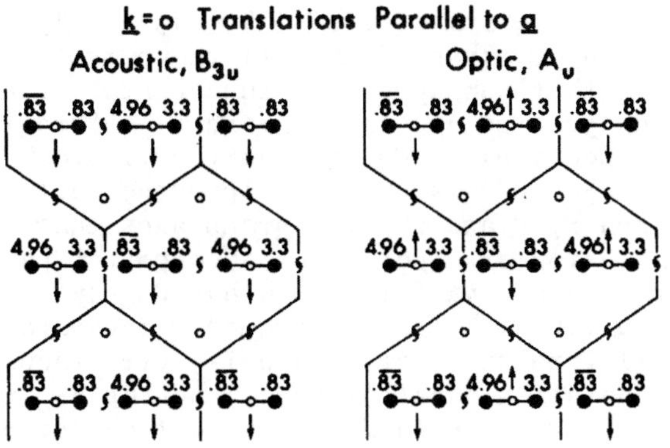

Fig. 11 The k = 0 vibrations of crystal chlorine due to translational displacements parallel to the a-axis.

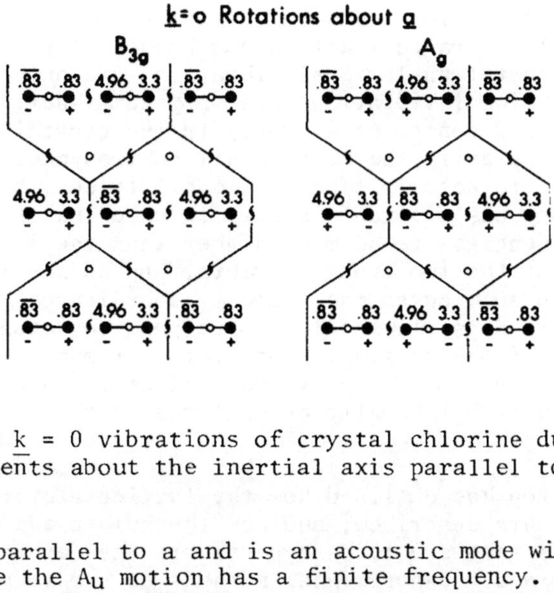

Fig. 12 The k = 0 vibrations of crystal chlorine due to rotational displacements about the inertial axis parallel to the a-axis.

the crystal parallel to a and is an acoustic mode with zero frequency, while the A_u motion has a finite frequency. The character table for D_{2h} reveals that the translations parallel to x, y and z, (the a, b and c axes of chlorine) have the symmetries B_{3u}, B_{2u} and B_{1u}, respectively, so the vibrations underlined in Table 1 are the three acoustic modes. Thus, the three translational modes with finite frequency are the A_u, B_{1u} and B_{2u} modes, while the A_g, B_{3g}, B_{1g} and B_{2g} rotational modes all have (Fig. 12) finite frequency.

A vibration whose fundamental transition is to be infrared active must have [14-16,23] the same symmetry as the dipole moment which, in turn, has the symmetries of T_x, T_y and T_z, i.e. B_{1u}, B_{2u} and B_{3u} under D_{2h}. Thus, (Table 1) the only lattice vibrations of chlorine that can appear as fundamental transitions in the far infrared spectrum are the B_{1u} and B_{2u} vibrations based on T_b and T_c, respectively. Under a unit cell group that is isomorphous with D_{2h} and has the correspondence between a, b and c crystal axes and x, y and z point group axes that is given in Table 1, a translation or dipole moment component parallel to c has B_{1u} symmetry and one parallel to b has B_{2u} symmetry. Conversely, the B_{1u} vibration with finite frequency causes a dipole moment change parallel to the c crystal axis, even though it is due to molecular translation parallel to b, and the B_{2u} vibration with finite frequency causes a dipole moment change parallel to the b crystal axis, even though it is due to translational motion parallel to c. If a single crystal is sliced perpendicular to the a axis, only the B_{1u} vibration will be observed with far infrared radiation whose electric vector is polarized parallel to the c axis and only the B_{2u} vibration will be seen with radiation polarized parallel to the b axis.

The rotational vibrations in crystal chlorine are all gerade, and are all infrared inactive but Raman-active. It is always the case that translational vibrations are only infrared-active and rotational vibrations are only Raman-active if the molecules are on a center of symmetry in the crystal. Note that this is not necessarily the case in all centrosymmetric crystals. A further point to note is that only one lattice vibration of finite frequency has a given symmetry in chlorine, and the intramolecular frequencies are so much higher than the intermolecular frequencies that the two types of vibrations do not interact significantly. In such cases the symmetry coordinates, i.e. the motions drawn in Figs. 11 and 12, correspond to actual $\underline{k} = 0$ normal vibrations of the crystal. But if two or more symmetry coordinates have the same symmetry and similar frequencies, the normal vibrations will usually be mixtures of the symmetry coordinates.

This section has outlined how the lattice vibrations of ordered crystals are described, and how the number and type of lattice vibrations that are to be expected in the far infrared spectrum is determined. Experimental methods of sample preparation and results for some ordered crystals are given in the next section.

4. SPECTRA OF ORDERED CRYSTALS

The vast majority of far infrared spectra of ordered molecular crystals that has been reported to date have been transmission spectra, either on percent transmission, $(I_t/I_o) \times 100\%$, or absorbance, $\log_{10}(I_o/I_t)$, scales. These spectra measure the energy that is lost by reflection or scattering from the surfaces of the crystals as well as that lost by absorption, and are only an approximation to the absorption spectrum that one wants to compare with theory and with the Raman spectrum. This is illustrated by the poor spectrum of crystalline oxetane in Fig. 4, which shows lines that are unsymmetrical and often tail to low frequency with a sharp dip on the high-frequency side. This arose because the sample froze into small crystals with many reflecting surfaces, so that reflection effects contributed markedly to the transmission spectrum. The reflectance at the surface of a crystal in air increases as the refractive index increases, and the anomalous dispersion of the refractive index through an absorption band [31] makes the refractive index decrease on the high frequency side of a band, increase sharply near the band center, and decrease slowly on the low-frequency side of the band, causing the effects shown in Fig. 4. Effects due to the frequency dependence of the refractive index are called dispersion effects, and because of them absorption bands of solid and liquid samples only appear to be symmetrical in the graph of the dielectric loss, ε'' (the imaginary part of the dielectric constant), against frequency and the band maximum is only at the vibration frequency in these graphs [15,32] (Fahrenfort [32] says that the frequency of the band maximum is at the resonant frequency in the graph of $\nu\varepsilon''$ against ν, not that of ε'' against ν, but for the fairly narrow peaks expected for ordered crystals the graphs of $\nu\varepsilon''$ and ε'' against ν differ very little in shape near the peak).

To obtain dielectric loss spectra one must measure the optical constants of a solid, that is the real, n, and imaginary, k, refractive indices, from which the real, ε', and imaginary, ε'', dielectric constants can be calculated via the relations

$$\varepsilon' = n^2 - k^2$$

$$\varepsilon'' = 2nk.$$

The imaginary refractive index, k, is related to the quantity K, which is sometimes called the absorptivity [33] (against I.U.P.A.C. recommendations) and which is obtained from the true absorbance spectrum (corrected for reflection and scattering losses) via the relation

$$2.303 \log_{10} I_o/I = K\ell$$

where ℓ is the thickness of the solid or liquid film. The relation between k and K is

$$K = 4\pi k\nu$$

where ν is the wavenumber of the radiation. The theory and practice of such measurements is discussed elsewhere, [14-16,32,34-38] and is not necessary in this chapter for reasons that will be given shortly.

The great difference that can exist between dielectric loss spectra and transmission spectra due to dispersion effects is well illustrated by the lattice vibrations of the ionic crystal potassium bifluoride [39,40]. The far infrared transmission spectrum of a thin transparent film of KHF_2 on a polyethylene window, or of powdered KHF_2 dispersed in a polyethylene disk, both at ~295°K, was reported [39] to contain two broad unsymmetrical bands with peak frequencies of ~130 and ~178 cm^{-1} and full widths at half height (FWHH) of 30 and 60 cm^{-1}, respectively. At 100°K the bands were not significantly sharper but had shifted to ~134 and ~191 cm^{-1}. It was reported [39] that the relative intensities of the two bands varied with the method of sample preparation, which is usually a sign of interference by dispersion effects. The reflection spectrum of KHF_2 [40,41] contains reflection peaks at 145 and 187 cm^{-1}, ±~5 cm^{-1} in each case, which resemble the peaks in the transmission spectrum [39]. The dielectric loss spectrum [40], calculated from the reflection spectrum, shows two lines at 132 and 170 cm^{-1} at 300°K and 139 and 181 cm^{-1} at 100°K, the width and shape of the features being comparable to those in the Raman spectrum [40], i.e. about ten times narrower than in the infrared transmission spectrum, and essentially symmetrical.

The frequency dependence of far infrared optical constants of molecular crystals in regions where lattice vibrations absorb appears to have been measured only for some polymers, namely polyethylene, TPX, mylar, and others [42]. The refractive index changes are small, because the absorption intensities are small, so dispersion effects are small and transmission spectra are a good qualitative approximation to the dielectric loss spectra. This has generally been assumed to be true for all molecular crystals and, because it is orders of magnitude easier to obtain transmission spectra than dielectric loss spectra of solids, practically all far infrared spectra of lattice vibrations of molecular crystals have been transmission spectra. Occasionally, as in Fig. 4, there is evidence of error due to dispersion effects, but it can usually be removed by more careful freezing of the liquid or by dispersing the powdered sample in a liquid or glassy medium, as in a nujol mull.

Even if dispersion effects are not a problem, one must still be careful about sample preparation to obtain a good transmission spectrum of lattice vibrations of molecular crystals. This is illustrated for crystal chlorine in Fig. 13. Curve A is the spectrum at 2 cm^{-1} resolution of powdered chlorine [7] that was dis-

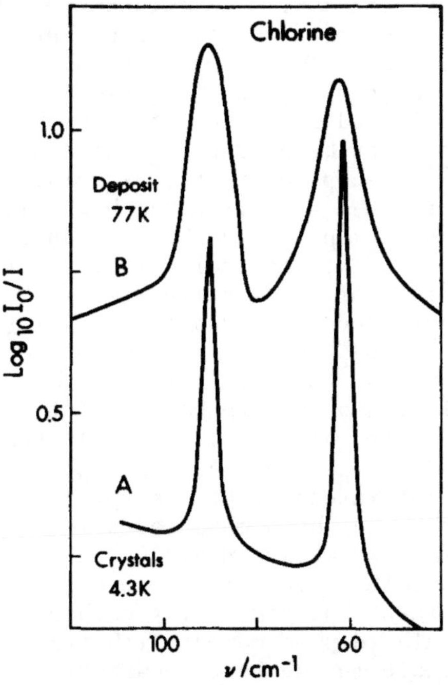

Fig. 13 The far infrared spectrum of chlorine. A) Powdered crystals dispersed in 3-methyl pentane glass at 4.3°K, 2 cm^{-1} resolution. B) Deposited from gas onto a copper plate at 77°K and not annealed; 77°K and 4 cm^{-1} resolution. Used by permission [44].

persed in 3-methyl pentane at ~120°K, cooled to ~90°K to transform the 3-methyl pentane into a rigid glass, loaded into a Janis Research Corporation model 8DT cryostat, and cooled to 4.3°K. The experimental details of this low-temperature mulling technique have been described [2,43]. Curve B of Fig. 13 is a spectrum at 4 cm^{-1} resolution of chlorine that was condensed from the gas phase onto a copper block at 77°K [44], producing, according to the authors [44], a deposit that was partly amorphous and partly crystalline. The much greater breadth of the absorption by the semi-amorphous deposit is clear.

At 4.3°K (Fig. 13, Curve A) the peak frequencies of solid chlorine are 61.2 and 90.4 cm^{-1}, both $\pm$~0.3 cm^{-1}, and the FWHH of the peaks is 3.5 cm^{-1}. At 90°K, the frequencies of crystalline samples of Cl_2 have been given as 60.0 and 88.0 cm^{-1} [45] and as 58.8 $\pm$ 0.2 and 86.8 $\pm$ 0.3 cm^{-1} [46], in both cases with the FWHH being 4.8 $\pm$ 0.2 and 2.9 $\pm$ 0.2 cm^{-1} for the low- and high-frequency peaks, respectively. The frequencies thus increase significantly on cooling from 100 to 4.3°K, but the effect on the FWHH is more complex. To assign these observed lines to the predicted (Sec. 3, Table 1) B_{1u} and B_{2u} vibrations, one needs to record spectra of oriented single crystals with polarized incident light. This has not been done for chlorine, but the assignment of the ~60 cm^{-1} absorption to the B_{1u} vibration and the ~90 cm^{-1} absorption to the B_{2u} vibration has been made [47] on the basis of calculations which fitted the frequencies of the Raman-active vibrations to an intermolecular force field and predicted the frequencies of the infrared-active vibrations. The Raman-active vibrations have been observed [47] at 87 (B_{1g}), 103 (B_{2g}), 120 (A_g) and 154 cm^{-1} (B_{3g}) at 4.3°K, none of which coincide with the infrared frequencies, thus illustrating the effect of the rule of mutual exclusion on the spectra of lattice vibrations. The existence of the band at 154 cm^{-1} is disputed and a more recent assignment is 83 (B_{1g}), 100 (A_g and B_{2g}) and 118 cm^{-1} (B_{3g}) at ~16°K [48,49]. The intensity of absorption by the lattice vibrations of chlorine has been measured twice [45,46]. The agreement between the two measurements is only about a factor of two, but even the smaller values [46] are a factor of about thirty-five larger than expected from the theory [50] that the change in the molecular dipole moment with translational displacement of a molecule in the lattice arises from the field due to the quadrupole moments of neighboring molecules. It has been concluded [46] that the high intensity is due to a weak type of bond between molecules, i.e. to particularly strong intermolecular forces. Whalley and Wong [46] reported that a 35 μm film of Cl_2 has peak absorbances ($\log_{10} I_o/I$) at 58.8 and 86.8 cm^{-1} under 1.5 to 2 cm^{-1} resolution of 0.6 and 0.46, yielding absorptivities, $K = \ell^{-1}\log_{10} I_o/I$, of 395 and 303 cm^{-1}, respectively.

The far infrared spectra of the lattice vibrations in a few more ordered molecular crystals will be briefly discussed in the remainder of this section.

Oxirane ($(CH_2)_2O$, ethylene oxide) is believed [43] to have a unit cell group or factor group that is isomorphous with C_{2h} and to have four molecules on general postions (site group C_1, i.e. no symmetry elements other than the identity) in the primitive unit cell. The Correlation Table is in Table 2. Each displacement of an oxirane molecule, the fifteen intramolecular vibrational displacements and the three translational and the three rotational displacements, yields a vibration of each symmetry under C_{2h}. Thus, the translational vibrations yield $3A_g + 3A_u + 3B_g + 3B_u$ and the rotational vibrations yield the same, $3A_g + 3A_u + 3B_g + 3Bu$. The three acoustic modes have the symmetry $A_u + 2B_u$, so the twenty-one intermolecular vibrations that have finite frequencies form the representation $6A_g + 6B_g + 5A_u + 4B_u$ under the unit cell group C_{2h}. The crystal is centrosymmetric but the

Table 2 Correlation Table for the Lattice Vibrations of Oxirane

Point group of free Molecule C_{2v}		Site group C_1	Unit Cell Group C_{2h}
T_x	B_1		$6A_g$
T_y	B_2		
T_z	A_1	A	$6B_g$
R_x	B_2		
R_y	B_1		$6A_u$
R_z	A_2		
			$6B_u$

molecules are not. Thus the twelve gerade vibrations are Raman active and the nine ungerade ones are infrared active, but translational and rotational displacements can contribute to each vibration.

The far infrared spectrum of oxirane [43] is shown in Fig. 14. Seven peaks due to the nine infrared-active fundamentals are clearly visible and a shoulder can be detected at 102 cm^{-1}. The spectrum depended critically on the time taken to crystallize the oxirane and to cool it from its melting point, 162°K, to 100°K. If this was done too quickly (e.g. ten to fifteen minutes) the lines were much broader, presumably due to the formation of a semi-vitreous solid.

For comparison, Fig. 15 shows the Raman spectrum of oxirane [43] at 22, 80 and 125°K. In the 22°K spectrum ten features are clearly visible, and polarized spectra of near-single crystals clearly showed [43] that the broad-topped feature at 92.5 cm^{-1} is the superposition of two bands, and confirmed the existence of two bands at 100 and 101 cm^{-1}. Thus, eleven of the twelve Raman active fundamentals are located, as well as eight of the nine infrared-active ones. Their frequencies are in Table 3, showing the centrosymmetry via the rule of mutual exclusion. Figure 15 also shows the importance of cooling the sample as far as possible, since at 125°K only four Raman features are evident, while eight can be easily seen at 80°K and ten at 22°K, and no phase transition occurs in this temperature range.

To proceed further with the assignment of the lattice spectra of oxetane one needs further experimental and theoretical information. If the spectra were recorded, with polarized light, of crystals for which the orientation of the crystallographic axes was known, the infrared-active bands could be assigned to vibrations of A_u or B_u symmetry and the Raman-active bands to vi-

Table 3. Frequencies of the Lattice Vibrations of Oxirane at 4.3°K

Infrared in cm^{-1}	Raman in cm^{-1}
	50.7
42.7	63.6
78.1	79.3
97.7	91.4
	93.1
102 sh	100.2
	101.3
106.2	110.0
117.7	115.2
123.2	130.5
139.7	137

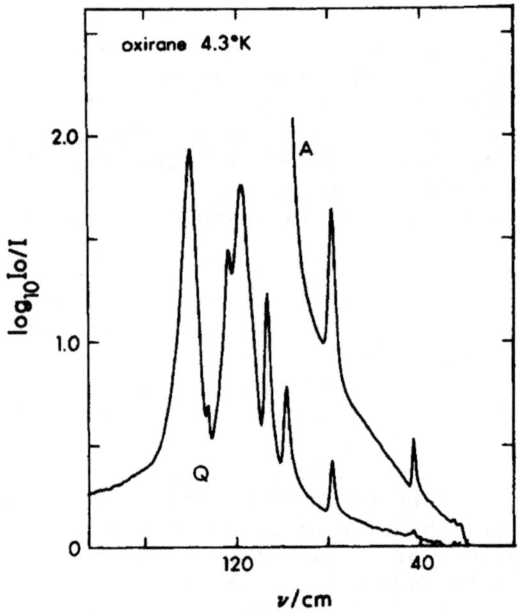

Fig. 14 Far-infrared spectrum of oxirane at 4.3°K. A) Neat solid film, B) powdered solid dispersed in 3-methyl pentane. Resolution 2 cm^{-1}. Used by permission [43].

brations of A_g or B_g symmetry. To proceed further, to associate

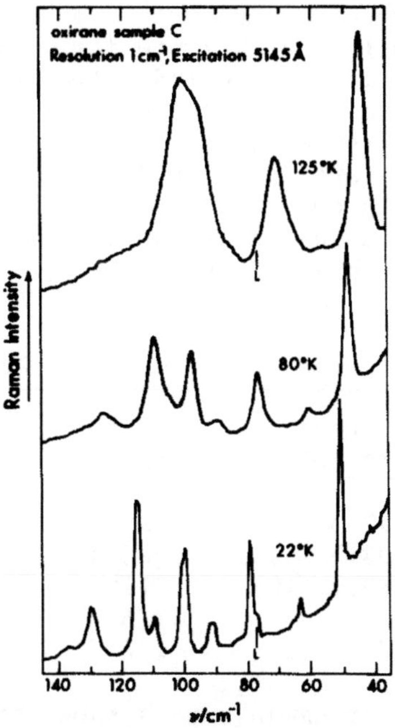

Fig. 15 Raman spectra of oriented crystals of oxirane at 1 cm^{-1} resolution. Used by permission [43].

particular molecular motions with each vibration, the detailed crystal structure must be known and a normal-coordinate calculation must be done using a model intermolecular force field. Such force fields are frequently based on atom-atom potential functions, when these are known reliably. Such calculations have been made for benzene for example [51-53]. It is probable that no simple interpretation would result for oxirane, because each translational and each rotational mode of motion can contribute to each vibration. It can be noted, however, from the lower intensity of the infrared bands near 43 and 78 cm^{-1}, that these two vibrations probably do not involve much reorientation of the molecular dipole.

A molecular crystal whose structure has been investigated spectroscopically for eighteen years is cyclopropane. Since the classic study by mid infrared spectroscopy of Halford and his coworkers [54] many studies have been made of this crystal and the work to 1974 is well summarized by Shurvell et al. [55]. Shurvell noted that the lattice vibration frequencies were not sufficiently well known to distinguish between two structures, a structure in the C_{2v} crystal class with two molecules per unit cell and a centrosymmetric structure in the D_{2h} class with four molecules per unit cell. In 1977 new far infrared and Raman spectra were reported [56] which indicated that the structure is not centrosymmetric and is, therefore, in the C_{2v} class. The far infrared and Raman spectra of cyclopropane are shown in Fig. 16 and the frequencies are in Table 4. The infrared and Raman frequencies coincide except that the Raman band at 126 cm^{-1} has no infrared counterpart and is undoubtedly a Raman-active but infrared-inactive A_2 vibration under the C_{2v} group.

Table 4. Lattice Vibration Frequencies of Cyclopropane at 4°K

Infrared[a] in cm^{-1}	Raman[a] in cm^{-1}
97.3 (3)	98.1
105.3 (6)	106.4
115.0 (10)	~116
	126.2
138.2 (3)	138.5

[a]Accuracy is ±1 cm^{-1} for Raman and is shown in parentheses for the infrared by the uncertainty in the last figure.

There are two points to note about the far infrared spectra of cyclopropane (Fig. 16). The first is the increased breadth of the features at higher temperatures, which is analogous to that shown for Raman spectra in Fig. 15. The second is that the relative intensities of the peaks differ markedly from those in the only earlier study [57]. It was shown, by preparing spectroscopic samples in a variety of ways [56],that crystalline films of cyclopropane tend to grow with a certain orientation which yields the pattern of intensities seen in the earlier work [57], and that the relative intensities of Fig. 16 are correct for a sample which consists of randomly oriented crystallites. A similar effect has been seen in the far infrared spectrum of crystalline benzene [58,59], and one should be aware of this effect if one is trying to identify a crystal by its far infrared spectrum.

Some molecular crystals neither absorb nor Raman scatter at low frequencies, so their lattice vibrations can not be studied directly by spectroscopic means. An example is hexamethylenetetramine (HMT), which crystallizes in the body-centered space group $I\bar{4}3m$, T_d^3, with two molecules per unit cell [60]. There is, therefore, one molecule in the primitive unit cell and the site group must be T_d, the same as the unit cell group. There are only three k = 0 translational lattice vibrations, they have

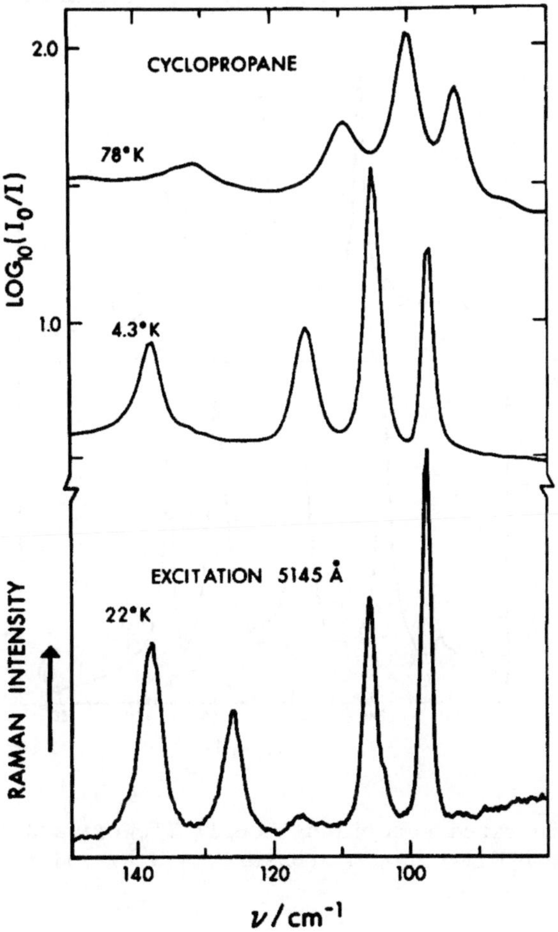

Fig. 16 Infrared and Raman (bottom) spectra of cyclopropane at 2 and 1 cm^{-1} resolution, respectively. The shoulder near 85 cm^{-1} in the top curve is spurious. Used by permission [56].

symmetry F_2 under T_d [23] and they must be the three zero-frequency acoustic vibrations. There are three k = 0 rotational vibrations. The character table for the group T_d shows [23] that they are triply degenerate and of symmetry species F_1, and are inactive in the infrared and Raman spectra. Thus, the only absorption or scattering that is seen to low-frequency of the lowest-frequency intramolecular mode, at 514 cm^{-1} (infrared) or 463 cm^{-1} (Raman), is very weak and due to overtone and combination transitions of the lattice vibrations.

It is generally believed that the spectra of hydrogen-bonded

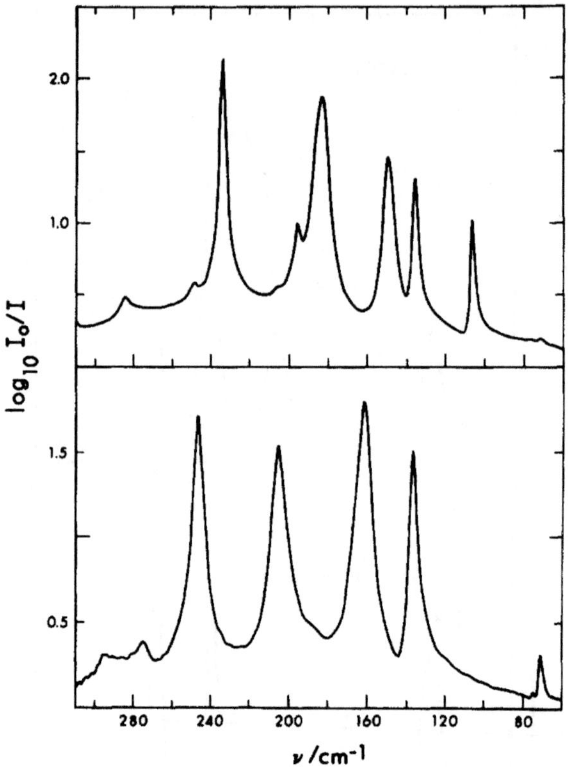

Fig. 17 Far-infrared spectra of ice IX (lower) and ice II (upper), both made from D_2O, at 1 atm pressure, 4.3°K, and 1.8 cm^{-1} resolution.

crystals contain primarily broad bands, but this is not necessarily the case. Figure 17 shows far infrared spectra of the ordered, high-pressure forms of deuterated ice, ices II and IX at 4.3°K [7] and spectra at 100°K have been published [61] and summarized [62] elsewhere. The features shown are due to essentially pure translational vibrations of the molecules and, no doubt because of the large oscillating mass, show none of the breadth that is usually associated with spectra of hydrogen-bonded molecules and which is generally attributed to large anharmonic effects.

One of the most surprising far infrared spectra of a fully ordered molecular crystal is that of acetic acid. The spectrum obtained in my laboratory a few years ago [63] of powdered acetic acid mulled in liquid propane at 100°K is shown in Fig. 18. Broad continuous absorption with some broad structure superimposed on

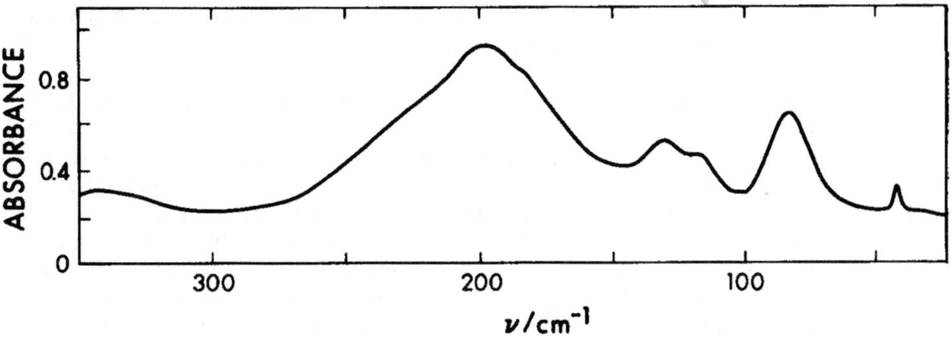

Fig. 18 Far infrared spectrum of acetic acid I at 90°K, 2.5 cm^{-1} resolution. The sample was powder mulled in liquid propane.

it can be seen, and it is typical of the spectra expected for orientationally disordered crystals (next Sec. and Sec. 2). But by neutron diffraction studies, Jonsson [64] has shown acetic acid to be fully ordered. The vibrations involved are relative displacements of the monomers that form the long hydrogen-bonded chains, and inter-chain vibrations, and all involve large inertial masses. The origin of the breadth is unexplained and we are currently studying the low-frequency Raman spectra of acetic acid, and other carboxylic acids, to seek further information about the vibrations.

5. ORIENTATIONALLY DISORDERED CRYSTALS

It was mentioned in Sec. 2 that orientationally disordered crystals have their molecules centered on positions that form a three-dimensional ordered lattice, i.e. on positions that show translational symmetry, but the molecular orientations do not show translational symmetry. Orientationally disordered crystals can be broadly classified into two types, those in which the molecules reorient very rapidly and those in which the molecules are fixed into orientationally disordered configurations. The first type is largely composed of the so-called plastic crystals [65-67], such as the high-temperature phases of the t-butyl halides, carbon tetrahalides, methane, cyclohexane, camphor, and other solids; while the disordered phases of ice [9] and the clathrate hydrates [1] are the classic examples of the second type.

The plastic crystals yield far infrared and Raman bands due to the lattice modes that are very similar to those of the liquid phase, being very broad and featureless. Examples are shown in Refs. [68-70] for t-butyl bromide, acetylene and hydrogen chloride. The theoretical interpretation of such spectra is complicated by the molecular reorientation rates being nearly as high as the lattice vibration frequencies.

The far infrared spectra of statically orientationally disordered crystals have been discussed for those cases in which the translational vibrations are well separated in frequency from the rotational vibrations. In such cases it has been argued [71] that the translational vibrations are to first approximation those of an ordered crystal which has the same diffraction structure as the disordered crystal of interest. Diffraction methods yield a structure that is the average over the whole crystal of the actual structure so that for ice, for example, they report that one-half of a hydrogen atom is situated at each end of each O---O hydrogen bond, instead of the actual case which is that the hydrogen atom is at one end of the bond or the other essentially at random. This argument leads to the conclusion that the density of states curve for the translational vibrations, that is the curve of the number of vibrations with each frequency plotted against frequency, will show considerable structure even though it is basically broad, in contrast to that for a fully disordered solid which shows little structure. It follows from the lack of symmetry in such a phase that there are no selection rules, and that all of the 18×10^{23} translational vibrations per mole of molecules are infrared and Raman active. Thus, the spectrum consists in part of the density of states curve multiplied by an intensity function. The intensity function is probably a slowly varying function of frequency so that features in the spectrum reflect features in the density of states curve. Thus, the peaks, shoulders inflections, etc. in the far infrared absorption by the water molecules in the clathrate hydrates and ice (Figs. 5 and 6) have been assigned to features in the density of states curve of the translational vibrations of the water molecules [2,3,10,12]. Various models for the intensity function have been proposed, and they have recently been reviewed and related [72].

It has also been argued [71] that the absorption or scattering by translational vibrations in orientationally-disordered crystals can show intense, sharp lines due to $k = 0$ vibrations which are allowed by the selection rules for ordered crystals. The relative intensity of this sharp, order-allowed absorption to that of the broad, disorder-allowed absorption depends upon the extent to which the dipole moment change due to the displacement of one molecule in the lattice depends upon which diffraction unit cell the molecule is in, i.e. upon the orientation of the molecule with respect to those of its neighbors. If the dipole moment change is independent of the relative orientations of the molecules, only order-allowed absorption will be seen and the spectrum will be like that of an ordered crystal. If the dipole moment change depends totally on the relative orientations of the molecules, only broad, disorder-allowed absorption will be seen. In the general case, sharp order-allowed absorption may be superimposed on broad disorder-allowed absorption. Similar arguments hold for the Raman spectra. Order-allowed bands have been assigned in the Raman spectrum of ice Ih [73] and, tentatively, in the far infrared spectrum [74] of hexamethylene tetramine hexahydrate which is partially orientationally disordered.

The situation for statically orientationally disordered crystals in which the rotational vibrations can not be separated from the translational vibrations is more complicated, because it is not the case that the rotational vibrations are an approximation to those of a crystal with the same diffraction structure.

6. CONCLUDING NOTE

The examples used in this chapter are, for convenience, largely from my own work. Many people have contributed to our knowledge of the lattice vibrations of molecular crystals, including A. Hadni, who has contributed to far infrared spectroscopy for over twenty years and who has two chapters in this volume, and two of the organizers of this conference, J. R. Durig and W. G. Fateley. To these people, and to all others whose often major contributions have not been mentioned, I apologize and note that people who go on to read the literature of this subject will soon learn of your work. Readers who wish to see a more quantitative and theoretical discussion of lattice vibrations of molecular crystals are referred to two reviews by O. Schnepp [75,76].

REFERENCES

1. D. W. Davidson, Chapter 3 in Volume 2 of Water - A Comprehensive Treatise , (F. Franks, ed.), Plenum Press, New York-London, 1973.

2. J. E. Bertie and S. M. Jacobs, Can. J. Chem., 55, 1777 (1977) and citations therein.

3. J. E. Bertie and S. M. Jacobs, J. Chem. Phys., 69, 4105 (1978) and citations therein.

4. R. K. McMullan and G. A. Jeffrey, J. Chem. Phys., 42, 2725 (1965).

5. F. Hollander and G. A. Jeffrey, J. Chem. Phys., 66, 4699 (1977).

6. T. C. W. Mak and R. K. McMullan, J. Chem. Phys., 42, 2732 (1964).

7. J. E. Bertie and S. M. Jacobs, unpublished results.

8. H. Wieser, M. Danyluk, and R. A. Kydd, J. Mol. Spectrosc., 43, 382 (1972) and citations therein.

9. P. V. Hobbs, Ice Physics, Oxford Univ. Press, Oxford, 1974.

10. J. E. Bertie and E. Whalley, J. Chem. Phys., 46, 1271 (1967).

11. Discussion report, J. E. Bertie, J. W. Glen and G. P. Johari, J. Glaciol., 21, 672 (1978).

12. J. E. Bertie and S. M. Jacobs, J. Chem. Phys., 67, 2445 (1977).

13. L. Brillouin. Wave Propagation in Periodic Lattices, Dover Publications, New York, 1953.

14. G. Turrell, Infrared and Raman Spectra of Crystals, Academic Press, London and New York, 1972.

15. J. C. Decius and R. M. Hexter, Molecular Vibrations in Crystals, McGraw-Hill, New York, 1977.

16. H. Poulet and J. P. Mathieu, Spectres de Vibration et Symétrie des Cristaux, Gordon and Breach, New York, 1970 (also published in English by same publisher).

17. C. Kittel, Introduction to Solid State Physics, J. Wiley, New York, 1967.

18. R. L. Collin, Acta Cryst., 5, 431 (1952); 9, 537 (1956).

19. E. W. Nuffield, X-ray Diffraction Methods, J. Wiley, New York, 1966.

20. M. J. Buerger, Elementary Crystallography, J. Wiley, New York, 1963.

21. International Tables for X-ray Crystallography, (N. F. M. Henry and K. Lonsdale, eds.), The Kynoch Press, Birmingham, England, 1965, Vol. 1.

22. E. Wigner and F. Seitz, Phys. Rev., 43, 804 (1933).

23. E. B. Wilson, J. C. Decius, and P. C. Cross, Molecular Vibrations. McGraw-Hill, New York, 1955.

24. W. G. Fateley, F. R. Dollish, N. T. McDevitt, and F. F. Bentley. Infrared and Raman Selection Rules for Molecular and Lattice Vibrations: The Correlation Method, Wiley-Interscience, New York, 1972.

25. W. G. Fateley, N. T. McDevitt, and F. F. Bentley, Appl. Spectrosc., 25, 155 (1971); W. G. Fateley, Appl. Spectrosc., 27, 395 (1973).

26. R. L. Carter, J. Chem. Ed., 48, 297 (1971).

27. S. Bhagavantam and T. Venkatarayudu, Proc. Ind. Acad. Sci., A9, 224 (1939).

28. D. F. Hornig, J. Chem. Phys., 16, 1063 (1948).

29. R. S. Halford, J. Chem. Phys., 14, 8 (1946); H. Winston and R. S. Halford, J. Chem. Phys., 17, 607 (1949).

30. L. A. Woodward, Introduction to the Theory of Molecular Vibrations and Vibrational Spectroscopy, Oxford University Press, Oxford, 1972.

31. F. A. Jenkins and H. E. White, Fundamentals of Optics, McGraw-Hill, New York, 1957.

32. J. Fahrenfort, Chapter XI in Infrared Spectroscopy and Molecular Structure, (M. Davies, ed.), Elsevier, Amsterdam, 1963.

33. J. E. Bertie, H. J. Labbe and E. Whalley, J. Chem. Phys., 50, 4501 (1969).

34. E. E. Bell, Proc. Aspen International Conference on Fourier Spectroscopy, Aspen, 1970, (G. A. Vanasse, A. T. Stair and D. J. Baker, eds.), AFCRL.

35. J. Chamberlain, The Principles of Interferometric Spectroscopy, (G. W. Chantry and N. W. B. Stone, eds.), Wiley-Interscience, Chichester, England, 1979.

36. N. W. B. Stone and G. W. Chantry, Chapter 2 in Adv. Infrared and Raman Spectroscopy, Vol. 3, (R. J. H. Clark and R. E. Hester, eds.), Heyden, London, 1977.

37. R. J. Bell, Introductory Fourier Transform Spectroscopy, Academic Press, New York, 1972.

38. K. D. Möller and W. G. Rothschild, Far-Infrared Spectroscopy, Wiley-Interscience, New York, 1971.

39. P. A. Giguère and K. Sathianandan, Can. J. Phys., 45, 2439 (1967).

40. P. Dawson, M. M. Hargreave, and G. R. Wilkinson, Spectrochim. Acta, 31A, 1055 (1975).

41. R. S. Singh, S. Trevino, and H. J. Prask, J. Chem. Phys., 58, 4703 (1973).

42. D. R. Smith and E. V. Loewenstein, Appl. Optics, 14, 1335 (1975).

43. J. E. Bertie and S. M. Jacobs, J. Chem. Phys., 68, 97 (1978).

44. S. H. Walmsley and A. Anderson, Mol. Phys., 7, 411 (1964).

45. J. G. David and Willis B. Person, J. Chem. Phys., 48, 510 (1968).

46. P. T. T. Wong and E. Whalley, Can. J. Phys., 50, 1856 (1972).

47. M. Suzuki, T. Yokoyama, and M. Ito, J. Chem. Phys., 50, 3392 (1969).

48. A. Anderson and T. S. Sun, Chem. Phys. Lett., 6, 611 (1970).

49. J. E. Cahill and G. E. Leroi, J. Chem. Phys., 51, 4514 (1969).

50. O. Schnepp, J. Chem. Phys., 46, 3983 (1967).

51. I. Harada and T. Shimanouchi, J. Chem. Phys., 46, 2708 (1967).

52. D. A. Oliver and S. H. Walmsley, Mol. Phys., 17, 617 (1969).

53. G. Taddei, H. Bonadeo, M. P. Marzocchi, and S. Califano, J. Chem. Phys., 58, 966 (1973).

54. C. Brecher, E. Krikorian, J. Blanc, and R. S. Halford, J. Chem. Phys., 35, 1097 (1961).

55. H. F. Shurvell, S. J. Daunt, and D. W. James, J. Mol. Spectrosc., 53, 77 (1974).

56. J. E. Bertie and S. M. Jacobs, J. Chem. Phys., 67, 4981 (1977).

57. J. B. Bates, D. E. Sands, and W. H. Smith, J. Chem. Phys., 51, 105 (1969).

58. Y. A. Sataty, A. Ron, and M. Brith, Chem. Phys. Lett., 23, 500 (1973).

59. I. Harada and T. Shimanouchi, J. Chem. Phys., 55, 3605 (1971).

60. J. A. K. Duckworth, B. T. M. Willis, and G. S. Pawley, Acta Cryst., A26, 263 (1970) and citations therein.

61. J. E. Bertie, H. J. Labbé, and E. Whalley, J. Chem. Phys., 49, 775 (1968).

62. J. E. Bertie, Appl. Spectrosc., 22, 634 (1968).

63. J. E. Bertie and R. Wilton, unpublished work.

64. P-G Jonsson, Acta Cryst., B27, 893 (1971).

65. J. G. Aston, Chapter 9 in Physics and Chemistry of the Organic Solid State, Vol. 1, Interscience, New York, 1963.

66. Phys. Chem. Solids, 18, issue 1, (1961).

67. A. Cabana, Chapter 2 in Vibrational Spectra and Structure, Vol. 4, (J. R. Durig, ed.), Elsevier, Amsterdam, 1975.

68. J. E. Bertie and S. Sunder, J. Chem. Phys., 59, 498 (1973).

69. Y. A. Schwartz, A. Ron, and S. Kimel, J. Chem. Phys., 51, 1666 (1969).

70. M. Ito, M. Suzuki, and T. Yokoyama, J. Chem. Phys., 50, 2949 (1969).

71. E. Whalley and J. E. Bertie, J. Chem. Phys., 46, 1264 (1967).

72. D. D. Klug and E. Whalley, J. Chem. Phys., 71, 2903 (1979).

73. P. T. T. Wong and E. Whalley, J. Chem. Phys., 65, 829 (1976).

74. J. E. Bertie and M. Solinas, Can. J. Chem., 53, 2642 (1975).

75. O. Schnepp, Adv. At. Mol. Phys., 5, 155 (1969).

76. O. Schnepp and N. Jacobi, Adv. Chem. Phys., 22, 205 (1972).

INFRARED SPECTRA OF SOLIDS AT NORMAL AND HIGH PRESSURES

John E. Bertie

Department of Chemistry
University of Alberta
Edmonton, Alberta, Canada

1. INTRODUCTION

The purpose of this chapter is to give an introduction to molecular vibrations in crystals and to the information that has been obtained by changing the pressure of the crystal. The emphasis will be on intramolecular vibrations, since lattice vibrations were discussed in a previous article and will be discussed again later by Dr. Hadni, and on infrared spectra, in keeping with the topic of this institute. Raman spectra will, however, be included where appropriate, and an example of the effect of pressure on lattice vibrations will also be given.

2. INTRAMOLECULAR VIBRATIONS IN CRYSTALS

This subject is treated in detail from several viewpoints in texts and reviews [1-10]. The following account omits much of the detail, for which the reader is referred to an appropriate source.

The potential energy of a molecule in a solid [1,3,5] can be regarded as the sum of the potential energy of the isolated molecule, V_i, plus the potential energy, V_s, due to the static forces exerted on the molecule by the surrounding molecules, plus the potential energy, V_d, due to the dynamic intermolecular forces that cause each molecule to know the state of vibration of other molecules. The two contributions, V_s and V_d, due to crystal forces cause the main differences that exist between the intramolecular vibration frequencies of a molecule in the gas phase and those in the crystal. These differences are: (1) the frequency of a vi-

J. R. Durig (ed.), Analytical Applications of FT-IR to Molecular and Biological Systems, 467–500.

bration that is not coupled to those of the other molecules in the crystal is not the same as the frequency of the same vibration in the gas-phase, because of V_s; (2) A vibration that is degenerate in the gas-phase can have its degeneracy reduced in the crystal if the site symmetry is lower than the symmetry of the gaseous molecule, so that two or more uncoupled vibration frequencies in the solid correspond to one frequency in the gas-phase; this is again due to V_s; (3) The vibrations of different molecules are all coupled by V_d, so that each uncoupled vibration in the solid yields a band of vibration frequencies by intermolecular vibrational coupling.

The first of these three differences is usually called a site shift and the second is called site splitting. But site splitting can also result from the first difference if the same chemical species occupies two non-equivalent sites in the crystal. In this case, the V_s term at one site can not be identical to that at the second site, so the uncoupled frequencies of the molecules at the first site differ from those at the second site, and site splitting results even if the degeneracy of the vibration has not been reduced in the crystal.

To introduce the language used to describe the third difference, due to intermolecular vibrational coupling, between the frequencies of molecules in the gas and solid phases, it is necessary to consider the effect of the crystal symmetry on the selection rules and, hence, on the spectra that can be observed. The principles are those outlined in the previous chapter on Lattice Vibrations in Molecular Crystals. Each uncoupled vibration of a molecule in the solid yields branches of dispersion curves, in the same way as each degree of translational or rotational freedom yields such branches except that the branches due to intramolecular vibrations are all optic branches. Thus, if there is just one molecule per primitive unit cell and it has 3N-6 intramolecular vibrations, there are 3N-6 optic branches due to intramolecular vibrations in addition to the three optic branches due to rotational vibrations and the three acoustic branches due to translational vibrations. If there are four equivalent molecules per primitive unit cell, which means that the four molecules are on sites that are related by the symmetry elements of the crystal and are, thus, symmetrically equivalent sites with a multiplicity of four [8,11,12], then there are 12N-24 intramolecular optic branches in addition to the twelve optic branches due to rotational vibrations and nine optic and three acoustic branches of translational vibrations. Further, if there are four molecules on one type of site and two molecules of the same species on a different type of site in each primitive unit cell, each of the 3N-6 intramolecular vibrations yields six optic branches. If the uncoupled frequency of the set of molecules on the fourfold site is close to that of the molecules on the twofold site, and if

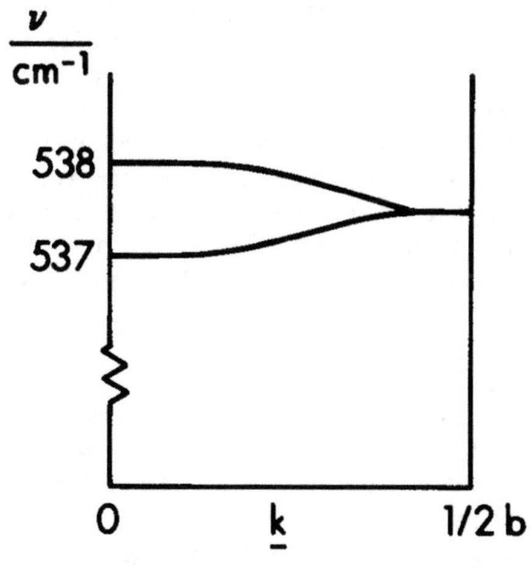

Fig. 1 Approximate dispersion curves for Cl-Cl stretching vibrations in solid chlorine which propagate in the b-crystallographic direction.

symmetry allows it, each normal vibration in each of the six branches in general involves motion of both sets of molecules. This is the usual case, but cases can exist in which hydrogen-bonding causes one set of molecules to have uncoupled frequencies that are greatly different from those of another set; in such cases each normal vibration of the crystal involves very largely one set or the other, so that, say, four branches of dispersion curves are due to one set and two branches are due to the second set.

Figure 1 shows approximate dispersion curves for the Cl-Cl stretching vibrations of crystalline chlorine, space group Cmca, D_{2h}^{18}, four molecules on C_{2h} sites per centered cell, which was discussed in the previous article. The two molecules in the primitive cell are on equivalent sites and yield two branches of dispersion curves for the Cl-Cl stretching vibration. The spectrum of solid chlorine is complicated by the mixture of isotopes, but the two branches of dispersion curves are believed [13] to be at 537 and 538 cm^{-1} at $\underline{k} = 0$, and they are coincident at the boundary of the Brillouin zone [14] for waves propagating parallel to the b crystallographic axis. (The Brillouin zone boundary parallel to $\underline{b}$ is the point where the wave vector in this direction has magnitude 1/2b, where b is the length of the primitive cell - and centered cell in this case - in this direction). The dispersion curves for chlorine are clearly almost flat, and this is usually the case for intramolecular modes [9], reflecting the fact that the intermolecular forces are usually much weaker than the intramolecular forces.

As was described in the previous article, the only vibrations that can absorb (or Raman-scatter) radiation in an ordered crystal are those whose wave-vector $\underline{k}$ equals the wave-vector $\underline{K}$ of the radiation. The approximation $\underline{k} = 0$ is used for intramolecular vibrations for the same reasons as for intermolecular ones and, while it greatly simplifies the theoretical treatment, it is important to be aware that it is an approximation that can lead one to predict a smaller number of bands in the spectrum than one would predict from $\underline{k} = \underline{K}$. This difficulty will be discussed explicitly later.

The contents of the last two paragraphs, taken with section 3 of the previous article, indicate that a molecular crystal which contains m N-atomic molecules per primitive unit cell has only m(3N-6) intramolecular vibrations, 3m rotational vibrations and 3m-3 translational vibrations which have $\underline{k} = 0$ and non-zero frequencies and which may, therefore, absorb or Raman-scatter radiation. If the m molecules are on equivalent sites, each uncoupled intramolecular vibration yields m $\underline{k} = 0$ vibrations which have different frequencies because of the intermolecular vibrational coupling. Thus one frequency splits into m because of intermolecular vibrational coupling, a phenomenon that is usually called Davydov splitting or correlation splitting. The combination of Davydov splitting and site splitting, that is the total splitting due to static and dynamic intermolecular forces, is called here factor group splitting or unit cell group splitting, but note that one text says that some people use the term 'factor group coupling' to describe intermolecular vibrational coupling (p. 251, Ref. [9]), so some confusion about the meaning of these terms does exist.

Not all of the $\underline{k} = 0$ vibrations can absorb or Raman-scatter radiation through fundamental transitions. To determine which ones are active, the factor group symmetries and selection rules must be found, usually by the correlation method [11,12]. That is, one must determine which vibrations have the same symmetry as the dipole moment or the polarizability under the factor group or unit cell group [8,9,15], which is isomorphous to the point group of the crystal class.

The $\underline{k} = 0$, Cl-Cl stretching vibrations of solid chlorine are shown in Fig. 2. The Davydov splitting is believed to be 1 cm^{-1}, with the B_{3g} vibration at 538 cm^{-1} and A_g at 537 cm^{-1} [13]. Neither vibration is infrared-active but both are Raman-active.

The Davydov components may or may not be centered on the frequency of the uncoupled vibration. If the unit cell is large they probably are centered and if it is small they probably are not. The displacement of the average of the frequencies of the Davydov components from the uncoupled frequency is due to the interaction between the vibrations of one molecule and those of the transla-

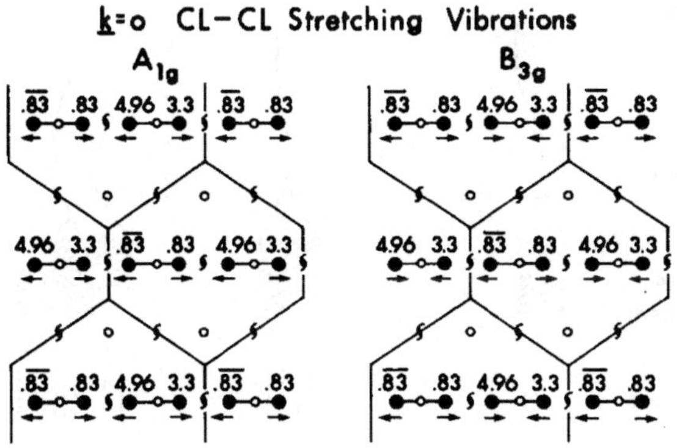

Fig. 2 The k = 0, Cl-Cl stretching modes of chlorine. The frequencies of the A_g and B_{3g} modes differ by 1 cm^{-1} due to intermolecular coupling.

tionally equivalent molecules in other unit cells [3]. If there is just one molecule per unit cell, the same considerations indicate that the frequency observed for a vibration in the pure crystal is probably not the uncoupled frequency. The uncoupled frequency can be observed to a good approximation as the frequency of a 'vibrationally isolated' molecule, i.e. one that is surrounded by isotopic molecules whose frequencies are quite different, so that the dynamic intermolecular forces do not significantly shift the frequencies of the isolated molecule or of its host.

Figure 3 summarizes schematically the frequency shifts and splittings that may occur when a gas is condensed to a crystal. It should be noted that for a disordered crystal one will not observe a discrete number of lines as shown but will observe the bands of frequencies shown on the third row of the figure. Further, the vibrations in the two overlapping bands in the right-hand column interact if symmetry permits.

The crystal symmetry can cause a further, important difference between the spectra of molecules in the gas phase and in crystals. The symmetry of the occupied site is determined by the symmetry of the static potential energy term, V_S, and can be lower than that of the molecule in the gas phase. As mentioned above, this can cause degeneracies to be reduced, thus causing site splitting, and it can also cause vibrations that are inactive in the gas phase to be active under the site group. At least one of the Davydov-split components of a vibration that is active under the site group is active under the unit cell group. Thus vibrations that are inactive in the gas phase but active under the site group do yield at least one line in the spectrum of the solid, if the crystal-activated intensity with which they absorb or Raman-scatter

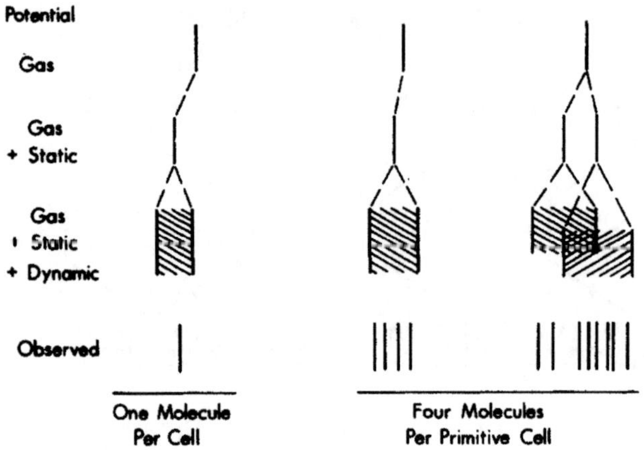

Fig. 3 Diagram showing the effects of static and dynamic crystal forces on intramolecular vibration frequencies. The observed lines may not be centered on the uncoupled frequencies shown in the 'gas + static' line.

is sufficiently high to be observed.

The spectrum of a crystal shows the unit cell group splittings or factor group splittings, that is the combined effects of static and dynamic crystal forces. It is necessary to next discuss what is observed experimentally, how one can determine whether observed splittings are due to site effects or intermolecular coupling, and what complications can confuse matters.

3. INTRAMOLECULAR VIBRATIONS IN CRYSTALS - PRACTICE

The simplest molecular crystals contain just one molecule per primitive unit cell, which means that the site group must be the same as the point group of the crystal class [16]. For primitive (P or R) space groups this requires that the space group be symmorphic, i.e. that the space group contain no screw-axes or glide-planes [8-10], but this is not the case for centered (A, B, C, F, or I) space groups; these must be non-symmorphic since the centering is generated by screw-axes or glide planes. The site group must also be either the point group of the gaseous molecule or a sub-group of it. Three examples of such crystals are: (1) the low-temperature phase of HCN [17], space group Imm2, C_{2v}^{20}, two molecules per body-centered cell or one molecule per primitive cell; the site symmetry is C_{2v}, a sub-group of the $C_{\infty v}$ point group of the gaseous molecule; (2) Hexamethylene tetramine, $(CH_2)_6N_4$, abbreviated hereafter to HMT, space group [18] I$\bar{4}$3m, T_d^3, two

Table 1 The Vibrational Representation of Hexamethylene Tetramine $N_4(CH_2)_6$

CH_2 stretch, symmetric	A_1		+ E		+ F_2
CH_2 stretch asymmetric				F_1	+ F_2
CH_2 deformation	A_1		+ E		+ F_2
CH_2 wag				F_1	+ F_2
CH_2 rock				F_1	+ F_2
CH_2 twist		A_2	+ E	+ F_1	
CN stretch	A_1		+ E	+ F_1	+ $2F_2$
CNC deformation	A_1		+ E	+ F_1	+ $2F_2$
	$4A_1$	$+A_2$	+5E	$+6F_1$	+ $9F_2$

Infrared active: F_2 Raman active: A_1, E, F_2

molecules per body-centered cell, one per primitive cell; the site symmetry is T_d, the same as that of the gaseous molecule; (3) Hexamethylene tetramine hexahydrate [19], $HMT{\cdot}6H_2O$, space group R3m, C_{3v}^5, one formula unit per primitive unit cell, yielding one HMT molecule per unit cell on a site of symmetry C_{3v}, a sub-group of the T_d point group of the free molecule. In such crystals there is only one $\underline{k} = 0$ vibration for each vibration of the free molecule, so Davÿdov splitting does not occur. The spectrum of the solid should just reflect site shifts and site splittings.

Since the site group of HMT is T_d, the same as the point group of the gaseous molecule, the vibrational representation formed by the $\underline{k} = 0$ vibrations is the same as that of the free molecule. This is analyzed in Table 1 and is $4A_1 + A_2 + 5E + 6F_1 + 9F_2$. The $9F_2$ vibrations are infrared-active and $4A_1 + 5E + 9F_2$ are Raman-active. The infrared spectrum of polycrystalline HMT at 300 and 100°K shows eight regions where the absorption is about ten times more intense than any other and is due to fundamental transitions [20,21]. These regions are shown in Fig. 4. Fundamentals can be identified at 513, 674, 812, 1007, 1238, 1372, 1458, 2873 and 2955 cm^{-1} by choosing the most intense bands [21]. However even in this relatively simple case, the bands at 812, 1007 and 1238 cm^{-1} show pronounced shoulders or side-peaks, and other peaks of comparable intensity to the fundamentals exist near 1440 cm^{-1} and between the two C-H stretching fundamentals near 2900 cm^{-1}. The Raman spectra of single crystals of HMT confirmed [21] the assignments of the F_2 fundamental vibrations, and yielded the frequencies of the A_1 and E vibrations which allowed the residual peaks near 1440 and 2900 cm^{-1} to be assigned with confidence to binary and ternary combination transitions or first overtones. But a lot of information is needed before this can be done and, without a good prediction of how many fundamental transitions are expected in each frequency region, it can be very difficult to decide which bands in a spectrum are due to fundamentals.

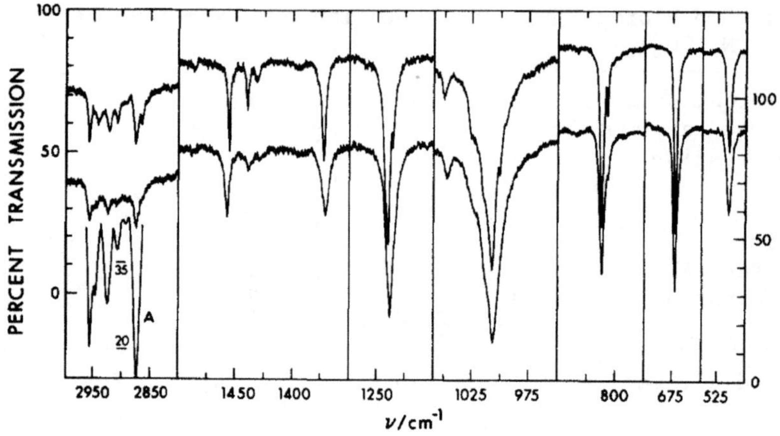

Fig. 4 The regions of strong absorption by polycrystalline hexamethylene tetramine at 100°K (upper) and 300°K (lower and curve A). Resolution 1.3 to 2 cm^{-1}. The left hand scale is for the upper curve, the other is for the lower curve and the numbers near curve A indicate its transmission scale. From Ref. [21], used by permission.

There are three more points to make about this 'simplest' case. First, the spectrum of gaseous HMT has been reported [22], but with the rather low frequency accuracy that was unavoidable in 1955. The frequencies were within 4 cm^{-1} of those reported above for the solid [21], except for differences up to 17 cm^{-1} at 2890 and 2967 cm^{-1}. This indicates that the site shifts are small in this case but they will not be well known until the spectrum of gaseous HMT is re-examined. Second, the frequencies of the F_2 modes in the infrared spectrum of pure HMT are not, theoretically, the frequencies of the uncoupled vibrations, even though there is no Davydov splitting. There are two reasons for this, first dispersion effects, which will be discussed as the next point, and second the interaction of the vibrations of one molecule with those of equivalent molecules in neighboring unit cells [3]. Some of the frequencies of the uncoupled vibrations could be measured from the spectrum of HMT-h_{12} molecules isolated in a lattice of HMT-d_{12}, at about 1 to 5 mole percent, but this experiment has not been done.

The third point to be made about this 'simplest' case carries with it a warning. The factor group analysis predicts that $9F_2$ modes are infrared and Raman-active, and the infrared spectrum is shown in Fig. 4. Part of the Raman-scattering by a single crystal of HMT-h_{12} is shown in Fig. 5, with the F_2 modes more intense in the lower curves. It is clear that the F_2 modes at 675, 812,

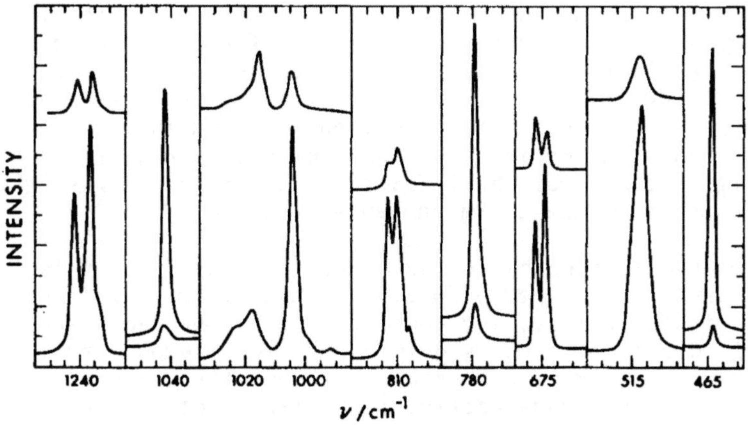

Fig. 5 Raman spectra of a single crystal of HMT at 300°K under x(zz)y (upper) and x(zx)y (lower) scattering geometries. The F_2 modes are more intense under x(zx)y.

~1010 and ~1240 cm^{-1} all appear as doublets, while that at 513 cm^{-1} is clearly an unresolved doublet. It is a universal phenomenon that any infrared-active vibration will yield two lines in the Raman spectrum if the infrared absorption intensity is sufficiently high. It arises because the active vibration has wave-vector $\underline{k} = \underline{K}$, not zero, which means that the displacement wave of the vibration is not a standing wave but propagates through the crystal in a specific direction. The dipole moment change that accompanies the vibration can be aligned parallel to this direction of propagation, in what is called a longitudinal optic vibration, or it can be aligned perpendicular to the propagation direction, in what is called a transverse optic vibration [7-10]. The longitudinal optic frequency is higher than the transverse optic frequency, the relation being

$$\nu_L^2 - \nu_T^2 = \frac{N}{\pi c^2} \left(\frac{n^2 + 2}{3n}\right)^2 \cdot (\partial\mu/\partial Q)^2,$$

where N is the number density of oscillators whose dipole moment derivative is $\partial\mu/\partial Q$, c is the velocity of light, and n is the refractive index of the solid near the frequencies ν_L and ν_T due to all oscillators other than the one under consideration [23]. The important point is that the separation between the LO and TO frequencies increases as the infrared intensity increases. To a good approximation the LO vibration is inactive in the infrared, because its dipole moment change is orthogonal to the electric vector of the radiation with which it must interact, but both the LO and TO vibrations are Raman-active if the factor-group selection rules, which assume that $\underline{k} = 0$, predict the vibration to be active [24].

The effect of the LO-TO splitting can often be detected in the infrared, however, because the crystal reflects radiation strongly between the LO and TO modes [7-10], because of the anomalous dispersion of the refractive index (previous article, section 4). Such dispersion effects are undoubtedly responsible for the breadth and side bands of the infrared bands of HMT at 812, 1007 and 1238 cm^{-1} (Fig. 4), and for the infrared band maxima being slightly to high frequency of the TO Raman bands [21].

The complications which have been discussed for this 'simplest' case of one molecule per primitive unit cell exist in the infrared spectra of all solids. The complication of the Raman spectrum by longitudinal optic modes only occurs for piezoelectric crystals, i.e. non-centrosymmetric crystals, because the infrared-active vibrations are not Raman-active in centrosymmetric crystals.

The second example of a crystal with one molecule per primitive unit cell that I wish to discuss is HMT · $6H_2O$. The vibrations of the HMT molecule are, empirically [25], essentially independent of those of the water molecule, and appear to obey factor group selection rules in spite of the fact that the water molecules are partly orientationally disordered so that the crystal has no symmetry [19]. Thus the vibrations of HMT can be analyzed based on the diffraction symmetry [19], using one molecule on a C_{3v} site per primitive (rhombohedral) unit cell.

The correlation table between the point group of HMT in the gas and pure solid, T_d, and C_{3v} is given as Table 2, with the infrared-active species underlined. The lower site symmetry in the hydrate causes the infrared-inactive A_1, E, and F_1 vibrations under T_d to become active in the hydrate, and causes the triply degenerate F_2 vibrations to yield two active vibrations of a_1 and e symmetry under C_{3v}. The infrared spectrum from 200 to 1500 cm^{-1} of HMT · $6H_2O$ at 95°K is shown in Fig. 6. By comparison with Fig. 4, the site splitting of the F_2 modes is clearly evident at 503/513, 671/692, 999/1012 and 1237/1245 cm^{-1} and is less clear but can be identified [25] for the remainder of the F_2 modes. Inactive modes of the T_d molecule are evident at 372 (F_1 under T_d), 784(A_1), 934(F_1), 1027(E), 1042(A_1), 1351(E), and 1442(A_1) cm^{-1}. These are shifted by 0 to +12 cm^{-1} from their frequencies in pure solid HMT [21,25].

Molecular crystals in which all of the molecules are on the same site, i.e. are all on symmetrically equivalent sites, and have no loss of degeneracy under the site group, yield spectra in which each intramolecular vibration appears as a multiplet through Davydov splitting. A simple example of this occurs in the Raman spectrum of solid I_2 [13], which has the same crystal structure and factor group selection rules as solid chlorine and whose spectrum is not confused by the presence of several isotopic forms of

Table 2 Correlation Table[a] Between the Point Groups T_d and C_{3v}

T_d	C_{3v}
A_1	a_1
A_2	a_2
E	e
F_1	a_2 + e
F_2	a_1 + e

[a]The infrared-active vibrations are underlined.

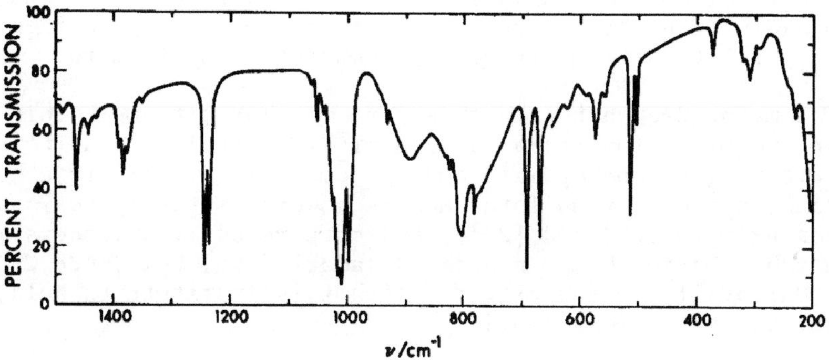

Fig. 6 The infrared spectrum of hexamethylene tetramine hexahydrate at 95°K and ~2 cm^{-1} resolution.

I_2. As discussed earlier for chlorine (Fig. 2), the $\underline{k}$ = 0, I-I stretching vibrations are of A_g and B_{3g} symmetry under the D_{2h} unit cell group and have different frequencies because of intermolecular coupling. Anderson and Sun [13] observed these vibrations in the Raman spectrum of solid I_2 at ~30°K, at 180 and 188 cm^{-1} with relative intensities 5 and 1, respectively. No other bands were in the vicinity, making it a particularly easy spectrum to interpret. They assigned the A_g band at 180 cm^{-1} on the basis that the greater intensity results from this mode in which the two molecules in the cell stretch simultaneously (Fig. 2).

The most complex situation that can exist in the vibrational spectra of solids is when site splitting and Davydov splitting both occur and perhaps are further confused by the occurrence of combination or overtone transitions at similar frequencies. Each system of bands in the spectrum must be examined independently, because they all usually have different appearances and it is quite possible for the major source of splitting to be site effects for one band system and intermolecular coupling for another. Some work of this type has been reviewed by Schnepp [5], while Decius and Hexter [9] discuss other systems. One important test is to study the spectrum of a single crystal with polarized light. The Davydov-split vibrations must have different symmetries under the unit cell group and, thus, different Davydov components absorb infrared light that is polarized in different directions. If the relation between the crystal axes and the direction of polarization is known, the symmetry of the absorbing component can be identified, but the key point for their identification as Davydov components is that all of the bands are not observed under all polarizations. On the other hand, if the splitting is predominantly site splitting, with all of the Davydov components essentially coincident, all of the bands are seen under all polarizations, but probably with varying relative intensities. Only accidentally, however, will the intensity of any component drop to zero.

The major drawback with this test is that extremely thin crystals are required for infrared transmission studies, and it is difficult to grow appropriate ones. Some work of this type has been done, notably by Halford and his coworkers twenty years ago (see for example [26] and [27]), and more recently by others [28-31], notably Marzocchi, Bonadeo and Taddei (e.g. Refs. [28,29]) at the University in the city at which this Institute is held, Firenze, and Katon (e.g. [31]).

Figure 7 shows a nice example of such work, parts of the polarized transmission spectrum of a single crystal of benzene, which is taken from a paper by Marzocchi and Bonadeo [28]. Benzene crystallizes in the orthorhombic space group D_{2h}^{15} (Pbca) with four molecules on equivalent C_i sites per unit cell. The vibrational representation of benzene is shown in Table 3 with the correlation table. Each b_{1u} and b_{2u} mode of gaseous benzene yields $A_u + B_{1u} + B_{2u} + B_{3u}$ Davydov-split components in the solid. The face of the single crystal was the ac face, so the B_{2u} Davydov component is polarized parallel to the propagation direction of the light and should not absorb radiation. The spectra with the electric vector polarized parallel to the a-axis show the B_{1u} components clearly at 1142.5 and 1311.0 cm^{-1} while the c-polarized spectra show the B_{3u} components at 1148.0 and 1312.5 cm^{-1}. The peak at 1306 cm^{-1} is not polarized, so it is not a Davydov component and can be assigned to $^{13}C^{12}C_5H_6$ molecules, which have the ^{13}C positioned at random among the six positions in the molecule and can, thus, yield no polarization phenomena.

Infrared reflection spectra of single crystals have been quite extensively used to identify Davydov components and for other purposes [9,32-34], and Raman spectroscopy of single or oriented crystals, which is much easier than infrared spectroscopy

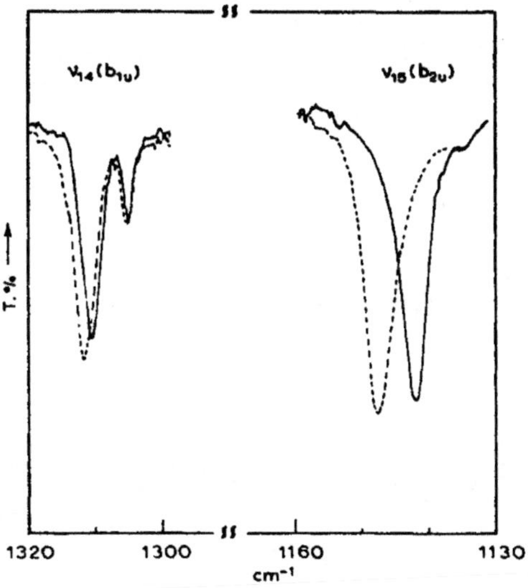

Fig. 7 Polarized infrared spectra of a single crystal of benzene ~80 μm thick at 80°K. Solid and dashed lines indicate radiation polarized parallel to the a- and c- crystal axes, respectively. From Ref. [28], used by permission.

of single crystals, has been quite extensively used since the advent of lasers (e.g. [21,30,35]).

A further important method of determining the origin of observed splittings is the use of isotopically mixed crystals. This technique was introduced in 1951-2 by Hrostowski and Pimentel [36] and by Hiebert and Hornig [37], and has been extensively used for the study of ice and related compounds [25,38], naphthalene [39], benzene [40], carbon dioxide [41] and other solids [42]. The principle is that a dilute isotopic impurity occupies at random all possible sites in the crystal. If its vibrational frequencies are well removed from those of the host isotopic species, they will be the uncoupled frequencies which reflect the site shift and site splitting only. The value of this technique is shown in Figs. 8 and 9.

Figure 8 shows parts of the infrared spectrum of crystalline benzene [43] (curves A and C) and of C_6H_6 as a very dilute ($\lesssim$1%) isotopic impurity in C_6D_6 [40]. The e_{1u} mode of benzene is split into two A_u modes under the site group and the a_{2u} mode should yield a single vibration under the site group. This is clearly shown by curves B and D, where the sharp feature at 694 cm^{-1}, on the side of the main peak at 697 cm^{-1}, is due to $^{13}C^{12}C_5H_6$ which is a natural, 6.7 mole percent, impurity in benzene. In the pure crys-

Table 3 Vibrational Representation of Gaseous Benzene and the Correlation Table to the D_{2h} Unit Cell Group

Point Group	Site Group	Unit Cell Group
D_{6h}	C_i	D_{2h}
2 A_{1g} +1 A_{2g} +0 B_{1g} +2 B_{2g} +1 E_{1g} +4 E_{2g}	A_g	A_g, B_{1g}, B_{2g}, B_{3g}
+0 A_{1u} +1 A_{2u} +2 B_{1u} +2 B_{2u} +3 E_{1u} +2 E_{2u}	A_u	A_u, B_{1u} (a)[a], B_{2u} (b)[a], B_{3u} (c)[a]

[a]The B_{1u}, B_{2u}, and B_{3u} modes are polarized parallel to the a, b and c crystal axes, respectively.

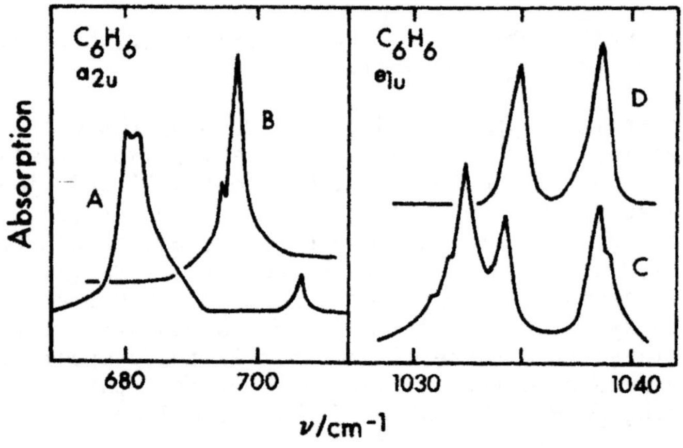

Fig. 8 Infrared absorption by C_6H_6 at 80°K, as the neat crystal (curves A and C) and as a dilute isotopic impurity in C_6D_6 (curves B and D). Curves A and C are on absorbance scales, re-drawn from [43] and curves B and D are on transmission scales, re-drawn from [40].

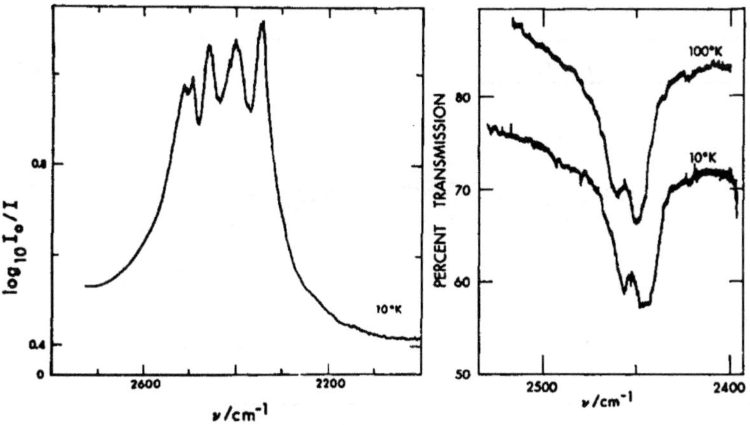

Fig. 9 O-D stretching bands of 99% D_2O ice IX (left) and of 10 mole % HDO in H_2O ice IX (right) at atmospheric pressure and 10 or 100°K. Note that the right-hand frequency scale is four times more expanded than the other.

tal - often called the neat crystal to recognize that it contains 6.7 mole percent of $^{13}C^{12}C_5H_6$ and is not pure $^{12}C_6H_6$ - the e_{1u} mode should yield six infrared-active Davydov components while the a_{2u} mode should yield three. These are visible at 681, 682 and 707 cm^{-1} for the a_{2u} mode in curve A of Fig. 8, while the six components of the e_{1u} mode have been shown by spectra of single crystals of benzene [28,29] to contribute to the three main peaks at ~1032, 1034 and 1039 cm^{-1} in curve C of Fig. 8, their frequencies being [29] 1032.5, 1033.6, 1034.5, 1038.5, 1039.0 and 1039.5 cm^{-1}. Thus the Davydov splitting appears to be fairly small for this e_{1u} mode (an in-plane H-C-C deformation) while the site splitting is quite large at ~4 cm^{-1}, whereas for the a_{2u} mode (the out-of-plane 'butterfly' vibration of benzene) the site splitting is zero by symmetry while the Davydov splitting is large at ~28 cm^{-1}.

Figure 9 shows the absorption by the OD stretching vibrations of D_2O ice IX on the left, and of 10 mole percent of HDO vibrationally isolated in H_2O ice IX on the right [44]. Ice IX crystallizes in the space group $P4_12_12$, $D_4{}^4$, with the twelve molecules per unit cell on 8-fold general positions and a 4-fold site of C_2 symmetry (the C_2'' axes of D_4). The correlation table for the OD stretching vibrations is in Table 4. Thus the eight D_2O molecules on general positions yield $2A_1 + \underline{2A_2} + 2B_1 + 2B_2 + \underline{4E}$ vibrations under the unit cell group, and the four molecules on the two-fold axis yield $A_1 + B_2 + \underline{E}$ from the symmetric stretching vibration and $\underline{A_2} + B_1 + \underline{E}$ from the antisymmetric stretching vibration. The total representation formed by the O-D stretching vibrations is, thus,

Table 4 Correlation Table for Ice IX

	Point Group	Site Group	Unit Cell Group[a]
	C_{2v}		D_4
molecule 1		C_1	
	A_1, B_2	A	A_1, $\underline{A_2}$, B_1, B_2, $\underline{E}$
molecule 2		C_2	
	A_1	A	$A_1 + B_2 + \underline{E}$
	B_2	B	$\underline{A_2} + B_1 + \underline{E}$

[a]The infrared-active species are underlined.

$3A_1 + \underline{3A_2} + 3B_1 + 3B_2 + 6\underline{E}$. Thus nine, $\underline{k} = 0$, O-D stretching vibrations are predicted to be infrared-active, and they yield the broad band with five distinct peaks at 2517, 2499, 2462, 2403 and 2348 cm^{-1} at 10°K that is shown in the left-hand box of Fig. 9. It has not proved possible to measure the frequencies of a D_2O molecule isolated in H_2O ice IX, but it is easy to measure the O-D stretching frequencies of HDO molecules isolated in ice IX. The structure indicates that the D atom can occupy three non-equivalent positions with equal probability. It can be in either of the two inequivalent positions on the eight water molecules on general positions, or in either of the two equivalent positions on the four water molecules on the 2-fold axes. Thus three isolated O-D stretching frequencies are predicted (the site splitting) and observed (Fig. 9, right box) at 2455, 2446 and 2444 cm^{-1} at 10°K. Thus the site splitting is quite small for ice IX and the Davydov splitting is very large. It is noteworthy that for ice II [44] the site splitting of the O-D stretching vibrations is much larger, about 65 cm^{-1}, but that the combination of Davydov and site splitting is approximately the same as in ice IX, about 150 cm^{-1}. These results for ice IX have been reproduced well by calculations [44] which yielded the splitting pattern shown in Fig. 10 and which reproduced the infrared and Raman frequencies and intensities extremely well [44,45]. The unit cell vibrations (lower-line of Fig. 10) are in general extensive mixtures of the symmetric and antisymmetric stretching vibrations of both types of water molecule, although the lowest-frequency vibration is an almost pure in-phase symmetric stretching of all water molecules in the cell.

A variant of the isotopic substitution technique described above is to compare the spectra of heavily isotopically substituted crystals with those of the host crystal. The heavy isotopic

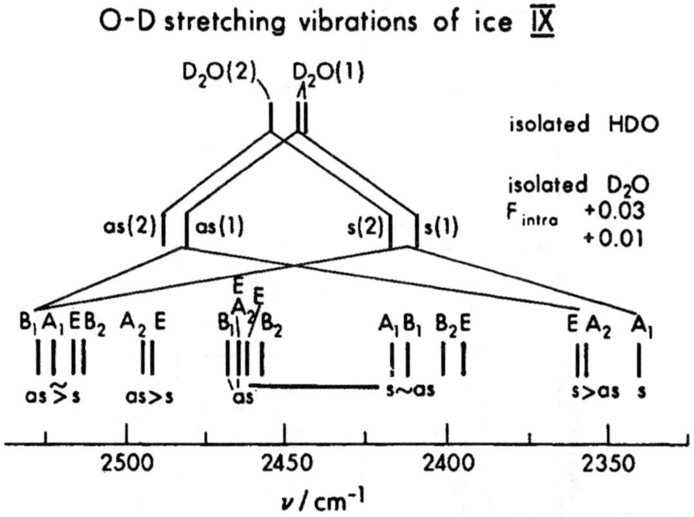

Fig. 10 Schematic diagram of the site splittings of O-D stretching vibrations of HDO in D_2O ice IX and the calculated site splittings of D_2O and the calculated [44,45] unit cell vibration frequencies.

substitution modifies the intermolecular vibrational coupling if the frequencies of the guest isotope are sufficiently different from those of the host, and destroys the selection rules, including the $\underline{k} = 0$ selection rule. All vibrations in the crystal become active, and the sharp lines due to the Davydov components are replaced by broad absorption within the band of frequencies which results from intermolecular coupling. But if the Davydov splitting is small and the observed splitting is largely due to site splitting, or to overtone or combination transitions, or to a further isotopic impurity, the spectrum of the heavily isotopically substituted crystal closely resembles that of the host lattice. This technique has been used by Kopelman and his coworkers and Fig. 11 shows its application to the Raman spectrum of naphthalene [39]. The left boxes show that the 392 cm^{-1} vibration of naphthalene yields a doublet in the neat crystal (center), a singlet for isotopically dilute $C_{10}H_8$ in $C_{10}D_8$ (bottom), and a broad absorption for a 50-50 mixture of $C_{10}H_8$ and $C_{10}D_8$ (top). There is clearly no site splitting and the doublet in the neat crystal is due to Davydov splitting. The right hand boxes show that the doublet at 3051/3057 cm^{-1} is not due to Davydov splitting, since it is invariant to isotopic substitution. It has not been assigned yet to my knowledge.

I have tried in this section to give an overview of the methods

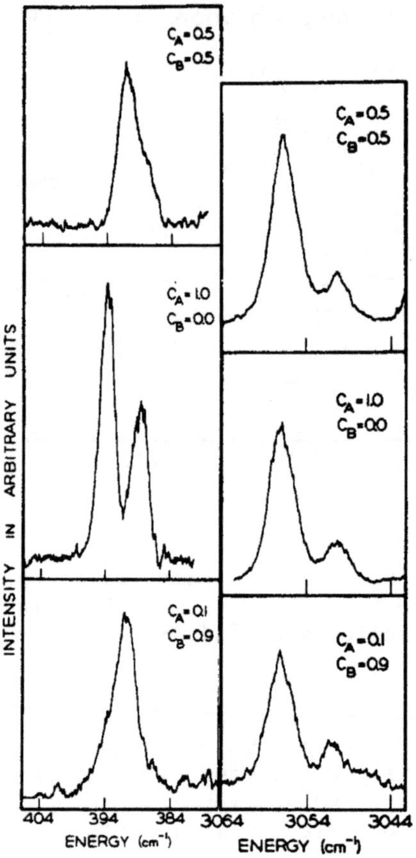

Fig. 11 Raman spectrum of naphthalene at 100°K. Middle boxes, neat $C_{10}H_8$; top boxes, 50% $C_{10}H_8$, 50% $C_{10}D_8$; bottom boxes, 90% $C_{10}D_8$, 10% $C_{10}H_8$. From Ref. [39], used by permission.

that must be used to understand the infrared absorption by the intramolecular vibrations in molecular crystals. It should be clear that a large amount of work is usually necessary before a spectrum is understood. Accordingly, the deduction of structural information from spectroscopic data also requires a lot of work and yields results that can not be regarded as definitive even in apparently straightforward cases [46]. One must also be prepared to work hard to deduce definitive results from the study of spectra of compounds under high-pressure, although useful information is sometimes obtained quite easily, as is discussed in the next section.

4. THE EFFECT OF HIGH PRESSURE ON THE VIBRATIONAL SPECTRA OF SOLIDS

The study of the effect of pressure on vibrational spectra is relatively young, being about twenty years old for infrared spectra and about twelve years old for Raman spectra, although pioneering papers appeared before these times. It is a flourishing field that has led to a large number of papers of varying scientific importance. Thus, in a recent review [47] that is confined to the effect of pressure on vibrational spectra of solid inorganic and coordination compounds, Ferraro has listed three hundred and twenty-four references. It is clear from the earlier sections of this chapter that it is difficult to understand the spectra of a solid in detail. Further, it is difficult experimentally to obtain any spectrum under high pressure, if the pressure is to be reasonably hydrostatic and reasonably well known. The combination of these two difficulties has made it inevitable that many studies of vibrational spectra under pressure have yielded results that are of little lasting scientific significance, while many other studies require the chemist's traditional use of logic, intuition, and likely approximation, to yield conclusions that are not definitively proved but are probably correct, and, as such, are extremely useful. The field is healthy, and the quality of the results has improved so much that the results of the early work, which was responsible for the growth of the field, are of little value today. To obtain a more complete overview of the field than can be given here, recent reviews [47-50] should be consulted.

Two broad objectives can be given of the study of systems under high pressure: first, to see the effect of pressure on the properties of the phases that exist at atmospheric pressure or are formed at high pressure; second, to determine how the high-pressure phases differ from the low-pressure phase. The first objective can only be reached by making the desired measurements while the sample is under pressure. The second objective can be reached in this way, but can sometimes be reached with better results by preparing the high-pressure phase at the appropriate temperature and pressure then quenching it to a sufficiently low temperature that the pressure can be released to atmospheric without loss of the desired phase, and measuring the desired properties of the quenched material.

To measure the infrared transmission spectrum of a solid or liquid under high-pressure a very thin sample must be compressed to the desired pressure, which should be perfectly hydrostatic, and two infrared-transmitting windows must withstand the applied pressure. The primary difficulties are obtaining very thin samples in the pressure vessel, exerting near-hydrostatic pressure on them, and finding sufficiently strong infrared-transmitting windows.

Most high-pressure infrared studies have used one of two cell designs. The first is the cell invented by Drickamer [51], in which the sample is enclosed by sodium chloride. The pressure is generated by a press which acts on two pistons which are orthogonal to the light beam. The pistons transmit the pressure to the sodium chloride which, in turn, transmits it to the sample. The pressure-transmitting windows are sodium chloride, which is forced under high-pressure into long, narrow, shaped, openings in the cell, to form a window that is about ten times longer than its average diameter. Thus, the sodium chloride is the pressure transmitting fluid as well as the pressure retaining medium. The sample is in intimate contact with an alkali halide, and the light must traverse a considerable thickness of sodium chloride through an opening of small diameter, but the pressure in the sample is quite hydrostatic and the sample diameter can be reasonably large (~0.75 to 1mm). This cell has been used ([47], [50] and references therein) at temperatures from 77°K to 500°C and to pressures over 15 GPa[15 GPa = 150 kbar ≈ 150,000 atmospheres].

The second cell is much simpler, but probably requires just as much skill as the Drickamer cell if one is to obtain results of high quality. It is the diamond anvil cell [52,53], universally called the DAC, which basically consists of two diamonds, each with a flat face of diameter ~0.5 mm between which the sample is placed and squeezed. The surface area is so small that relatively little force is required to generate a high pressure. For example a 2 Kg force yields a pressure of ~1 kbar (0.1 GPa). The diamonds are securely mounted in metal and the pressure is applied by compressing a spring in the basic design of the cell. If this cell is used without a gasket to contain the sample, the pressure gradients across the sample are very large, so shear stresses are large and the pressure is poorly known and not at all hydrostatic. When a gasket is used, however, the aperture of the cell is even smaller than 0.5 mm, and a means of measuring the sample pressure is necessary. This is achieved (also in the Drickamer cell) by including a ruby chip in the sample and observing the wavelengths of the R fluorescence lines of ruby (694.2 and 692.8 nm at 1 atm.), which change linearly with pressure at a rate of +0.365 nm GPa^{-1} (= 0.365 Å $kbar^{-1}$, [47], [50], and references therein). Compounds other than ruby have also been used [47,50]. The early studies with the DAC tended to be rather qualitative, but improved designs [54,55], and increased attention to optically matching it to spectrometers (e.g. [56]), have led to quantitative results that compare well with those obtained with cells that have much larger samples and apertures but which are less versatile. The DAC has been used for high and low temperatures and pressures up to 30 GPa, and its great advantage is that mid infrared, far infrared, and Raman spectra can be obtained [57] from the same sample under the same conditions. The DAC is also useful for small samples and for growing single crystals [58] even when pressure changes are not of primary interest. Other, more specialized, cells have been described [47,50] for infrared [59] and Raman [60,61] spectroscopy under hydrostatic pressure.

FT-IR spectrometers are admirably suited to the energy-limited spectroscopy that is inevitable through a high-pressure cell (see earlier chapter on "Energy-Limited Systems"). Slow-scanning, far-infrared interferometers have been used for this purpose for years [57,62] but I am aware of only one such study with a rapid-scan FT-IR spectrometer [63]. The field will benefit greatly from the use of mid infrared FT-IR spectrometers.

The general effect of increasing the pressure is to increase vibrational frequencies, except in the case of hydrogen-bonded X-H stretching vibrations which decrease with increasing pressure. For lattice vibrations and hydrogen-bonded X-H stretching vibrations the shift with increasing pressure is, thus, in the same direction as the shift with decreasing temperature, which is the direction expected from increased intermolecular forces. The magnitude of pressure shifts can, however, be much larger than the magnitude of temperature shifts, if a sufficiently large pressure range is available. Not all vibrations shift to the same extent, and it is sometimes possible to separate overlapping bands, to distinguish lattice vibrations from intramolecular vibrations by the greater pressure sensitivity of the former, and to identify Fermi resonance doublets from the relative intensity changes which occur with changing pressure.

Some results taken with little selection from the literature are in Table 5. They are all results of Raman spectroscopy under pressure. They show that lattice modes of molecular crystals have a normalized pressure shift (NPS = $(\nu^{-1}(\partial\nu/\partial P)_T)$) of ~0.1 to 0.4 GPa^{-1} while intramolecular modes of benzene and the lattice mode of diamond have NPS ~0.001 to 0.005 per GPa. The NPS for translational vibrations in ice 1h is +0.22 GPa^{-1} while that for hydrogen-bonded O-H and O-D stretching vibrations of water ranges from -0.044 GPa^{-1} in ice 1h to -0.023 in ice VI, -0.009 in ice VII and -0.007 to -0.009 in liquid water. The metal-ligand vibrations of $ZnCl_4^=$ in Cs_2ZnCl_4 and of dichloro bis(pyridine)zinc(II) have NPS values in the range 0.03 to 0.10 and 0.02 to 0.08 GPa^{-1}, respectively. The lattice modes in the latter compound have NPS values of 0.08 and 0.3 and, although coincident at atmospheric pressure, are well separated at high pressure [69]. Thus lattice modes tend to shift far more than intramolecular modes although cases do exist in which similar shifts occur for the two types of vibration.

Interesting changes in Fermi resonance diads have been observed with increasing pressure [72-74]. Wu and Jura [72] and Lewis and Sherman [73] have used the changes in Fermi diads with

Table 5 Pressure Dependence of Vibrational Frequencies

Species	Ref.	Temp. /°K	Pressure Range[a]/GPa	Wavenumber Range[b]/cm^{-1}	Dependence[a]	$\frac{1}{\nu}(\frac{\partial\nu}{\partial P})_T \cdot 10^3$ / GPa^{-1}
Diamond	[60]	295	0 - 2.4	1333 - 1340	L	2.0
Benzene I	[64]	77	0.5 - 2(?)	16 bands between 608 and 3080 cm^{-1}	?	1.6 to 5.9
		77	0 - 5.0	7 bands between 59 - 110 and 132 - 260	NL	≳170 ≳190
		295	0 - 2.5	7 bands between 42 - 83 and 105 - 200	NL	≳400 ≳360
Benzene II	[64]	295	1.25 - 3.0	86 - 108	NL	≳150
				130 - 175	NL	≳200
				155 - 205	NL	≳180
H_2O and HDO in $H_2O(\ell)$	[65]	298	0 - 1.0	2513 - 2490	L	-9.2
				3402 - 3379	L	-6.8
H_2O in ice VI	[66]	294	0.9 - 2.12	3213 - 3124	L	-23
H_2O in ice VII	[67]	295	2.5 - 7.3	3300 - 3150	?	-9.5
H_2O in ice Ih	[70]	255	0 - 0.16	3141 - 3119	L	-44
	[71]	255	0 - 0.16	214.5 - 222.2	L	+220
Cs_2ZnCl_4	[68]	293	0 - 1.2	5 bands between 298 - 316 and 131 - 135	L	30 to 50
				117.9 - 130	L	84
				111.3 - 125	L	100
Dichloro-bis(pyridine) zinc(II)	[69]	293	0 - 1.2	326.3 - 335	L	22
				292.9 - 302	L	26
				222.1 - 244.5	L	84

Table 5 (continued)

Species	Ref.	Temp. /°K	Pressure Range[a]/GPa	Wavenumber Range[b]/cm^{-1}	Dependence[c]	$\frac{1}{\nu}(\frac{\partial \nu}{\partial P})_T \cdot 10^3$ GPa^{-1}
				204.7 - 222.5	L	70
				197.0 - 204	L	29
				151.5 - 161.5	L	56
				29.6 - 40.5	L	290
				29.6 - 32.4	L	84

[a] 1 GPa = 10 kbar.

[b] where two numbers are separated by a hyphen, the first number applies to the lower limit of the pressure range and the second to the higher limit.

[c] 'L' means linear, 'NL' means non-linear, '?' means unknown.

pressure to analyze Fermi resonances, and thus to extract information on the unperturbed transitions, for solid naphthalene [72] and the sulfate ion in alkali halide matrices [73]. Both groups used Drickamer cells but Lewis also used a cell in which He or N_2 gas generated hydrostatic pressure to 1.5 GPa. Figure 12 shows three of the bands reported by Lewis and Sherman [73] for $CaSO_4$ in KBr at 80°K. The relative intensities of the Fermi diad change with pressure due to the different pressure dependences of the energies of the upper states. Such intensity dependence indicates rather clearly the existence of a Fermi diad, rather than splitting by some other source, and its analysis provides useful information on the unperturbed states [72,73]. Hamann [74] has observed a different manifestation of Fermi resonance, the creation of an Evan's hole [75] by change of pressure. An Evan's hole is caused by a sharp, weak transition in Fermi-resonance with a broad, strong transition. The resonance removes the intensity of the strong transition from the vicinity of the weak one, thus creating a small region of high transmission, i.e. a 'transmission window' or an 'absorption hole'. Figure 13 shows the infrared transmission spectrum of nicotinic acid in a DAC at ~295°K at atmospheric pressure (solid curve) and at 4 GPa (dashed curve) [74]. At atmospheric pressure, distinct absorption peaks exist at 640 and 1035/1045 cm^{-1}, marked B and C in the figure. At 4 GPa these features have turned into transmission windows, the one at 1040 cm^{-1} being particularly pronounced. Although Hamann does not comment upon it, the reverse effect may occur at ~1470 cm^{-1} where a mild Evan's hole at 1 atmosphere seems to disappear at the higher pressure. The changes occur gradually as the pressure is increased, undoubtedly due to the relative displacement with pressure of the upper states of the transitions in question. These results are interesting because it is not at all obvious why the weak transition should yield an absorption peak when it is superimposed on one part of the broad absorption but yield an Evan's hole when it is very slightly shifted.

The measurement of vibrational spectra under pressure enables one to determine how much of the change in vibrational frequencies with changing temperature is due to the change in volume of the solid and how much is a purely temperature effect. Medina and Daniels [76] have shown that the frequencies of the Raman bands due to rotational vibrations in α-nitrogen shift with temperature at constant volume by about one-fifth of the shift at constant pressure. Johari [71] has shown that the Raman peak due to translational vibrations of ice Ih shifts with temperature at 1bar and 255°K by -0.073 cm^{-1} $°K^{-1}$ at constant volume but by 0.134 cm^{-1} $°K^{-1}$ at constant pressure, so in this case only about one half of the frequency shift is due to the volume change. For the OH stretching vibrations in ice Ih at 255°K and 1 bar the temperature dependences are [70] 0.63 cm^{-1} $°K^{-1}$ at constant pressure and 0.45 cm^{-1} $°K^{-1}$ at constant volume, so in this case only about 30% of the shift is due to the volume change.

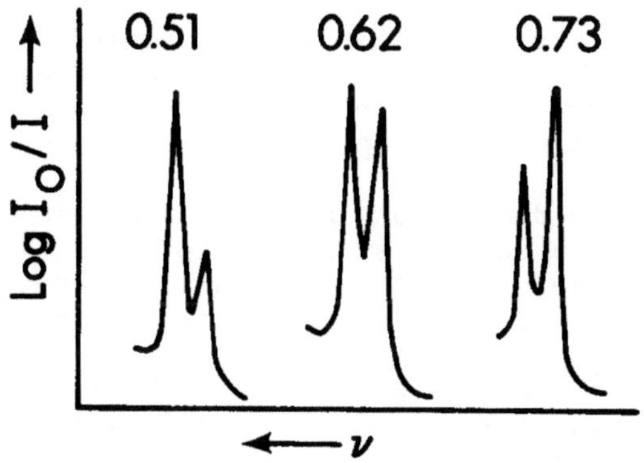

Fig. 12 The 1090-1070 cm^{-1} region of the infrared absorption by $CaSO_4$ in a KBr matrix at 80°K and pressures of 0.51, 0.62, and 0.73 GPa. Re-drawn from Ref. [73].

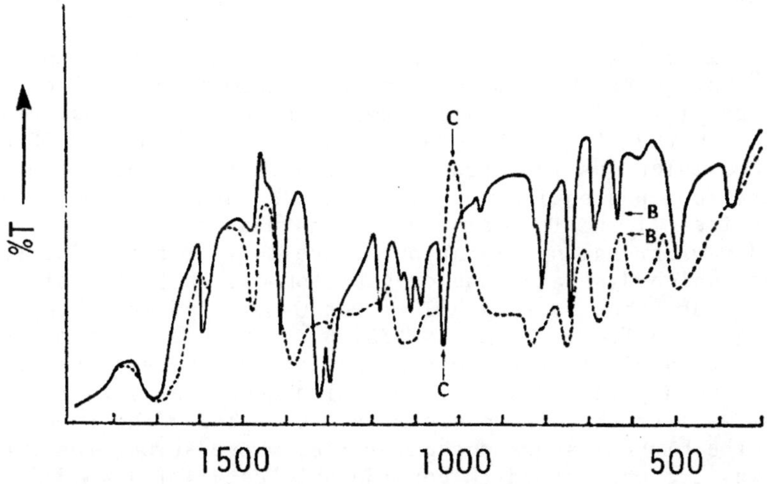

Fig. 13 Infrared transmission spectrum of nicotinic acid at 295°K and atmospheric (solid curve) or 4 GPa (dashed curve) pressure. From Ref. [74], used by permission.

Nearly all crystals undergo solid-solid phase transitions under pressure [77]. Vibrational spectroscopy has been extensively used to detect phase transitions, to study the vibrational properties of high-pressure phases, and to try to determine how the high-pressure phase differs from the low-pressure phase. Further, the spectra of a high-pressure phase can give valuable clues to the correct interpretation of the spectra of the low-pressure phase.

There is little doubt that the substance that has been most extensively studied under pressure is water, and the infrared spectrum of the phase of ice that is stable at atmospheric pressure, called ice Ih, was very poorly understood until the spectra of high-pressure phases were found to contain clues [38,78-80]. Figures 6 and 17 of the chapter on Lattice Vibrations show the far infrared spectra of ice I and of ices II and IX, respectively, the latter being high-pressure phases which were quenched to 77°K before the spectra were recorded. The sharp-line spectra of ices II and IX made it clear [79,80] that the absorption by ice I is very broad because of the disorder in the structure - an idea that had not been discussed before the spectra of ices II and IX were available - and that ices II and IX are essentially fully ordered, which has since been confirmed by diffraction methods. Further, the absorption by the O-D stretching vibration of HDO isolated in H_2O ice IX (Fig. 9) contains sharp features, since two lines separated by 2 cm^{-1} are nearly resolved at 10°K and lines separated by 9 cm^{-1} are clearly resolved, and this absorption in ice II [38,44] consists of four sharp lines. These absorptions, contrasted with the single featureless band of F.W.H.H. ~20-30 cm^{-1} due to HDO molecules isolated in H_2O ice Ih [78], indicated that there are four non-equivalent sites for the hydrogen atoms in ice II, that ices II and IX are ordered, and that the breadth of the isolated O-D stretching vibration of ice Ih is largely due to the disorder, which displaces the oxygen atoms slightly off of regular lattice sites, thus causing all of the O-D--O hydrogen bonds to have slightly different lengths [38,78]. In this way the vibrational spectra of high-pressure phases of ice whose structures were not known gave valuable information about the low-pressure phase as well as the high-pressure phases. Today the spectra of some of the high-pressure phases of ice, whose structures are now all known, are understood in considerable detail [44,45,81].

Ice is generally regarded as an atypical substance, so I will close this chapter with an account of the changes that are observed in the infrared [57] and Raman [64] spectra of a more normal solid, benzene, as the pressure is increased. As is often the case today, rather more detailed information was obtained from the Raman spectrum [64] than from the infrared spectrum, but the infrared spectrum did give evidence of a phase transformation that was not observed in the Raman. The Raman spectra were obtained with a modified [64] Drickamer-type cell, and the spectra of the lattice vibrations is shown in Fig. 14. The mid and far infrared spectra were obtained with a DAC and the Raman spectra were also recorded with the DAC of the same samples at the same temperatures and

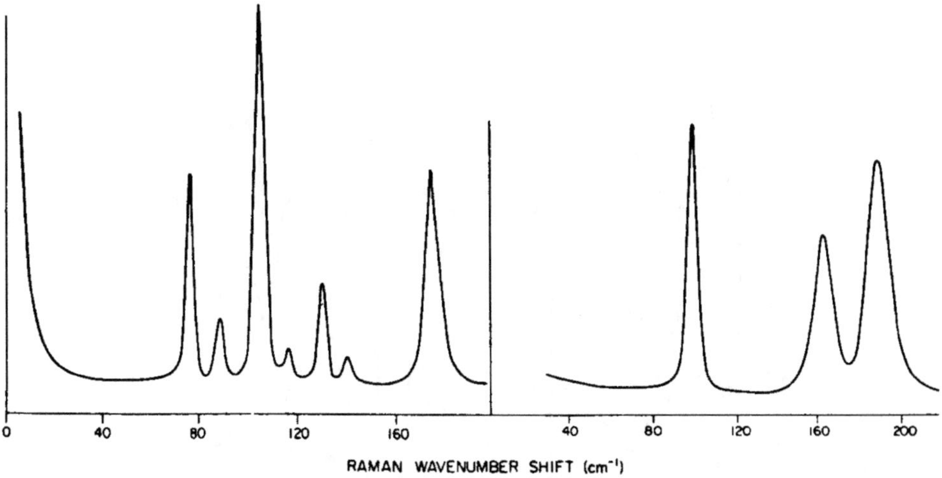

Fig. 14 Raman spectra of C_6H_6 I at 77°K and 1.1 GPa (left) and C_6H_6 II at 77°K and 2.3 GPa. From Ref. [64], used by permission.

pressure as were used for the infrared spectra [57]. The infrared spectra are shown in Fig. 15.

As the pressure is increased on benzene I, the solid phase at atmospheric pressure, the frequencies increase, as is shown in Table 5. Only three lattice bands are visible in the Raman spectrum at 295°K and low pressures, but as the pressure increases the bands separate until six bands are visible at 1.1 GPa and a seventh is resolved at $\geq$1.6 GPa [64]. This seventh band had not been observed previously, even in studies with polarized light of benzene at low temperature. At 77°K all of these seven bands are visible at pressures $\geq$0.45 GPa, and at atmospheric pressure a weak band at 90 cm^{-1} is visible but merges with a stronger band at 92 cm^{-1} as the pressure increases. Thus, increasing the pressure can cause bands to merge or separate. The bands due to intramolecular vibrations change as the pressure increases but the multiplet splitting does not increase substantially with pressure.

When the pressure exceeds ~1.7 GPa at 295°K, new bands appear in the spectrum to clearly indicate that a phase transition has started. However, the transition is very sluggish, which is common for solid-solid transitions [77], and the pressure can be raised to 3 GPa without the transformation proceeding far. Similar effects are observed at 77°K where the transition is not complete even at 4.5 GPa. To make the transformation to benzene II go completely, the temperature can be raised at 1.5 GPa. At this pressure the rate of transformation increases markedly above 320°K and the transition is complete before 373°K is reached [57,64].

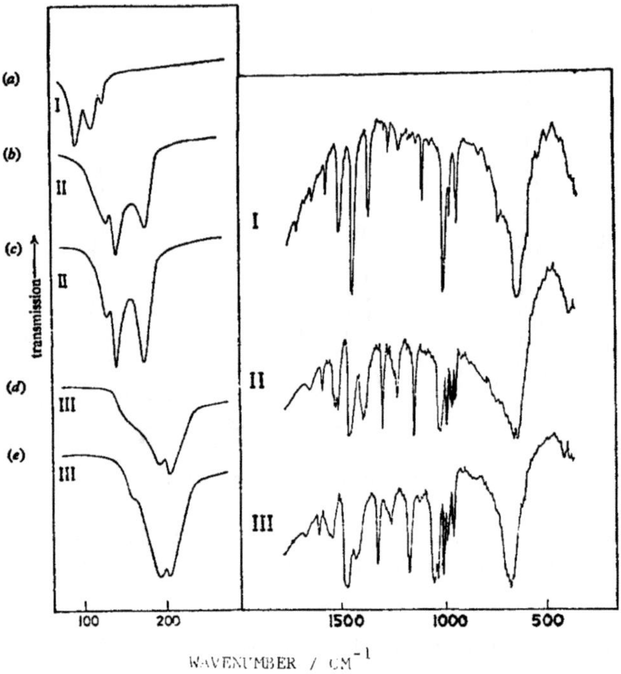

Fig. 15 Infrared spectra of C_6H_6 in phases I (295°K and 0.41 GPa (left) or 0.16 GPa (right)), II (2.03 GPa and 295°K or 188°K (curve c, left box)) and III (3.44 GPa and 295°K or 188°K (curve e, left box)). Resolution 3.7 cm^{-1} above 400 cm^{-1}.

The Raman and infrared spectra of benzene II are quite different from those of benzene I (Figs. 14 and 15). Fewer lattice vibrations are seen in the Raman, and the multiplet splitting of some of the intramolecular modes is much greater than in benzene I, both in the Raman and the infrared. For example, the splitting of ν_9 of C_6H_6 near 1180 cm^{-1} in the Raman spectrum is ~15 cm^{-1} in phase I but ~45 cm^{-1} in phase II. Again, increasing the pressure shifts the bands to high frequency. The infrared spectra of benzene II show no evidence of further phase transition up to 4.0 GPa at 295°K [57]. But if the sample is left at 373°K and 4.0 GPa overnight and cooled to 295°K, the far-infrared (Fig. 15) and Raman frequencies change markedly, although there is little change in band shape. The mid infrared spectrum shows changes in band-shape and frequency (Fig. 15). The volume of the sample also decreases by about 15%, and these observations support previous suggestions that there is a third phase of solid benzene, benzene III.

The crystal structure of benzene II is known [82]. The space group is the very common $P2_1/C$ (C_{2h}^5) with two molecules on sites

Table 6 Correlation Table for the Lattice Vibrations in the Two Phases of Benzene

Benzene I: Orthorhombic, Pbca, D_{2h}^{15}, 4 molecules per cell

	Molecule	Site	Unit Cell
	D_{6h}	C_i	D_{2h}
T_z	A_{2u}	A_u	A_u
			$\underline{B_{1u}}$
			$\underline{B_{2u}}$
T_x, T_y	E_{1u}		$\underline{B_{3u}}$
R_z	A_{2g}	A_g	A_g
			B_{1g}
			B_{2g}
R_x, R_y	E_{1g}		B_{3g}

Benzene II: Monoclinic, $P2_1/c$, C_{2h}^5, 2 molecules per cell

	Molecule	Site	Unit Cell
	D_{6h}	C_i	C_{2h}
T_z	A_{2u}	A_u	$\underline{A_u}$
T_x, T_y	E_{1u}		$\underline{B_u}$
R_z	A_{2g}	A_g	A_g
R_x, R_y	E_{1g}		B_g

[a]The infrared active species under the unit cell group are underlined. All gerade species are Raman active.

of symmetry C_i per unit cell - the same space and site groups as for napthalene. The spectra of benzene II are understood remarkably well. The correlation tables for the lattice vibrations in benzene I and II are shown in Table 6. Thus, $3A_g + 3B_{1g} + 3B_{2g} + 3B_{3g}$ rotational lattice vibrations are Raman-active for benzene I while $2B_{1u} + 2B_{2u} + 2B_{3u}$ translational lattice vibrations are infrared-active (excluding the $B_{1u} + B_{2u} + B_{3u}$ acoustic modes), and $3A_g + 3B_g$ rotational vibrations are Raman-active and $2A_u + B_u$ translational vibrations are infrared-active for benzene II. Only eight of the Raman-active modes of benzene I and three of benzene II have been observed. Calculations of the lattice frequencies, based on atom-atom potentials, and calculations based on the oriented gas model for the Raman intensities, have explained [64] the infrared and Raman frequencies and Raman absences extremely well for both phases, and have enabled the low-frequency band in the far infrared spectrum of benzene II to be assigned [57] to the B_u mode. Not surprisingly, more work is required before the splittings of the intramolecular modes are understood, but the facts that the splittings in benzene I do not increase markedly with pressure and that those of benzene II are generally greater than those of benzene I provide experimental evidence about the intermolecular forces. It is to be expected that an FT-IR spectrometer will enable the mid infrared spectra of benzene under pressure to be measured with sufficiently high resolution and signal-to-noise ratio to enable the splittings to be precisely measured and, with luck, interpreted.

A chapter such as this is finished with the realization that it merely scratches the surface of its subject. I hope that the principles have been given with sufficient examples and references to enable the interested reader to explore the field more deeply.

REFERENCES

1. D. F. Hornig, Disc. Farad. Soc., 9, 115 (1950).

2. W. Vedder and D. F. Hornig., Advan. Spectry., 2, 189 (1961).

3. D. A. Dows, in Physics and Chemistry of the Organic Solid State, (D. Fox, M. M. Labes and A. Weissberger, Ed.), Wiley-Interscience, New York, 1963, Vol. 1, p. 658.

4. S. S. Mitra and P. J. Gielisse, Progress in Infrared Spectroscopy, (H. A. Szymanski, Ed.), Plenum, New York, 1964, Vol. 2, p. 47.

5. O. Schnepp, Adv. At. Molec. Phys., 5, 155 (1969).

6. G. W. Robinson, Ann. Rev. Phys. Chem., 21, 429 (1970).

7. P. M. A. Sherwood, Vibrational Spectroscopy of Solids, Cambridge University Press, Cambridge, 1972.

8. G. Turrell, Infrared and Raman Spectra of Crystals, Academic Press, London, (1972).

9. J. C. Decius and R. M. Hexter, Molecular Vibrations in Crystals, McGraw-Hill, New York, 1977.

10. H. Poulet and J. P. Mathieu, Spectres de Vibration et Symétrie des Cristaux, Gordon and Breach, New York, 1970. Also published in English by the same publishers.

11. W. G. Fateley, F. R. Dollish, N. T. McDevitt, and F. F. Bentley, Infrared and Raman Selection Rules for Molecular and

Lattice Vibrations: The Correlation Method, Wiley-Interscience, New York, 1972.

12. W. G. Fateley, N. T. McDevitt, and F. F. Bentley, Appl. Spectrosc., 25, 155 (1971). W. G. Fateley, Appl. Spectrosc., 27, 395 (1973).

13. A. Anderson and T. S. Sun, Chem. Phys. Lett., 6, 611 (1970).

14. L. Brillouin, Wave Propagation in Periodic Lattices, Dover Publications, (1953).

15. J. E. Bertie and J. W. Bell, J. Chem. Phys., 54, 160 (1971); J. E. Bertie and R. Kopelman, J. Chem. Phys., 55, 3613 (1971).

16. M. J. Buerger, Elementary Crystallography, J. Wiley, New York, 1956.

17. W. J. Dulmage and W. N. Lipscomb, Acta Cryst., 4, 330 (1951).

18. J. A. K. Duckworth, B. T. M. Willis, and G. S. Pawley, Acta Cryst., A26, 263 (1970) and references cited therein.

19. T. C. W. Mak, J. Chem. Phys., 43, 2799 (1965).

20. R. Mecke and H. Spiesecke, Spectrochim. Acta, 7, 387 (1957).

21. J. E. Bertie and M. Solinas, J. Chem. Phys., 61, 1666 (1974).

22. R. Mecke and H. Spiesecke, Chem. Ber., 88, 1997 (1955).

23. C. Haas and D. F. Hornig, J. Chem. Phys., 26, 707 (1957).

24. R. Loudon, Advan. Phys., 13, 423 (1964); 14, 621 (1965).

25. J. E. Bertie and M. Solinas, Can. J. Chem., 53, 2624 (1975).

26. C. Brecher, E. Krikorian, J. Blanc, and R. S. Halford, J. Chem. Phys., 35, 1097 (1961).

27. S. Zwerdling and R. S. Halford, J. Chem. Phys., 23, 2221 (1955).

28. M. P. Marzocchi and H. Bonadeo, Chem. Phys. Lett., 14, 282 (1972).

29. M. P. Marzocchi, H. Bonadeo, and G. Taddei, J. Chem. Phys., 53, 867 (1970).

30. J. E. Bertie, A. M. Heyns, and O. Oehler. Can. J. Chem., 51, 2275 (1973).

31. P. F. Krause, J. E. Katon, and R. W. Mason, J. Phys. Chem., 82, 690 (1978).

32. H. Yamada and K. Suzuki, Spectrochim. Acta, 23A, 1735 (1967); K. Tsuji and H. Yamada, J. Phys. Chem., 76, 260 (1972).

33. R. D. Cooke, C. Pastorek, R. E. Carlson, and J. C. Decius, J. Chem. Phys., 69, 5 (1978).

34. P. Dawson, M. M. Hargreave, and G. R. Wilkinson, Spectrochim. Acta, 31A, 1055 (1975).

35. H. Bonadeo, M. P. Marzocchi, E. Castellucci, and S. Califano, J. Chem. Phys., 57, 4299 (1972).

36. H. J. Hrostowski and G. C. Pimentel, J. Chem. Phys., 19, 661 (1951).

37. G. L. Hiebert and D. F. Hornig, J. Chem. Phys., 20, 918 (1952).

38. J. E. Bertie and E. Whalley, J. Chem. Phys., 40, 1646 (1964) and earlier papers cited therein.

39. P. N. Prasad and R. Kopelman, J. Chem. Phys., 57, 856 (1972).

40. E. R. Bernstein, J. Chem. Phys., 50, 4842 (1969).

41. J. E. Cahill, J. Chem. Phys., 66, 4847 (1977).

42. M. Falk and O. Knop, Water - a Comprehensive Treatise, (F. Franks, Ed.), Plenum, New York-London, 1973, Vol. 2, Ch. 2.

43. N. Rich and D. Dows, Mol. Cryst., 5, 111 (1968).

44. J. E. Bertie and F. E. Bates, J. Chem. Phys., 67, 1511 (1977).

45. J. E. Bertie and B. F. Francis, J. Chem. Phys., In press (1980).

46. J. E. Bertie and S. M. Jacobs, J. Chem. Phys., 68, 97 (1978).

47. J. R. Ferraro, Coordination Chem. Rev., 29, 1 (1979).

48. E. Whalley, in High Pressure Chemistry, (H. Kelm, Ed.), D. Reidel, Pub., Dordrecht, 1978, p. 127.

49. E. Whalley in "Proc. 4th Int'l Conf. High Pressure," Kyoto, 1974, (J. Osugi, Ed.), The Physico-Chemical Society of Japan, Kyoto, 1975, p. 35.

50. J. R. Ferraro and L. J. Basile, Appl. Spectrosc., 28, 505 (1974).

51. R. A. Fitch, T. E. Slykhouse, and H. G. Drickamer, J. Opt. Soc. Amer., 47, 1015 (1957).

52. C. E. Weir, E. R. Lippincott, A. van Valkenburg, and E. N. Bunting, J. Res. Nat'l Bur. Stds., A63, 55 (1959).

53. E. R. Lippincott, F. E. Welsh, and C. E. Weir, Anal. Chem., 33, 137 (1961).

54. D. M. Adams, S. J. Payne, and K. Martin, Appl. Spectrosc., 27, 377 (1973).

55. J. D. Barnett, S. Block, and G. J. Piermarini, Rev. Sci, Instr., 44, 1 (1973).

56. D. M. Adams, S. K. Sharma, and R. Appleby, Appl. Opt., 16, 2572 (1977); D. M. Adams and S. K. Sharma, Appl. Opt., 18, 595 (1979).

57. D. M. Adams and R. Appleby, J. Chem. Soc. Faraday II, 73, 1896 (1977).

58. J. W. Brasch, Spectrochim. Acta, 21, 1183 (1965); C. W. Brown, R. J. Obremski, J. R. Allkins, and E. R. Lippincott, J. Chem. Phys., 49, 4806 (1968); A. van Valkenburg, Appl. Opt., 9, 1 (1970); A. J. Melveger, J. W. Brasch, and E. R. Lippincott, Appl. Opt., 9, 11 (1970); R. J. Jakobsen and Y. Mikawa, Appl. Opt., 9, 17 (1970).

59. R. P. Lowndes, Page 805 in Ref. [49], D. B. Fitchen , Rev. Sci. Instr., 34, 673, (1963).

60. E. Whalley, A. Lavergne, and P. T. T. Wong., Rev. Sci. Instr., 47, 845 (1976).

61. R. S. Hawke, K. Syassen, and W. B. Holzapfel, Rev. Sci. Instr., 45, 1598 (1974); H. J. Jodl and W. B. Holzapfel, Rev. Sci. Instr., 50, 340 (1979).

62. C. C. Bradley, H. A. Gebbie, V.V. Kechin, and J. H. King, Nature, 211, 839 (1966).

63. P. Klaeboe and T. Woldback, Appl. Spectrosc., 32, 589 (1978).

64. W. D. Ellenson and M. Nicol, J. Chem. Phys., 61, 1380 (1974).

65. G. E. Walrafen, J. Sol. Chem., 2, 159 (1973).

66. M. Abebe and G. E. Walrafen, J. Chem. Phys., 71, 4167 (1979).

67. W. B. Holzapfel, R. S. Hawke, and K. Syassen, Page 344 in Ref. [49].

68. P. T. T. Wong, J. Chem. Phys., 66, 2347 (1977).

69. P. T. T. Wong, J. Chem. Phys., 63, 5108 (1975).

70. T. C. Sivakumar, H. A. M. Chew, and G. P. Johari, Nature, 275, 524 (1978).

71. G. P. Johari and T. C. Sivakumar, J. Chem. Phys., 69, 5557 (1978).

72. Che-Kuang Wu and G. Jura, Spectrochim. Acta, 30A, 797 (1974).

73. S. Lewis and W. F. Sherman, Spectrochim.Acta, 35A, 613 (1979).

74. S. D. Hamann, Aust. J. Chem., 30, 71 (1977).

75. J. C. Evans, Spectrochim. Acta, 16, 994 (1960); 17, 129 (1961); 18, 507 (1962).

76. F. D. Medina and W. B. Daniels, Phys. Rev. Lett., 32, 167 (1974).

77. P. W. Bridgman, The Physics of High Pressure, Dover Pub., New York, 1970.

78. J. E. Bertie and E. Whalley, J. Chem. Phys., 40, 1637 (1964).

79. J. E. Bertie and E. Whalley, J. Chem. Phys., 46, 1271 (1967).

80. J. E. Bertie, H. J. Labbé, and E. Whalley, J. Chem. Phys., 49, 775 (1968).

81. P. T. T. Wong and E. Whalley, J. Chem. Phys., 64, 2359 (1976).

82. G. J. Piermarini, A. D. Mighell, C. E. Weir, and S. Block, Science, 165, 1250 (1969).

SPECIAL APPLICATIONS OF FAR INFRARED INTERFEROMETRY - INTERNAL ROTATION OF MOLECULES WITH TWO C_{3v} ROTORS

J. R. Durig and S. D. Hudson

Department of Chemistry
University of South Carolina
Columbia, South Carolina, U. S. A.

1. INTRODUCTION

Internal rotation was once thought to be essentially "free". Thermodynamic studies gradually provided stronger evidence for the fact that internal rotation is not "free". Despite the fact that internal rotational barriers are relatively small when compared to the total energy of a molecule or even of a bond, heat capacity, entropy, equilibrium constants, etc. contain appreciable contributions from internal rotation. From a chemist's viewpoint it is then important that information be obtained on barriers to internal rotation. The prediction of barriers to internal rotation from factors such as substituent effects, electronegativities and nonbonded interactions have also proven to be useful.

Torsional vibrations were first observed in the mid-infrared region as combination and difference bands with other normal modes. Observed in this manner, torsional vibrations lead to complicated spectra which are not resolvable and/or difficult to interpret. In an earlier chapter, we discussed the utility of far infrared data for the determination of the barriers to internal rotation for molecules containing one rotor. In this chapter we shall cover the rotation of molecules with two C_{3v} rotors.

About twenty years ago Taylor and Vidale [1] reported the Raman spectrum of gaseous dimethylether using a Toronto arc source. Five years later, Fateley and Miller [2] initiated the far infrared studies of such two-top molecules and reported the spectra of a series of gaseous dimethyl compounds. Many molecules with two internal C_{3v} rotors were shown to have rich far infrared

J. R. Durig (ed.), Analytical Applications of FT-IR to Molecular and Biological Systems, 501–522.

spectra [2-6]. Barriers calculated in such early studies [2-6] used approximation methods derived from theoretical work done in the field of microwave spectroscopy. At most, only two Fourier coefficients of the potential function could be determined which allowed interpretation and assignment of only a few of the usually large number of spectral features found in the far infrared region. It is apparent that this approach of treating the semi-rigid model, based on the use of tabulated perturbation sums, is not at all sufficient to explain the commonly very "rich" torsional far infrared spectra. These early methods were not capable of extracting all the information about the potential functions of these molecules. Meanwhile, instrumentation has become more powerful in far infrared and Raman spectroscopy. Since the application of interferometry and the introduction of high-powered continuous lasers as excitation sources for Raman spectroscopy, the amount and quality of the information available has been drastically increased.

Since torsional vibrations are low energy vibrations, the Boltzmann distribution allows the population of several excited states at ambient temperatures. This excited state population often allows a "series" of torsional transitions to be observed. These "hot bands" provide information not only about the height of the barrier but also can provide information leading to the determination of other coefficients in the potential function, namely the sine-sine and cosine-cosine coupling terms.

Far infrared spectroscopy has some distinct advantages over other methods for the determination of barriers and potential coefficients. For example, barrier heights are obtained from microwave studies indirectly by the splitting of the pure rotational levels or by the measurement of the relative intensities of the torsional excited state lines. The A-E splittings observed in the microwave spectra for the ground state lines only provide information on the barrier and the sine-sine coupling terms. In order to obtain the cosine-cosine coupling terms, splitting must be observed in the pure rotational levels of the torsional excited states. It should be pointed out that barriers obtained from the splittings in the torsional excited states may differ from those obtained in the ground state, but barriers obtained by the splitting method still should be good to two or three percent. However barriers obtained by the intensity method are only good to about ten percent. Also, since molecules with a center of symmetry have no microwave spectrum, no theoretical work had been carried out to treat such cases. In contrast to microwave spectroscopy, low frequency vibrational spectra give the actual vibrational energy level separations which in turn provide information on the shape of the potential function as well as values for the coupling terms.

Far infrared studies of two-top molecules are not, however,

trouble-free. It is frequently found that two (or more where conformers are present) series of torsional transitions overlap, and this requires the synthesis of an isotopic species with a $-d_3$ moiety in order to sort out the torsional transitions. In most cases the torsional transitions are very weak due to the extremely small dipole moment change. This often requires the use of long pathlength cells and high sample pressures (if attainable). Additionally, overtones are often high enough in frequency to be obscured by other fundamentals, or interact through Fermi resonance with other vibrational modes. Also one of the more frustrating and challenging problems is the elimination of trace amounts of water. Minute quantities of water from the cell walls or in the sample can totally obliterate torsional transitions in the far infrared region.

In a recent series of papers, Durig and co-workers [7-19] have demonstrated how far infrared interferometry and laser Raman spectroscopy may be combined with modern computing techniques to determine three or more Fourier coefficients of the potential function in two variables and to interpret most or all of the features observed in the far infrared spectra of such two-top molecules. The theoretical background of this method was derived by Groner and Durig [7] and the approach is heavily based on the concept of the internal isometric group introduced by Günthard and co-workers [20].

2. THEORY AND POTENTIAL FUNCTIONS

The discussion of the theory follows essentially the work of Groner and Durig [7]. In contrast to the customary procedure of first setting up the Hamiltonian and then deriving its symmetry properties, they first thoroughly investigated the internal isometric group of any model before setting up the Hamiltonian. This method allows the relation between the different models to always persist and new insight is gained into the nature of these models.

The Hamiltonian for the most general case of two nonequivalent rotors may be set up according to the semi-rigid model $C_{3v}(T) - C_1(F) - C_{3v}(\bar{T})$ where T = top, F = frame and where the bar indicates that the tops are nonequivalent, such as in the case of CD_3PHCH_3. The internal Hamiltonian for this model as obtained by Groner and Durig [7] is expressed by

$$H = \tfrac{1}{2}[g^{44}\hat{p}_o^{\,2} + 2g^{45}\hat{p}_o\hat{p}_1 + g^{55}\hat{p}_1^{\,2}] + V(\tau_o,\tau_1)$$

where

$$V(\tau_o,\tau_1) = \tfrac{1}{2}[V_{30}(1 - \cos 3\tau_o) + V_{30}' \sin 3\tau_o + V_{60}(1 - \cos 6\tau_o)$$

$$+ V'_{60} \sin 6\tau_o + V_{03}(1 - \cos 3\tau_1) + V'_{03} \sin 3\tau_1$$

$$+ V_{06}(1 - \cos 6\tau_1)$$

$$+ V'_{06} \sin 6\tau_1 + V_{33}(\cos 3\tau_o \cos 3\tau_1 - 1)$$

$$+ V'_{33} \sin 3\tau_o \sin 3\tau_1 + V''_{33} \sin 3\tau_o \cos 3\tau_1 + V'''_{33} \cos 3\tau_o$$

$$\sin 3\tau_1].$$

The symmetry group of this Hamiltonian is of order 9 and is isomorphous to $C_3 \otimes C_3$, the direct product of the cyclic group of order 3 with itself. Its character table is shown in Table 1 as an example. In the high barrier case, the energy levels may be labeled by the symmetry species Γ^{mn} and by the limiting vibrational quantum numbers v and $\bar{v}$ of the torsional modes. Each torsional state ($v\bar{v}$) splits into 5 sublevels according to

$$\Gamma^{00} \oplus \Gamma^{10} \oplus \Gamma^{01} \oplus \Gamma^{11} \oplus \Gamma^{12}$$

where Γ^{00} is the totally symmetric one-dimensional representation. The other symbols Γ^{mn} each represent a complex conjugate pair of one-dimensional representations whose energy matrices are the complex conjugates of each other and contain the same eigenvalues (pseudodegenerate representations).

Models with symmetry higher than $C_3 \otimes C_3$ are obtained as soon as the molecular frame has symmetry higher than C_1. To obtain the potential function for models with higher symmetry, all linear operators $\hat{P}_{FJ}$ of the model must be applied to the potential function of the $C_{3v}(T) - C_1(F) - C_{3v}(\bar{T})$ model. In order to keep the potential function invariant with respect to these operators, certain conditions must be placed on the potential coefficients. These conditions are given in Table 2.

The internal Hamiltonian for the $C_{3v}(T) - C_s(F) - C_{3v}(T)$ model, as for $(CH_3)_2NH$, may therefore be obtained by introducing the restrictions

$$g^{44} = g^{55},\ V_{30} = V_{03},\ V_{60} = V_{06},\ V'_{60} = -V'_{06},\ V''_{33} = -V'''_{33}$$

and

$$V'_{30} = V'_{03} = 0.$$

Its symmetry group is of order 18 and is isomorphous to $C_3 \otimes \theta_3$ where C_3 is the cyclic group of order 3 and θ_3 is the dihedral group of order 6. The energy levels of each torsional state ($v\bar{v}$) split in the high barrier case into the sublevels

$$\Gamma^{00\sigma} \oplus \Gamma^{10} \oplus \Gamma^{01} \oplus \Gamma^{11} \oplus \Gamma^{12\sigma}$$

Table 1 Character Table of the Group $C_3 \otimes C_3$

$C_3 \otimes C_3$	E	V_1	V_1^{-1}	V_2	V_2^{-1}	$V_1^{-1}V_2^{-1}$	V_1V_2	$V_1^{-1}V_2$	$V_1V_2^{-1}$	$\phi_{mn}(\tau_0,\tau_1)$[a]
Γ^{00}	1	1	1	1	1	1	1	1	1	m=3m' and n=3n'
Γ^{11}	1	ε^*	ε	ε^*	ε	ε^*	ε	1	1	m=σ(3m'+1) and n=σ(3n'+1)
	1	ε	ε^*	ε	ε^*	ε	ε^*	1	1	
Γ^{12}	1	ε^*	ε	ε	ε^*	1	1	ε^*	ε	m=σ(3m'+1) and n=-σ(3n'+1)
	1	ε	ε^*	ε^*	ε	1	1	ε	ε^*	
Γ^{10}	1	ε^*	ε	1	1	ε	ε^*	ε	ε^*	m=σ(3m'+1) and n=σ3n'
	1	ε	ε^*	1	1	ε^*	ε	ε^*	ε	
Γ^{01}	1	1	1	ε^*	ε	ε	ε^*	ε^*	ε	m=σ3m' and n=σ(3n'+1)
	1	1	1	ε	ε^*	ε^*	ε	ε	ε^*	

$\varepsilon = \exp(i2\pi/3)$; $\varepsilon^* = \exp(-i2\pi/3)$

[a] Transformation properties of basis functions $\phi_{mn}(\tau_0,\tau_1) = (2\pi)^{-1} \exp(im\tau_0) \exp(in\tau_1)$. $m',n' = 0, \pm1, \pm2, \pm3, \ldots$ The transformation properties of the one-stop basis function $\phi_{0,m}(\tau_0)$ and $\phi_{1,n}(\tau_1)$ are obtained with the relations $(2\pi)^{\frac{1}{2}}\phi_{m0}(\tau_0,\tau_1) = \phi_{0,m}(\tau_0)$ and $(2\pi)^{\frac{1}{2}}\phi_{0n}(\tau_0,\tau_1) = \phi_{1,n}(\tau_1)$, respectively. Functions with $\sigma = +1$ and $\sigma = -1$ transform like the first and second of each pair of pseudodegenerate representations, respectively.

Table 2 Potential Coefficients for Two-Top Models (Standard Form Potential Function)

Model[a]	$V_s : V_{30}$ / $V_t : V_{03}$	V'_{30} / V'_{03}	V_{60} / V_{06}	V'_{60} / V'_{06}	V_{33} / V'_{33}	V''_{33} / V'''_{33}
C_{2v}(e)	P	0	P	0	X	0
C_{2h}(e)	P	0	P	0	X	0
C_s(n)	X	0	X	0	X	0
C_2(e)	P	0	P	P	X	P
C_s(e)	P	0	P	M	X	M
C_i(e)	P	0	P	M	X	M
C_1(n)	X	0	X	X	X	X

[a]The point group symbol refers to the covering symmetry of the molecular frame. (e) and (n) refer to equivalent and nonequivalent tops, respectively. Code for V_s:/V_t:, 0: $V_s = V_t = 0$; P: $V_s = V_t$; M: $V_s = -V_t$; X: no conditions on V_s or V_t.

where sigma is + or - for $\bar{v}$ even or odd, respectively, with $\bar{v}$ as the limiting vibrational quantum number of the torsional mode which is antisymmetric with respect to the symmetry plane of the molecular frame. $\Gamma^{00\sigma}$ are one-dimensional representations and Γ^{11} is two-dimensional. $\Gamma^{12\sigma}$(σ = + or -) are the symbols for two complex conjugate pairs of one-dimensional representations, whereas Γ^{10} represents a complex conjugate pair of two-dimensional representations thus leading to fourfold degenerate energy levels.

The internal Hamiltonian for the C_{3v}(T) - C_2(F) - C_{3v}(T) model, e.g., trans-1,2-dimethylcyclopropane, may be obtained by introducing the symmetry restrictions

$$V_{30} = V_{03},\ V_{60} = V_{06},\ V'_{60} = V'_{06},\ V''_{33} = V'''_{33},\ g^{44} = g^{55}.$$

Its symmetry group is of order 18 and is isomorphous to $C_3 \otimes \theta_3$ as is the C_{3v}(T) - C_s(F) - C_{3v}(T).

Now, by applying two covering operators simultaneously to the molecular frame i.e. combining any two of the three previously discussed models, it is possible to derive higher symmetry models. The isometric groups, orders of these groups, and some examples are given in Table 3. The selection rules for infrared and Raman activity of several two-top models are summarized in Table 4.

Table 3 Two-Top Models

Model[a]	Isometric group[b]	Order	Example
$D_{2h}(e)$	$\theta_3 \otimes \theta_3 \otimes \nu_2$	72	p-xylene
$C_{2v}(n)$	$G_{18} \otimes \nu_2$	36	p-silyltoluene
$D_2(e)$	$C_3 \otimes \theta_3 \otimes \nu_2$	36	
$C'_{2h}(e)$	$C_3 \otimes \theta_3 \otimes \nu_2$	36	
$C_2(n)$	$C_3 \otimes C_3 \otimes \nu_2$	18	
$C_{2v}(e)$	$\theta_3 \otimes \theta_3$	36	dimethylether
$C_{2h}(e)$	$\theta_3 \otimes \theta_3$	36	trans-2-butene
$C_s(n)$	G_{18}	18	
$C_2(e)$	$C_3 \otimes \theta_3$	18	trans-1,2-dimethylcyclo-propane
$C_s(e)$	$C_3 \otimes \theta_3$	18	dimethylamine
$C_i(e)$	$C_3 \otimes \theta_3$	18	
$C_1(n)$	$C_3 \otimes C_3$	9	N-silylmethylamine

[a]The point group symbol refers to the covering symmetry of the molecular frame. (e) and (n) refer to equivalent and nonequivalent tops, respectively.

[b]ν_2, C_3, θ_3, and G_{18} denote the two group, the cyclic group of order 3, the dihedral group of order 6 and group 18, respectively.

3. EXAMPLES

It should now be useful to consider some representative examples. We shall show some typical spectra and then explain their assignments. In many cases the spectra are very rich and complex and additional data were needed from the Raman spectra of the gases before confident assignments could be made. The relative intensities of the Q branches can also aid in their assignment. In this regard we have found that it is very important to record the spectra with a resolution of at least 0.25 cm^{-1} since the Q branches are very sharp and the relative intensities may change with a resolution of 0.5 cm^{-1}. Also, because of the limited number of points in the interferogram, we have found that a Q

Table 4 Selection Rules for Two-Top Transitions of Some Models

Model	Active Transitions		Comments
$C_1(n)$	A $\Gamma^{00} \leftrightarrow \Gamma^{00}$ $\Gamma^{10} \leftrightarrow \Gamma^{10}$ $\Gamma^{01} \leftrightarrow \Gamma^{01}$ $\Gamma^{11} \leftrightarrow \Gamma^{11}$ $\Gamma^{12} \leftrightarrow \Gamma^{12}$		IR active and Raman polarized
$C_s(n)$	A' $\Gamma^{00\sigma} \leftrightarrow \Gamma^{00\sigma}$ $\Gamma^{10} \leftrightarrow \Gamma^{10}$ $\Gamma^{01} \leftrightarrow \Gamma^{01}$ $\Gamma^{11} \leftrightarrow \Gamma^{11}$ $\Gamma^{12} \leftrightarrow \Gamma^{12}$	A'' $\Gamma^{00+} \leftrightarrow \Gamma^{00-}$ $\Gamma^{10} \leftrightarrow \Gamma^{10}$ $\Gamma^{01} \leftrightarrow \Gamma^{01}$ $\Gamma^{11} \leftrightarrow \Gamma^{11}$ $\Gamma^{12} \leftrightarrow \Gamma^{12}$	All are IR active; A' are Raman polarized; A'' are Raman. depolarized
C_i	A_g	A_u	
$C_s(e)$	A' $\Gamma^{00\sigma_2} \leftrightarrow \Gamma^{00\sigma_2}$ $\underline{\Gamma^{10} \leftrightarrow \Gamma^{10}}$ $\underline{\Gamma^{11} \leftrightarrow \Gamma^{11}}$ $\Gamma^{12\sigma_2} \leftrightarrow \Gamma^{12\sigma_2}$	A'' $\Gamma^{00+} \leftrightarrow \Gamma^{00-}$ $\underline{\Gamma^{10} \leftrightarrow \Gamma^{10}}$ $\underline{\Gamma^{11} \leftrightarrow \Gamma^{11}}$ $\Gamma^{12+} \leftrightarrow \Gamma^{12-}$	A_g and A' are Raman polarized; A'' are Raman depolarized; A', A'', A_u are IR active
C_2	A $\Gamma^{00\sigma_1} \leftrightarrow \Gamma^{00\sigma_1}$ $\Gamma^{10} \leftrightarrow \Gamma^{10}$ $\Gamma^{11\sigma_1} \leftrightarrow \Gamma^{11\sigma_1}$ $\Gamma^{12} \leftrightarrow \Gamma^{12}$	B $\Gamma^{00+} \leftrightarrow \Gamma^{00-}$ $\Gamma^{10} \leftrightarrow \Gamma^{10}$ $\Gamma^{11+} \leftrightarrow \Gamma^{11+}$ $\Gamma^{12} \leftrightarrow \Gamma^{12}$	A, B are IR active; A are Raman polarized; B are Raman depolarized
C_{2h}	A_g	B_u	
C_{2v}	A_1 $\Gamma^{00\sigma_1\sigma_2} \leftrightarrow \Gamma^{00\sigma_1\sigma_2}$ $\underline{\Gamma^{10} \leftrightarrow \Gamma^{10}}$ $\underline{\Gamma^{11\sigma_1} \leftrightarrow \Gamma^{11\sigma_1}}$ $\Gamma^{12\sigma_2} \leftrightarrow \Gamma^{12\sigma_2}$	B_2 or B_1 $\Gamma^{00++} \leftrightarrow \Gamma^{00--}$ $\Gamma^{10+-} \leftrightarrow \Gamma^{00+-}$ $\underline{\Gamma^{10} \leftrightarrow \Gamma^{10}}$ $\underline{\Gamma^{11+} \leftrightarrow \Gamma^{11-}}$ $\Gamma^{12+} \leftrightarrow \Gamma^{12-}$	A_1, B_u, B_1, B_2, A_u are IR active; A_1, A_g are Raman polarized; B_1, B_2, A_2, B_g are Raman depolarized
	A_u A_2 $\Gamma^{00\sigma_1+} \leftrightarrow \Gamma^{00\sigma_1-}$ $\underline{\Gamma^{10} \leftrightarrow \Gamma^{10}}$ $\underline{\Gamma^{11\sigma_1} \leftrightarrow \Gamma^{11\sigma_1}}$ $\underline{\Gamma^{12+} \leftrightarrow \Gamma^{12-}}$	B_g B_1 or B_2 $\Gamma^{00+\sigma_2} \leftrightarrow \Gamma^{00-\sigma_2}$ $\underline{\Gamma^{10} \leftrightarrow \Gamma^{10}}$ $\underline{\Gamma^{11+} \leftrightarrow \Gamma^{11-}}$ $\Gamma^{12\sigma_2} \leftrightarrow \Gamma^{12\sigma_2}$	

—— Underlined transitions may become weakly active.

---- Underlined transitions may become weakly active only in the C_{2v} case.

branch may be "cut off" if the point does not come at the maximum of the absorption peak. This may cause the Q branch to appear "weaker" than it really is and therefore the relative intensities will be incorrect. One further point should be made. As mentioned earlier, water vapor is a serious problem in the far infrared spectral region and at times we have attempted to subtract it from the sample spectrum. We have found that it is almost impossible to distinguish between the "spectral artifacts" from the improper canceling of the water bands and the torsional transitions. Therefore we should like to stress that it is very important to have both a dry sample and a dry cell simultaneously, which is no small task, if good torsional data are to be obtained.

A. Dimethylphosphine-d_3 ($C_1(n)$)

The internal Hamiltonian for CH_3PHCD_3 may be set up according to the $C_{3v}(T)$ - $C_1(F)$ - $C_{3v}(\bar{T})$ model. Details of this model were previously discussed and will not be repeated here. From the far infrared spectrum of this molecule, Fig. 1, it is clearly evident that two distinct series of torsional transitions exist. The higher frequency series begins at 183.2 cm^{-1} and the lower frequency series begins at 137.2 cm^{-1}. An initial assignment was made using the first three transitions in each series. The transitions (01) ← (00), (02) ← (01), and (03) ← (02) were assigned to bands at 183.2, 172.6 and 158.6 cm^{-1}, respectively, and the transitions (10) ← (00), (20) ← (10), and (30) ← (20) were assigned to bands at 137.2, 131.2 and 124.2 cm^{-1}, respectively. This assignment was used as the basis for the more complete assignment shown in Table 5. It should be noted that splitting of several of the torsional transitions was observed. For example, the (50) ← (40) transition was predicted to be split by 6.4 cm^{-1} whereas a splitting of 8.3 cm^{-1} was observed. Similarly, the (40) ← (30) transition was split by 3.4 cm^{-1} between the Γ^{10} and Γ^{11} levels whereas the calculated splitting was 2.4 cm^{-1}. A better fit of this splitting could not be obtained without drastically affecting the fit for the transitions lower in the well. The potential and kinetic coefficients used for the final frequency fit are shown in Table 6. The average barrier of 832 cm^{-1} agrees extremely well with the value of 811 cm^{-1} obtained by the microwave splitting method [22]. However, it should be pointed out that the barrier of 711 cm^{-1} which was obtained in the initial far infrared study [10] of dimethylphosphine was in error because of the misassignment of the "second" set of transitions. Spectral "artifacts" resulting from the subtraction of water led to the misassignment.

B. Dimethylamine ($C_s(e)$)

The internal Hamiltonian for CH_3NHCH_3 may be set up according to the $C_{3v}(T)$ - $C_s(F)$ - $C_{3v}(T)$ model. Since this model was

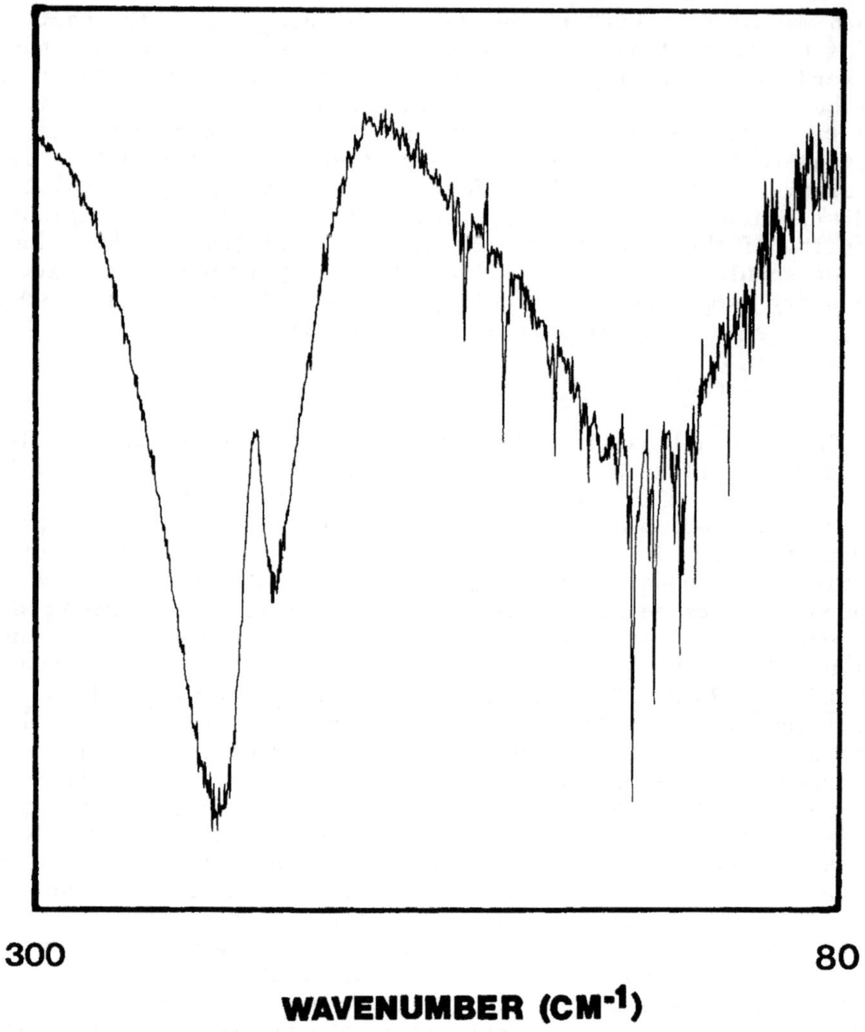

Fig. 1 Far infrared spectrum of gaseous dimethylphosphine-d_3 at 0.25 cm^{-1} resolution.

discussed in a previous section of this work the symmetry restrictions, splitting of the torsional energy levels, etc. will not be restated.

The far infrared spectrum of dimethylamine-d_0 is reproduced in Fig. 2. The assignment of the observed Q branches is relatively difficult until one compares their frequency sums with the frequencies observed in the "double-jump" region of the Raman

Table 5 Observed Torsional Data, Calculated Frequencies and Assignment for Dimethylphosphine-d_3[a]

Observed		Assignment	Calculated	
Raman (Δcm^{-1})	Infrared (cm^{-1})	$v'\bar{v}' \leftarrow v\bar{v}$	Γ^{10}	Γ^{11}
183	183.2	0,1 ← 0,0	182.1	182.1
172	172.6	0,2 ← 0,1	173.0	173.2
	158.6	0,3 ← 0,2	--	157.8
	149.5	5,0 ← 4,0	--	148.4
	141.2	5,0 ← 4,0	142.0	--
	138.5	hot band	--	--
	137.2	1,0 ← 0,0	137.1	137.1
	132.8	hot band	--	--
	131.2	2,0 ← 1,0	132.0	132.0
	124.9	hot band	--	--
	124.2	3,0 ← 2,0	125.8	125.8
	123.4	4,0 ← 3,0	122.5	--
	120.0	4,0 ← 3,0	--	120.1
117		?	--	--
	110.6	?	--	--

[a]Taken from Ref. [21].

Table 6 Torsional Potential Constants (cm^{-1}) from the Low Frequency Vibrational Data and Kinetic Coefficients (cm^{-1}) for Dimethylphosphine-d_3

Parameter	Value	Dispersion
V_{30}[a]	828.7	10.6
V_{03}[a]	835.8	6.8
V_{60}	-28.5	5.9
V_{06}	-37.2	4.7
$\hbar^2 g^{44} = 2F_1$	6.26312	
$\hbar^2 g^{55} = 2F_2$	11.47490	
$\hbar^2 g^{45} = 2F'$	-0.47727	

Standard deviation of frequency fit is 1.08 cm^{-1}.

[a]V_{30} represents the barrier for the CD_3 group and V_{03} represents the barrier for the CH_3 group.

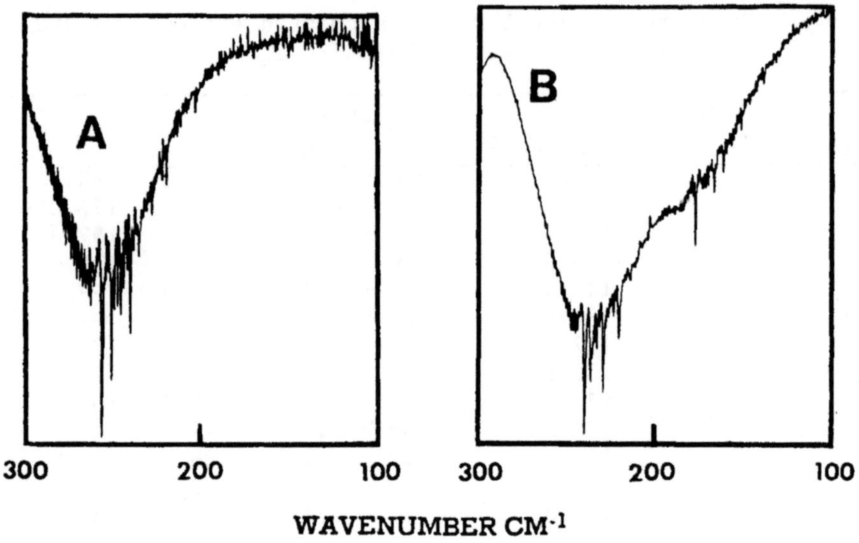

Fig. 2 Far infrared spectra of gaseous dimethylamine-d_0 (A) and dimethylamine-d_3 (B). Used by permission [8].

spectrum (see Fig. 3). The first five "double-jumps" of the high-energy series and the first four "double-jumps" of the low-energy series are readily observable in the Raman spectrum. With these data the far infrared spectrum can be easily assigned (see Table 7). However, it should be noted that these overtone transitions are considerably more intense in the Raman spectrum than are the fundamental or "hot band" transitions (see the spectral region of 250 cm^{-1} in Fig. 3). Additionally it should be noted that the relative intensities of these "double-jumps" do not decrease according to the Boltzmann distribution. It appears that in the higher excited states of these methyl torsional modes, the increased mechanical as well as the increased electrical anharmonicity give rise to successive double-jumps which are much more intense than one would expect from the Boltzmann factor. Thus, the Raman spectrum has proven to be extremely important in not only providing a check of the assignment of the far infrared Q branches but it frequently is the only means of obtaining frequencies for the lower energy transitions. This is particularly true for the molecules of $C_{2v}(e)$ symmetry.

The initial fitting of the torsional potential function for dimethylamine was conducted using three potential constants, $V_{30} = V_{03}$, V_{33} and V_{33}'. Attempts to expand the potential function and improve the fit were futile. V_{60}, V_{06}, V_{60}', V_{06}', V_{33}'' and V_{33}''' were therefore set to zero for the final computations. V_{60}

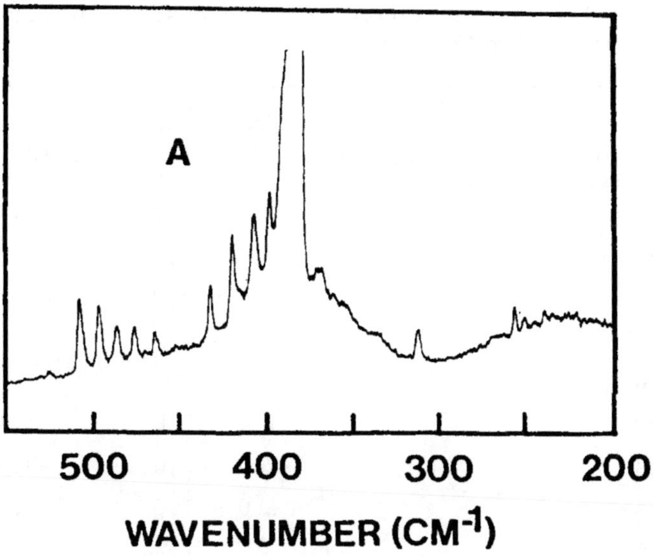

Fig. 3 Raman spectrum of gaseous dimethylamine-d_0. Used by permission [8].

and V_{06} were found to have such a small value for dimethylamine that they were excluded. V_{60}', V_{06}', V_{33}'' and V_{33}''' were found to have little effect on the calculated values of the transitions and were, therefore, very ill-determined so they were also excluded.

As shown by Groner and Durig [7], the parameter set commonly used for the potential function is not the best one in terms of determinability because some of the coefficients are strongly correlated. They introduced the linear combinations

$$\begin{bmatrix} V_{3s} \\ V_{3d} \\ V_{eff} \end{bmatrix} = \begin{bmatrix} 1/2 & 1/2 & 0 \\ 1/2 & -1/2 & 0 \\ 1/2 & 1/2 & -1 \end{bmatrix} \begin{bmatrix} V_{30} \\ V_{03} \\ V_{33} \end{bmatrix}$$

as determinable and less correlated parameters in place of V_{30}, V_{03} and V_{33}. These new parameters have generally lower dispersions than the original ones. In the case of a potential function for equivalent tops, $V_{3s} = V_{03} = V_{30}$ and $V_{3d} = 0$ holds, of course. The value of the effective barrier which is given directly by V_{eff} is listed in Table 10 and the value is 1054 cm^{-1}. This value is about 71 cm^{-1} lower than the V_3 calculated from the average of the microwave splittings measured in the first excited state for

Table 7 The Observed Torsional Data and Assignment for the Vapor of Dimethylamine-d_0[a]

RAMAN		FIR		ASSIGNMENT	
Δcm^{-1}	rel. int.[b]	cm^{-1}	rel. int.[b]	$v'_1v'_2 \leftarrow v_1v_2$	Obsd.-Calc.
507	vs			0,2 ← 0,0	0.98
496	vs			0,3 ← 0,1	-1.91
486	m			0,4 ← 0,2	0.77
476	m				
464	w				
452	vw			1,1 ← 0,0	-1.43
432	ms			2,0 ← 0,0	-0.33
419	vs			3,0 ← 1,0	1.85
407	s				
398	s				
383	*			δ CNC	
370	w				
362	vw				
312	w				
256	w	256.3	vs	0,1 ← 0,0	1.83
251	w	250.8	s	0,2 ← 0,1	-0.75
245	vvw	245.3	ms	0,3 ← 0,2	-1.06
240	vw	239.8	s	0,4 ← 0,3	0.94
		235.0	mw	1,1 ← 1,0	0.61
		227.6	mw	(0,1 ← 0,0)'	hot band of δ CNC
		222.0	mw	(0,2 ← 0,1)'	hot band of δ CNC
		219.4	m	1,0 ← 0,0	0.36
		213.0	w	2,0 ← 1,0	-0.29
		211.1	w	2,1 ← 2,0	-0.37
		202.7	w	3,0 ← 2,0	-1.17

[a] Data taken from Ref. [8].
[b] vs=very strong, s=strong, m=medium, w=weak, sh=shoulder, b=broad. The intensities listed are relative to the other torsional data, not to the rest of the vibrational data.

each torsion [23]. The microwave analysis was complicated by the inversion splitting which was also averaged. In view of the compromises made in the calculations based on the microwave data, it is reasonable that the V_3 thus computed (1125 cm^{-1}) is not very different from the "effective" V_3 (1054 cm^{-1}) reported from the vibrational study [8].

C. Dimethylether ($C_{2v}(e)$)

The internal Hamiltonian for CH_3OCH_3 may be set up according to the $C_{3v}(T)$ - $C_{2v}(F)$ $\dot{-}$ $C_{3v}(T)$ model. The symmetry restrictions, isometric group and selection rules for this model may be found in Tables 2, 3 and 4, respectively.

The torsional modes of dimethylether have been extensively studied [1-4] by both far infrared and Raman spectroscopy. In the far infrared spectrum of this molecule, Fateley and Miller [2] assigned the C-type Q branch at 241 cm^{-1} as the infrared active torsional fundamental, (01) ← (00). This assignment was supported by more recent studies [3, 4]. However Groner and Durig [7] recently showed by a study of the low frequency spectrum with a Fourier transform spectrometer (0.25 cm^{-1} resolution) that this band is composed of at least six sharp Q branches and contains not only the (01) ← (00) transition but also the (02) ← (01) and (03) ← (02) transitions. In the Raman spectrum of the overtone region, two relatively strong bands are apparent at 481 and 450 cm^{-1} and were previously assigned by Taylor and Vidale [1] to the overtones of both torsional fundamentals. Indeed, since the band complex at 241 cm^{-1} is assigned to the first three transitions of the series $(0,\bar{v}+1) \leftarrow (0,\bar{v})$ where $\bar{v}$ = 0, 1, 2, the 481 cm^{-1} Raman line must be assigned to the (02) ← (00) and (03) ← (01) transitions. However, from the microwave data [24], the second torsional fundamental was predicted at 197 cm^{-1} so the Raman bands at 395.5 and 385.0 cm^{-1} were assigned to the (20) ← (00) and (30) ← (10) transitions, respectively, of the A_2 torsional fundamental. With this assignment the 450 cm^{-1} Raman line remains unexplained. Observed frequencies, assignments and calculated values are found in Table 8 and the potential and kinetic coefficients are found in Table 9. The energy level diagram for dimethylether is shown in Fig. 4. From these torsional data, the potential constants V_{03}, V_{33}, V_{33}' and V_{eff} were calculated to have values of 1215, 273, 22 and 943 cm^{-1}, respectively. The effective barrier agrees well with the value (909 cm^{-1}) obtained from microwave splittings [24]. It should be pointed out that the term V_{33} is surely quite different from V_{33}' and in contrast to some earlier proposals [25, 26] that these coefficients might have the same values but opposite signs.

Table 8 Observed and Calculated Frequencies[a] of Torsional Transitions of CH_3OCH_3 in cm^{-1}.

Observed		Assignment	Calculated				
Infrared	Raman	$(v'\bar{v}') \leftarrow (v\bar{v})$	ν_{00}	ν_{10}	ν_{11}	ν_{12}	$\nu_{obs}-\nu_{10}$
	481.2	(01) ← (00)	481.63	481.67	481.71	481.71	-0.47
	479.1 sh	(03) ← (01)	478.99	478.79	478.60	478.60	0.31
	450.5						
		(12) ← (10)	448.22	447.63	447.03	447.03	
	438.0						
	430.0						
	426.0						
	424.0						
	412.0	COC bending					
	395.5	(20) ← (00)	394.85	394.92	395.00	395.00	0.58
	385.0	(30) ← (10)	385.77	385.83	385.08	385.07	-0.83
	381.5						
	363.0						
241.0		(01) ← (00)	240.84	240.84	240.83	240.83	0.16
240.5							
240.2		(02) ← (01)	240.79	240.83	240.88	240.88	-0.63
239.8							
238.9		(03) ← (02)	238.21	237.96	237.72	237.72	0.94
237.8							
226.1							
225.5							
224.8		(12) ← (11)	225.84	225.13	224.41	224.41	-0.33
223.1		(11) ← (10)	222.38	222.50	222.63	222.63	0.60
200.7		(21) ← (20)	201.96	200.67	199.21	199.21	0.03
199.0							

[a]Data taken from Ref. [7].

Table 9 Kinetic Coefficients[a] and Potential Parameters[b] in cm^{-1}.

	CH_3OCH_3	CD_3OCH_3
$\hbar^2 g^{44} = 2F_1$	13.56441	8.19286
$\hbar^2 g^{55} = 2F_2$		13.10636
$\hbar^2 g^{45} = 2F'$	-2.63808	-2.47808
V_{30}	1215.28 ± 16.58	1250.08 ± 34.23
V_{03}		1335.12 ± 25.50
V_{60}	-20.75 ± 4.28	-6.32 ± 6.19
V_{06}		-19.25 ± 8.80
V_{33}	272.68 ± 18.23	397.38 ± 35.64
V'_{33}	21.64 ± 4.63	41.56 ± 10.30
$\frac{1}{2}(V_{30} + V_{03} - 2V_{33})$	942.60 ± 8.65	895.22 ± 14.20
$V_{30} - V_{33}$		852.70 ± 12.90
$V_{03} - V_{33}$		937.74 ± 23.44
$\frac{1}{2}(V_{30} + V_{03})$		1292.60 ± 27.47
$\frac{1}{2}(V_{03} + V_{03})$		45.52 ± 12.51
σ[c]	0.720	0.846

(In CH_3OCH_3, the values 13.56441, 1215.28 ± 16.58 and -20.75 ± 4.28 are bracketed to apply to both F_1/F_2, V_{30}/V_{03} and V_{60}/V_{06} respectively.)

[a] g^{mn} taken from Ref. [4].

[b] Error limit of a parameter is twice its dispersion.

[c] σ is the standard deviation of the fitted frequencies in cm^{-1}.

4. SUMMARY AND CONCLUSIONS

For many of the molecules studied so far (i.e. dimethylphosphine [10], propane [9], and isobutene [12]), rather large, negative sine-sine coupling terms, V'_{33}, have been obtained (see Table 10). Differences in frequency between that of the geared and antigeared torsions (an A_2 and a B_2 in the case of a C_{2v} molecular frame) are determined by the g^{45} and V'_{33}. The effects of the kinetic term far outweigh those of the sine-sine coupling

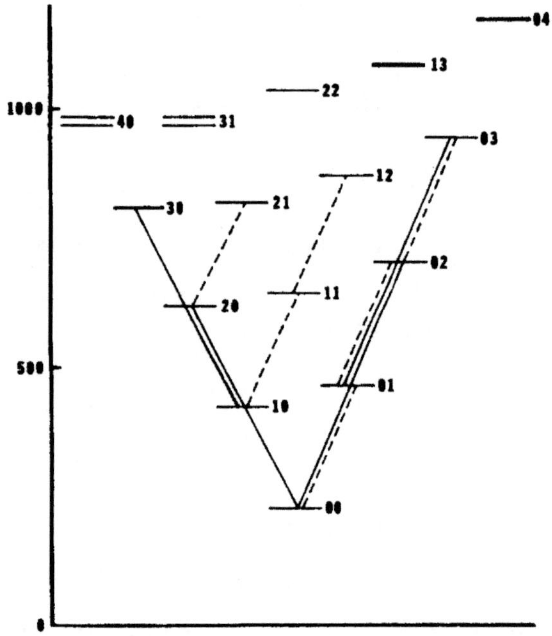

Fig. 4 Energy level diagram for CH_3OCH_3, obtained with the data in Table 8. Ordinate in cm^{-1}. The numbers to the right of the energy levels denote the limiting vibrational quantum numbers ($v\bar{v}$). Any energy level consists of four sublevels which are partly resolved for the (40) and (31) level, whereas "broadening" occurs for the levels (30), (21), (13), and (04).

term and the latter can actually be thought of as a small perturbation of g^{45}. Potential functions which have relatively large negative V'_{33} terms have been obtained for molecules which have rather small negative g^{45} terms. On the other hand, when large negative g^{45} terms were obtained from the structural parameters as in dimethylether [7] and dimethylamine [8], the sine-sine coupling term, V_{33}', was relatively small in magnitude and in fact for these two molecules it was positive.

The terms V'_{60}, V'_{06}, V''_{33} and V'''_{33} in the potential function have been found to have little effect on the calculated values of the torsional transitions and are, therefore, ill-determined. These terms have been excluded in all of the calculations for the molecules cited above. Coefficients of this type have been considered in the potential function of a semi-rigid $C_{3v}(T)$ - $C_1(F)$-$C_{3v}(\bar{T})$model by Grant et al. [25] but they had not been considered for semi-rigid molecules, $C_{3v}(T)$ - $C_s(F)$ - $C_{3v}(T)$, such as

Table 10 Torsional Potential Constants and Kinetic Coefficients (cm^{-1}) for Some Internal C_{3v} Rotors

Parameter	$(CH_3)_2NH$	$(CH_3)_2PH$	$(CH_3)_2O$	$(CH_3)_2S$	$CH_3CH_2CH_3$	$(CH_3)_2C{=}CH_2$
V_{30}	1281.7	940.8	1215.28	804.3	1323.4	893.7
V_{03}	1281.7	940.8	1215.28	804.3	1323.4	893.7
V_{60}	--	--	-20.75	--	--	-4.0
V_{06}	--	--	-20.75	--	--	-4.0
V_{33}	227.9	240.0	272.68	145.8	176.6	165.7
V'_{33}	30.2	-55.8	21.64	-22.4	-132.0	-135.1
V_{eff}	1053.8	700.8	942.60	658.6	1146.8	728.0
$\hbar^2 g^{44}=2F_1$	13.243	11.455	13.564	11.922	12.204	11.533
$\hbar^2 g^{45}=2F'$	-2.230	-0.509	-2.738	-0.638	-1.433	-0.353
$\hbar^2 g^{55}=2F_2$	13.243	11.455	13.564	11.922	12.204	11.533
σ^a	1.25	0.56	0.720	0.89	0.81	0.82
Ref.[b]	[8]	[21]	[7]	[11]	[9]	[12]
V_3^c	1125	811	909	762	1152	773
V'^{c}_{33}	--	29	-5.1	12.6	-55	--
Ref.[d]	[23]	[22]	[24]	[27]	[27]	[28]

[a] σ is the standard deviation in the frequency fit in cm^{-1}
[b] References for vibrational studies
[c] Barrier and V'_{33} from microwave splittings
[d] References for microwave studies

$V_{eff} = [(V_{30} + V_{03})/2] - V_{33}$

dimethylamine and dimethylphosphine, until Groner and Durig [7] introduced such terms with these coefficients into the potential function of certain types of semi-rigid two-top models. Using the fact that the potential function is invariant with respect to operations of the internal isometric symmetry group, they proved [7] that these coefficients may be non-zero by symmetry arguments. Without introducing these terms, the potential function would have the same symmetry for $(CH_3)_2NH$ as for $(CH_3)_2O$ and this seems unreasonable from a chemist's viewpoint.

It appears that vibrational spectroscopic data provide excellent promise for the determination of the potential coupling terms for molecules with two C_{3v} rotors. However, it should be emphasized that usually the Raman data for the two-quantum jumps are needed for a confident assignment of the far infrared transitions. There has been some difficulty of an energy level being perturbed by interaction with the bending mode in several of the molecules studied so far. However, there are usually eight to twelve transitions observed for these two-top molecules, so the perturbed level can be readily identified.

The cosine-cosine coupling terms, V_{33}, are usually taken to be zero in microwave studies but it has been found that these have values of the order of 15% of the V_3 term. This coupling term is obtained only from excited torsional states in the microwave studies, and frequently there are not sufficient splitting data to determine simultaneously both the sine-sine and cosine-cosine coupling parameters. The sine-sine coupling term is very much dependent on the g^{45} term, so a good structural determination is necessary for the evaluation of this coupling term. It is suspected that part of the differences between the microwave barrier values and those obtained by vibrational spectroscopy can be attributed to poor structural parameter values. Further work is needed on more two-top molecules before trends in the coupling parameters can be ascertained, if indeed they do exist. A combination of infrared and Raman spectroscopy studies appears to provide the best method for continuing such investigations.

REFERENCES

1. R. C. Taylor and G. L. Vidale, J. Chem. Phys., 26, 122 (1957).

2. W. G. Fateley and F. A. Miller, Spectrochim. Acta, 18, 977 (1962).

3. K. D. Möller, A. R. DeMeo, D. R. Smith, and L. H. London, J. Chem. Phys., 47, 2609 (1967).

4. E. C. Tuazon and W. G. Fateley, J. Chem. Phys., 54, 4450 (1971).

5. D. R. Smith, B. M. McKenna, and K. D. Möller, J. Chem. Phys., 45, 1904 (1966).

6. S. Weiss and G. E. Leroi, Spectrochim. Acta, 25A, 1759 (1969).

7. P. Groner and J. R. Durig, J. Chem. Phys., 66, 1856 (1977).

8. J. R. Durig, M. G. Griffin, and P. Groner, J. Phys. Chem., 81, 554 (1977).

9. J. R. Durig, P. Groner, and M. G. Griffin, J. Chem. Phys., 66, 3061 (1977).

10. J. R. Durig, M. G. Griffin, and W. J. Natter, J. Phys. Chem., 81, 1588 (1977).

11. J. R. Durig and M. G. Griffin, J. Chem. Phys., 67, 2220 (1977).

12. J. R. Durig, W. J. Natter, and P. Groner, J. Chem. Phys., 67, 4948 (1977).

13. J. R. Durig and D. A. C. Compton, J. Chem. Phys., 69, 4713 (1978).

14. J. R. Durig, D. A. C. Compton, and M. R. Jalilian, J. Phys. Chem., 83, 511 (1979).

15. J. R. Durig and D. A. C. Compton, J. Phys. Chem., 83, 265 (1979).

16. J. R. Durig, G. A. Guirgis, and D. A. C. Compton, J. Phys. Chem., 83, 1313 (1979).

17. J. R. Durig, S. D. Hudson, and W. J. Natter, J. Chem. Phys., 70, 5747 (1979).

18. J. R. Durig and D. A. C. Compton, J. Phys. Chem., 83, 2879 (1979).

19. J. R. Durig and G. A. Guirgis, Chem. Phys., 44, 309 (1979).

20. A. Bauder, R. Meyer, and Hs. H. Günthard, Mol. Phys., 28, 1305 (1974); 31, 647 (1976).

21. S. D. Hudson, Ph.D. Thesis, University of South Carolina, Columbia, SC (1980).

22. J. R. Durig, P. Groner and Y. S. Li, J. Chem. Phys., 67, 2216 (1977).

23. J. E. Wollrab and V. W. Laurie, J. Chem. Phys., 54, 532 (1971).

24. J. R. Durig, Y. S. Li, and P. Groner, J. Mol. Spectrosc., 62, 159 (1976).

25. D. M. Grant, R. J. Pugmire, R. C. Livingston, K. A. Strong, H. L. McMurry, and R. M. Brugger, J. Chem. Phys., 52, 4424 (1976).

26. J. R. Hoyland, J. Chem. Phys., 49, 1908 (1968).

27. A. Trinkaus, H. Dreizler and H. D. Rudolph, Z. Naturforsch., 28A, 750 (1973).

28. V. W. Laurie, J. Chem. Phys., 34, 1516 (1961).

SYMMETRY DETERMINATIONS BY VIBRATIONAL SPECTROSCOPY

J. R. Durig and S. A. Johnston

Department of Chemistry
University of South Carolina
Columbia, South Carolina, U. S. A.

1. INTRODUCTION

Vibrational spectroscopy has often been used to determine the symmetry of molecules. However, a confident determination of the molecular symmetry necessitates a study of both the infrared and Raman spectra, and earlier determinations by one technique alone have led, in some cases, to incorrect conclusions. Additionally, vibrational data obtained by one technique for one physical phase (eg. infrared data of the gas) have frequently been compared with vibrational data taken by another technique for a different phase (eg. Raman data of the liquid). Also it should be pointed out that the low frequency skeletal stretching and bending motions are usually the most sensitive for symmetry determinations, but this region of the spectrum was frequently inaccessible for many of the early spectroscopic studies. These facts have lowered confidence in the vibrational spectroscopic technique for the determination of molecular symmetries. However when molecules have a symmetry center, the results are usually quite definitive.

A center of symmetry or center of inversion is a symmetry element which allows the inversion of all atoms through a center of symmetry. That is, if a molecule can be brought to an equivalent configuration by changing the coordinates (x, y, z) of every atom, where the origin of the coordinate system lies at the center of mass of the molecule, into (-x, -y, -z), then the point at the origin is said to be a center of symmetry or center of inversion. When this center exists, restrictions are placed on a number of atoms in the molecule. Only one atom may be at the center since it is a point. This atom at the center is unique since it does not shift when the inversion is performed. All other atoms must

J. R. Durig (ed.), Analytical Applications of FT-IR to Molecular and Biological Systems, 523–536.

have a "twin" with which it can exchange when inversion is performed.

A molecule which contains a center of symmetry follows the rule of mutual exclusion with respect to its fundamental modes. That is, a normal mode can not be both infrared and Raman active. The mutual exclusion rule holds for the following molecular symmetries: C_i, C_{2h}, C_{4h}, C_{6h}, D_{2h}, D_{4h}, D_{6h}, D_{3d}, D_{5d}, S_6, $D_{\infty h}$, T_n and O_h. The rule may hold for the whole molecule or perhaps only the skeletal modes. Nonetheless, the rule of mutual exclusion makes the assignment of these symmetries fairly definitive. Some obvious examples of molecules with an inversion center, i, are: carbon dioxide, carbon suboxide, acetylene, ethylene, square planar XeF_4, benzene, and octahedral PtF_6. When checking to see if the rule of mutual exclusion holds, one must compare spectra that were recorded in the same physical state. Initially some incorrect conclusions were drawn on many molecules due to the fact that infrared spectrum of the gas and Raman data on the liquid were compared. Intermolecular interactions in the liquid can often cause difficulties with such comparisons.

It is possible to obtain information on the relative amounts of two different isomers in the fluid states using vibrational spectroscopy and to determine, by temperature variations, the relative difference in energy between the two isomers. Also, all three physical states can often be studied. It is therefore possible in many cases to determine the symmetry in the solid, liquid, and gaseous states and to follow a change in symmetry if one occurs. However it is usually necessary to obtain vibrational data on the solid phase for the determination of two isomers and their relative stabilities as well as to observe possible changes in symmetry with a change in physical state.

The fundamental vibrational modes encountered below 600 cm^{-1} are somewhat different from those occurring at higher frequency. These low frequency modes which are usually the skeletal stretching and bending motions are often greatly influenced by changes in the structure or conformation of the molecule. They also make the largest contribution to the thermodynamic function of a molecule. Since the low frequency modes are generally the most sensitive to changes in molecular structure and conformation, it is in this spectral region one must search for changes in vibrational activity.

Vibrational data is usually easily obtainable, and the interpretation is frequently straightforward. This method is applicable to a large variety of molecules and usually only a short period of time is needed to collect the data. Therefore a larger variety of molecules can be studied in more physical states by vibrational spectroscopy than by any other technique.

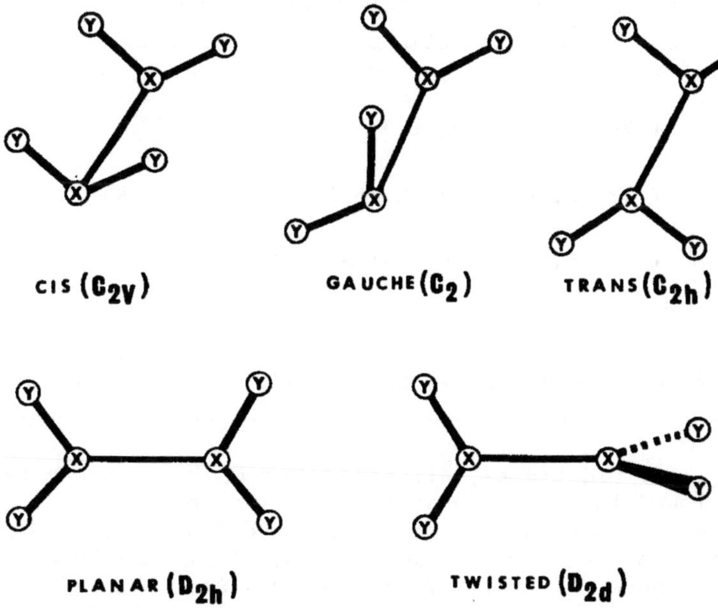

Fig. 1 The possible structures for X_2Y_4-type molecules.

2. MOLECULAR SYMMETRIES OF SOME X_2Y_4 MOLECULES

The molecules that have the general formula X_2Y_4 exhibit a wide variety of molecular symmetries. The five possible structures for these X_2Y_4 molecules include planar (D_{2h}), staggered with a twist angle of 90° and (D_{2d}), and three resulting from a non-planar arrangement of the XY_2moiety: trans (C_{2h}), gauche (C_2) and cis (C_{2v}). These five symmetries are illustrated in Fig. 1. The representations for the normal modes for each molecular symmetry group are given in Table 1.

If X is a Group VA element such as nitrogen, a pyramidal arrangement of the neighboring atoms and nonbonded electron pairs about the nitrogen atom leads to structures which have been determined to be either gauche or trans. Several factors may influence the relative amounts of the trans or gauche isomers including interactions of the two non-bonded electronic clouds, exchange forces, the steric repulsion of the substituent atoms, and others. By comparing the Raman and infrared data, one can easily distinquish between C_{2h} (trans) and C_2 (gauche) symmetry. The rule of mutual exclusion holds for C_{2h} symmetry, and a comparison of the infrared and Raman frequencies usually provides conclusive evidence for or against the presence of the symmetry center. When both isomers are

Table 1 Representations of Normal Modes of Molecules with Molecular Formula X_2Y_4

D_{2h}	$3A_g(R,p) + 2B_{1g}(R,dp) + 1B_{2g}(R,dp) + 0B_{3g}(R,dp) + 1A_u +$ $1B_{1u}(IR) + 2B_{2u}(IR) + 2B_{3u}(IR)$
D_{2d}	$3A_1(r,p) + 1B_1(R,dp) + 2B_2(R,dp,IR) + 3E(R,dp,IR)$
C_{2h}	$4A_g(R,p) + 2B_g(R,dp) + 3A_u(IR) + 3B_u(IR)$
C_2	$7A(R,p,IR) + 5B(R,dp,IR)$
C_{2v}	$4A_1(R,p,IR) + 3A_2(R,dp) + 2B_1(R,dp,IR) + 3B_2(R,dp,IR)$

present in the fluid states, all eighteen possible fundamental frequencies allowed in the Raman or infrared are not usually observed. In fact, usually only the X-X stretching and XY_2 bending modes have different frequencies for the two isomers. If isomers are present, usually only one will persist in the solid state. Hence, it is absolutely necessary to record the vibrational spectrum for the solid state, so that the bands that arise from the second isomer in the fluid states will disappear upon solidification of the sample. Since, in general, Raman spectra have fewer overtones and combination bands than infrared spectra, a comparison of the Raman spectra of the liquid and solid provides a more convenient method for the detection of the presence of isomers than the infrared since it is frequently characterized by broad, ill-defined band contours in the gas phase. However, a study of both Raman and infrared spectra is necessary for a confident determination of the molecular symmetries. Some earlier determinations of the molecular symmetries by one technique alone led to incorrect conclusions. To illustrate these concepts, several examples will be considered.

Tetrafluorohydrazine exists as an equilibrium mixture of the _gauche_ (C_2) and _trans_ (C_{2h}) isomers. The microwave spectrum [1] was obtained and the observed rotational constants indicated that the molecule has a dihedral angle of 65°, and therefore it has C_2 symmetry. From an investigation of the Raman spectrum of the gas phase [2], it was concluded that no mutual exclusion exists between the infrared [3] and Raman frequencies in the region from 700-100 cm^{-1} and that significant amounts of the _trans_ form was improbable. Appreciable error limits could be assumed in the Raman frequencies resulting from measuring weak bands against a strong background and having a spectrometer with a spectral slit width of only 16 cm^{-1}. More complete infrared studies [4,5] showed that the data could be interpreted on the basis of C_2 symmetry if the above assumption was valid. Colburn _et al._ [6] concluded from a low temperature NMR study that two rotational isomers were present and that they differed in energy by 100-200 cal/mole. The presence of a significant amount of the trans form was excluded by Bohn and Bauer [7] on the basis of an electron diffraction study of N_2F_4. Although the presence of the gauche form was well established, the reality of a trans form was in doubt. Therefore Durig and Clark [8] initiated a variable temperature Raman study of liquid N_2F_4. The presence of seventeen pronounced Raman lines suggested that a second conformer exists since only twelve lines are expected for the gauche form. The mutual exclusion principle was operative for at least eight bands below 800 cm^{-1}. It was concluded that the trans N_2F_4 isomer is present at ambient and lower temperatures giving rise to the mutually exclusive bands. Later infrared studies [9,10] were in agreement although the precise assignment of some of the normal modes differed [11].

Tetramethylhydrazine has been the subject of two vibrational studies. McKean [12] measured the infrared spectra from 400-4500 cm^{-1} of the gaseous and condensed states. In the more recent study, Durig et al. [13] recorded the Raman spectra of the gas, liquid, and solid and measured the depolarization values. They also recorded the far infrared spectrum from 33 to 450 cm^{-1}. Only the skeletal stretching and bending frequencies were expected to show significant variations with changes in molecular structure [13]. The Raman spectrum of solid tetramethylhydrazine shows eleven distinct lines between 300 and 1050 cm^{-1}. This is exactly the number expected for skeletal modes, excluding the torsion, if the molecule has C_2 symmetry. If the trans form were present only six lines would be expected and mutual exclusion would be operative. However excellent correspondence between the infrared and Raman bands in the solid is found. Therefore, the molecule has C_2 symmetry in the solid. There were no bands observed in the liquid which disappear with solidification. In the Raman spectrum of the gas, the Q branches for the symmetric modes were well pronounced and the broad lines of the asymmetric modes could still be identified. Thus, it was concluded that tetramethylhydrazine exists in only one isomeric form in all states and this form is the gauche (C_2) isomer.

For the X_2Y_4 molecules that have a boron for the X atom rather than a Group VA element, a decrease from 38 to 34 valence electrons occurs, and the two most probable structures for molecules of this type are either planar (D_{2h}) or staggered with a 90° twist angle (D_{2d}). Since the rule of mutual exclusion holds for D_{2h} symmetry, it is easy to distinguish between D_{2d} and D_{2h} molecular symmetry by comparing the infrared and Raman data for the same physical state. The representations for these two symmetries are given in Table 1. In some cases the Raman data are sufficient, since six bands are expected for D_{2h} symmetry, but nine are possible for D_{2d} symmetry. However, it is not possible to determine the symmetry from the infrared data alone, since five bands are expected for both symmetries.

Diboron tetrachloride, B_2Cl_4, was shown by x-ray diffraction [14] to be planar (D_{2h}) in the crystal at -165°C. An infrared spectrum of the gas to 200 cm^{-1} and a Raman study of the liquid were reported by Mann and Fano [15]. They concluded that the molecule has D_{2d} symmetry with the BCl_2 groups oriented perpendicularly in the fluid states. Hedberg and Ryan [16,17] also concluded that the "free" molecule has D_{2d} symmetry based on their electron diffraction studies. Nimon and co-workers [18] reported an infrared study of matrix-isolated B_2Cl_4. They also reported the ^{11}B-^{10}B isotopic shifts of the fundamental vibrations for the Raman spectrum of the liquid. The molecular symmetry of the species in the liquid phase and in the argon matrix was confidently established as D_{2d}. This conclusion was based on the infrared activity of the ^{11}B-^{10}B isotopic species in the matrix and the presence of four coincidences between the infrared and Raman bands. Durig et al. [19] reported the infrared and Raman spectra of B_2Cl_4 in the fluid and crystalline states with particular emphasis on the spectra of the solid phase. It was clearly demonstrated that the molecules convert from the 90° staggered D_{2d} form in the liquid to the planar D_{2h} form in the solid. Particularly large shifts can be noted for the BCl_2 bending modes with the change of symmetry. This example vividly demonstrates that the low frequency skeletal bending modes are usually the most sensitive to changes in molecular symmetry.

Durig et al. [11] published an overview of the symmetry of these X_2Y_4 molecules. It can be shown that the molecular symmetry is affected by bond lengths, steric effects, electronegativity, the substituent (Y), the nature of X, and so on. Trends due to these effects are discussed in more detail and are perhaps easier to understand when a larger group of molecules are considered. Due to limited space, only a few have been discussed and anyone interested in these molecules should consult the overview [11] and the references therein. In Table 2, the structural parameters and molecular symmetries of several X_2Y_4 molecules are summarized.

3. MOLECULAR SYMMETRIES OF SOME $C_2O_2X_2$ MOLECULES

The molecules with the general formula $X_2Y_2Z_2$ exhibit several possible molecular symmetries. The possible structures for $C_2O_2X_2$ molecules include trans (C_{2h}), a centrosymmetric nonplanar (C_i), cis (C_{2v}) and gauche (C_2). The representations for the normal modes for each molecular symmetry group is given in Table 3. It is important to remember that a comparison of the Raman spectra of the liquid and solid provides a convenient method for the detection of the presence of isomers. However, a study of both Raman and infrared spectra is necessary for a confident determination of molecular symmetries.

Table 2 Structural Parameters and Molecular Symmetries of Some X_2Y_4 Molecules

Molecule	Symmetry	State[a]	X-X distance (Å)	X-Y distance (Å)	X-X-Y-angle (deg)	Y-X-Y angle (deg)	Dihedral Angle	Ref.
B_2F_4	planar D_{2h}	s	1.67±0.04	1.32±0.03	-	120±2.5	0	20
	planar D_{2h}	g	1.720±0.004	1.317±0.002	-	117.3±0.2	0	21
B_2Cl_4	planar D_{2h}	s	1.75±0.05	1.73±0.02	-	120.5±1.3	0	14
	twisted D_{2d}	g	1.70±0.03	1.750±0.004	-	118.6±0.3	90	16,17
B_2Br_4	twisted D_{2d}	g,s	1.70-1.80	1.87-1.90	-	120-124	90	11
N_2H_4	gauche C_2	g,s	1.449±0.004	1.022±0.006	112.0±1.5	-	90	22-27
N_2F_4	trans C_{2h}	g	1.492±0.007	1.372±0.002	100.6±0.6	-	180	28,29
N_2F_4	gauche C_2	g	1.492±0.007	1.372±0.002	100.1±1.0	-	64.2±3.7	28,29
N_2O_4	planar D_{2h}	g	1.782±0.008	1.190±0.002	112.3±0.6	-	0	30
$N_2(CH_3)_4$	gauche C_2	g						12,13
$N_2(CF_3)_4$	gauche C_2	g	1.402±0.002	1.433±0.001	119.0±1.5	-	88±4	31,32
$N_2(SiH_3)_4$	twisted D_{2d}	g	1.457±0.016	1.731±0.004	115.5±0.7	-	82.5±0.8	33,34
P_2H_4	gauche C_2	g	2.218±0.004	1.451±0.004	95.2±0.6	91.3±1.4	81	35,36
P_2F_4	trans C_{2h}	g	2.281±0.003	1.587±0.002	95.4±0.2	99.1±0.2	180	37
P_2Cl_4	trans C_{2h}	g[b],s						38,39
P_2I_4	trans C_{2h}	s	2.21±0.6	2.475±0.028	94	102	180	40
$P_2(CH_3)_4$	trans C_{2h}	g,s	2.192±0.009	1.853±0.003	101.1±0.7	99.6±1.0	164±23	41
$P_2(CH_3)_4$	gauche C_2	g						42
$P_2(CF_3)_4$	trans C_{2h}	g,s	2.16	1.912	107	104	-	43
$P_2(CF_3)_4$	gauche C_2	g						44

[a]Abbreviations used: g, gas and s, solid.

[b]P_2Cl_4 has about 5% gauche isomer at ambient temperature.

Table 3 Representations of Normal Modes of Molecules with Molecular Formula $C_2O_2X_2$

C_{2h}	$5A_g(R,p) + 2A_u(IR) + 1B_g(R,dp) + 4B_u(IR)$
C_{2v}	$5A_1(R,p,IR) + 2A_2(R,dp) + 4B_1(R,dp,IR) + 1B_2(R,dp,IR)$
C_i	$6A_g(R,p) + 6A_u(IR)$
C_2	$6A(R,p,IR) + 6B(R,dp,IR)$

Oxalylchloride is one of the molecules having the formula $C_2O_2Cl_2$ and has been the subject of many vibrational studies. Since the frequencies for the skeletal bending modes fall in the far-infrared region of the spectrum studies in this region are more complete. An x-ray investigation [45] of crystalline oxalylchloride has shown that the molecule exists as a centrosymmetric isomer in the solid with space group D_{2h}^{15} and four molecules per primitive cell. A confident assignment for all the normal modes can be obtained from a comprehensive study of the vibrational spectrum of the solid. Durig and Hannum [46] undertook a complete investigation of the vibrational spectra of oxalylshloride in the solid, liquid, and gaseous states, emphasizing the solid state studies. For the sake of simplicity, the isomer present in the solid is called <u>trans</u>. It is not necessarily truly the planar <u>trans</u> form; in fact, it is suggested [45] that the molecule is not strictly planar. The centrosymmetric nonplanar C_i symmetry differs from C_{2h} symmetry in that the B_g mode under C_{2h} goes to A_g under C_i. Since gerade modes remain gerade and ungerade remain ungerade, and since it is not possible to determine depolarization ratios for the bands in the spectrum of a polycrystalline solid, no distinction can be made between a B_g and A_g mode in the solid. Mutual exclusion is present in the spectra of the solid phase [46]. However, there are more bands present in the infrared and Raman spectra of liquid oxalylchloride than in the solid. Many of these "extra" Raman lines were shown to be highly temperature sensitive [46,47]. Also, Kagarise [48] showed how the infrared spectrum changed drastically upon solidification. It is not possible to assign these "extra" bands as difference tones [46]. Therefore, a second isomer is present and since the <u>cis</u> is favored on theoretical grounds over the <u>gauche</u>, the bands were assigned as the normal modes for the <u>cis</u> isomer. Assuming that the C=O stretching modes have approximately the same frequency in both isomers, the majority of the remaining modes have been assigned [46]. Even though the assignment given by Durig and Hannum [46] may be considered tentative, there is no doubt that a second isomer is present in the liquid.

Oxalylbromide, $C_2O_2Br_2$ is another molecule that has been the subject of many studies. Early studies incorrectly concluded that the molecule exists as only one conformer in all phases. An x-ray investigation [45] of crystalline oxalylbromide has shown that the molecule exists as the <u>trans</u> isomer in the solid with space group

C_{2h}^5 and two molecules per unit cell. A confident assignment for all the normal modes of the trans isomer results from a comprehensive study of the vibrational spectrum by Durig et al. [49]. Variable temperature studies of the Raman spectrum of oxalylbromide by Durig et al. [49] in the 300-450 cm^{-1} region showed significant changes in the intensities of several bands in this region. Since these bands could not be assigned to difference bands of the trans fundamentals and since they also varied with temperature, the fluid states of oxalylbromide are characterized by the presence of a second isomer. The remaining bands were tentatively assigned to the cis isomer (C_{2v}) although the second isomer may be the gauche isomer (C_2). As in oxalylchloride, the cis is chosen as the second isomer on theoretical grounds; however, the presence of a second isomer is not in doubt.

Oxalylfluoride, $C_2O_2F_2$ is the last in this series of halogenated oxalic acids to be considered. A comparison of the infrared spectra of oxalylfluoride vapor and solid [50] shows the disappearance of three major bands upon solidification. Therefore, it was concluded that there must be two isomeric forms in the vapor. The frequencies in the solid phase were assigned to the trans isomer and bands unique to the fluid phases were assigned to the cis isomer [50]. A comparison of the infrared and Raman bands in the crystal show that the mutual exclusion principle is operative. From the alternate forbiddance, Durig et al. concluded that the molecule has a centrosymmetric structure in the crystal and that each molecule occupies a C_i site.

Glyoxal, $C_2O_2H_2$, is an interesting molecule to conclude this section on $X_2Y_2Z_2$ molecules. It has been clearly shown that the molecule exists in the trans form in the solid [51,52] since the principle of mutual exclusion is operative. The molecule has a centrosymmetric structure in the crystal and each molecule occupies a C_i site. The factor group is believed to be either $C_{2h}(2/m)$ or D_{2h} (mmm) with two or four molecules per unit cell [51,52]. Cole and Durig recorded the spectra of glyoxal-d_1 and d_2. The loss of the symmetry center in the d_1 compound (C_s) means that coincidences will occur in the Raman and infrared active modes, while the dideutero compound (C_{2h}) should exhibit mutual exclusion. Eight clear coincidences were observed in the spectra of the solid monodeutero compound [52]. The Raman and infrared spectra of solid di-deuteroglyoxal clearly demonstrate the C_{2h} molecular symmetry of the trans isomer [52]. Unfortunately, it was not possible to record spectra of the liquid phase to determine if a second conformer existed [51]. Currie and Ramsay [53] reported an electronic band which was attributed to the cis isomer. Durig et al. [54] recorded the microwave spectrum of glyoxal. They [54] determined that in addition to the trans isomer there was a second isomer present at ambient temperature. This isomer was determined to be the cis isomer by inertial defect calculations confirming a planar structure. Further studies by Cole *et al.* [55] on the mono-deutero compounds confirmed the presence of the *cis* isomer. The torsional potential function of glyoxal was calculated by Durig *et al.* [56] using the infrared data of the *trans* conformer and the microwave intensity data of the *cis* isomer.

4. SUMMARY AND CONCLUSIONS

The limited examples given above should demonstrate to the reader that a combination of infrared and Raman data can provide a powerful method for the determination of the molecular symmetries of a large variety of molecules. For the X_2Y_4 molecules it has been demonstrated that the conformer present in the fluid phases may not be the one present in the solid state, i.e., B_2Cl_4. Also it should be noted that, for the N_2Y_4 molecules, four different symmetries have been found. For the most part, the X-X stretch or the NY_2 bending motions are the most sensitive to conformational changes. Thus, it is very important that the vibrational spectrum be recorded over the entire spectral range expected for the fundamentals if symmetry determinations are going to be made. Additionally, it is very important to compare infrared and Raman data which are from the same physical state, i.e., Raman data from the liquid state should not be compared to infrared data from the vapor state.

If one expects that conformers are present in the fluid states, then it is necessary to record the vibrational spectrum of the solid. Usually only one conformer is present in the solid state and the disappearance of bands initially observed in the fluid states is a strong indication of the presence of conformers or a symmetry change with crystallization. However, it should be noted that if the conformers have about the same energy it is possible that they may persist in the solid state, i.e., N_2F_4.

The examples of the $C_2O_2X_2$ molecules demonstrate clearly that there are two conformers present at ambient temperature and that the most stable one has C_{2h} symmetry. However the symmetry of the second conformer has only confidently been determined for the glyoxal molecule (C_{2v}). For the halogen containing molecules the symmetry may either be C_2 (*gauche*) or C_{2v} (*cis*). Usually it is quite difficult to distinguish between two low symmetries such as C_2 or C_{2v} by vibrational techniques.

Although only two types of examples have been given as illustrations, there are of course many others which could have been used. For example, the XY_3 molecules may have D_{3h}, C_{3v} or C_{2v} symmetry and vibrational spectroscopy is a very convenient technique to distinguish among these symmetries. However, the authors' research interests more or less dictated the examples which would be utilized.

Finally, the advantages of the vibrational techniques for symmetry determinations should be mentioned. First, these techniques are applicable to all three physical states. Thus, it is possible in most cases to determine the symmetry in the solid, liquid, and gaseous states and to follow a change in symmetry, if it should occur with a change of physical state. Second, it is possible to obtain the relative amounts of two different isomers in the fluid states and by temperature variations to determine the relative differences in energy between the isomers. Third, by using the frequencies for the internal torsional modes for the two isomers along with the enthalpy difference, it is possible, in principle, to determine the complete potential function for the interconversion of the two isomers. Fourth, the vibrational data are usually easily obtainable, and the interpretation is straightforward; the method is applicable to a large variety of molecules, and only a short period of time is needed to collect the data. Thus, a larger variety of molecules can be studied in more physical states by vibrational spectroscopy than by any other physical technique.

ACKNOWLEDGMENT

The authors gratefully acknowledge the financial support of our research studies by the National Science Foundation, Grant CHE-79-20763.

REFERENCES

1. D. R. Lide, Jr. and D. E. Mann, J. Chem. Phys., 31, 1129 (1959).

2. Y. I. Koton and V. M. Tatenskii, Opt. Spektrosk., 14, 443 (1963).

3. C. B. Colburn and A. Kennedy, J. Am. Chem. Soc., 80, 5004 (1958).

4. J. R. Durig and R. C. Lord, Spectrochim. Acta, 16, 1471 (1960).

5. E. N. Moskvitina, Y. Y. Kuzyakov, N. A. Knyazeva, and V. M. Tatevaskii, Opt. Spektrosk., 16, 768 (1964).

6. C. B. Colburn, F. A. Johnson, and C. Haney, J. Chem. Phys., 43, 4526 (1965).

7. R. K. Bohn and S. H. Bauer, Inorg. Chem., 6, 304 (1967).

8. J. R. Durig and J. W. Clark, J. Chem. Phys., 48, 3216 (1968).

9. A. Oskam, R. Elst, and J. C. Dunker, Spectrochim Acta, 26A, 2021 (1970).

10. D. F. Kostor and F. A. Miller, Spectrochim Acta, 24A, 1487 (1968).

11. J. R. Durig, B. M. Gimarc, and J. D. Odom in Vibrational Spectra and Structure, Vol. II, (J. R. Durig, ed.), Marcel Dekker, New York, 1975, Chapter 1.

12. D. C. McKean, Spectrochim. Acta, 10, 161 (1957).

13. J. R. Durig, R. W. MacNamee, L. B. Knight, and W. C. Harris, Inorg. Chem., 12, 804 (1973).

14. M. Atoji, P. J. Wheatley, and W. N. Lipscomb, J. Chem. Phys., 27, 196 (1957).

15. D. E. Mann and L. Fano, J. Chem. Phys., 26, 1665 (1957).

16. K. Hedberg and R. Ryan, J. Chem. Phys., 41, 2214 (1964).

17. R. R. Ryan and K. Hedberg, J. Chem. Phys., 50, 4986 (1969).

18. L. A. Nimon, K. S. Seshardi, R. C. Taylor, and D. White, J. Chem. Phys., 53, 2416 (1970).

19. J. R. Durig, J. E. Saunders, and J. D. Odom, J. Chem. Phys., 54, 5285 (1971).

20. L. Trefonas and W. N. Lipscomb, J. Chem. Phys., 28, 54 (1958).

21. D. D. Danielson, J. V. Patton and K. Hedberg, J. Am. Chem. Soc., 99, 6484 (1977).

22. Y. Morino, T. Iijima, and Y. Murata, Bull. Chem. Soc. Japan, 33, 46 (1960).

23. A. Yamaguchi, I. Ichishima, T. Shimanouchi, and S. Mizushima, J. Chem. Phys., 31, 843 (1959).

24. T. Kasuya and T. Kojima, J. Phys. Soc. Japan, 18, 364 (1963).

25. J. R. Durig, M. G. Griffin and R. W. MacNamee, J. Raman Spectrosc., 3, 133 (1975).

26. R. L. Collin and W. N. Lipscomb, J. Chem. Phys., 18, 566 (1950).

27. M. Guay and R. Savoie, Can. J. Chem., 47, 201 (1969).

28. M. Cardillo and S. H. Bauer, Inorg. Chem., 8, 2086 (1969).

29. M. M. Gilbert, G. Gundersen, and K. Hedberg, J. Chem. Phys., 56, 1691 (1972).

30. B. W. McClelland, G. Gunderson and K. Hedberg, J. Chem. Phys., 56, 4541 (1972).

31. L. S. Bartell and H. K. Higgenbotham, Inorg. Chem., 4, 1346 (1965).

32. J. R. Durig, R. W. MacNamee, L. B. Knight, and W. C. Harris, Inorg, Chem., 12, 804 (1973).

33. C. Glidewell, D. W. Rankin, A. G. Robiette, and G. M. Sheldrick, J. Chem. Soc. A, 318 (1970).

34. J. R. Durig, K. S. Kalasinsky, and V. F. Kalasinsky, J. Mol. Struct., 35, 201 (1976).

35. B. Beagley, A. R. Conrad, J. M. Freeman, J. J. Monaghan, B. G. Norton, and G. C. Holywell, J. Mol. Struct., 11, 371 (1972).

36. J. R. Durig, L. A. Carreira, and J. D. Odom, J. Am. Chem. Soc., 96, 2688 (1974).

37. L. Hodges and L. S. Bartell, Fourth Austin Meeting on Molecular Structure, Austin, Texas, 1972.

38. J. D. Odom, J. E. Saunders and J. R. Durig, J. Cryst. Mol. Struct., 2, 169 (1972).

39. S. G. Frankiss and F. A. Miller, Spectrochim. Acta, 21, 1235 (1965).

40. Y. C. Leung and J. Van Waser, J. Phys. Chem., 60, 539 (1956).

41. A. McAdam, B. Beagley, and T. G. Hewitt, Trans. Faraday Soc., 66, 2732 (1970).

42. J. R. Durig and R. W. MacNamee, J. Mol. Struct., 17, 426 (1973).

43. L. S. Su and L. S. Bartell, private communication.

44. J. D. Witt, J. W. Thompson and J. R. Durig, Inorg. Chem., 12, 811 (1973).

45. P. Groth and O. Hassel, Acta Chem. Scand., 16, 2311 (1962).

46. J. R. Durig and S. E. Hannum, J. Chem. Phys., 52, 6089 (1970).

47. B. D. Saksena and R. E. Kagarise, J. Chem. Phys., 19, 987 (1951).

48. R. E. Kagarise, J. Chem. Phys., 21, 1615 (1953).

49. J. R. Durig, S. E. Hannum, and F. G. Baglin, J. Chem. Phys., 54, 2367 (1971).

50. J. R. Durig, S. C. Brown, and S. E. Hannum, J. Chem. Phys., 54, 4428 (1971).

51. J. R. Durig and S. E. Hannum, J. Cryst. Mol. Struct., 1, 131 (1971).

52. A. R. H. Cole and J. R. Durig, J. Raman Spectrosc., 4, 31 (1975).

53. G. N. Currie and D. A. Ramsay, Can. J. Phys., 49, 317 (1971).

54. J. R. Durig, C. C. Tong, and Y. S. Li, J. Chem. Phys., 57, 4425 (1972).

55. A. R. H. Cole, Y. S. Li, and J. R. Durig, J. Mol. Spectrosc., 61, 346 (1976).

56. J. R. Durig, W. E. Bucy, and A. R. H. Cole, Can. J. Phys., 53, 1832 (1975).

COMPARISON OF FT-IR AND RAMAN SPECTROSCOPY

W. Kiefer and J. Laane*

Physikalisches Institut
Universität der Bayreuth
Bayreuth, West Germany

1. INTRODUCTION

Paralleling the evolution of Fourier Transform Infrared Spectroscopy during the 1960's and 1970's there have been rapid developments in the area of Raman spectroscopy. The laser serves only in a secondary role for FT-IR spectroscopy, but as a source for excitation it has greatly revolutionized conventional Raman spectroscopy and led to the advent of new techniques including coherent anti-Stokes Raman scattering (CARS), Raman gain, inverse or hyper-Raman spectroscopy. Both FT-IR and Raman techniques have also benefited from the improvements in opto-electronic and data-processing systems. While a moving mirror with high mechanical precision and a high quality beam splitter are basic to an FT-IR instrument, a double monochromator with precision gratings, which may be holographic, and a laser as light source are intrinsic to a Raman spectrometer. Improved detector systems (pyroelectric bolometers, such as triglycine sulfate, TGS, for FT-IR or photon counting for Raman), not available a decade ago now serve vital functions. Utilization of an on-line mini-computer is all but essential for practical FT-IR spectroscopy in order to facilitate the data handling and computation; it can also serve as a valuable accessory for the signal averaging and manipulation of Raman data.

In the following sections we will compare basic concepts, instrumentation, and applications of Raman spectroscopy to FT-IR spectroscopy. Raman techniques will be emphasized since FT-IR

*Humboldt U. S. Senior Scientist Awardee on leave from Texas A & M University, College Station, Texas.

J. R. Durig (ed.), Analytical Applications of FT-IR to Molecular and Biological Systems, 537–577.

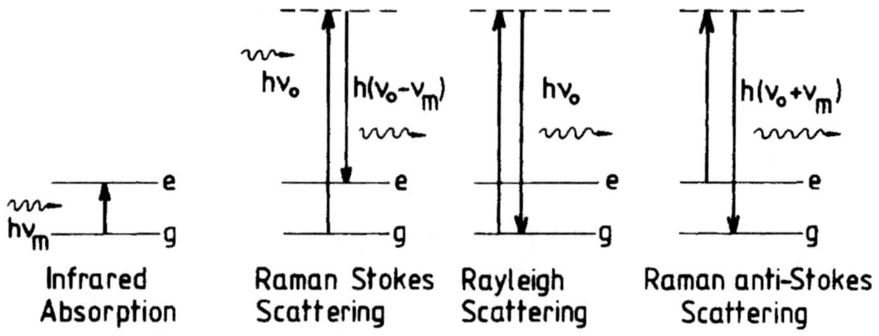

Fig. 1 Comparison of energy transitions between two molecular states (e: excited state; g: ground state); ν_m: molecular frequency (vibrational and/or rotational); ν_o: incident light frequency.

methods are well covered in the other articles. It may be advantageous to first compare generally infrared and Raman spectroscopy.

2. BASIC CONCEPTS IN INFRARED AND RAMAN SPECTROSCOPY

Raman and infrared spectroscopy are complementary techniques for studying the vibrational and rotational energy spacings in molecules. Since the two methods depend on different quantum mechanical processes and very often provide different information, both techniques should ideally be used to obtain the maximum information about molecular vibrations and/or rotations. Although the infrared spectrum alone may suffice to solve a particular problem, for a structural determination it is necessary for the chemist to measure both kinds of spectra. For complete knowledge of the dynamics of the molecule under investigations and hence of its structure, additional and complementary information through Raman spectroscopy is needed.

Let us first point out the basic concepts for these two molecular spectroscopic methods. Figure 1 compares the energy transitions associated with infrared absorption (one-photon process) and Raman scattering (two-photon process). For the former the molecular frequencies ν_m^i at which light absorption occurs in the infrared correspond to the energy spacings ΔE^i according to $\nu_m^i = \Delta E^i/h$. In a Raman experiment, monochromatic light (in most cases from a laser) is used as the exciting source, and the frequencies and intensities of the scattered light are then measured. Generally more than 99.99% of the scattered light is in the form of Rayleigh radiation which has the same frequency as that of the laser (ν_o). However, a small fraction of the light is scattered inelastically with shifted frequencies $\nu_S = \nu_o - \nu_m^i$ (Stokes lines) or $\nu_{AS} = \nu_o + \nu_m^i$ (anti-Stokes lines). Because Stokes lines originate from the generally more populated ground state, they have greater intensity than the anti-Stokes lines originating from molecular excited states at higher energies. In each experiment ν_o is well known and the Raman frequency $\nu_R = \nu_o - \nu_S = \nu_m$, which can be interpreted in the same way as an infrared absorption frequency, is readily determined from the scattered Stokes frequency ν_S. The quantum mechanical selection rules for infrared and Raman processes are different; consequently, a Raman spectrum for a particular molecular sample will be different from its infrared spectrum, even though there will likely be frequency coincidences between the two. While infrared absorption (or emission) requires a dipole moment change during the course of the energy transition, Raman scattering occurs when a polarizability change (a distortion of the molecular electron distribution) is involved. By use of character tables the infrared or Raman activity of a particular transition is readily determined [1-3]. Infrared transitions have the same symmetry species as molecular translations (T_x, T_y, T_z) while the Raman transitions have symmetry species which result from the direct product of these (α_{xy}, α_{yz}, α_{zz}). In general, less symmetric vibrations produce greater intensity in infrared spectra but low or negligible intensity in Raman spectra. On the other hand, the more symmetric motions result in the strongest Raman and the weakest infrared intensities. Moreover, "totally symmetric vibrations", which are symmetric with respect to all the symmetry operations of a molecule, can be identified not only by their strong Raman intensities but also by the fact that their Raman lines are polarized. It is an especially useful feature of Raman scattering that bands can be determined to be "polarized" or "depolarized". In most Raman experiments a rectangular scattering geometry is used (see Fig. 2), where a polarized laser beam (e.g. x-polarized) is incident on the sample (y-direction). Light which is scattered at 90° (z-direction) is then analyzed by means of a polarization analyzer. This allows the measurement of the "parallel" component ($I_{\parallel}$, polarized in the x-direction), which is composed of isotropically and anisotropically scattered Raman light, as well as the "perpendicular" component ($I_{\perp}$, y-direction polarized), which contains only anisotropically scattered Raman light. The ratio of these measured intensities is defined as the depolarization ratio $\rho = I_{\perp}/I_{\parallel}$. Simple theory shows that ρ is equal to 3/4 for Raman lines which correspond to non-symmetric or degenerate vibrational modes ("depolarized" Raman lines) whereas for symmetric vibrations $\rho < 3/4$ ("polarized" Raman lines; often $\rho < 0.2$). A detailed discussion of the polarization properties of Raman scattering may be found elsewhere [4].

As an example of how the infrared and Raman spectra differ for a typical organic molecule, Fig. 3 shows the infrared and Raman spectra [5] of liquid 3-cyclopenten-1-one ($O{=}\overline{C}CH_2CH{=}CH\overline{C}H_2$),

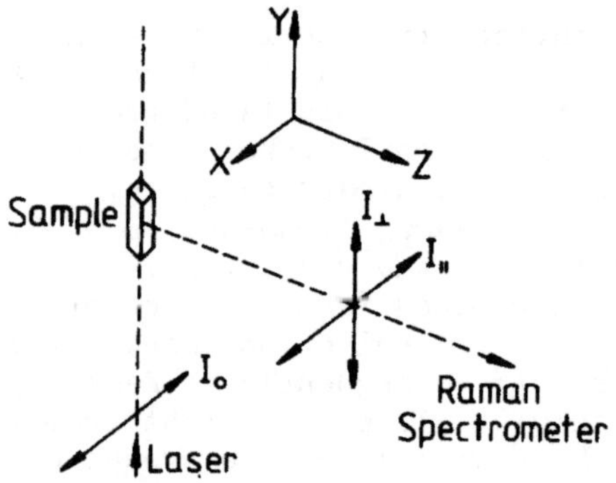

Fig. 2 90° Raman scattering arrangement and the definition of $I_\parallel$ and $I_\perp$.

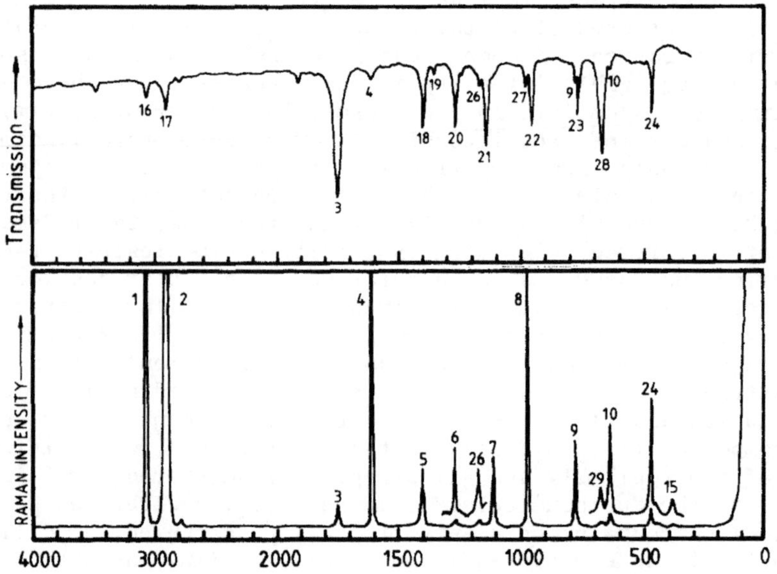

Fig. 3 Infrared (upper field) and Raman (lower field) spectra of liquid 3-cyclopenten-1-one (adapted from Ref. [5]).

which is planar and has C_{2v} symmetry. This molecule has thirty vibrations with symmetry species $10A_1 + 6A_2 + 9B_1 + 5B_2$. In principle, all vibrations are Raman active and the the ten A_1 motions are polarized in the Raman spectrum; all but the six A_2 vibrations are infrared active. Vibrations numbered 1 to 10 are the A_1 modes, 11 to 16 are A_2, 17 to 25 are B_1, and 26 to 30 are B_2; the most intense spectral bands in Fig. 3 have been labeled with their corresponding vibration number. It can be seen that the eight most intense Raman lines arise from the totally symmetric A_1 modes. The other two A_1 vibrations also result in appreciable Raman intensity while only four of the other twenty A_2, B_1, or B_2 motions are readily seen in the Raman spectrum. However, the only vibration of A_1 symmetry to have substantial intensity in the infrared is the stretching of the polar C=O bond. The other strong infrared bands all arise from the less symmetric B_1 or B_2 motions.

Figure 3 also demonstrates another important feature of infrared and Raman spectra: certain functional groups give rise to strong Raman lines while others produce strong infrared bands. The C=O stretch has strong infrared but weak Raman intensity whereas the C=C stretch at 1614 cm^{-1} has strong Raman but weak infrared intensity. The strongest infrared bands, in general, occur for stretching of groups with strong dipolar character: C=O, O-H, Si-H, Si-O, Si-Cl, N-O, C-O, C-H, S-H, S-O, and P-H. The largest Raman intensities are found for functional groups such as C≡C, C=C, C-C, C=S, C≡N, C=N, S-S, C-Cl, C-Br, C-H, and N-H. Listings of group frequencies and intensities are available from several sources [6,7].

Quantitative intensity measurements by Raman spectroscopy are based on different principles from those for infrared spectroscopy. For IR the absorbance A, which is a logarithmic function of per cent transmittance, is linearly related to sample concentration c, path length b, and the absorptivity a:

$$A = -\log_{10}(T) = abc \quad (IR). \tag{1}$$

This can be quite inconvenient since strong bands (with large a) may often have little transmittance and can go "off scale".

For Raman spectroscopy, a linear relationship exists between Raman intensity I_R and concentration. The signal is also proportional to laser power P, scattering frequency to the fourth power ν_S^4, and the differential scattering cross section $(d\sigma/d\Omega)$. Unlike with double beam infrared techniques which cancel instrumental parameters, the intensity measured with a single channel Raman spectrometer depends also on the signal throughput efficiency $T(\nu)$, and the focusing and collection optics as well as the monochromator slit width used. The dependence on these instrumental parameters we simply express by the function F. We then have

$$I_R = (d\sigma/d\Omega)cPT(\nu)F\nu_S^4 \quad (Raman). \tag{2}$$

In the rare case when the Raman signal becomes too great it

may easily be reduced by lowering the laser power or by narrowing the slit width. For quantitative Raman investigations utilizing the intensity of one band for measuring concentration, the instrumental parameters P, F, and T are fixed as is the scattering frequency and the differential scattering cross section. The Raman intensity and concentration (number of scattering molecules in the region of the laser focus) then are related by a simple proportionality constant $K = (d\sigma/d\Omega)PFT\nu_S^4$ according to

$$I_R = Kc. \tag{3}$$

This simple relationship of Raman intensity and concentration is valid only for the linear Raman effect. For nonlinear coherent Raman scattering spectroscopy (e.g. CARS) the relationship to concentration is more complicated.

Until now, we have discussed only a few fundamental aspects while comparing infrared with Raman spectroscopy. Since FT-IR represents an elegant but only a special technical way of performing absorption spectroscopy in the infrared spectral region, everything said above for infrared spectroscopy holds clearly also for FT-IR spectroscopy.

In the following we will direct our discussion to more or less technical problems. We will especially discuss how the spectroscopic data are collected with both methods and how the two techniques can be applied to various spectroscopic problems.

3. DATA COLLECTION AND INSTRUMENTATION

A. FT-IR Spectroscopy

The data collected from a Michelson interferometer for FT-IR spectroscopy represents spectral intensity I vs. path difference δ between the fixed and moving mirrors. Figure 4 shows the type of an interferogram typically obtained by a Fourier transform spectrophotometer. Under optimal conditions or after computational correction, the interferogram is symmetric about zero-path difference $\delta = 0$. Therefore, it is generally sufficient to utilize only half of the data to transform the interferogram into a frequency spectrum. The advantages of interferometry over conventional grating infrared spectroscopy are well known [8]:

(1) All of the frequencies of the spectrum are gathered simultaneously resulting in the multiplex (Fellgett's) advantage.

(2) The optical throughput is considerably greater for an interferometer than for a grating monochromator (Jacquinot's advantage).

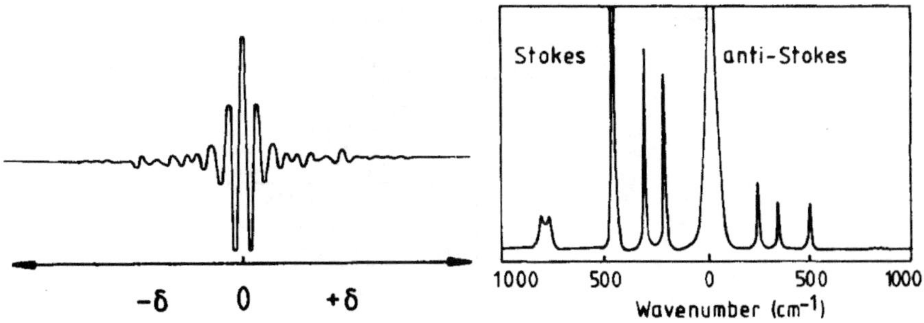

Fig. 4 Examples of data collected for FT-IR (left) and Raman (right) spectroscopy.

(3) The laser interferometer used to reference the position of the moving mirror in the main interferometer results in very high frequency accuracy in the calculated spectrum (Connes' advantage).

(4) There is no equivalent of stray light for rapid scanning interferometers.

(5) The frequency data obtained from the FT calculation are in digital form and allow ready manipulation by the mini-computer.

B. Raman Spectroscopy

1. General. Like conventional (pre-FT-IR) infrared spectra, Raman spectra are obtained using a dispersive system which yields intensity vs. frequency data. Figure 4 shows a typical Raman spectrum (of CCl_4) which is surprisingly similar to an infrared interferogram in appearance. The large spectral intensity in the center at frequency ν_o corresponds to the Rayleigh radiation, and the Stokes and anti-Stokes lines are symmetrically distributed about this frequency. Because the Stokes lines are more intense than the anti-Stokes lines

$$I_{AS}/I_S = (\nu_{AS}/\nu_S)^4 \exp(-h\nu_m/kT), \tag{4}$$

only the former are generally scanned to obtain the Raman spectrum. The Raman frequencies, which may be compared directly to infrared values, are determined from the scattered Stokes frequencies ν_S from $\nu_m = \nu_o - \nu_S$.

Conventional Raman spectroscopy utilizing a grating monochromator does not enjoy the advantages discussed for FT-IR techniques. However, use of photographic detection or of an optical multi-

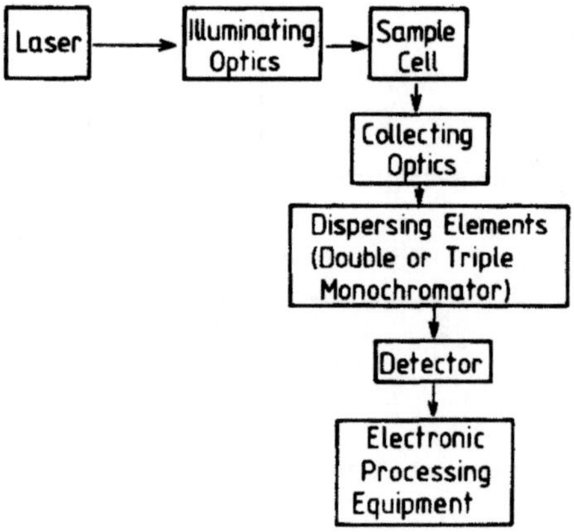

Fig. 5 Schematic diagram of a laser Raman spectrometer.

channel analyzer (OMA) permits the simultaneous detection of a wide range of frequencies. Triple monochromators can be used to minimize stray light, and on-line mini-computers may be used to signal average and manipulate Raman spectra. Various instrumental techniques used for Raman spectroscopy will now be discussed.

2. Conventional Raman Spectroscopy. The essential parts of any Raman apparatus are: a monochromatic light source, a sampling area, a monochromator (double or triple) to separate the scattered light according to frequency, and a radiation detection system. In Fig. 5 the main features of a laser Raman spectrometer are displayed.

Most commonly used lasers for Raman spectroscopy are the continuous wave (cw) argon ion and krypton ion lasers with emission lines in the blue to red spectral region (the strongest laser lines are at 488 and 514.5 nm for the argon and at 647 nm for the krypton ion laser). Lasers provide not only a wide variety of useful excitation lines, but the additional properties of a high degree of polarization, narrow line width, and high flux density.

In the simplest arrangement the laser beam is focused with a lens with focal length of the order of 3-10 cm into the sample. The scattered (Raman + Rayleigh) light is collected with a high speed objective (preferably f/1 or even better) and focused onto the slit of a double or triple monochromator. It is important that the dispersing element has high stray light rejection. With triple monochromators equipped with holographic gratings, stray light rejections of the order of 10^{-15} or better can be achieved nowadays.

The sensitivity for the detection of weak light levels, as is the case for many Raman signals, could be improved by the development of good detectors with high quantum efficiency (up to 50%) and low dark current (typically less than 10 c.p.s.) and by the straightforward potentiality of the photon counting technique and digital data processing equipment. Many articles describing these techniques have appeared in the past and demonstrate the usefulness of their application in laser Raman spectroscopy [9].

Typical commercial monochromators used in Raman spectroscopy are one-meter grating instruments with maximum resolution of typically 0.2 cm^{-1}. Higher resolution (~0.05 cm^{-1}) can be achieved with homemade instrumentation [10]. However, routine scans in Raman spectroscopy are taken typically at 1-5 cm^{-1} resolution. Although the Raman process is extremely inefficient (a good Raman scatterer converts only one photon out of about 10^7-10^8 incident photons into a frequency shifted Raman photon) state-of-the-art Raman instruments give signals of up to 10^6 counts per second for medium to strong Raman lines at a resolution of approximately 3 cm^{-1}. The dynamic range, however, is wide enough to detect one single Raman photon. The detectable Raman signal may be increased in some cases by multiple passing of the laser beam through the sample, either in or out of the laser cavity, or by increasing the laser power up to the point of sample decomposition. The latter can be held fairly high if special sample techniques [11] are used which avoid thermal decompositions.

3. Computer Interfaced Systems. With the availability and relatively low cost of mini- and micro-computers, more and more of these systems have been interfaced to Raman spectrometers. The signal from a phototube can be fed directly to the computer where it is stored in digital form as a function of frequency, which is determined by a stepping motor that governs the grating drive. Such systems have been described in the literature [12].

A computerized Raman spectrometer is capable of carrying out various tasks otherwise not possible. When several spectra of the same sample are recorded and stored in the computer, signal averaging can be performed to improve the signal-to-noise ratio. Figure 6 demonstrates the improvement that is possible with signal averaging [13]. An on-line computer system also facilitates the computation of Raman difference spectra (see Sec. 4.F.) which are useful for analyzing spectra without solvent interference and also for the determination of small solvent or isotopic shifts. Difference spectra combined with polarization techniques permit the direct presentation of the isotropic and anisotropic components of the Raman signal [11].

Digital techniques also permit instrumental corrections (for example, for changes in phototube sensitivity and grating reflec-

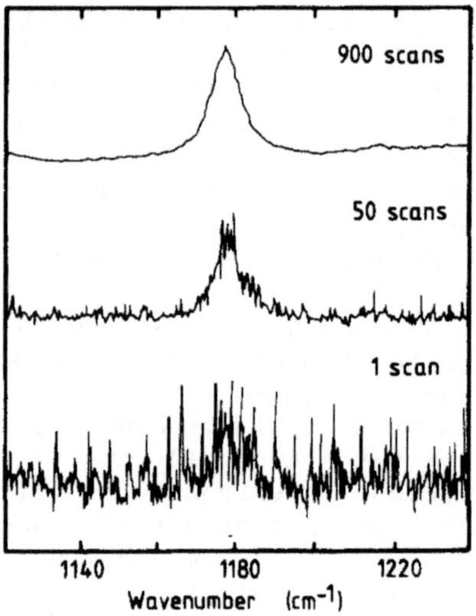

Fig. 6 Increase in S/N via computer averaging demonstrated for the 1178 cm^{-1} line of C_6H_6 in CCl_4 ($3.5 \cdot 10^{-3}$M) (adapted from Ref. [13]).

tivity as a function of frequency), peak area determinations, spectral smoothing, curve resolution, calculation of derivative spectra, and also experimental control [13].

4. Multichannel and Time-Resolved Raman Spectroscopy. In conventional Raman spectroscopy a single radiation detector (photomultiplier) is used to analyze the spectrum by means of a scanning monochromator. Depending on resolution, only an extremely small amount of the total Raman radiation reaches the detector through the exit slit of the dispersing element. The main part of Raman scattered light is wasted. Similar to the improvement in infrared spectroscopy through FT-IR, advances have been made in Raman spectroscopy by the multichannel technique, where the whole Raman spectrum is collected simultaneously by means of a photoelectric detector having a large number of channels. These gather independently the information from small spectral regions. Multichannel Raman spectroscopy was pioneered by Delhaye [14] and his group in France already a long time ago and has now reached a fairly high standard [15]. In principle, the photographic emulsion could serve as a multichannel detector. However, due to the very low quantum efficiency of the photographic material it was replaced by highly sensitive photoelectric image devices such as image intensifiers, microchannel plates, or semiconductor diode-arrays. The sensitivity has now been brought up to be comparable to that of single-channel Raman spectroscopy using photomultipliers,

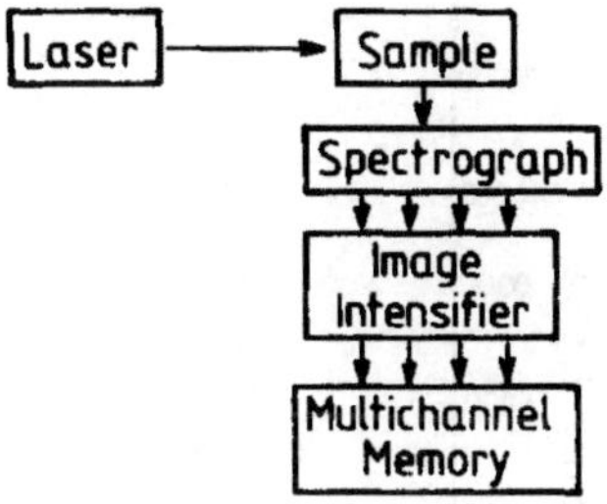

Fig. 7 Schematic diagram of a multichannel Raman spectrometer.

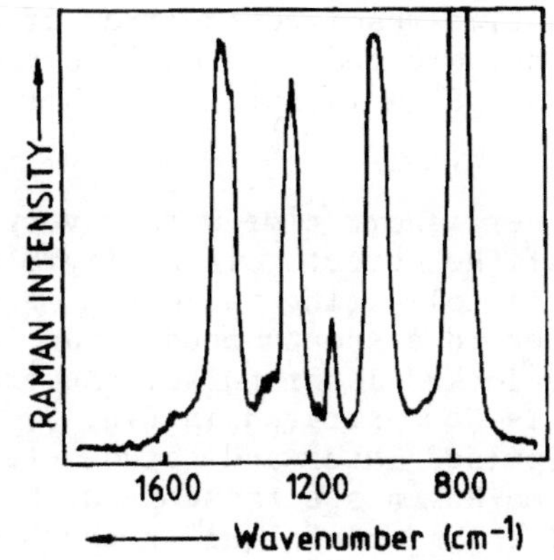

Fig. 8 Part of the Raman spectrum of cyclohexane excited by a single laser pulse of 25 ps duration (adapted from Ref. [16]).

so it is possible today to detect single Raman photons simultaneously emitted at various frequencies.

A schematic diagram of a multichannel Raman spectrometer is shown in Fig. 7. The total Raman spectrum intensified by up to 10^6 is transferred on to a secondary detector where it can be stored. This is accomplished by a low light-level television camera tube (SEC or SIT tube) which has an electrical storage target with integration and storage capability. With this it is possible to directly read the spectrum onto a magnetic tape or into the memory of a computer.

When a pulsed laser is used, it becomes possible to record

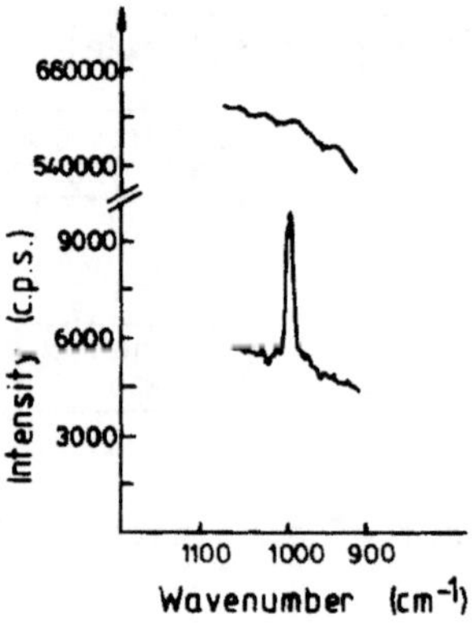

Fig. 9 Rejection of fluorescence from a Raman spectrum by sub-nano-second time-resolved spectroscopy; upper field: cw experiment; lower field: time-resolved spectrum of a 10^{-4}M solution of rubrene in benzene, excited by a mode-locked cavity-dumped argon ion laser at 514.5 nm (adapted from Ref. [18]).

Raman spectra in a very short time with a very good S/N ratio, because the noise of the detector is negligible for the short period of time one is collecting the data. Bridoux et al. [16] have been able to record a spontaneous Raman spectrum excited by a single 25 ps mode-locked laser pulse. An example of this most advanced technique is demonstrated in Fig. 8. With a similar experimental set-up (pulsed laser and gated detector), one can also perform time-resolved Raman spectroscopy with subnanosecond time-resolution [17]. The same technique is capable of suppressing disturbing fluorescence which, in many cases, is emitted slightly delayed in time from the sample [18]. The dramatic gain in S/N, when the technique of the time-resolved rejection of fluorescence from Raman spectra is applied, is demonstrated in Fig. 9, where the upper spectrum (cw-experiment) only shows a high fluorescence background with nearly no detectable Raman signal, whereas the lower (time-resolved) spectrum clearly indicates a Raman band at 992 cm^{-1}.

In Table 1 we compare the two techniques assuming state-of-the-art designs.

4. APPLICATIONS

A. General

Because infrared and Raman spectroscopy are essentially based on different phenomena, one or the other is often better

Table 1 Comparison of Raman and FT-IR Instrumentation

	Raman	FT-IR
System Complexity	moderate	higher
Efficiency	low	high
Reliability	good	good
Expenses	appr. \$60,000	appr. \$100,000
Source	laser	blackbody
Selective Element	double or triple monochromator	Michelson interferometer
Detector	PM tube	Golay, TGS
Electronics	photon counting	computer
Resolution:		
commercial instruments	$0.15\ cm^{-1}$*	$0.06\ cm^{-1}$
special home-made instruments	$0.05\ cm^{-1}$	$0.005\ cm^{-1}$
Principal Limitation of S/N	sample decomposition	energy (emission of blackbody; NEP of detector)
Spectral Range	$0 - \pm 10{,}000\ cm^{-1}$	$6 - 20{,}000\ cm^{-1}$
Time to Obtain Single Spectrum (Routine)	appr. 10 min.	appr. 0.5 s (excluding computation; appr. 10 s with comp.)
Time Resolution	10^{-12} s	10^{-4} s
Special Requirements	laser safety	purge or evacuate to overcome water vapor absorption
Cell Materials	glass; quartz	salt plates
Normal Sampling Volume	appr. 1 mm^3	0.1 to $10^5\ cm^3$

* For linear Raman spectroscopy; for "active coherent Raman spectroscopy" resolution of 10^{-4} to $10^{-5}\ cm^{-1}$ possible.

suited for a particular application. For example, aqueous solutions are more readily investigated using Raman spectroscopy whereas matrix-isolated samples are less difficult to examine with infrared instrumentation. However, as Fourier methods have extended the sensitivity of infrared measurements and as new developments have advanced Raman spectroscopy, both techniques can now be used for numerous applications which heretofore were impossible for one or both. Thus, aqueous studies with FT-IR and matrix work with Raman spectroscopy are today relatively commonplace.

Low frequency (0-400 cm^{-1}) vibrational spectroscopy has, in particular, benefited from the developments in FT-IR and laser Raman spectroscopy. Before FT-IR only specially constructed (and very expensive) far infrared spectrometers were capable of examining absorption spectra in this region due to severe problems with inadequate source intensity and detector capability resulting in weak signals. The substantial improvements in sensitivity available with FT-IR, however, have made studies in this region considerably more accessible. Even pre-laser Raman spectroscopy was capable of examining spectra to below 100 cm^{-1}. However, gases and crystals, whose spectra are of special interest in this region, were exceedingly difficult to examine with the earlier excitation sources. Modern Raman spectrometers with laser sources readily permit the examination of crystal lattice modes and pure rotational lines of gases, even at values below 1 cm^{-1}.

In the following sections we will briefly examine the applications of vibrational techniques, with particular emphasis on Raman methods.

B. Gas-Phase Molecules

1. Principles. Both the infrared and Raman spectra of gases and vapors are about two to three orders of magnitude weaker than the corresponding spectra of condensed phases, due to the lower molecular density in gases. Nonetheless, many samples are preferentially examined as vapors so as to minimize the intermolecular interactions. For a normal sample about 20 to 50 mbar of pressure in a 10 cm glass cell with salt windows produces a satisfactory infrared spectrum. While certain bands, such as those for Si-H stretching modes, may go off scale even at 3 mbar, others (such as ring-puckering vibrations) require maximum vapor pressure and path lengths of several meters in order to be observed. Vibrational Raman spectra of gases are generally more difficult to obtain than the infrared spectra, and the laser power (2 to 10 watts) and photon counting sensitivity (1000 counts/s or less) must be quite high. Usually a multipass system (see below) is utilized for concentrating the laser power at the sample in order to increase the Raman signal. Glass or quartz is generally used

for the window material of Raman gas cells, which may be quite small since effective scattering only occurs from a very tiny (≈ 1 mm^3) portion of the sample. Since glass, quartz, and other materials which transmit visible light are thermally much more stable than the infrared transmitting salt windows, high and low temperature samples are more readily investigated by Raman than infrared spectroscopy. Vibrational bands are quite weak in Raman spectra, but the pure-rotational spectra are considerably stronger and are readily examined down to less than 1 cm^{-1}.

The vibrational spectra of gases also always show the energy contributions from the rotational states. Thus, the infrared bands of vapors show P, Q, and R branches corresponding to changes of -1, 0, +1 in the rotational quantum number J. The actual band shapes observed in infrared spectra depend on the symmetry and the relative values of the moments of inertia of a molecule as well as the instrumental resolution. Virtually all spectroscopic texts have discussed these band types [1]. Spherical top molecules (with all moments of inertia equal) have only one band type with a strong spectral Q branch between the P and R branches; symmetric top molecules (with two equal moments of inertia) produce parallel bands, similar in appearance to those of the spherical tops, and also perpendicular bands, which have many Q branches and rotational fine structure that is generally merged into a broad background. The parallel bands of linear molecules, including diatomics such as HCl and CO, are unusual in that the Q branch is forbidden and no band maximum occurs at the actual vibrational frequency. Most molecules have less symmetry and are classified as asymmetric tops, which have bands described as types A, B, or C. Type A and C bands have central Q branches (stronger in C bands) surrounded by the P and R branches, but the type B bands show a spectral minimum at the vibrational frequency since the Q branch corresponding to $\Delta J = 0$ is forbidden.

The vibrational bands in Raman spectra are simpler in the sense that often a very intense Q branch is present at the vibrational frequency and the bands from the rotational contributions (bands with $\Delta J \neq 0$) are much weaker. On the other hand, the rotational vibrational bands in the Raman spectrum often have O and S branches, corresponding to $\Delta J = -2$ and $+2$, in addition to the P, Q, and R structure. Similarly, the pure rotational Raman spectra of diatomics are based on the selection rule $\Delta J = 2$, so that the band spacing is double that of infrared spectra for which $\Delta J = 1$. Moreover, the pure rotational spectra of all molecules (except spherical top molecules such as SF_6, CCl_4, etc., if one neglects centrifugal distortion) are Raman active whereas the pure rotations of the homonuclear diatomics (N_2, O_2, Cl_2, etc.) and the spherical top molecules can not be investigated by infrared spectroscopy. As an example of a pure rotational spectrum which is inactive in IR we show in Fig. 10 the Stokes Raman spectrum of

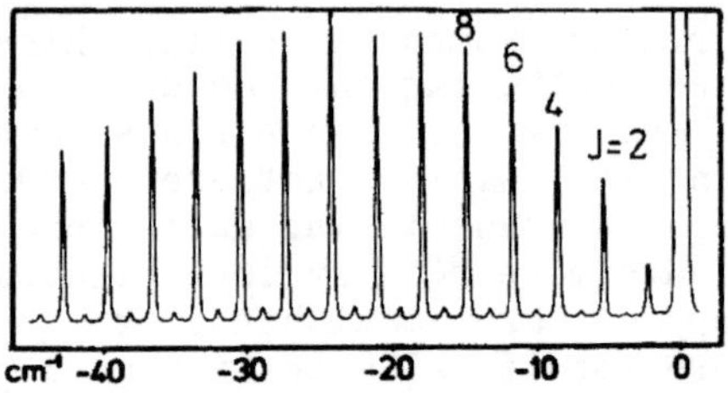

Fig. 10 Pure rotational Raman spectrum of CO_2 (adapted from Ref. [19]).

CO_2. The weak lines between the even J values are due to molecules in the first vibrationally excited state [19].

In studies of anharmonic large-amplitude vibrations, different transitions generally dominate Raman spectra than those in the infrared. For example, for ring-puckering and internal rotation vibrations the vibrational quantum number v generally changes by two ($\Delta v = 2$) for Raman transitions and by one ($\Delta v = 1$) for infrared transitions. Such complementary behavior greatly facilitates data interpretation. Figure 11 shows the infrared and Raman ring-puckering spectra for trimethylene oxide [20,21] demonstrating the differences in the two type of transitions. Comprehensive reviews on large-amplitude vibrations (ring-puckering and pseudorotation) may be found in Ref. [22].

Whereas a vapor phase sample has at most three infrared band types, the same sample may have up to four different types of Raman band contours for various types, the identification of which can be useful for proper band assignment. Very often, however, the band structure may be complicated by various effects including Coriolis coupling, Fermi resonance, hot bands, and the overlap of different Raman lines. Nonetheless, some generalizations can be made [23]. All totally symmetric vibrations have strong Q branches whereas for the other modes this central peak will be much weaker and often unobservable. The totally symmetric vibrations of spherical top molecules will produce only Q branches while such vibrations for linear molecules produce only O, Q, and S branches. In principle, all other vibrations for all other symmetry types will show O, P, Q, R, and S contributions, although the relative intensities of the branches may differ markedly. Figure 12 shows a vapor phase Raman spectrum of the E_{2g} fundamental of benzene at 1178 cm^{-1}. Computed Q, P, R, O and S branches and the total computed contour help assign the various peaks in the observed spectrum [24].

Vapor phase Raman spectra have proved useful for both temper-

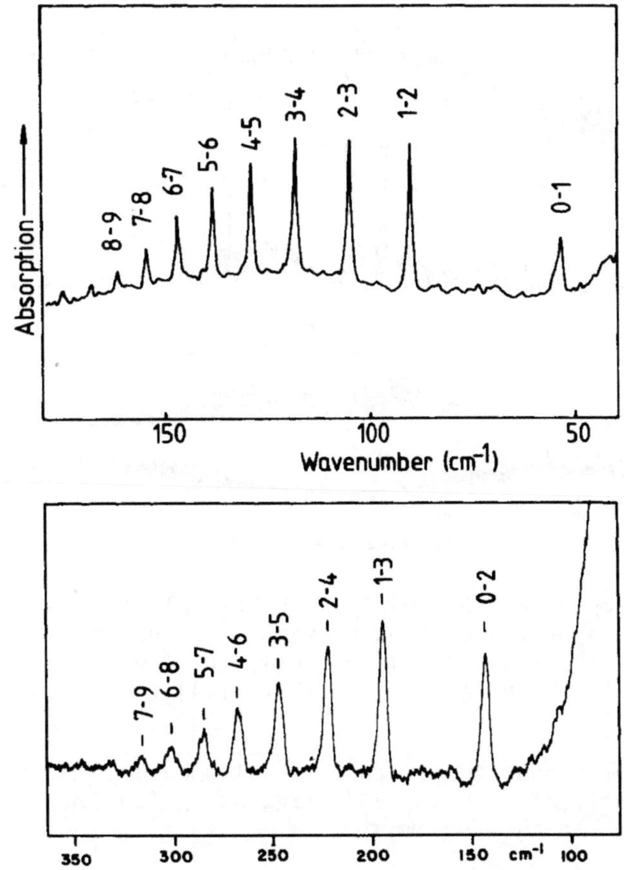

Fig. 11 Pure puckering transitions of trimethylene oxide observed in the far-infrared (upper field, $\Delta v = 1$ transitions) and in the Raman (lower field, $\Delta v = 2$ transitions) spectrum (adapted from Refs. [20] and [21]).

ature measurements and remote sensing. An entire text [23] which emphasizes such techniques has been published. Rotational temperatures of molecules can be measured by examining the relative intensities of the various J lines in a pure rotational Raman spectrum. Such a procedure is described in Ref. [25]. The vibrational temperature T_V which is often different from the rotational temperature T_R, may be obtained by several different techniques. First, the anti-Stokes to Stokes intensity ratio I_{AS}/I_S, which decreases with temperature, can be measured and T_V can be determined from Eq. (4) when I_{AS} and I_S are measured.

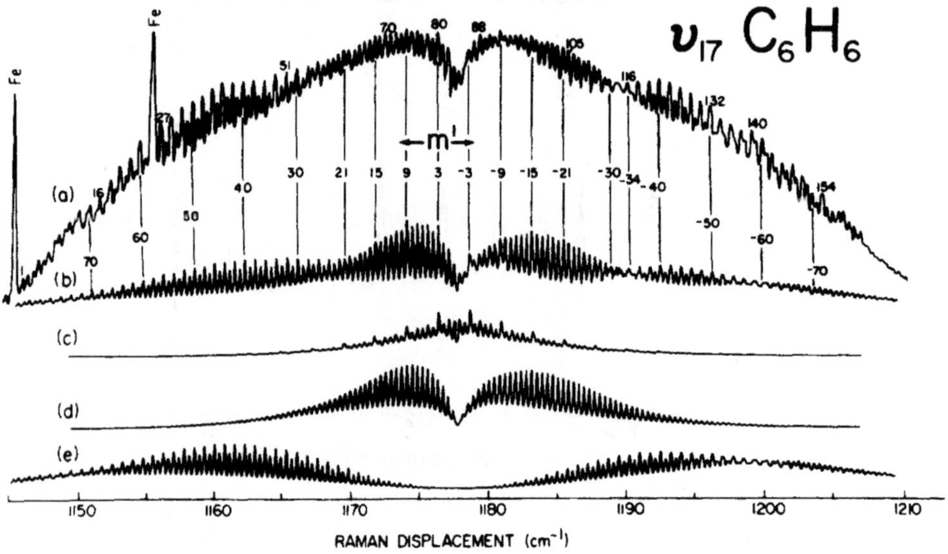

Fig. 12 Example for the complex vibrational-rotational structure in the Raman spectrum of benzene vapor; (a) observed spectrum, (b) computed spectrum (total contour), (c) Q branches, (d) P and R branches, (e) O and S branches (Ref. [24]).

The vibrational temperature may also be determined from the intensity ratio of a "hot band" originating from an excited vibrational state, to the fundamental:

$$I_H/I_F = \exp(-E_H/kT_V) \qquad (5)$$

where H and F refer to the hot band and fundamental, respectively. A third general technique which can be used to determine both rotational and vibrational temperatures involves contour analysis. That is, the entire calculated spectrum is computationally adjusted by varying T_R and T_V until the observed spectrum is attained.

Such temperature measurements, and also other remote sensing applications, are feasible since Raman systems can be set up to be single ended when back-scattering is used. That is, both the exciting source (laser) and detection system may be in the same location. Remote sensing by FT-IR emission spectroscopy is also possible since in this case the sample serves as the energy source; however, infrared absorption techniques are not practical for remote detection. With the Raman technique, scattering from N_2 up to 40 km altitude has been observed, and detection approaching ppm levels at distances of 1 km is possible [23].

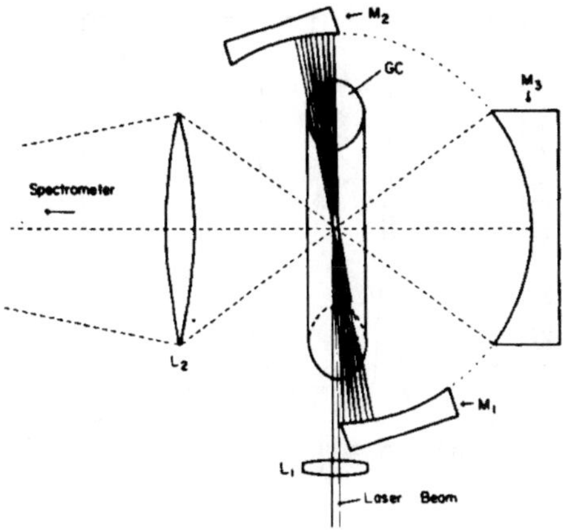

Fig. 13 Schematic diagram of Raman multipass system (Ref. [26]).

2. Multiple-Reflection Systems. Infrared spectroscopy, including FT-IR, has long relied on multiple-reflection optics to achieve a longer path length through the gas sample of interest. This increases the detectability of weak bands since in Eq. (1) A is directly proportional to path length b. "White cells" of effective path length of more than 1 km have been designed.

The Raman spectroscopy of gases also benefits from multiple-reflection systems; not by increasing the sample path length, but by increasing the laser power P at the sampling region (see Eq. 2). Various multiple reflection systems have been proposed and realized to increase the Raman signal collected from a gas [26,27]. A schematic diagram of a very compact and efficient multipass system [27] is shown in Fig. 13. The principle is an off-axis external resonator which is formed by two concentric spherical mirrors M_1 and M_2 with dielectric coatings (reflectivity at 488 nm is 99.9%). The incident laser beam is focused very close to the center of M_1 and M_2 and is then reflected and refocused by M_2. The beam reflects alternately from the two mirrors and passes two focal points on each side of the center of the two spheres. The gas cell is a cylinder with Brewster-angle windows to minimize reflection losses. The scattered light is collected by lens L_2 and the backing mirror. A gain factor of about 20 over a single focused beam arrangement has been realized by this technique. A similar system is available commercially.

3. High Resolution Studies. The high resolution studies of the pure rotation and rotation-vibration Raman bands of many molecules have been performed by various research groups [28]. Much of the success of high resolution Raman spectroscopy is due to

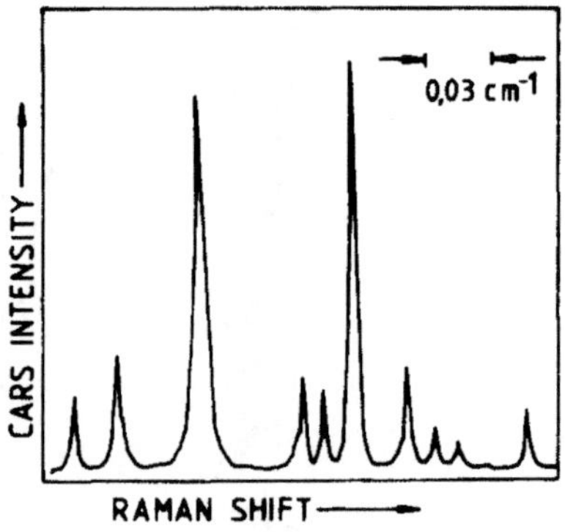

Fig. 14 Part of the high resolution CARS spectrum of the $\nu_1(A_1)$ Q branch of CH_4 at 2917 cm^{-1}; pressure 0.5 Torr; temperature: 80 K (adapted from Ref. [32]).

the fact that single mode lasers and high dispersive, mainly home-made, monochromators are available. Many problems can be solved which range from the investigation of Dicke narrowing to detailed studies of the fine structure of the Raman transitions in simple gases. Space does not allow the discussion of any of these studies in greater detail. For the interested reader we refer to the various review and book articles published in this research field [10,23,26,28-30]. Non-linear coherent Raman spectroscopy of gases, in particular the CARS technique, Raman gain, or inverse Raman spectroscopy promise major advances in the area of high resolution, molecular structure studies and applications. By these techniques resolution of better than 10^{-3} cm^{-1} has been achieved [31]. A high resolution CARS spectrum of part of the ν_1 band of CH_4 at 0.5 Torr pressure and 80 K temperature with resolution of 0.002 cm^{-1} is shown in Fig. 14, which demonstrates the high state of development of this technique [32]. Considerable improvements are to be expected for various kinds of these non-linear coherent Raman techniques.

C. Liquids and Solutions

1. Principles. Liquids and solutions, especially colorless ones, are easily examined by Raman spectroscopy, perhaps even more readily than by infrared methods. While samples for infrared studies must be contained at small path lengths between salt plates, which are subject to cracking and deterioration, almost any glass container serves as a suitable Raman cell. Even reaction flasks, bottles, test tubes, vials, and capillary tubes can be used. Unlike infrared spectroscopy where too much sample or too long a path length often cause bands to go off scale, the

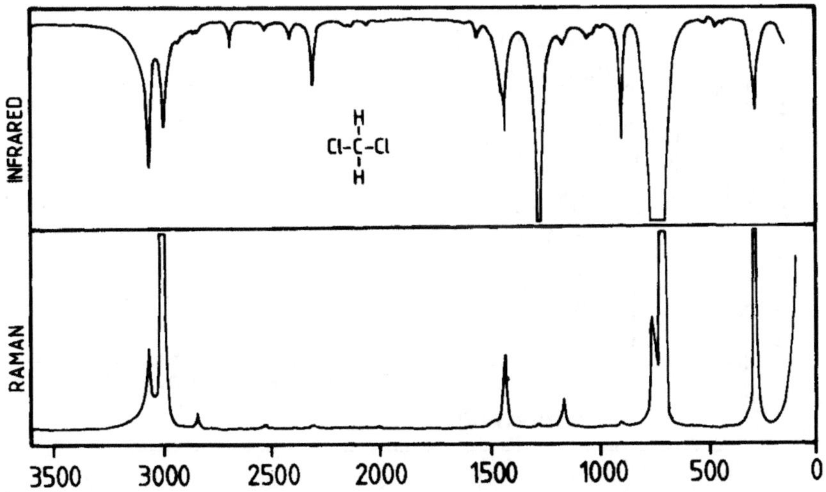

Fig. 15 Infrared and Raman spectra of liquid methylene chloride (CH_2Cl_2).

amount of sample and path length for Raman spectroscopy generally are not important. Spectra can be obtained from micro- or nanogram quantities or from large containers holding kilograms or more of material.

The Raman spectra of solvents, of course, differ markedly from their infrared spectra. This means that spectral regions inaccesible by infrared spectroscopy for a particular solvent due to strong solvent absorption may be well suited for Raman studies. Similarly, poor regions for Raman spectra may be good for infrared work. Figure 15 shows both the IR and Raman solvent spectra for methylene chloride CH_2Cl_2 demonstrating this point.

Infrared spectra of solutions are generally run in a double-beam or double channel mode so that the solute and solvent spectrum may be collected in one and the pure solvent spectrum in the other. In this manner, weak to moderate infrared absorptions of the solvent can be compensated for; however, in regions of very strong solvent bands where negligible radiation is transmitted no meaningful spectra can be obtained. Conventional Raman spectra are recorded "single beam" so that solvent bands are always present in the spectra. However, utilization of an on-line computer or an electronic Raman difference system (see Sec. 4.F.) allows a second solvent spectrum to be recorded which may be subtracted from that of the solution.

Perhaps the most valuable contributions of Raman spectra of

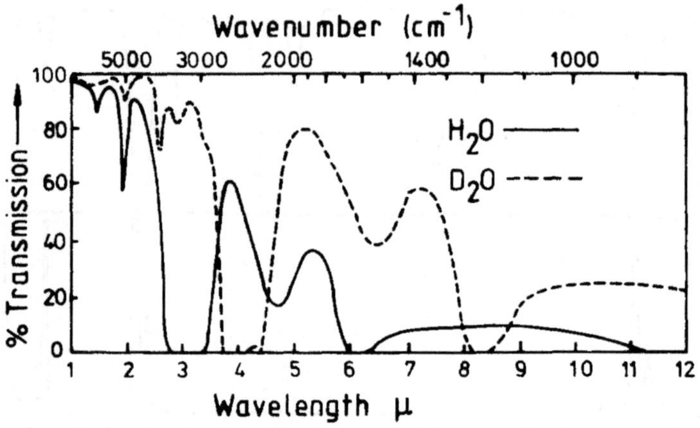

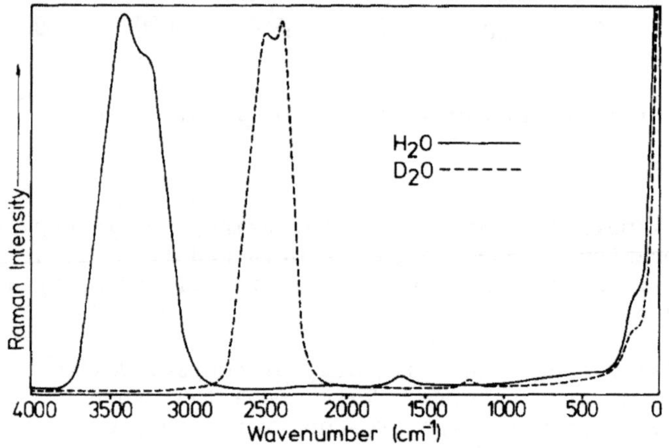

Fig. 16 Infrared and Raman spectra of liquid H_2O and D_2O.

solutions have come from the studies of aqueous systems containing inorganic or biologically important molecules. Some examples of these applications will be considered next.

2. Aqueous Solutions. Because of its highly polar character, liquid water is a very strong infrared absorber and consequently a poor solvent for infrared studies. Furthermore, all but a few optical materials (BaF_2, CaF_2) which transmit infrared radiation dissolve in water and thus cannot be used. On the other hand, Raman scattering from water is quite weak and the cell material (glass), of course, is not affected by the solvent. Figure 16 compares the infrared spectra of liquid H_2O and D_2O to their Raman spectra. As can be seen, except for the broad band between 3000 and 3700 cm^{-1} (2300 - 2700 cm^{-1} for D_2O), the Raman spectrum of water is essentially clear of any interference. While the in-

frared spectra of aqueous systems are much more difficult to obtain, there are cases where both IR and Raman data are desirable. For example, the planar, undistorted nitrate ion NO_3^- in aqueous solution has D_{3h} symmetry with vibrational species $A_1' + A_2'' + 2E'$. The symmetric NO_3 stretching (ν_1) vibration is only active in the Raman spectra and can be observed near 1040 cm^{-1}. The out-of-plane bending motion (ν_2) is only infrared active and occurs near 820 cm^{-1}.

With pre-FT-IR instrumentation only molecules with intense absorptions in the relatively clean 1800-2800 cm^{-1} region (e.g. C≡N groups) could be easily investigated. However, the ability of an FT-IR system to rapidly collect data and to signal average has made it possible to examine spectral regions where less than 1% transmittance is present. Consequently, by using either H_2O or D_2O to get maximum transmittance, the complete infrared spectra of most aqueous samples can be recorded from 800 cm^{-1} on up. It should be noted, however, that the spectra of the dissolved material can only be measured if the absorption intensities of its bands are greater than that of the (displaced) water in the spectral regions of interest. Furthermore, aqueous infrared studies are hindered by the very small path lengths (~50 μm) that must be used as well as by the expense of all materials.

Because water is a natural solvent for many inorganic and biological material and because it is one of the solvents best suited for Raman work, Raman studies of aqueous solutions have found great applicability. When colored materials are present, the resonance Raman effect can further enhance the technique, but even under normal conditions excellent spectra are readily obtained even for complex systems. Ionic equilibria involving numerous different chemical species may be better understood by examining the Raman intensities of the bands for each species as the chemical conditions (concentrations, pH, etc.) are changed. It is clear that water coordination plays a significant role in such systems and that Raman spectra are invaluable for sorting out the nature of the species. For example, from the symmetric stretching M-O frequencies in $M(H_2O)_6^{+n}$ complexes (n = 2 or 3 and M = Mg, Co, Ni, Cu, Zn, Cd, Hg, Cr, Fe, etc.), which are observed in the 350-550 cm^{-1} range, the strength of the hydration can be evaluated. A recent review on the infrared and Raman spectra of electrolytes in solution has been published [33].

As an example of the study of a radical in aqueous solution, Fig. 17 shows the spectra from a mixture involving the following reaction [34]:

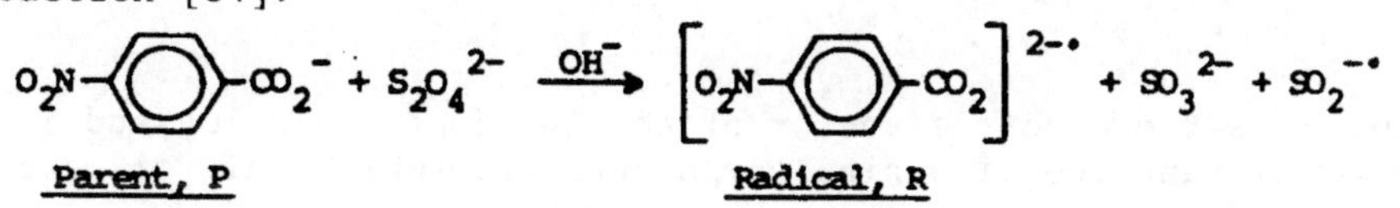

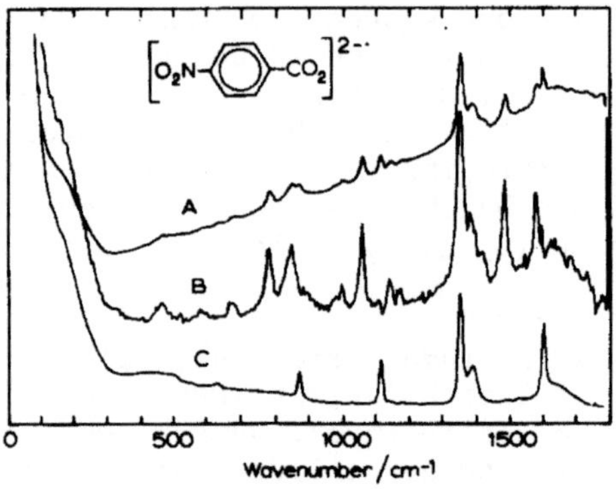

Fig. 17 Raman spectra from (A) a reacting mixture of 4-nitrobenzoate and $S_2O_4^{2-}$, and (C) 4-nitrobenzoate, both in NaOH solution. Spectrum B resulted from computer processing of A and C. (Ref. [34]).

Spectrum A represents that of the reaction mixture while C is for the starting material. The difference spectrum (B = A - C, with scaling) reveals many bands of the free radical.

Numerous studies of proteins and nucleic acids in aqueous solutions have been carried out using Raman spectroscopy [35-40]. It has been demonstrated that structure-frequency correlations may be formulated from which the secondary structure of proteins may be determined. For example, the "Amide I" Raman line occurs in the 1655-1659 cm^{-1} region for an alpha helix, from 1667-1677 cm^{-1} for an antiparallel sheet, and between 1665 and 1675 cm^{-1} for a disordered protein. Similar correlations exist for the "Amide III" band near 1250 cm^{-1} as well as for selected skeletal modes. Many other kinds of biological Raman studies [41-43], including the investigation of viruses, have been carried out. Figure 18 shows as an example the investigation of oxy- and deoxyhemoglobin demonstrating that Raman spectroscopy distinguishes between these species [44].

D. Solids

1. Single Crystals. As generally known, a normal mode of vibration will be active in the Raman effect if it involves a change in the polarizability of the molecule. The induced moment is given by

$$\mu = \alpha E$$

where E is the electric vector of the incident radiation and α a tensor of rank two with nine components from which only six are

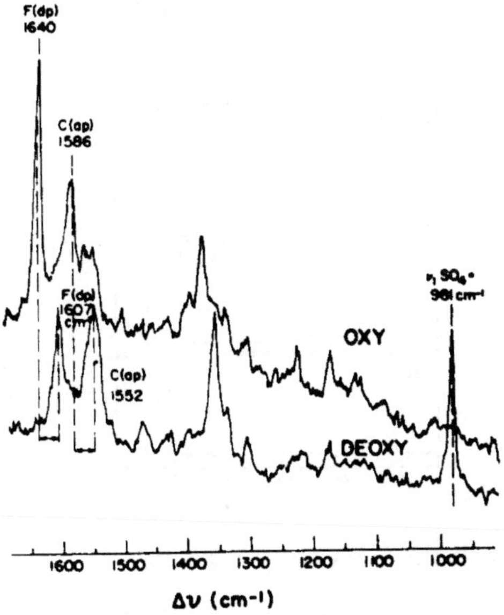

Fig. 18 Part of the Raman spectra of oxy- and deoxyhemoglobin (adapted from Ref. [44]).

unique, because α is symmetric (except for resonance Raman or electronic Raman scattering). Unlike in gases and liquids, in the solid state the polarizability tensor can be oriented since the positions of all atoms are fixed with respect to the crystallographic axis. This fact suggests a powerful tool for interpreting Raman spectra of anisotropic substances. It is only necessary to orient the axis of the polarizability tensor with respect to the directions of propagation, polarity of the incident, and the scattered radiation in order to carry out polarized Raman spectra of single crystals. The Raman experiment then permits identification of the polarizability tensor components associated with each normal mode of vibration in the oriented crystal. Thus it becomes easy to correlate the observed modes to the various symmetry species and hence to the particular vibrations. It should be mentioned that useful information in this respect is only derivable when the propagation direction of the incident light, the electric vectors of the incident and scattered radiation, and the faces of the crystal are parallel or perpendicular to the polarizability ellipsoid. Hence, to perform a polarized Raman single crystal experiment, it is necessary to cut and orient (with X-rays) the crystal before useful Raman spectra are observed. For Raman spectroscopy, the size of the single crystal may be as small as the dimensions of a focused laser beam (e.g. a crystal of volume 1 mm^3 can be investigated). Because of this and because most laser

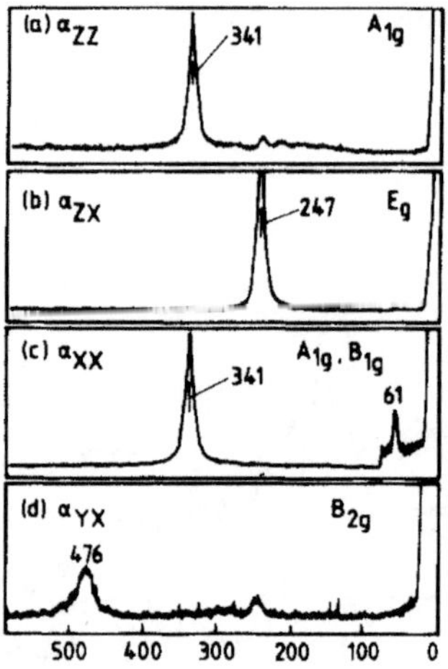

Fig. 19 Polarized Raman spectra of an MnF_2 single crystal (adapted from Ref. [45]).

beams are highly polarized, polarized single crystal Raman spectroscopy is straightforward.

Figure 19 shows a typical example for single crystal Raman work (MnF_2). Group theory predicts four Raman active modes: $1A_{1g}$ (with α_{xx}, α_{yy}, $\alpha_{zz} \neq 0$), $1B_{1g}$ (α_{xx}, $\alpha_{yy} \neq 0$), $1E_g$ (α_{yz}, α_{xz}, α_{zy}, $\alpha_{zx} \neq 0$) and $1B_{2g}$ (α_{xy}, $\alpha_{yx} \neq 0$). The figure shows the spectra observed for the different α_{ij} spectra in complete agreement with the predictions from group theory [45]. For further details on single crystal Raman spectroscopy we refer to various monographs [46-50].

In principle, it is possible (at least for special cases) to elucidate the kind of vibration by the use of polarized infrared radiation and an oriented single crystal. For instance, in an anisotropic medium, an infrared-active vibration involves an oscillating dipole, μ, which occurs along a specific direction, resolvable into only one or two components (instead of three components $\mu = \mu_x + \mu_y + \mu_z$; x, y, z: Cartesian coordinates). In such a medium, plane polarized infrared radiation with its electric vector parallel to μ will result in maximum absorption at the frequency of the normal mode, while radiation polarized in a plane perpendicular to μ will not give rise to any absorption. If normal modes are non-degenerate, μ will be perpendicular to z (principal symmetry axis) and therefore lies in the xy-plane. So, in principle, polarized infrared radiation can be used to identify the orientation of μ and, consequently, the symmetry species of each infrared-active vibration. However, in contrast to Raman

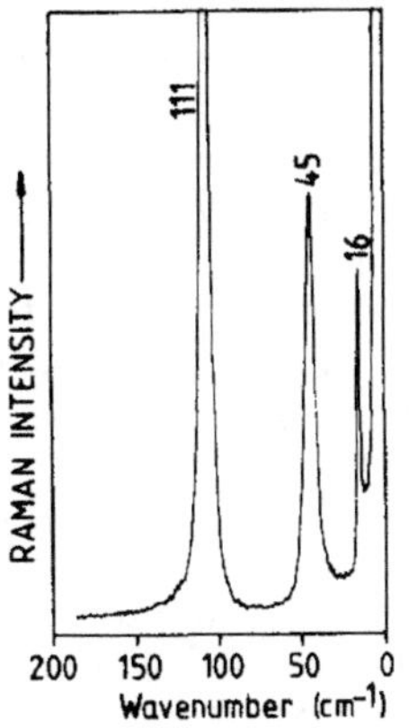

Fig. 20 Raman spectrum of CdI_2 crystal powder (Ref. [52]).

spectroscopy, single crystals are far more difficult to handle experimentally for IR spectroscopy. Ideally, one should have a crystal whose surface area is roughly that of the infrared beam at the sample point (perhaps 10 mm^2 or more) and a thickness of about 10μ. Clearly, it is virtually impossible to prepare single crystals of such shapes. Sometimes reflection techniques are employed, but the observed reflection spectrum is usually complex and therefore, it is necessary to accomplish a mathematical transformation in order to derive the desired absorption spectrum [51].

2. Crystal Powders. Powders are, in principle, easy to scan in the Raman when no fluorescence is involved. They can be placed, either held in a transparent glass capillary tube or pressed into a sample holder (as pellet), in front of the spectrometer at the sample location. The high (non-frequency-shifted) stray light, the major limitation in earlier days of Raman spectroscopy, can be effectively separated by means of triple monochromators. Very often, double monochromators are also capable of scanning crystal powder Raman spectra even at low frequencies. An example is given in Fig. 20 where a low frequency Raman spectrum of a pressed pellet of CdI_2 crystal powder is displayed [52]. The band at 16 cm^{-1}, which looks rather like an instrumental artifact (e.g. reflection from a slit jaw), was shown to be a real Raman transition [53].

Earlier problems of sample decomposition due to absorption of laser light may be overcome by dynamical Raman techniques such as the rotating sample or the surface scanning techniques [11]. When employing these techniques, it is possible to scan Raman spectra even from black crystal powders.

In the infrared, insoluble powders are usually prepared as mulls (nujol or fluorolube) or pressed alkali halide pellets (KBr, NaCl, etc.). The latter offers a number of advantages over nujol mulls, because there are no interference solvent bands and the sample can be stored indefinitely for future reference. However, with inorganic samples there arise difficulties due to the fact that the salt sometimes is not inert with respect to the sample. Moreover, frequency shifts and additional bands may occur due to effects of the host lattice system.

3. Matrix Isolation Studies. Starting in the 1950's the technique of matrix isolation infrared spectroscopy was developed primarily by G. Pimentel [54] and co-workers. Typically, an inert gas such as argon, krypton, or nitrogen is mixed with a small amount (less than 1%) of sample. The mixture may be frozen directly onto an infrared window, which is maintained at a low temperature (2 to 40 K), or a photolysis, pyrolysis, or electric discharge may precede the matrix deposition. The sample molecules (or the newly formed chemical species) are then trapped between inert matrix molecules and no further intermolecular reactions or interactions may occur unless the system is intentionally warmed or photolyzed, for example. The matrix isolation technique thus allows spectra to be obtained for species which, because of their reactivity, otherwise could not be studied. This method also permits the low temperature study of unreactive molecules which have been isolated from each other by the matrix, and thus the perturbations from the normal crystal forces have been eliminated. Because rare gas to sample ratios are often 2000 or more, very little sample is actually available for spectral examination and thus the signals are weak. The development of FT-IR, by allowing improved signal-to-noise ratios to be achieved through signal averaging, has made it possible to detect weak signals from matrix isolated species which otherwise would have been buried in the noise. As a result, infrared investigations of matrix isolated species are relatively routine, aside from the need for good vacuum and low-temperature equipment.

Matrix isolation with Raman spectroscopy, applied extensively by L. Andrews [55-57], G. Ozin [58-60], and other workers, is more difficult to persue than with FT-IR. The main problem arises from sample warming by the laser beam, resulting in collapse of the matrix. If the laser power is reduced or the laser defocused, then the signal from the highly dilute sample may be difficult to attain. Nonetheless, by careful selection of the rare gas and the experimental parameters such as rare gas to sample ratio, laser frequency, laser power, and operating temperature, quality spectra can be obtained. Also a good approach to this problem would be the application of the rotating surface scanning technique [11].

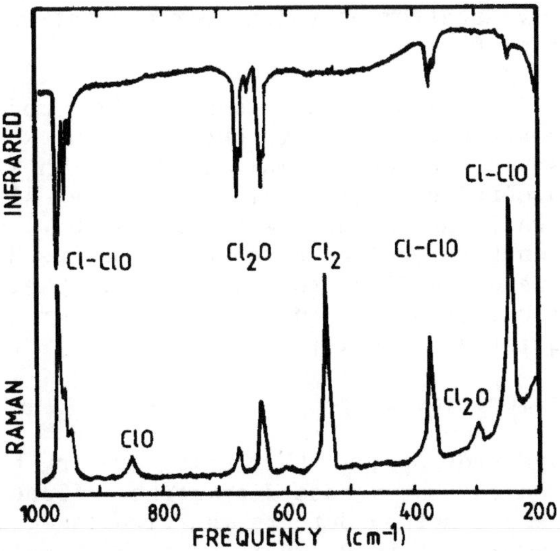

Fig. 21 Matrix isolated infrared and Raman spectra of Cl_2O and its photolysis products (adapted from Ref. [61]).

Figure 21 shows the matrix isolated infrared and Raman spectra of Cl-O-Cl and its photolysis products in an argon matrix [61]. In addition to the starting material (Cl_2O) bands can be seen for the Cl-Cl-O isomer, the diatomic Cl-O species, and for a $(ClO)_2$ dimer (at 995 cm^{-1}) as well as for diatomic chlorine Cl_2. It should be evident that the production of Cl-O and Cl_2 were only determined from the Raman data.

E. Microsampling

The greater sensitivity brought to infrared spectroscopy by FT-IR is most useful for examining samples whose absorptions would be lost into background noise with conventional instruments. The spectra of only about 11 ng of sample can now be obtained. One area of analytical chemistry which has rapidly developed because of the small sample detectability is GC-IR (gas chromatography-infrared spectroscopy). This approach to identifying minor or major components of a mixture utilizes normal GC methods to separate the components which are passed to a lightpipe and then detected by the FT-IR instrument. More than fifty components in microgram sized samples may thus be identified.

Raman spectroscopy is also inherently an excellent microsampling technique since the actual scattering volume, which is determined by the diameter and length of the region of the laser focus, can be kept extremely small. Barrett and Adams [62] showed that the effective scattering volume for excitation of gases can be of the order of 0.01 nl.

More recently, Rosasco and co-workers [63] modified a conventional laser Raman spectrometer to obtain useful Raman spectra from discrete solid samples as small as 0.7 μm in linear dimensions. Spectra obtained from single, micrometer-sized particles of several inorganic and organic materials were reported [63]. Freeman and co-workers [64] have reported routine Raman analyses of liquid samples as small as two nanoliters. Useful spectra from an effective volume of 0.1 nl in a 50 μm diameter capillary were obtained by Nyquist and Kagel [65].

F. Difference Spectroscopy

Infrared spectroscopy, whether using grating or FT instruments, has long utilized the ratio recording method. Generally a "double-beam" or two-channel method detects the signals I_S and I_R through two different samples S and R, where S and R normally represent the sample and reference, respectively. The transmittance through the sample relative to the reference is given by

$$T = I_S/I_R. \qquad (6)$$

T can be related logarithmically to the difference of the absorptions (which are proportional to sample concentrations according to Eq. (1)) by the two samples:

$$A_S - A_R = -\log(I_S/I_o) + \log(I_R/I_o) = -\log(I_S/I_R) = -\log T.$$

Since FT-IR instruments readily permit conversions of a percent T spectrum to an absorption scale, a difference spectrum of two absorbance curves is routinely determined. The purpose of such ratio recording techniques in the past has been almost exclusively to subtract out the undesired elements (solvent bands, instrumental effects) in the reference spectrum.

Raman spectroscopy is intrinsically a "single-beam" technique, but methods have been developed for subtracting data in one channel from another. While this is also useful for canceling solvent or interfering effects in some cases, Raman difference spectroscopy can also be valuable for determining (1) small frequency shifts resulting from solvent or isotopic effects; or (2) the isotropic and anisotropic components of the Raman signal. The theory for calculating the frequency shifts between two similar spectra has recently been fully developed [66]. For most cases a small frequency shift results in a Raman difference signal resembling a first derivative curve. From the measurement of the peak to peak height (d) of this difference curve and by ratioing it against the intensity of the normal Raman band (I_o) the frequency shift may be calculated when the bandwidth Γ is known.

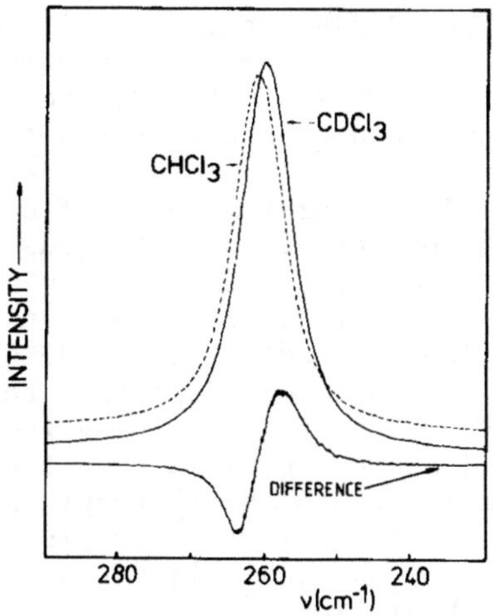

Fig. 22 Raman spectra of the ν_6(E) vibration of $CHCl_3$ and $CDCl_3$ and their difference spectrum (Ref. [66]).

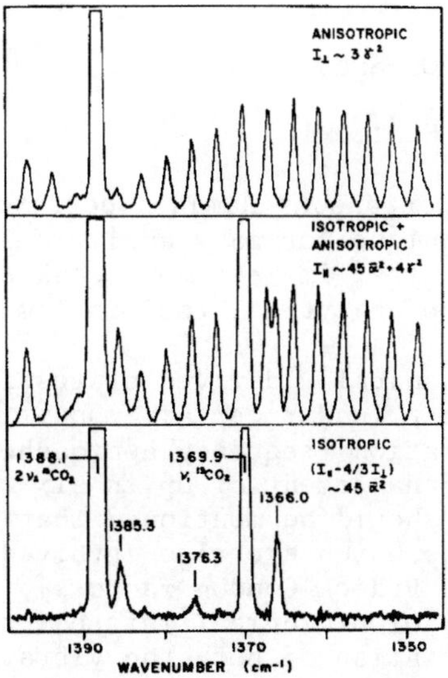

Fig. 23 Part of the vibrational-rotational spectrum of CO_2. All three spectra were recorded simultaneously (Ref. [67]).

Figure 22 shows the Raman spectra of liquid $CHCl_3$ and $CDCl_3$ near 260 cm^{-1} along with the difference spectrum. From the measured values of $d/I_o = 0.34$ and $\Gamma = 8.4\ cm^{-1}$, the isotopic frequency shift of 1.1 cm^{-1} can be accurately determined ($\pm 0.1\ cm^{-1}$).

The Raman signal is normally made up of both isotropic ($\bar{\alpha}^2$) and anisotropic ($\bar{\gamma}^2$) components. Simultaneous recording of parallel ($I_{\parallel} \propto 45\bar{\alpha}^2 + 4\bar{\gamma}^2$) and perpendicular ($I_{\perp} \propto 3\bar{\gamma}^2$) polarized Raman spectra permits the direct determination of the isotropic component since $I_{\parallel} - 4/3 I_{\perp} \propto 45\bar{\alpha}^2$. Figure 23 clearly demonstrates this application [67]. The middle spectrum represents the normal Raman spectrum of CO_2 vapor (containing isotropically and anisotropically scattered Raman light) whereas the top and bottom spectra are for the anisotropic and isotropic parts, respectively. Since the rotational-vibrational structure is anisotropic, its elimination clearly makes the analysis of the CO_2 pure vibrational bands (Q branches) much simpler.

G. Summary of Applicability of Raman and FT-IR Spectroscopy

Table 2 summarizes the applicability of both techniques to various spectroscopic problems.

5. SPECIAL RAMAN PROCESSES

A. Resonance Raman Scattering

Conventional, spontaneous Raman spectroscopy has been an extremely useful technique for many analytical applications. However, due to the fact that it is a very weak phenomenon, its usefulness as a sensitive analytical tool is limited.

The quantum mechanically derived expression for the Raman intensity I_R shows that (under certain conditions) I_R will increase manyfold as the excitation frequency approaches that of an electronic absorption. Enhancement of up to six orders of magnitude can be achieved. It should be mentioned that only specific vibrations, namely those which are also involved in the absorption process (having large Franck-Condon factors), will be enhanced. For complex molecules of biological interest this may be of great help for the interpretation of both the vibrational and absorption spectrum. Thus, via resonance Raman scattering also excited electronic state properties may be studied [68]. For an excellent review on resonance Raman scattering see Ref. [68].

The technique is, of course, useful only for substances that absorb in the region of excitation. With the availability of tunable lasers which operate between 1000 nm and 350 nm (with

Table 2 Applicability of Raman and FT-IR Spectroscopy

	Raman	FT-IR
General Applicability	90-95%	95-99%
Limitations	short-lived fluorescence; extreme absorption	aqueous solutions; single crystals; metals
Specific Applicability		
gases	simple	simple
liquids	very simple	very simple
single crystals	very simple	very difficult
powders	simple	moderate
matrix-isolation	difficult	good
low temperature	simple	simple
high temperature	moderate	difficult
microsampling	good (<1 ng)	good (<1 ng)
difference spectroscopy	easy and common	easy but not common
polarization measurements	easy with valuable information	moderate with limited information
quantitative analysis	more difficult	good
Applicability to Molecular Spectroscopy		
group frequencies	very good	very good
low frequencies	very good	more difficult
group intensities	good	good
assignment	excellent	very good
future outlook	promising (non-linear techniques; resolution)	promising

frequency doubling even down to about 200 nm) resonance Raman spectroscopy can be applied to a great variety of samples.

If a molecule does not absorb at an accessible wavelength, then it is also possible to attach "resonance Raman labels" (molecules with useful resonance Raman spectral characteristics) to specific sites of biological function, to report back structural information. This label-technique, started by Carey et al.[69-71], is quite promising. For various reviews on the application of resonance Raman scattering on molecules of biological interest we refer to Refs. [43,72,73].

B. Hyper-Raman Scattering

Hyper-Raman scattering [74] arises from the dipoles induced by a non-linear interaction of the scattering system with the electric field of the incident radiation. Via the hyperpolarizability, β, Stokes- and anti-Stokes-shifted photons with respect to the second harmonic of the exciting laser frequency are generated at frequencies $2\nu_o \pm \nu_m$. The theory for hyper-Raman scattering has been worked out in great detail [4,74,75] and several reviews of this research field have been published [76,77]. One of the important results is that the vibrational selection rules for hyper-Raman scattering are less restrictive than for normal Raman scattering and many vibrational modes which are both Raman and infrared inactive are hyper-Raman active. For structural determination via vibrational analysis this effect therefore complements normal Raman and infrared data.

Unfortunately, the efficiency of this scattering process is extremely small, even by comparison with normal Raman spectroscopy. Hence, great experimental effort is needed. Nevertheless, the method has been developed to high standard by Long and co-workers [78] at the University of Bradford, England. The hyper-Raman spectra of liquids and solids reported from this laboratory are now comparable in quality to spontaneous Raman spectra recorded in the early days of laser Raman spectroscopy.

C. Active Raman Spectroscopy

In the last few years, a new method in Raman spectroscopy, namely the active Raman spectroscopy (as christened by Akhmanov and Koroteev [79]) was developed, which has sort of an intermediate position between spontaneous and stimulated Raman spectroscopy. This method combines the various spectroscopic possibilities of the spontaneous Raman effect with the most important characteristic of stimulated Raman scattering, the strong and coherent excitation of Raman active modes, which causes strong signal levels.

The method of active Raman spectroscopy is based on the coherent excitation of vibrations in the field of two strong laser beams with frequencies ω_1 and ω_2 in such a way that their difference corresponds to the frequency of the molecular vibration ω_m. Hereby, the excitation occurs via the third-order non-linear term in the polarizability of the molecule in a field with two frequencies. The existence of the coherent molecular vibrations and hence information on eigenstates of the molecules under investigation can be monitored by employing a probe wave with frequency ω, which is modulated by the coherent molecular vibrations leading, for instance, to a strong coherent signal at frequency $\omega_a = \omega + \omega_m$. Instead of a different third wave (the probe wave) with frequency ω, the wave with frequency ω_1 can be used for this purpose. In this case, a scattered radiation at the anti-Stokes frequency $\omega_a = 2\omega_1 - \omega_2$ is generated. As commonly known, this technique is called coherent anti-Stokes Raman spectroscopy (CARS) [80,81]. But this is not the only possibility for manifesting coherent molecular vibrations. The interaction of the electric fields of the two lasers with frequencies ω_1 and ω_2 with the electric field associated with the molecular vibrations can yield also to a gain or a loss in the power of the lasers. The method where the gain at the Stokes frequency is measured is generally referred [82] to as "Raman gain spectroscopy", whereas the "inverse Raman effect" is the terminology commonly used [83] to designate the induced loss at the anti-Stokes frequency. Also higher order Raman scattering effects (HORSES) can occur.

A schematic diagram for the different methods mentioned above is illustrated in Fig. 24. From an experimental point of view, it is evident that in active Raman spectroscopy one either measures the radiation generated at anti-Stokes frequency (CARS) or the change ΔS in the laser beam power (ω_1: inverse Raman effect; ω_2: Raman gain). In order to get full information of the medium, it is necessary to tune the frequency difference $\omega_1 - \omega_2$; then, successively all Raman-active vibrations (or rotations) will be excited and a complete Raman spectrum is obtained. Active Raman spectroscopy can, of course, only be performed since frequency tunable laser sources are available.

One of the major advantages of active Raman spectroscopy over conventional Raman spectroscopy is the ability to obtain Raman spectra with very high resolution which is solely determined by the line width of the input sources. The experimental requirements for high resolution measurements utilizing these techniques preclude the necessity of using monochromators, interferometers, or other apparatus usually used in conventional Raman spectroscopy. Since precisely controlled, tunable, narrow line-width laser sources with appreciable power are available commercially, high resolution molecular spectroscopy [31], which for quite a long time was only possible by infrared absorption, has become feasible now also in the visible region through active Raman spectroscopy.

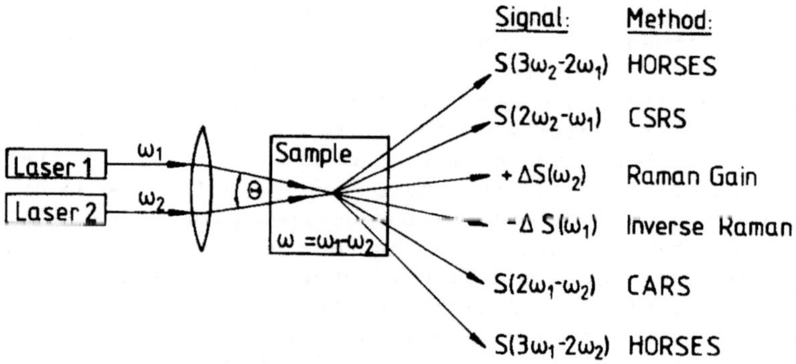

Fig. 24 Schematic diagram for various methods in active Raman spectroscopy.

Although these techniques are (at the moment) not easy to perform and also the equipment needed to carry out an experiment is very expensive, considerable developments and applications in molecular spectroscopy are to be expected in the future.

REFERENCES

1. G. Herzberg, Molecular Spectra and Molecular Structure; I. Spectra of Diatomic Molecules, 2nd Ed., Van Nostrand Reinhold, New York, 1950; II. Infrared and Raman Spectra of Polyatomic Molecules, D. Van Nostrand, Princeton, New Jersey, 1945.

2. E. B. Wilson, Jr., J. C. Decius, and P. C. Cross, Molecular Vibrations, McGraw-Hill, New York, 1955.

3. F. A. Cotton, Chemical Applications of Group Theory, Wiley-Interscience, New York, 1971.

4. D. A. Long, Raman Spectroscopy, McGraw-Hill, London, 1977.

5. J. D. Lewis and J. Laane, Spectrochim. Acta, 21A, 755 (1975).

6. N. B. Colthup, L. H. Daly,and S. E. Wiberley, Introduction to Infrared and Raman Spectroscopy, Academic Press, New York, 1974.

7. L. J. Bellamy, Advances in Infrared Group Frequencies, Methner, London, 1968.

8. P. R. Griffiths, Chemical Infrared Fourier Transform Spectroscopy, John Wiley & Sons, New York, 1975.

9. C. E. Hathaway, in The Raman Effect, Vol. 1, (A. Anderson, ed.), Marcel Dekker, New York, 1971.

10. A. Weber, Raman Spectroscopy of Gases and Liquids, Topics in Current Physics, Springer, Berlin, 1979.

11. W. Kiefer, in Advances in Infrared and Raman Spectroscopy, Vol. 3, p. 1, (R. J. H. Clark and R. E. Hester, eds.), Heyden & Son, London, 1977.

12. J. R. Downey and G. J. Janz, in Advances in Infrared and Raman Spectroscopy, Vol. 1, p. 1, (R. J. H. Clark and R. E. Hester, eds.), Heyden & Son, London, 1975.

13. E. D. Schmid, G. Berthold, H. Berthold, and B. Brosa, Ber. Bunsenges. Phys. Chemie, 75, 149 (1971).

14. M. Delhaye, Appl. Optics, 7, 2195 (1968).

15. M. Bridoux and M. Delhaye, in Advances in Infrared and Raman Spectroscopy, Vol. 2, p. 140, (R. J. H. Clark and R. E. Hester, eds.), Heyden & Son, London, 1976.

16. M. Bridoux, A. Deffontaine, and C. Reiss, C. R. A. S., 282, 771 (1976).

17. M. Bridoux, A. Deffontaine, M. Delhaye, and C. Reiss, Proc. Fifth Int. Conf. Raman Spectrosc.,(E. D. Schmid, J. Brandmuller, W. Kiefer, B. Schrader, and H. W. Schrotter, eds.), H. F. Schulz-Verlag, Freiburg 1976, p. 760.

18. J. M. Harris, R. W. Chrisman, F. E. Lyttle, and R. S. Tobias, Anal. Chem., 48, 1937 (1976).

19. W. Klockner, Ph.D. Dissertation, Munich, 1977.

20. H. Wieser, M. Danyluk, and R. A. Kydd, J. Mol. Spectrosc., 43, 382 (1972).

21. W. Kiefer, H. J. Bernstein, M. Danyluk, and H. Wieser, Chem. Phys. Lett., 12, 605 (1972).

22. Vibrational Spectra and Structure,(J. R. Durig, ed.), Vol. 1, Marcel Dekker, New York, 1974; Vol 5, Elsevier, Amsterdam, 1976.

23. M. Lapp and C. M. Penney, Laser Raman Gas Diagnostics, Plenum Press, New York, 1974.

24. A. Hollinger and H. L. Welsh: Meeting of the Canadian Association of Physicists, June 1977, paper CE 5; picture taken

from A. Weber, Raman Spectroscopy of Gases and Liquids, Springer, Berlin, 1979.

25. R. S. Hickman and L. H. Liang, Rev. Sci. Instr., 43, 796 (1972).

26. J. M. Cherlow and S. P. S. Porto in Laser Spectroscopy of Atoms and Molecules, Topics in Applied Physics, Vol. 2, (H. Walther, ed.), Springer, Berlin 1976.

27. W. Kiefer, H. J. Bernstein, H. Wieser, and M. Danyluk, J. Mol. Spectrosc., 43, 393 (1972).

28. A. Weber, Proc. Fifth Int. Conf. Raman Spectrosc., (E. D. Schmid et al., eds.), H. F. Schulz-Verlag, Freiburg, 1976, p. 391 and references therein.

29. A. Weber, in The Raman Effect, (A. Anderson, ed.), Marcel Dekker, New York, 1973, Vol. 2, p. 543.

30. H. G. M. Edwards, in Essays in Structural Chemistry, (A. J. Downs, D. A. Long, and L. A. K. Staveley, eds.), Plenum, New York, 1971.

31. A. Owyoung, in Laser Spectroscopy IV, (H. Walther and K. W. Rothe, eds.), Springer Series in Optical Sciences, Vol. 21, Springer, Berlin, 1979, p. 175.

32. D. N. Kozlov, A. M. Prokhorov, and V. V. Smirnov, J. Mol. Spectrosc. 77, 21 (1979).

33. D. E. Irish and M. H. Brooker in Advances in Infrared and Raman Spectroscopy, Vol. 2, p. 212, (R. J. H. Clark and R. E. Hester, eds.), Heyden & Son, London, 1976.

34. R. E. Hester, W. E. L. Grossman, and E. Ernstbrunner, Proc. Fifth Int. Conf. Raman Spectrosc., (E. D. Schmid et al., eds.), H. F. Schulz-Verlag, Freiburg, 1976, p. 13.

35. B. G. Frushour and J. L. Koenig, in Advances in Infrared and Raman Spectroscopy, (R. J. H. Clark and R. E. Hester, eds.), Vol. 1, p. 35, Heyden & Son, London, 1975.

36. T. G. Spiro and T. M. Loehr, ibid., Vol. 1, P. 98.

37. G. Vergoten, G. Fleury, and Y. Moschetto, ibid. Vol. 4, 1978, p. 195.

38. G. J. Thomas, Jr., Proc. Sixth Int. Conf. Raman Spectrosc., (E. D. Schmid, R. S. Krishnan, W. Kiefer, and H. W. Schrotter, eds.), Heyden & Son, London,1978, Vol. 2, p. 91.

39. P. R. Carey, ibid. p. 105.

40. T. Theophanides, ibid. p. 115.

41. G. J. Thomas, Applied Spectrosc., 30, 483 (1976).

42. G. J. Thomas in Vibrational Spectra and Structure, (J. R. Durig, ed.), Vol. 3, p. 239, Marcel Dekker, New York, 1975.

43. T. G. Spiro, in Vibrational Spectra and Sturcture, (J. R. Durig, ed.), Vol. 5, p. 101, Elsevier, Amsterdam, 1976.

44. T. G. Spiro and T. C. Strekas, J. Am. Chem. Soc., 96, 338 (1974).

45. S. P. S. Porto, P. A. Fleury, and T. C. Damen, Phys. Rev., 154, 522 (1967).

46. H. Poulet and J. P. Mathieu, Spectres de Vibration et Symetrie des Crystaux, Gordon and Breach, Paris, 1970.

47. G. Turrell, Infrared and Raman Spectra of Crystals, Academic Press, New York, 1972.

48. P. M. A. Sherwood, Vibrational Spectroscopy of Solids, Cambridge University Press, Cambridge, 1972.

49. G. R. Wilkerson, in The Raman Effect, Vol. 2, (A. Anderson, ed.), Marcel Dekker, New York, 1973.

50. M. Cardona (Ed.), Light Scattering in Solids, Topics in Applied Physics, Vol. 8, Springer, Berlin, 1975.

51. K. D. Moller and W. G. Rothschild, Far-Infrared Spectroscopy, Wiley-Interscience, New York, 1971.

52. W. Kiefer, Ph.D. dissertation, Munich 1970.

53. S. Montero and W. Kiefer, J. Raman Spectrosc.,1, 565 (1973).

54. G. C. Pimentel, in Formation and Trapping of Free Radicals, (A. M. Brass and H. P. Broida, eds.), p. 69, Academic Press, New York, 1960.

55. L. Andrews, J. Chem. Phys., 48, 972 (1968).

56. L. Andrews, Appl. Spectrosc., 33, 205 (1979).

57. L. Andrews, in Vibrational Spectra and Structure, (J. R. Durig, ed.), Vol. 4, p. 1, Elsevier, Amsterdam, 1975.

58. G. A. Ozin, Spex Speaker, Vol. XVI (1971).

59. G. A. Ozin and A. Van der Volt, in Progress in Inorganic Chemistry, Vol. 19, (S. J. Lippard, ed.), Wiley, New York, 1975.

60. M. Moskovits and G. A. Ozin in Vibrational Spectra and Structure, (J. R. Durig, ed.), Vol. 4, p. 187, Elsevier, Amsterdam, 1975.

61. F. K. Chi and L. Andrews, J. Chem. Phys., 77, 3062 (1973).

62. J. J. Barrett and N. I. Adams III, J. Opt. Soc. Am., 58, 311 (1968).

63. G. J. Rosasco and J. H. Simmons, Am. Ceram. Soc. Bull., 53, 626 (1974); Appl. Spectrosc., 29, 396 (1975) and 32, 292 (1978).

64. S. K. Freeman, P. R. Reed, Jr., and D. O. Landon, Mikrochim. Acta (Wien) 288 (1972).

65. R. A. Nyquist and R. O. Kagel, in Practical Spectroscopy, (E. G. Brame and J. S. Grasselli, eds.), Dekker, New York, 1975, Vol. 1, p. 441.

66. J. Laane and W. Kiefer, J. Chem. Phys., May 1980, in press; J. Laane and W. Kiefer, Proc. Seventh Int. Conf. Raman Spectrosc., to be published (1980).

67. W. Klockner, W. Kiefer, and H. W. Schrotter, Appl. Spectrosc., 31, 52 (1977).

68. D. L. Rousseau, J. M. Friedman, and P. F. Williams, in Raman Spectroscopy of Gases and Liquids, (A. Weber, ed.), Springer, Berlin, 1979, p. 203.

69. P. R. Carey, H. Schneider, and H. J. Bernstein, Biochem. Biophys. Res. Comm., 47, 588 (1972).

70. P. R. Carey, A. Froese, and H. Schneider, Biochemistry, 12, 2198 (1973).

71. P. R. Carey and H. Schneider, Accounts of Chem. Res., 11, 122 (1978).

72. T. G. Spiro, in Chemical and Biochemical Applications of Lasers, Vol. 1, (C. B. Moore, ed.), Academic Press, New York, 1974.

73. T. M. Theophanides (Ed.), Infrared and Raman Spectroscopy of

Biological Molecules, Reidel, Dordrecht, 1979.

74. S. J. Cyvin, J. E. Rauch, and J. C. Decius, J. Chem. Phys., 43, 4083 (1965).

75. J. H. Christie and D. J. Lockwood, J. Chem. Phys., 54, 1141 (1971).

76. D. A. Long, in Essays in Structural Chemistry, (A. J. Downs, D. A. Long and L. A. K. Staveley, eds.), Macmillan, London, 1971.

77. M. J. French and D. A. Long, in Molecular Spectroscopy (Specialist Periodical Report No. 29), (R. F. Barrow, D. A. Long and J. Sheridan, eds.), The Chemical Society, London, 1976, Vol. 4, p. 225.

78. T. J. Dines, M. J. French, R. J. B. Hall, and D. A. Long, Proc. Fifth Int. Conf. Raman Spectrosc., (E. D. Schmid et al., eds.), Schulz-Verlag, Freiburg, 1976, p. 707.

79. S. A. Akhmanov and N. I. Koroteev, Zh. Eksp. Teor. Fiz., 67, 1306 (1974).

80. J. W. Nibler, W. M. Schaub, J. R. McDonald, and A. B. Harvey, in Vibrational Spectra and Structure, (J. R. Durig, ed.), Vol. 6, Elsevier, Amsterdam, 1977, p. 173.

81. J. W. Nibler and G. V. Knighten, in Raman Spectra of Gases and Liquids, (A. Weber, ed.), Springer, Berlin, 1978, p. 253.

82. A. Owyoung, IEEE J. Quantum Electron., QE14, 192 (1978).

83. A. Lau et al., Sov. J. Quantum Electron., 6, 402 (1976).

SUBJECT INDEX

AUTHOR INDEX

REFERENCE AUTHOR INDEX

Numbers in brackets are reference numbers and indicate that an author's work is referred to although his name is not cited in the text. Underlined numbers give the page on which the complete reference is listed.

GPSR Compliance

The European Union's (EU) General Product Safety Regulation (GPSR) is a set of rules that requires consumer products to be safe and our obligations to ensure this.

If you have any concerns about our products, you can contact us on ProductSafety@springernature.com

In case Publisher is established outside the EU, the EU authorized representative is:

Springer Nature Customer Service Center GmbH
Europaplatz 3
69115 Heidelberg, Germany

Batch number: 10815974

Printed by Printforce, the Netherlands